# Machine Learning and Wireless Communications

How can machine learning help the design of future communication networks, and how can future networks meet the demands of emerging machine learning applications? Discover the interactions between two of the most transformative and impactful technologies of our age in this comprehensive book.

First, learn how modern machine learning techniques, such as deep neural networks, can transform how we design and optimize future communication networks. Accessible introductions to concepts and tools are accompanied by numerous real-world examples, showing you how these techniques can be used to tackle longstanding problems. Next, explore the design of wireless networks as platforms for machine learning applications. An overview of modern machine learning techniques and communication protocols will help you to understand the challenges, while new methods and design approaches will be presented to handle wireless channel impairments such as noise and interference, to meet the demands of emerging machine learning applications at the wireless edge.

**Yonina C. Eldar** is a professor of electrical engineering at the Weizmann Institute of Science, where she heads the Center for Biomedical Engineering and Signal Processing. She is also a visiting professor at Massachusetts Institute of Technology and the Broad Institute and an adjunct professor at Duke University. She is a member of the Israel Academy of Sciences and Humanities, an IEEE fellow, and a EURASIP fellow.

**Andrea Goldsmith** is the Dean of Engineering and Applied Science and the Arthur LeGrand Doty Professor of Electrical Engineering at Princeton University. She is a member of the US National Academy of Engineering and the American Academy of Arts and Sciences. In 2020, she received the Marconi Prize.

**Deniz Gündüz** is a professor of information processing in the Electrical and Electronic Engineering Department of Imperial College London, where he serves as the Deputy Head of the Intelligent Systems and Networks Group. He is also a part-time faculty member at the University of Modena and Reggio Emilia.

**H. Vincent Poor** is the Michael Henry Strater University Professor at Princeton University. He is a member of the US National Academy of Engineering and the US National Academy of Sciences. In 2017, he received the IEEE Alexander Graham Bell Medal.

# Machine Learning and Wireless Communications

Edited by

**YONINA C. ELDAR**
*Weizmann Institute of Science*

**ANDREA GOLDSMITH**
*Princeton University*

**DENIZ GÜNDÜZ**
*Imperial College*

**H. VINCENT POOR**
*Princeton University*

# CAMBRIDGE
## UNIVERSITY PRESS

University Printing House, Cambridge CB2 8BS, United Kingdom

One Liberty Plaza, 20th Floor, New York, NY 10006, USA

477 Williamstown Road, Port Melbourne, VIC 3207, Australia

314–321, 3rd Floor, Plot 3, Splendor Forum, Jasola District Centre, New Delhi – 110025, India

103 Penang Road, #05–06/07, Visioncrest Commercial, Singapore 238467

Cambridge University Press is part of the University of Cambridge.

It furthers the University's mission by disseminating knowledge in the pursuit of education, learning, and research at the highest international levels of excellence.

www.cambridge.org
Information on this title: www.cambridge.org/9781108832984
DOI: 10.1017/9781108966559

First published 2022

Printed in the United Kingdom by TJ Books Limited, Padstow Cornwall

*A catalogue record for this publication is available from the British Library.*

*Library of Congress Cataloging-in-Publication Data*
Names: Eldar, Yonina C., editor.
Title: Machine learning and wireless communications / edited by Yonina C. Eldar,
    Weizmann Institute of Science, Andrea Goldsmith, Princeton University,
    Deniz Gündüz, Imperial College, H. Vincent Poor, Princeton University.
Description: First edition. | Cambridge, United Kingdom ; New York, NY :
    Cambridge University Press, 2022. | Includes bibliographical references and index.
Identifiers: LCCN 2021063108 (print) | LCCN 2021063109 (ebook) |
    ISBN 9781108832984 (hardback) | ISBN 9781108966559 (epub)
Subjects: LCSH: Wireless communication systems. | Machine learning. |
    BISAC: TECHNOLOGY & ENGINEERING / Signals & Signal Processing
Classification: LCC TK5103.2 .M3156 2022 (print) | LCC TK5103.2 (ebook) |
    DDC 621.382–dc23/eng/20220318
LC record available at https://lccn.loc.gov/2021063108
LC ebook record available at https://lccn.loc.gov/2021063109

ISBN 978-1-108-83298-4 Hardback

To our families.

# Contents

# Contributors

**Ziv Aharoni,**
Ben-Gurion University of the Negev

**Mohammad Mohammadi Amiri,**
Princeton University

**Mehdi Bennis,**
University of Oulu

**Mingzhe Chen,**
Princeton University

**Shuguang Cui,**
The Chinese University of Hong Kong, Shenzhen

**Mérouane Debbah,**
CentralSupélec

**Nariman Farsad,**
Stanford University

**Ziv Goldfeld,**
Cornell University

**Hengtao He,**
Southeast University

**Xiufeng Huang,**
Tsinghua University

**Shi Jin,**
Southeast University

**Hyeji Kim,**
The University of Texas

**Seong–Lyun Kim,**
Yonsei University

**Visa Koivunen,**
Aalto University

**David Burth Kurka,**
Imperial College London

**Hoon Lee,**
Pukyong National University

**Sang Hyun Lee,**
Korea University

**Kin K. Leung,**
Imperial College London

**Geoffrey Y. Li,**
Imperial College London

**Dongzhu Liu,**
King's College London

**Litian Liu,**
Massachusetts Institute of Technology

**Derya Malak,**
EURECOM

**Philippe Mary,**
Institut National des Sciences Appliquées

**Muriel Médard,**
Massachusetts Institute of Technology

**Christophe Moy,**
University of Rennes 1

**Zhisheng Niu,**
Tsinghua University

**Seungeun Oh,**
Yonsei University

**Mehmet Emre Ozfatura,**
Imperial College London

**Jihong Park,**
Deakin University

**Haim H. Permuter,**
Ben-Gurion University of the Negev

**Tony Q. S. Quek,**
Singapore University of Technology and Design

**Milind Rao,**
Stanford University

**Stefano Rini,**
National Chiao Tung University

**Salman Salamatian,**
Massachusetts Institute of Technology

**Hyowoon Seo,**
Kwangwoon University

**Wenqi Shi,**
Tsinghua University

**Nir Shlezinger,**
Ben-Gurion University of the Negev

**Osvaldo Simeone,**
King's College London

**Amit Solomon,**
Massachusetts Institute of Technology

**Amir Sonee,**
National Chiao Tung University

**Tiffany Tuor,**
Imperial College London

**Shiqiang Wang,**
IBM Thomas J. Watson Research Center

**Yue Xu,**
Beijing University of Posts and Telecommunications

**Hao Ye,**
Georgia Institute of Technology

**Feng Yin,**
The Chinese University of Hong Kong, Shenzhen

**Alessio Zappone,**
University of Cassino

**Sheng Zhou,**
Tsinghua University

**Dor Tsur,**
Ben-Gurion University of the Negev

# Preface

Machine learning (ML) and wireless communications are two of the most rapidly advancing technologies of our time. The main premise of ML is to enable computers to learn and perform certain tasks without being explicitly programmed to do so. This is achieved by training algorithms on data available for the task to be accomplished. Although the basic ideas and ambitions of ML go back to the 1950s, there has been a recent surge in interest and applications in this area, fueled by the availability of increasingly powerful computers, large amounts of data, and developments in new learning algorithms as well as their theoretical underpinnings. At the same time, wireless communication has evolved, through advances in both theory and supporting technologies, to encompass a variety of application areas, from high-performance data transmission tasks such as media distribution to the massive deployment of end-devices to enable Internet of Things (IoT) tasks such as sensing, inference, and control.

We are now witnessing the confluence of these two fields, with two primary aspects to this connection. One is the application of ML techniques to the optimization of wireless networks. This is a natural use of ML, as wireless networks involve many inferential and control tasks, which often must operate under dynamic or uncertain conditions, and create many examplars for learning because data transmissions take place at very high rates. The other aspect of this connection is the use of wireless networks as ML platforms. This again is a natural application of emerging wireless networks, such as those supporting IoT applications, because they involve sensing, inference, and control and provide edge devices with considerable processing power. Learning at the network edge has advantages in terms of latency and privacy, and it capitalizes on the fact that many learning tasks, such as those supporting automated driving, are locality specific.

To realize the promise of these opportunities, significant research in many dimensions is needed. Important issues include the adaptation of existing ML techniques to wireless system design and the design and development of new techniques that can meet the constraints and requirements of communication networks, including the capability to implement at least some of these techniques in low-power chips that can be used in mobile devices, as well as developing fundamental analytical techniques and bounds on the performance of distributed ML algorithms operating within the constraints of wireless connectivity. This book focuses on these research issues through a series of 18 chapters written by experts in the field, beginning with an introductory chapter providing a brief general overview of ML methodology. By presenting a systematic overview of the most promising aspects of the connection between ML in wireless networks, this book provides an entry point and a comprehensive overview of the state of the art for researchers in academia and industry who are interested in learning and contributing to this growing field.

This book is the culmination of the efforts of many people, including the chapter authors and the editorial and production staff at Cambridge University Press. We wish to express our deep gratitude for their contributions.

# 1 Machine Learning and Communications: An Introduction

Deniz Gündüz, Yonina C. Eldar, Andrea Goldsmith, and H. Vincent Poor

## 1.1 Introduction

Wireless communications is one of the most prominent and impactful technologies of the last few decades. Its transformative impact on our lives and society is immeasurable. The fifth generation (5G) of mobile technology has recently been standardised and is now being rolled out. In addition to its reduced latency and higher data rates compared to previous generations, one of the promises of 5G is to connect billions of heterogeneous devices to the network, supporting new applications and verticals under the banner of the Internet of Things (IoT). The number of connected IoT devices is expected to increase exponentially in the coming years, enabling new applications including mobile healthcare, virtual and augmented reality, and self-driving cars as well as smart buildings, factories, and infrastructures.

In parallel with recent advances in wireless communications, modern machine learning (ML) techniques have led to significant breakthroughs in all areas of science and technology. New ML techniques continue to emerge, paving the way for new research directions and applications, ranging from autonomous driving and finance to marketing and healthcare, just to name a few. It is only natural to expect that a tool as powerful as ML should have a transformative impact on wireless communication systems, similarly to other technology areas. On the other hand, wireless communication system design has traditionally followed a model-based approach, with great success. Indeed, in communication systems, we typically have a good understanding of the channel and network models, which follow fundamental physical laws, and some of the current systems and solutions already approach fundamental information theoretical limits. Moreover, communication devices typically follow a highly standardized set of rules; that is, standardization dictates what type of signals to transmit, as well as when and how to transmit them, with tight coordination across devices and even different networks. As a result, existing solutions are products of research and engineering design efforts that have been optimized over many decades and generations of standards, which are based upon theoretical foundations and extensive experiments and measurements. From this perspective, compared to other areas in which ML has made significant advances in recent years, communication systems can be considered more amenable to model-based solutions rather than generic ML approaches. Therefore, one can question what potential benefits ML can provide to wireless communications [1–3].

The chapters in the first part of this book provide many answers to this question. First, despite the impressive advances in coding and communication techniques over the last few decades, we are still far from approaching the theoretical limits in complex networks of heterogeneous devices, or even understanding and characterizing such limits. Current design approaches are built upon engineering heuristics: we try to avoid or minimize interference through various access technologies, which allows us to reduce a complex interconnected network into many parallel point-to-point links. We then divide the design of the point-to-point communication systems into many different blocks, such as channel estimation, feedback, equalization, modulation, and coding, which are easier to model and may even lend themselves to optimal solutions. Although this modular design is highly attractive from an engineering perspective, it is generally suboptimal.

With 5G, we aim to connect billions of IoT devices, which have significantly different requirements in terms of latency and energy efficiency compared to conventional mobile handsets. The networks will have to sustain a much more diverse set of devices and traffic types, each with different requirements and constraints. This will require significantly more flexibility and more adaptive and efficient use of the available resources. Therefore, the performance loss due to the aforementioned modular design and separate optimization of the individual blocks is becoming increasingly limiting. Cross-layer design as an alternative has been studied extensively, but this type of design still depends on a model-based approach that is often significantly more complex than the modular design. This complexity has led to ad-hoc solutions for each separate cross-layer design problem and scenario, with performance gains that were often limited or nonexistent relative to more traditional modular design. Data-driven ML techniques can provide an alternative solution, with potentially larger gains in performance and reasonable complexity in implementation.

Second, in current systems, even with the modular approach, the optimal solution may be known theoretically, yet remain prohibitive to attain computationally. For example, we can write down the maximum likelihood detection rule in a multiuser scenario, but the complexity of its solution grows exponentially with the number of users [4]. Similarly, in a communication network with a multi-antenna transmitter and multiple single-antenna receivers, the optimal beamforming vectors can be written as the solution of a well-defined optimization problem, albeit with no known polynomial complexity solution algorithm [5]. As a result, we either provide low-complexity yet suboptimal solution techniques to the problem [6] or resort to simple solutions that can be solved in closed-form [7]. Recent results have shown that machine learning techniques can provide more attractive complexity-performance trade-offs [8, 9].

There has already been a significant amount of research proposing data-driven ML solutions for different components of a communication system, such as channel estimation [10–12], channel state information compression and feedback [13, 14], channel equalization [15], channel decoding [16], and more, many of which will be discussed in detail in the later chapters of this book. These solutions are shown to outperform conventional designs in many scenarios, particularly when we do not have a simple model of the communication system (e.g., when the channel coefficients do not follow a distribution with a known covariance matrix). However, another potential benefit of the ML approaches is to go beyond the aforementioned modular design and learn the best communication scheme in an end-to-end fashion rather than targeting each module separately [17, 18]. This end-to-end ML approach facilitates attacking much more complex problems with solutions that were deemed elusive with conventional approaches. A good example is the joint source-channel coding problem, which is typically divided into the compression and channel coding subproblems. Despite the known suboptimality of this approach, and decades of efforts in designing joint source-channel coding schemes, these often resulted in ad-hoc solutions for different source and channel combinations with formidable complexity. However, recent results have shown that ML can provide significant improvements in the end-to-end performance of joint source-channel coding [19–22].

An additional benefit of ML is that is does not require knowledge of channel parameters or statistics. In particular, many known communication methods rely on knowledge of the channel parameters or statistics or the ability to estimate these statistics with minimal resources. In modern wireless networks, this is no longer the case. Thus, ML can offer efficient techniques to compensate for unknown and varying channel knowledge [23, 24]. In addition, hardware limitations, such as the use of low bit rate quantizers or nonlinear power amplifiers, can significantly increase the complexity of the underlying channel model. This renders a design based on exact channel knowledge difficult to implement [25].

Another dimension of the potential synergies between communications and ML is the design of communication networks to enable ML applications [26]. While a significant amount of information is collected and consumed by mobile devices, most of the learning is still carried out centrally at cloud servers. Such a centralized approach to learning at the edge has limitations. First, offloading all the data collected by edge devices to the cloud for processing is not scalable. For example, an autonomous car is expected to collect terrabytes of data every day. As the number of vehicles increases, the cellular network infrastructure cannot support such a huge rise in data traffic solely from autonomous cars. A similar surge in traffic is expected from other connected devices. The sheer volume of the collected data makes such centralized approaches highly unsustainable. This is particularly problematic when the collected data has large dimensions and has low *information density*, which refers to the information within the data relevant for the underlying task. For example, a surveillance camera may collect and offload hours of recording of a still background, which has little use for the task of detecting intruders. Therefore, it is essential to enable edge devices with some level of intelligence to extract and convey only the relevant information for the underlying learning task [27–29].

Having all the intelligence centralized also causes significant privacy and security concerns [30–32]. Most of the data collected from edge devices, such as smart meters [33, 34], autonomous cars [35], health sensors [36], or entertainment devices [37], collect highly sensitive information that reveals significant personal information about our lives and daily habits. Sharing this data in bulk with other parties is a growing concern for consumers, which can potentially hinder the adoption of some of these services. The alternative can be to carry out local learning at each individual device with the available data; however, the locally available data can be limited in quantity and variety, resulting in overfitting.

Fully centralized intelligence also introduces latency. In many applications that involve edge devices, inference and action need to be fast. For example, autonomous vehicles must detect and avoid pedestrians or other obstacles rapidly. Even though some of the inference tasks can be carried out on board, devices may not have the necessary processing capability, and some of the inference tasks may require fusing information distributed across multiple edge devices. For example, an autonomous car may benefit from camera or LIDAR data from other nearby cars or terrain information available at a nearby base station to make more accurate decisions about its trajectory and speed. In such scenarios, it is essential to enable distributed inference algorithms that can rapidly gather the most relevant information and make the most accurate decisions [28].

In light of these observations, an important objective in merging communications and ML is to bring intelligence to the network edge, rather than offloading the data to the cloud and relying on centralized ML algorithms. The two core ingredients behind the recent success of ML techniques are massive datasets and tremendous memory and processing power to efficiently and rapidly process the available data to train huge models, such as deep neural networks (DNNs). Both of these ingredients are plentiful at the network edge, yet in a highly distributed manner. The second part of this book is dedicated to distributed ML algorithms, particularly focusing on learning across wireless networks, addressing these challenges and highlighting some of the important research achievements and remaining challenges.

Before revieweing the individual chapters in the book, we provide a brief overview of basic ML techniques, particularly focusing on their potential applications in communication problems.

## 1.2     Taxonomy of Machine Learning Problems

The type of problems to which machine learning algorithms are applied can be grouped into three categories: *supervised learning, unsupervised learning*, and *reinforcement learning (RL)*. All three types of problems have found applications in wireless communications. Here we provide a general overview of these different categories and what types of problems they are used to solve and then highlight their applications in wireless communications.

### 1.2.1     Supervised Learning

In supervised learning the goal is to teach an algorithm the input-output relation of a function. This can be applied to a wide variety of functions and input/output types. Consider, for example, the input vector denoted by $\mathbf{x} \in \mathcal{X}$ and the associated vector of target variables $\mathbf{c} \in \mathcal{C}$. Supervised learning problems are classified into two groups: if the input is mapped to one of a finite number of discrete classes (i.e., $\mathcal{C}$ is a discrete set), then the learning task is called *classification*. If, instead, the output can take continuous values (i.e., $\mathcal{C}$ is a continuous set), then the learning task is called *regression*. The often cited toy problem of classifying images into dog and cat labels is a classical example of a classification problem. Most classification tasks have more than two classes, and other common examples that are often used to compare and benchmark supervised learning algorithms are classification of handwritten digits and spam detection.

In communications, the design of a receiver for a fixed transmission scheme is a classification task. For example, consider a simple modulation scheme mapping input bits to constellation points. The receiver trying to map each received noisy symbol to one of the constellation points can be considered a classification problem. This can also be extended to decoding of coded messages transmitted over a noisy channel, where the decoder function tries to map a vector of received symbols to a codeword [16, 38].

Another application of classification in communications is the detection of the type of wireless signals in the air, which is often required for military applications, or for cognitive radios [39]. This includes the detection of the transmitting device by identifying the particular hardware impairments of each individual transmitter [40], detection of the modulation type used by the device [41–43], or detection of the wireless technology employed by the transmitter [44]. A dataset of synthetic simulated channel effects on wireless modulated signals of 11 different modulation types has been released in [45], which has significantly helped the research community to test and compare proposed techniques on a benchmark dataset.

A well-known example of a regression problem encountered in wireless communication systems is channel estimation [10, 11, 46], where the goal is to estimate the channel coefficients from noisy received versions of known pilot signals. While traditional channel estimation methods assume a known channel model, and try to estimate the parameters of this model through least squared or minimum mean-squared error estimation techniques, a data-driven channel estimator does not make any assumptions on the channel model. It instead relies on training data generated from the underlying channel. In the context of wireless communications, while a data-driven approach is attractive when an accurate channel model is not available, the requirement of a large training dataset can mean that training needs to be done off-line.

Supervised learning can also be used as an alternative method to rapidly obtain reasonable suboptimal solutions to complex optimization problems. In wireless networks, we often face highly complex nonconvex or combinatorial optimization problems, for example, in scheduling or deciding transmit powers in a network of interfering transmitters. Many of these problems do not have low-complexity optimal solutions; however, in practice, we need a fast solution to be implemented within the coherence time of the channels. Therefore, we typically resort to some low-complexity suboptimal solution. An alternative would be to train a neural network using the optimal solution for supervision. This can result in a reasonable performance that can be rapidly obtained once the network is trained. Such methods have been extensively applied to wireless network optimization problems with promising results in terms of the performance-complexity trade-off [8, 47].

### 1.2.2    Unsupervised Learning

In unsupervised learning problems, we have training data without any output values, and the goal is to learn functions that describe the input data. Such functions may be useful for more efficient supervised learning, as they can be seen as a method for feature extraction. They may also be used to make the input data more amenable to human understanding and interpretation. Unsupervised learning is not as well-defined as supervised learning, since it is not clear what type of description of the data we are looking for. Moreover, it is often the case that the measure to use to compare different descriptions is not obvious and may highly depend on the type of data and application we have in mind. Common unsupervised learning problems are *clustering*, *dimension reduction*, and *density estimation*, all of which have been used extensively in communication systems.

Indeed, clustering is nothing but source compression or quantization, where the goal is to identify a small number of representatives that can adequately represent all possible input vectors. In density estimation, the goal is to determine the distribution of data as accurately as possible from a limited number of samples. Parameterized density estimation is often used in wireless communications when estimating the channel from a limited number of pilot signals. Dimensionality reduction is similar to clustering in the sense that we want a more efficient representation of data, but rather than limiting this representation to a finite number of clusters, we limit its dimension. Projecting a large-dimensional input data to two or three dimensions is used for visualization. A common technique for dimensionality reduction is principle component analysis (PCA), which is also known as the Karhunen-Loeve transform and is used for lossy data compression. More recently neural networks in the form of autoencoders have been employed for dimensionality reduction. Autoencoders play an important role in the data-driven design of compression and communication schemes.

In machine learning, some approaches first try to learn the distribution of data (or the data as well as the output in the context of supervised learning). These are called *generative models*, because once the underlying distribution is learned, one can generate new samples from this distribution. With the advances in deep learning, DNNs have also been used in *deep generative models*, which have shown remarkable performance in modeling complex distributions. Two popular architectures for deep generative modeling are variational autoencoders (VAEs) [48] and generative adversarial networks (GANs) [49]. Generative models have recently been employed in [50, 51] to model a communication channel from data; such models can be used for training other communication components when no channel model is available.

### 1.2.3    Reinforcement Learning (RL)

RL is another class of machine learning problems, where the goal is to learn how to interact in a random unknown environment based on feedback received in the form of costs or rewards following each action. In contrast to supervised learning, where the outputs corresponding to a dataset of input samples are given, in RL these need to be learned through interactions with the environment. The environment has a state, and the agent interacts with the environment by taking actions. The goal is to learn the right action to take at each state in order to maximize (minimize) the long-term reward (cost). The cost/reward acquired depends both on the state and the action taken and is typically random with a stationary distribution. Notable examples of RL algorithms are game playing agents that can beat human masters in chess and Go, or more recently in Atari games or more advanced multiplayer video games.

A fundamental aspect of RL is the trade-off between *exploration* and *exploitation*. Exploration refers to taking new unexplored actions to gather more information about the environment to potentially discover actions with higher rewards (lower costs). Exploitation, on the other hand, refers to exploiting the actions that are more likely to provide higher rewards (lower costs) based on past observations. A conservative agent that is more likely to exploit its current knowledge risks losing out on high reward actions that it has never tried. On the other hand, an agent that keeps exploring without considering its past experiences ends up with a low reward. Hence, the goal is to find the right balance between exploration and exploitation.

RL has found applications in wireless networks as early as in the 1990s [52, 53], including power optimization in the physical layer for energy efficient operation [54]. Similarly to other machine learning tools, RL algorithms can be used for two types of problems requiring interactions with an environment, which can model the wireless network environment a device is operating in: the device might have an accurate model of the environment, but the solution of an optimal operation policy for the device may be elusive. In such a case, RL techniques can be used as a numerical solution technique to characterize the optimal (or near optimal) strategy for the device. Note that this is similar to the application of RL in games such as chess or Go, which have well-understood rules but are still highly complex to study methodically. This type of problem typically appears in networking, where multiple devices are scheduled to share the limited spectral resources.

Alternatively, RL methods can be used in wireless networking problems for which the environment is known to be stationary, yet we do not have a good model to characterize its statistical behaviour. This might be the case, for example, when operating over unlicensed bands, where it is difficult or impossible to model the statistics of device activations, spectrum activity, traffic arrivals, and channel variations. In such scenarios, RL tools can be used to learn the best operation strategy through direct interactions with the environment in an online fashion. An example of the application of RL techniques to such a problem is channel access for cognitive users, where the probability of availability of each channel is unknown to the user. This problem is formulated in [55] as a multiarmed bandit (MAB) problem, which is a special case of RL with a single state. The name MAB is motivated by bandit machines in casinos: a gambler wants to play the arm that has the highest chance of winning, but this is unknown in advance. Hence, the gambler has to try as many different arms as possible to discover the best one, while also trying to exploit the estimated best arm as much as possible, since trying each arm has a cost. In [56], scheduling of multiple energy-harvesting wireless transmitters is considered. Due to the presence of finite-capacity batteries, this problem is formulated as a restless MAB, which refers to the fact that the rewards of the arms are governed by an underlying Markov process, and the state of the arms may change even if they are not played.

## 1.3        ML Tools Used in Communication System Design

In the previous section we classified ML problems into three main groups. There are many different formulations and approaches within each of these categories. Moreover, many different techniques are available to solve the same problem, with each method providing a different complexity-performance trade-off depending on the implementation constraints, availability of data, and uncertainty in the system parameters. While providing background on all existing tools is beyond the scope of this chapter, we briefly review here three fundamental tools that are commonly used in the design of communication networks (and many other engineering problems), which also appear in most of the later chapters of the book. In particular, we provide a brief introduction to neural networks and how they are trained, particularly for supervised learning tasks, then present the concept of an autoencoder for unsupervised learning, and finally discuss the main challenges of RL along with common solution approaches.

### 1.3.1        Deep Neural Networks (DNNs)

An artificial neural network is a popular ML model consisting of several layers of "neurons," which are processing units loosely inspired by biological neurons. In a *feedforward neural network*, neurons are organized into layers, and there are no feedback connections that would feed the output of any layer back into the model itself. Neural networks that incorporate feedback connections are called *recurrent neural networks*. Please see Fig. 1.1 for an illustration of a neural network architecture.

The first layer takes as input an affine function of the values of the input signal, while the following layers take as input affine functions of the outputs of the neurons in the previous layer. The left-most layer is typically considered the *input layer* and the right-most layer is the *output layer*, while the remaining layers in between are called the *hidden layers*; see Fig. 1.1. The number of hidden layers provides the *depth* of the network, which led to the terminology of "deep learning." Each hidden layer typically consists of many neurons, the number of which determines the *width* of the network. Let $d_k$ denote the width of the $k$th layer, where $d_0$ corresponds to the input dimension (e.g., the number of pixels in the input image). If we denote the output of neuron $i$ in layer $k$ by $x_i^{(k)}$, and its input by $y_i^{(k)}$, we have

$$y_i^{(k)} = w_{0,i}^{(k)} + \sum_{n=1}^{d_{k-1}} w_{n,i}^{(k)} x_n^{(k-1)}, \qquad (1.1)$$

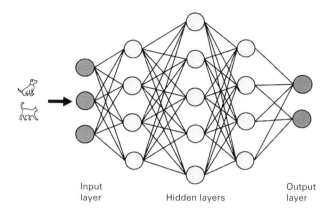

**Figure 1.1** Neural network architecture.

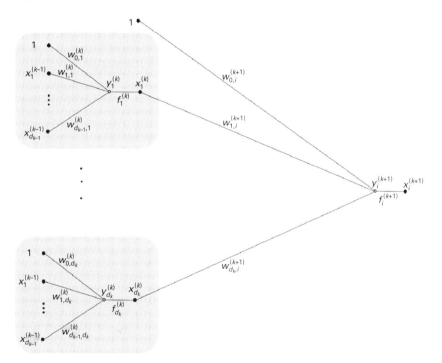

**Figure 1.2** One layer of a neural network, and the forward propagation of the neuron activations.

where $w_{n,i}^{(k)}$ denotes the weight of the connection from the output of the $n$th neuron in layer $k-1$ to the input of the $i$th neuron in layer $k$. Here, $w_{0,i}^{(k)}$ corresponds to the bias term for the $i$th neuron at layer $k$. Each neuron then applies a nonlinear "activation function" on its input, denoted by $f_i^{(k)}$ for the $i$th neuron at layer $k$, resulting in

$$x_i^{(k)} = f_i^{(k)}(y_i^{(k)}), \tag{1.2}$$

as illustrated in Fig. 1.2.

The evaluation of the output values of the neurons in a neural network by recursively computing Eqs. (1.1) and (1.2) is called *forward propagation*, as the information flows from the input layer toward the output layer.

Different activation functions can be used in Eq. (1.2). The most common activation functions are the identity function, step function, a rectified linear unit (ReLU), sigmoid function, hyperbolic tangent, and softmax. Softmax is a generalization of the sigmoid function, which represents a binary variable. Note that the model in Eq. (1.2) allows for employing a different activation function for each neuron in the network. Softmax is often used as the activation function in the output layer of a classifier network to represent a probability distribution over $n$ classes. On the other hand, ReLU is typically used as the activation function for hidden layers, although it is not possible to say which activation function will perform the best. This is often determined through trial and error.

Another aspect of a neural network that we must design is its *architecture*, which refers to the depth of the network and the width of each layer. Although a network with a single layer would be sufficient to fit the training dataset, deeper networks have been shown to generalize better to the test set. Deeper networks may also use less parameters in total; however, they are harder to train. Therefore, the network architecture is often determined through trial and error, taking into account any constraints on the training time and complexity.

Neural networks without an activation function, or an identity activation function, can only learn linear models; nonlinear activation functions are required to learn more complex nonlinear functions from the inputs to the outputs. Even though the choice of the activation function and the network architecture are important design parameters in practice, a network with a single hidden layer with a sigmoid activation function is sufficient to approximate (with arbitrary precision) any continuous function between any two Euclidean spaces [57, 58]. Note, however, that this is an existence result, and we still need to identify the right parameters that would provide the desired approximation. Identification of these parameters corresponds to the *training* stage of the neural network.

For a given neural network architecture, the output corresponding to each input data sample is determined by the weights, $w_{n,i}^{(k)}$. Therefore, the goal of the training process is to determine the weight values that will result in the best performance. Here, the performance measure will depend on the underlying problem. It can be the accuracy of detecting the correct label in a classification problem or determining the correct value in a regression problem. For example, in a multiclass classification problem with $C$ classes, the width of the output layer is chosen as $C$, and the normalized values of the output layer are interpreted as the likelihoods of the corresponding classes. Let $s_1, \ldots, s_N$ denote the data points in the training dataset, with corresponding labels $l_1, \ldots, l_N$ represented as one-hot encoded vectors of size $C$, whose elements are all 0 except for the entry corresponding to the correct class, which is set to 1. That is, if data sample $s_n$ in the training dataset belongs to class $c$, then we have $l_{n,c} = 1$, and $l_{n,m} = 0$ for $m \neq c$. In this case, the loss function will be given by

$$E(\mathbf{w}) = -\sum_{n=1}^{N}\sum_{c=1}^{C} l_{n,c} \ln x_c^{(K)}(\mathbf{s}_n, \mathbf{w}), \tag{1.3}$$

where $K$ is the number of layers in the network, $\mathbf{w}$ represents all the weights (including the bias terms) in the network, and we have written $x_k^{(K)}(\mathbf{s}_n, \mathbf{w})$ to represent the outputs of the neurons in the output layer to highlight the dependence of their values on the network weights as well as the input data. This error function is called the *cross entropy*, or the *log loss*, and is widely used as the objective/loss function when training parameterized classification algorithms.

The error function in neural networks is highly nonlinear and nonconvex, therefore it is difficult to find the globally optimal weight parameters that minimize Eq. (1.4). Instead, we resort to numerical approaches with the hope of finding reasonable, potentially locally optimal, solutions. A common numerical approach to solve such optimization problems is the *gradient descent* method, where we start from an initial value of the weight vector $\mathbf{w}$, and iteratively update it along the negative gradient direction:

$$\mathbf{w}(t+1) = \mathbf{w}(t) + \eta \nabla E(\mathbf{w}(t)), \tag{1.4}$$

where $\eta > 0$ is the *learning rate*. Since the error function is defined over the whole training dataset, the gradient at each iteration must be computed for all the data points. Alternatively, we can use an unbiased estimate of the gradient at each iteration by computing it only at a random data sample. This is called *stochastic gradient descent*, and it results in faster convergence in training DNNs.

In a neural network, the error function is a complex function of the weights, and how to compute the gradient with respect to the weight vector is not obvious at all. Fortunately, the recursive structure of the network allows for computing the gradients in a systematic and efficient manner using the chain rule of differentiation. This is known as the *backpropagation algorithm*, also referred to as *backprop*. There are excellent textbooks explaining the technical details of the backpropagation algorithm and other techniques to speed up neural network training. We refer the readers to [59, 60] for further details.

## 1.3.2    Autoencoders

Autoencoders play an important role in the application of neural networks in wireless communications. An autoencoder is a pair of neural networks, called the *encoder* and *decoder* networks, trained together in an unsupervised manner in order to recover the input signal at the output of the decoder network with minimal distortion (see Fig. 1.3). The output of the encoder network is called the *bottleneck layer*, which typically has lower dimension compared to the input signal. After training, this bottleneck layer recovers a low-dimensional representation of the input signal that still allows the decoder network to recover the input signal within some distortion. Equivalently, the encoder acts as a data-driven dimensionality reduction technique,

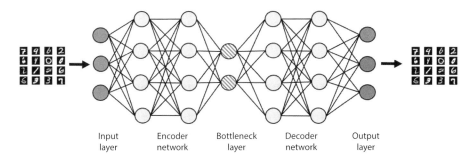

**Figure 1.3** Autoencoder neural network architecture.

similarly to principal component analysis (PCA). Autoencoders were introduced in the 1980s [61], but they achieved competitive or superior performance compared to other unsupervised learning techniques only recently with the advances in computational capabilities that allowed fast training of sufficiently complex autoencoder models on large datasets [62].

While autoencoders have been widely used for preprocessing in supervised learning and visualization, their potential for data compression was also recognized in the 1990s [63, 64]. In data compression, the low-dimensional representation generated by the encoder network must be converted into a bit sequence that can then be used by the decoder network to recover the input signal. Since we are typically limited by a finite bit budget, the output of the encoder network needs to be quantized, which introduces additional distortion at the reconstruction; hence, the autoencoder must be trained together with the quantization layer. This creates some difficulty as the quantization operation is not differentiable, which prevents backpropagation during training. Several methods have been introduced in the literature to deal with this, such as straight through estimation, which ignores the quantization operation in the backpropagation [65], or approximating the quantization operation with a differentiable function [66].

Denoising autoencoders were introduced in [67] in order to provide robustness in the generated low-dimensional representation to partial corruptions in the input signal. In a denoising autoencoder, some components of the input signal are randomly erased before being fed into the encoder network, and the autoencoder is trained to recover the uncorrupted input signal with the highest fidelity. This way, the denoising autoencoder learns not only to represent the input signal efficiently, but also to recover the random perturbation introduced on the input signal. If we treat the basic autoencoder as applying source compression, we can consider the setting in the denoising autoencoder as remote source coding [68].

Inspired by the denoising autoencoder, communication over a noisy channel is treated in [69], where the channel introduces random noise on the signal generated by the encoder output before it is fed into the decoder network. To the best of our knowledge, this was the first work to treat the design of the encoder and decoder in a channel communication problem as training of an autoencoder. In this model, the input signal is a sequence of equally likely bits, representing the input message, and the goal is to transmit these inputs bit sequence over the noisy channel as reliably as possible. Note that the input message has maximum entropy, so there is no structure to be learned; instead, the goal of the autoencoder in this model is to learn the set of input codewords that can best mitigate the impairments introduced by the channel. Here, we can remove the low-dimensionality requirement on the bottleneck layer, as we typically introduce additional redundancy against channel noise. The dimension of the bottleneck layer determines the code rate.

A major challenge in the design of channel coding schemes using autoencoders is the sheer number of input signals that must be recognized at the decoder. In particular, for a message size of $n$ bits, we have $2^n$ possible input messages; equivalently, the number of classes that must be recognized by the decoder network grows exponentially with $n$. This makes the training of such an autoencoder highly challenging for moderate to large blocklengths.

### 1.3.3      Deep Reinforcement Learning (DRL)

DRL refers to the particular family of RL techniques where DNNs are employed to approximate the agent's strategy or utility function. In RL techniques, actions are chosen at each state based on the expected rewards they will provide. The expected reward from each action is updated continuously based on the past experiences of the agent. In general, it is difficult to map these experiences to an accurate estimate of the reward function. In many complex scenarios, the agent will not be able to explore all possible state-action pairs a sufficient number of times to make accurate estimates. For such problems, DNNs are used to estimate the average reward for each state-action pair or directly predict the best action to take (more precisely, a distribution over actions). DNN parameters are updated as in neural network training, but based on the rewards accrued instead of the labels used in supervised training of DNNs. Thanks to the generalization power of DNNs, DRLs can solve highly complex RL problems even with continuous state and action spaces. However, similarly to supervised learning problems, to achieve a good performance level, DRLs require significant training data and training time. Hence, in the wireless context, it may be more appropriate for stationary scenarios where the training can be carried out off-line using the available data or the model of the underlying system, and the trained network is used afterward. They can also be effective when the variations in the system statistics are relatively slow, so that it is possible to track the optimal policy with limited additional training.

### 1.4      Overview of the Book

This book is aimed at postgraduate students, Ph.D. students, researchers, and engineers working on communication systems, as well as researchers in machine learning seeking an understanding of the potential applications of ML in wireless communications. To read and fully understand the content it is assumed that the reader has some background in communication systems and basic ML. The book consists of 17 technical chapters written by leading experts in their areas, organised into two parts. The first part, consisting of 10 chapters, considers the application of machine learning techniques in solving a variety of wireless communication problems. The remaining seven chapters of the book comprise the second part, which focuses on the design and optimization of wireless network architectures to carry out machine learning tasks in an efficient manner.

Each chapter is self-contained and the chapters are independent of each other. Therefore, there is flexibility in selecting material both for university courses and short seminars. A brief summary of each chapter is given next.

## 1.4.1    Part I: Machine Learning for Wireless Networks

In Chapter 2, the authors treat the problem of information transmission over a wireless channel in an end-to-end manner, considering jointly the source compression and channel coding problems. This chapter explores joint source-channel coding schemes based on an autoencoder architecture, where a pair of jointly trained DNNs act as the encoder and the decoder. The chapter focuses on image and text transmission problems and shows that a neural network aided design not only outperforms state of the art digital transmission schemes, but also provides graceful degradation with variations in channel quality.

In Chapter 3, the authors extend the joint source-channel coding problem treated in the previous chapter to the lossy transmission of correlated sources over a network of orthogonal links. Joint source-network coding is known to be an NP-hard problem, and no practical solution exists for general sources (e.g., images). This chapter explores neural network aided design of such codes for arbitrary network topologies.

In Chapter 4, the author focuses explicitly on the design of channel codes using deep learning. The chapter highlights that deep learning based design of channel coding results in a broader range of code structures compared to traditional code designs, which can then be optimized efficiently through gradient descent. However, identifying the right network architectures and efficiently training the network parameters are challenges that still need to be overcome. The chapter provides a range of neural network aided code designs and highlights many different scenarios where such designs outperform conventional codes.

Chapter 5 focuses on the applications of deep learning to the problems of channel estimation, feedback, and signal detection, particularly for orthogonal frequency division multiplexing (OFDM) and millimeter-wave (mmWave) systems. Convolutional neural networks (CNNs) are employed for these tasks and are shown to learn channel features and estimate the channel successfully, after being trained on a large dataset generated using the channel model. For the signal detection task, "deep unfolding" is employed to benefit from the structure of an iterative decoder to improve the speed and accuracy of the training process.

In Chapter 6, the authors contrast the traditional model-based design approach to communication systems with the data-driven approach based on ML. They demonstrate through various examples that a combination of the two approaches, as model-based ML, can benefit from the best of both worlds. In particular, such designs use ML to learn unknown aspects of the system model and reduce complexity, while benefiting from the lower training requirements and near-optimal performance of model-based approaches.

Chapter 7 tackles the challenging problem of distributed optimization in wireless networks, where the nodes have to make decisions based on their local view of the network, while their actions collaboratively decides the global reward or cost function. This chapter considers the framework in which the nodes can exchange limited amounts of information through backhaul links in order to coordinate their actions. This highly complicated nonconvex optimization problem is solved by employing DNNs trained in an unsupervised manner.

Chapter 8 also deals with the radio resource allocation problem in wireless networks employing neural networks, which are trained in an off-line manner and are employed for inference assuming the radio environment remains sufficiently stationary with respect to the training setting. This chapter focuses on the global energy efficiency of the network, defined as the ratio of the total throughput achieved over the network to the total power consumption. The authors show that neural networks can be trained to operate close to the optimal performance, while having much lower inference complexity, albeit at the expense of additional training cost.

Chapter 9 focuses on the applications of RL in physical layer wireless network problems. After a brief overview of Markov decision processes (MDPs), partially observable MDPs, and multiarmed bandits, the authors present some of the most widely used algorithms, in particular Q-learning, SARSA, and deep RL. They then provide a number of examples of how these techniques can be applied in a variety of wireless networking problems from power management and cache content optimization to channel access and real-world spectrum sharing problems.

In Chapter 10, the authors present scalable learning frameworks that are capable of processing and storing large amounts of data generated in a large network. This requires parallel learning algorithms that can decompose the global learning problem into smaller tasks that can be carried out locally. The authors propose Bayesian nonparametric learning and RL as potential tools for scalable learning in the wireless context. Finally, they present several practical use cases for the presented tools. The authors also touch upon the model interpretability and adaptivity aspects of these solutions.

Chapter 11 deals with the fundamental limits of communication over noisy channels. Shannon's channel coding theorem identifies the fundamental limit of communication over a memoryless noisy channel as a single-letter mutual information expression, which is maximized over input distributions. For certain channels with feedback, capacity can be formulated as a directed information. This chapter introduces RL tools for the identification and optimization of the capacity for both known and unknown channel models.

## 1.4.2    Part II: Wireless Networks for Machine Learning

Chapter 12 presents a general introduction to collaborative learning and how it can be implemented over wireless nodes. The authors first provide an overview of distributed learning algorithms, including federated and fully decentralized learning. They also present methods to reduce the communication load of these distributed learning algorithms, which measures the amount of information that must be exchanged among the devices that participate in the learning process. Then, resource allocation and scheduling problems are presented when nodes within physical proximity of each other carry out distributed learning over a shared wireless medium.

Chapter 13 focuses on federated learning at the wireless network edge. After a comprehensive introduction to the federated learning framework and existing algorithms, the authors focus on the implementation of federated learning with limited communication resources. The adaptation to available channel resources is achieved by adjusting the number of local iterations.

Chapter 14 also deals with federated learning, focusing on the communication bottleneck. The authors present a quantization theoretic framework to reduce the communication load from the devices to the parameter server in federated learning. In particular, scalar quantization, subtractive dithering, and universal vector quantization techniques are considered to reduce the communication load in federated learning.

While Chapter 15 also considers federated learning over wireless networks, the main focus of this chapter is to highlight how the signal superposition property of the wireless medium can be exploited to increase the speed and accuracy of the learning process. In this chapter, unlike in conventional digital communication techniques, the devices participating in the federated learning process are allowed to transmit their model updates in an uncoded/uncompressed fashion simultaneously over the same channel resources. With this approach the wireless medium becomes a computation tool instead of solely enabling the transmission of information from the transmitter to a receiver.

Chapter 16 presents federated distillation as a communication-efficient distributed learning framework. The authors first introduce the concepts of knowledge distillation and codistillation and then explain how these techniques can be exploited in the federated learning setting to reduce the communication load. The chapter concludes with the application of federated distillation to two selected applications.

Chapter 17 addresses the privacy aspects of federated learning in the wireless context. It is known that released model parameters in federated learning may leak sensitive information about the underlying datasets. Differential privacy has been introduced as a method to address this privacy leakage. This chapter studies how differential privacy can be adopted when federated learning is carried out across wireless nodes.

Chapter 18 focuses on the inference aspect of machine learning at the wireless network edge. The authors first overview techniques such as network splitting, joint source-channel coding, and inference-aware scheduling to improve the speed and efficiency of inference over wireless networks. Then, they provide a detailed analysis of pruning-based dynamic neural network splitting and dynamic compression ratio selection methods.

In conclusion, we would like thank all the authors for their contributions to this book and for their hard work in presenting the material in a unified and accessible fashion. We hope that you, the reader, will find these expositions to be useful.

## References

[1] D. Gündüz et al., "Machine learning in the air," *IEEE Journal on Selected Areas in Communications*, vol. 37, no. 10, pp. 2184–2199, 2019.

[2] N. Farsad et al., "Data-driven symbol detection via model-based machine learning," *Communications in Information and Systems*, 2020.

[3] M. Chen et al., "Communication efficient federated learning," in *Proc. National Academy of Sciences (PNAS)*, doi:10.1073/pnas.2024789118, April 2021.

[4] S. Verdu, *Multiuser Detection*, 1st ed. Cambridge University Press, 1998.

[5] Y. Liu, Y. Dai, and Z. Luo, "Coordinated beamforming for MISO interference channel: Complexity analysis and efficient algorithms," *IEEE Trans. on Signal Processing*, vol. 59, no. 3, pp. 1142–1157, 2011.

[6] Q. Shi et al., "An iteratively weighted MMSE approach to distributed sum-utility maximization for a MIMO interfering broadcast channel," *IEEE Trans. on Signal Processing*, vol. 59, no. 9, pp. 4331–4340, 2011.

[7] M. Joham, W. Utschick, and J. A. Nossek, "Linear transmit processing in MIMO communications systems," *IEEE Trans. on Signal Processing*, vol. 53, no. 8, pp. 2700–2712, 2005.

[8] H. Sun et al., "Learning to optimize: Training deep neural networks for interference management," *IEEE Trans. on Signal Processing*, vol. 66, no. 20, pp. 5438–5453, 2018.

[9] H. Huang et al., "Unsupervised learning-based fast beamforming design for downlink MIMO," *IEEE Access*, vol. 7, pp. 7599–7605, 2019.

[10] H. Ye, G. Y. Li, and B. Juang, "Power of deep learning for channel estimation and signal detection in OFDM systems," *IEEE Wireless Communications Letters*, vol. 7, no. 1, pp. 114–117, 2018.

[11] M. B. Mashhadi and D. Gündüz, "Pruning the Pilots: Deep Learning-Based Pilot Design and Channel Estimation for MIMO-OFDM Systems," in *IEEE Transactions on Wireless Communications*, vol. 20, no. 10, pp. 6315–6328, Oct. 2021.

[12] M. Boloursaz Mashhadi and D. Gündüz, "Deep learning for massive MIMO channel state acquisition and feedback," *Journal of Indian Institute of Sciences*, no. 100, p. 369–382, 2020.

[13] C. Wen, W. Shih, and S. Jin, "Deep learning for massive MIMO CSI feedback," *IEEE Wireless Communications Letters*, vol. 7, no. 5, pp. 748–751, 2018.

[14] M. B. Mashhadi, Q. Yang, and D. Gündüz, "Distributed deep convolutional compression for massive MIMO CSI feedback," in *IEEE Transactions on Wireless Communications*, vol. 20, no. 4, pp. 2621–2633, Apr. 2021.

[15] W. Xu et al., "Joint neural network equalizer and decoder," in *15th Int. Symp. on Wireless Communication Systems (ISWCS)*, pp. 1–5, 2018.

[16] T. Gruber et al., "On deep learning-based channel decoding," in *51st Annual Conf. on Information Sciences and Systems (CISS)*, pp. 1–6, 2017.

[17] S. Dorner et al., "On deep learning-based communication over the air," in *51st Asilomar Conf. on Signals, Systems, and Computers*, pp. 1791–1795, 2017.

[18] A. Felix et al., "OFDM-autoencoder for end-to-end learning of communications systems," in *IEEE Int. Workshop on Signal Processing Advances in Wireless Communications (SPAWC)*, pp. 1–5, 2018.

[19] E. Bourtsoulatze, D. Burth Kurka, and D. Gündüz, "Deep joint source-channel coding for wireless image transmission," *IEEE Trans. on Cognitive Communications and Networking*, vol. 5, no. 3, pp. 567–579, 2019.

[20] N. Farsad, M. Rao, and A. Goldsmith, "Deep learning for joint source-channel coding of text," in *IEEE Int. Conf. on Acoustics, Speech and Signal Processing (ICASSP)*, pp. 2326–2330, 2018.

[21] D. B. Kurka and D. Gündüz, "DeepJSCC-f: Deep joint source-channel coding of images with feedback," *IEEE Journal on Selected Areas in Information Theory*, vol. 1, no. 1, pp. 178–193, 2020.

[22] D. B. Kurka and D. Gündüz, "Bandwidth-Agile Image Transmission With Deep Joint Source-Channel Coding," in *IEEE Transactions on Wireless Communications*, vol. 20, no. 12, pp. 8081–8095, Dec. 2021.

[23] N. Shlezinger, R. Fu, and Y. C. Eldar, "Deepsic: Deep soft interference cancellation for multiuser MIMO detection," *IEEE Trans. on Wireless Communications*, pp. 1–1, 2020.

[24] N. Shlezinger et al., "ViterbiNet: A deep learning based Viterbi algorithm for symbol detection," *IEEE Trans. on Wireless Communications*, vol. 19, no. 5, pp. 3319–3331, 2020.

[25] N. Shlezinger and Y. C. Eldar, "Deep Task-Based Quantization", Entropy 2021, 23, 104, January 2021.

[26] D. Gündüz et al., "Communicate to learn at the edge," *IEEE Communications Magazine*, vol. 58, no. 12, pp. 14–19, Dec. 2020.

[27] N. Shlezinger, Y. C. Eldar, and M. R. D. Rodrigues, "Hardware-limited task-based quantization," *IEEE Trans. on Signal Processing*, vol. 67, no. 20, pp. 5223–5238, 2019.

[28] M. Jankowski, D. Gunduz, and K. Mikolajczyk, "Wireless image retrieval at the edge," *IEEE Journal on Selected Areas in Communications*, vol. 39, no. 1, pp. 89–100, Jan. 2021.

[29] P. Neuhaus et al., "Task-based analog-to-digital converters," *IEEE Trans. on Signal Processing*, vol. 69, pp. 5403–5418, 2021.

[30] C. Ma et al., "On safeguarding privacy and security in the framework of federated learning," *IEEE Network*, vol. 34, no. 4, pp. 242–248, 2020.

[31] K. Wei et al., "Federated learning with differential privacy: Algorithms and performance analysis," *IEEE Trans. on Information Forensics and Security*, vol. 15, pp. 3454–3469, 2020.

[32] B. Hasırcıoğlu and D. Gündüz, "Private Wireless Federated Learning with Anonymous Over-the-Air Computation," ICASSP 2021 – 2021 *IEEE International Conference on Acoustics, Speech and Signal Processing (ICASSP)*, 2021, pp. 5195–5199.

[33] O. Tan, J. Gómez-Vilardebó, and D. Gündüz, "Privacy-cost trade-offs in demand-side management with storage," *IEEE Trans. on Information Forensics and Security*, vol. 12, no. 6, pp. 1458–1469, 2017.

[34] G. Giaconi, D. Gunduz, and H. V. Poor, "Privacy-aware smart metering: Progress and challenges," *IEEE Signal Processing Magazine*, vol. 35, no. 6, pp. 59–78, 2018.

[35] S. Karnouskos and F. Kerschbaum, "Privacy and integrity considerations in hyperconnected autonomous vehicles," *Proc. IEEE*, vol. 106, no. 1, pp. 160–170, 2018.

[36] M. Malekzadeh et al., "Protecting sensory data against sensitive inferences," in *Proc. 1st Workshop on Privacy by Design in Distributed Systems*, 2018.

[37] V. Sivaraman et al., "Smart iot devices in the home: Security and privacy implications," *IEEE Technology and Society Magazine*, vol. 37, no. 2, pp. 71–79, 2018.

[38] E. Nachmani et al., "Deep learning methods for improved decoding of linear codes," *IEEE Journal of Selected Topics in Signal Processing*, vol. 12, no. 1, pp. 119–131, 2018.

[39] O. A. Dobre et al., "Survey of automatic modulation classification techniques: classical approaches and new trends," *IET Communications*, vol. 1, no. 2, pp. 137–156, 2007.

[40] L. J. Wong, W. C. Headley, and A. J. Michaels, "Specific emitter identification using convolutional neural network-based IQ imbalance estimators," *IEEE Access*, vol. 7, pp. 33 544–33 555, 2019.

[41] T. J. O'Shea, J. Corgan, and T. C. Clancy, "Convolutional radio modulation recognition networks," in *Engineering Applications of Neural Networks*, C. Jayne and L. Iliadis, eds. Springer International Publishing, pp. 213–226, 2016.

[42] T. J. O'Shea, T. Roy, and T. C. Clancy, "Over-the-air deep learning based radio signal classification," *IEEE Journal of Selected Topics in Signal Processing*, vol. 12, no. 1, pp. 168–179, 2018.

[43] P. Triantaris et al., "Automatic modulation classification in the presence of interference," in *European Conf. on Networks and Communications (EuCNC)*, pp. 549–553, 2019.

[44] M. Kulin et al., "End-to-end learning from spectrum data: A deep learning approach for wireless signal identification in spectrum monitoring applications," *IEEE Access*, vol. 6, pp. 18484–18501 2018.

[45] T. O'Shea and N. West, "Radio machine learning dataset generation with gnu radio," *Proc. GNU Radio Conf.*, vol. 1, no. 1, 2016.

[46] P. Dong et al., "Deep cnn-based channel estimation for mmwave massive MIMO systems," *IEEE Journal of Selected Topics in Signal Processing*, vol. 13, no. 5, pp. 989–1000, 2019.

[47] W. Cui, K. Shen, and W. Yu, "Spatial deep learning for wireless scheduling," *IEEE Journal on Selected Areas in Communications*, vol. 37, no. 6, pp. 1248–1261, 2019.

[48] C. Doersch, "Tutorial on variational autoencoders," *arXiv preprint*, arXiv stat.ML.1606.05908, 2016.

[49] I. Goodfellow et al., "Generative adversarial nets," in *Advances in Neural Information Processing Systems*, vol. 27, Z. Ghahramani, M. Welling, C. Cortes, N. Lawrence, and K. Q. Weinberger, eds. Curran Associates, Inc., 2014, pp. 2672–2680.

[50] H. Ye et al., "Channel agnostic end-to-end learning based communication systems with conditional gan," in *IEEE Globecom Workshops (GC Wkshps)*, pp. 1–5, 2018.

[51] T. J. O'Shea, T. Roy, and N. West, "Approximating the void: Learning stochastic channel models from observation with variational generative adversarial networks," in *Int. Conf. on Computing, Networking and Communications (ICNC)*, 2019, pp. 681–686.

[52] J. A. Boyan and M. L. Littman, "Packet routing in dynamically changing networks: A reinforcement learning approach," in *Proc. 6th Int. Conf. on Neural Information Processing Systems*, 1993, p. 671–678.

[53] S. Singh and D. Bertsekas, "Reinforcement learning for dynamic channel allocation in cellular telephone systems," in *Advances in Neural Information Processing Systems*, vol. 9, M. C. Mozer, M. Jordan, and T. Petsche, eds. MIT Press, pp. 974–980, 1997.

[54] T. X. Brown, "Low power wireless communication via reinforcement learning," in *Proc. 12th Int. Conf. on Neural Information Processing Systems*, pp. 893–899, 1999.

[55] L. Lai et al., "Cognitive medium access: Exploration, exploitation, and competition," *IEEE Trans. on Mobile Computing*, vol. 10, no. 2, pp. 239–253, 2011.

[56] P. Blasco and D. Gündüz, "Multi-access communications with energy harvesting: A multi-armed bandit model and the optimality of the myopic policy," *IEEE Journal on Selected Areas in Communications*, vol. 33, no. 3, pp. 585–597, 2015.

[57] G. Cybenko, "Approximation by superpositions of a sigmoidal function," *Math. Control Signal Systems 2*, pp. 303–314, 1989.

[58] K. Hornik, "Approximation capabilities of multilayer feedforward networks," *Neural Networks*, vol. 4, no. 2, pp. 251–257, 1991.

[59] C. M. Bishop, *Pattern Recognition and Machine Learning (Information Science and Statistics)*. Springer-Verlag, 2006.

[60] I. Goodfellow, Y. Bengio, and A. Courville, *Deep Learning*. MIT Press, 2016.

[61] D. E. Rumelhart, G. E. Hinton, and R. J. Williams, *Learning Internal Representations by Error Propagation*. MIT Press, pp. 318–362, 1986.

[62] G. E. Hinton and R. R. Salakhutdinov, "Reducing the dimensionality of data with neural networks," *Science*, vol. 313, no. 5786, pp. 504–507, 2006.

[63] Y. Yaginuma, T. Kimoto, and H. Yamakawa, "Multi-sensor fusion model for constructing internal representation using autoencoder neural networks," in *Proc. Int. Conf. on Neural Networks (ICNN'96)*, vol. 3, pp. 1646–1651, 1996.

[64] A. Steudel, S. Ortmann, and M. Glesner, "Medical image compression with neural nets," in *Proc. 3rd Int. Symp. on Uncertainty Modeling and Analysis and Annual Conf. of the North American Fuzzy Information Processing Society*, 1995, pp. 571–576.

[65] Y. Bengio, N. Léonard, and A. Courville, "Estimating or propagating gradients through stochastic neurons for conditional computation," *arXiv preprint*, arXiv cs.LG.1308.3432, 2013.

[66] R. Gong et al., "Differentiable soft quantization: Bridging full-precision and low-bit neural networks," in *IEEE/CVF Int. Conf. on Computer Vision (ICCV)*, 2019, pp. 4851–4860.

[67] P. Vincent et al., "Extracting and composing robust features with denoising autoencoders," in *Proc. 25th Int. Conf. on Machine Learning*, pp. 1096–1103, 2008.

[68] S. I. Gelfand and M. S. Pinsker, "Coding of sources on the basis of observations with incomplete information," *Problems of Information Transmission*, vol. 15, pp. 115–125, 1979.

[69] T. J. O'Shea, J. Corgan, and T. C. Clancy, "Unsupervised representation learning of structured radio communication signals," in *1st Int. Workshop on Sensing, Processing and Learning for Intelligent Machines (SPLINE)*, pp. 1–5, 2016.

# Part I

## Machine Learning for Wireless Networks

# 2 Deep Neural Networks for Joint Source-Channel Coding

David Burth Kurka, Milind Rao, Nariman Farsad, Deniz Gündüz, and Andrea Goldsmith

## 2.1 Introduction

Digital communication systems typically entail separate steps for source coding and channel coding. In source coding, the source signal, e.g., text, image, or video, is mapped to a sequence of symbols that compresses the data for efficient transmission while guaranteeing a certain reconstruction quality. This involves stripping the original source signal of any redundancies. In channel coding, redundant symbols are systematically added to the compressed source sequence prior to transmission. These redundant symbols enable the receiver to detect or correct any errors that may be introduced during the transmission of data over the channel. Surprisingly, Shannon showed that this two-step approach to data communication is optimal for ergodic sources and channels when infinite blocklength codes are allowed [1]. Known as *Shannon's separation theorem*, this has been extended to a larger class of source, channel, and network scenarios [2–4].

Optimality of separation in Shannon's theorem assumes no constraint on the complexity of the source and channel code design. However, in practice, having large blocklengths may not be possible due to computational complexity as well as delay constraints. Thus, in challenging communication scenarios with more stringent power or latency constraints, or in the presence of multiple users or rapidly changing channels, the limitations of this separation-based approach become more apparent, significantly limiting the performance with respect to its fundamental information theoretic limit [5–7]. The alternative is to design the mapping from the source signal directly to the channel input, which is called joint source-channel coding (JSCC). There have been significant research efforts on JSCC over the years; however, these have focused either on the theoretical analysis under some idealistic source and channel distributions, e.g., [8–13], or on the joint optimization of the component parameters (vector quantizer, index assignment, channel code, and modulator) of an inherently separate design, e.g., [14–21]. Nonetheless, despite the suboptimality of separate source and channel coding in many practical settings, the lack of powerful JSCC techniques with reasonable coding and decoding complexities has prevented the emergence of alternatives to the modular separation-based approach.

In this chapter, we show that deep neural networks (DNNs) can be used to design JSCC solutions with impressive results. We illustrate the potential of DNN-based JSCC through concrete examples for various source and channel distributions.

In particular, we show that DNN-based JSCC schemes (a) achieve performance comparable or superior to state-of-the-art separation-based schemes, (b) provide graceful degradation upon deterioration of channel conditions, (c) have the versatility to adapt to different channels and source domains, (d) allow successive refinement with almost no performance loss, (e) exploit channel output feedback, and (f) support variable-length encoding.

The rest of this chapter is organized as follows. First, we review separate source and channel coding in Section 2.2. We then present how DNNs have been used for compression and channel coding in Section 2.3. A neural network–based JSCC for the transmission of text or natural language over discrete channels is presented in Section 2.4, and Section 2.5 deals with transmission of images over continuous channels. Section 2.6 concludes the chapter.

## 2.2    Source and Channel Coding

Consider a source signal $x \in \mathcal{R}^n$ to be transmitted over $k$ uses of a noisy channel, where $k/n$ is denoted as the *bandwidth ratio*. Conventional wireless point-to-point communication systems follow a modular design approach (see Fig. 2.1(a)), consisting of two steps: a source encoder followed by a channel encoder. The source encoder $f_s: \mathcal{R}^n \to \mathcal{B}^m$, $\mathcal{B} = \{0, 1\}$ maps $x$ into as few bits as possible, while the source decoder $g_s: \mathcal{B}^m \to \mathcal{R}^n$ reconstructs the original source signal from the compressed bits. When designing the source encoder and decoder, the goal is to minimize $m$ by compressing the source signal while allowing for reconstruction of the original source within the allowed distortion under a prescribed distortion measure.

Let $b \in \mathcal{B}^m$ be the compressed bits. The channel encoder $f_c: \mathcal{B}^m \to \mathcal{Z}^k$ maps the compressed bit sequence into a sequence of symbols transmitted over the channel, where $\mathcal{Z}$ denotes the channel input alphabet. In principle, the channel encoder introduces structured redundancy to correct any errors that may be introduced during transmission over the channel. Typically, the set $\mathcal{Z}$ is finite in digital communication systems (i.e., the channel input is discrete). However, in this work we also consider continuous-input channels, where $\mathcal{Z}$ is the set of complex numbers, $\mathcal{C}$. The channel introduces errors and distortion, and the received symbol sequence at the receiver is denoted by $\hat{z} \in \mathcal{Z}$. The channel decoder $g_c: \mathcal{Z}^k \to \mathcal{B}^m$ estimates the original $b$, potentially correcting the errors that are introduced during transmission. When designing the channel encoder and decoder, the goal is to use the smallest $k$, or add the fewest number of redundant symbols, while guaranteeing reliable communication of the bit sequence $b$.

Alternatively, in JSCC, the source $x \in \mathcal{R}^n$ is directly mapped to the channel input vector using a JSCC encoder $f_\theta: \mathcal{R}^n \to \mathcal{Z}^k$. Similarly, the channel output is directly mapped to an estimate of the source using a JSCC decoder $g_\phi: \mathcal{Z}^k \to \mathcal{R}^n$ (see Fig. 2.1(b)).

Practical wireless communication systems today almost exclusively rely on a separate design of the source and channel codes. The separate design provides modularity;

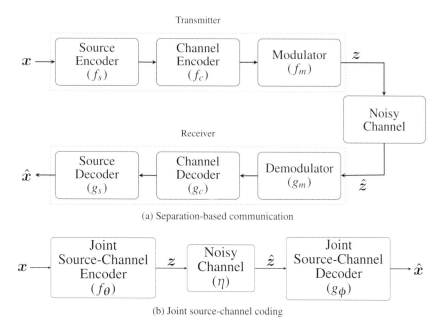

Figure 2.1 The components of (a) separation-based and (b) JSCC communication systems. In separation-based systems, each component is optimized independently. In JSCC, the input source is transformed directly into channel inputs that are transmitted and restored in a single step.

that is, the design and optimization of the source and channel components can be carried out independently, which simplifies the design process as the source and channel coding problems individually are much simpler to design and are better understood. Moreover, we have highly specialized source codes for different types of information sources, e.g., JPEG2000/ BPG for images, MPEG-4/ WMA for audio, or H.264 for video, which have been engineered by domain experts over many decades and many generations of standards. There are also universal source encoders such as gzip, which are designed to compress any type of data. Similarly, highly optimized channel coding techniques have been developed for additive white Gaussian noise (AWGN) channels, such as turbo, low-density parity-check (LDPC), and polar codes.

Shannon's Separation Theorem establishes the theoretical optimality of the separate design in the asymptotic infinite blocklength regime [22]. However, as we move toward less conventional communication paradigms, we are reaching the limits of this separate design. Particularly, for machine-type communications within the Internet of Things (IoT), this modular approach is increasingly limited in meeting the stringent transmission power and latency constraints of the devices and the underlying applications. For example, compression delay is currently the main bottleneck in ultra-low-latency communications, which is essential for many emerging applications such as virtual reality or tactile Internet. Moreover, due to the overcrowding of the wireless spectrum, communications increasingly take place over more challenging environments that do not follow traditional channel models. Fading and interference can have more degrading effects on the transmission compared to AWGN. Existing coding techniques perform poorly in such channel environments, and adapting their design to these complex channel statistics is extremely challenging, if not impossible.

Even within the scope of existing channel models, separation-based schemes are extremely sensitive to the channel parameters and can suffer severely when the channel conditions differ from those for which the codes have been optimized. For example, if the signal-to-noise ratio (SNR) in an AWGN channel is worse than the one for which the channel code rate is chosen, the error probability increases rapidly, provoking errors in both the channel and the source decoders, which can compromise the reconstruction quality significantly. Additionally, because the source and channel code rates are fixed, the reconstruction quality remains the same regardless of how much the channel SNR improves. These two characteristics are known as the "cliff effect" in digital communications. This also has implications when broadcasting to multiple receivers: those with worse channel conditions than the one targeted by the channel code are not able to reconstruct the source, while those with much better channel conditions do not obtain a better source reconstruction as a result.

In recent years, DNNs have been employed to improve both the source coding and the channel coding components of the conventional digital communication systems. In addition, the aforementioned limitations of the existing separate source and channel coding approach have motivated the use of deep learning (DL) to solve the JSCC problem. Before summarizing the proposed solution approaches in the next section, we will first give a brief overview of how DNNs have been used for source compression and channel coding.

## 2.3    DL-Based Source and Channel Coding

The design of source and channel codes has traditionally relied on human ingenuity. Several decades of intensive research have resulted in capacity achieving codes for the AWGN channel, such as turbo codes, LDPC codes, and polar codes. In parallel, numerous standards have been developed for specialized compression techniques for different types of information sources, e.g., JPEG2000 for images, as well as H.264 and H.265 for videos. However, with the recent developments in DL, a new data-driven approach for source and channel coding individually as well as for JSCC has emerged, as shown in Fig. 2.2. In this new paradigm, the encoders and decoders are replaced by DNNs, which are trained directly from data for source and channel coding, as we now describe in more detail.

- *DL for source coding*: Data-driven methods, such as principal component analysis (PCA), have long been used for dimensionality reduction and feature extraction. More recently, neural networks paired to form an autoencoder network have been shown to provide much better performance for dimensionality reduction and feature extraction [23]. This naturally led to employing autoencoders for source

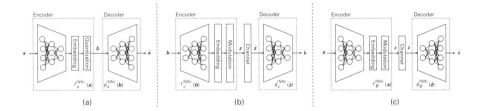

**Figure 2.2** Data-driven approach using DL for (a) source coding, (b) channel coding, and (c) JSCC.

compression by incorporating a quantization layer as shown in Fig. 2.2(a). In this architecture, the neural network encoder and decoder are jointly trained as autoencoders together with the quantization component, using sample source data (e.g., image, audio, or video samples).

- *DL for channel coding*: For a fixed channel encoder, the decoding operation at the receiver is a standard classification problem, where the received noisy channel output is classified into one of the input channel codewords. Many recent works have shown that DNN-based decoders can outperform existing conventional decoder architectures, e.g., the belief propagation decoder [24, 25]. On the other hand, both the encoder and the decoder can also be replaced by DNNs, which results in an autoencoder architecture similar to source compression. However, in the case of channel coding, we need to introduce the modulation and the channel as an untrainable neural network layer between the encoder and the decoder. This is illustrated in Fig. 2.2(b). Hence, as opposed to source coding, where the goal of the autoencoder is to learn the most efficient representation of the source sequence, the goal here is to learn channel codewords that can be recovered reliably despite channel impairments.

- *DL for JSCC*: Neural networks can also be used to design the mapping from the source signals directly to the channel inputs and the reconstruction of the source signal directly from the noisy channel output, i.e., $f_\theta$, and $g_\phi$ in Fig. 2.1(b), jointly. This method is similar to the channel code design: it incorporates the channel as an untrainable layer in the autoencoder architecture but uses sample source data for training the network. This architecture is illustrated in Fig. 2.2(c).

Next, we discuss each of these approaches in more detail.

## 2.3.1　Source Coding Using DL

Lossy source coding using DL has primarily relied on an autoencoder-type architecture. In this architecture, an encoder neural network compresses the input into a lower-dimensional embedding and a decoder reconstructs the input from this lower-dimensional embedding. For lossy source coding, the lower-dimensional embedding needs to be quantized. This process is shown in Fig. 2.2(a).

For end-to-end training of lossy source coding methods using DL, prior works have typically relied on the optimization of a rate-distortion loss given by

$$\mathcal{L} = - \underbrace{\log_2 P(\lceil f_s(\boldsymbol{x}) \rceil)}_{\text{Number of bits}} + \lambda \underbrace{d(\boldsymbol{x}, g_s(\lceil f_s(\boldsymbol{x}) \rceil))}_{\text{Distortion}}, \qquad (2.1)$$

where $P$ is a probability mass function that depends on the network parameters, distortion $d(\cdot, \cdot)$ measures the discrepancy between the input signal and its reconstruction at the decoder, and the parameter $\lambda > 0$ controls the trade-off between the distortion and the number of bits. Here $\lceil \cdot \rceil$ is the rounding function used for quantization.

Since quantization is not a smooth function, it can suppress the backpropagation of gradients during training. Several methods have been proposed to deal with this challenge. In [26], a stochastic binarization method is proposed for quantization. This stochastic binarization was used in [27] along with recurrent neural networks to achieve a reconstruction performance that outperforms BPG, WebP, JPEG2000, and JPEG. Later in [28], stochastic binarization was generalized to stochastic rounding to the nearest integer. In this method, during the forward pass, the following random variable is used:

$$b = \lfloor f_s(\boldsymbol{x}) \rfloor + \epsilon, \qquad \epsilon \in \{0, 1\}, \qquad P(\epsilon = 1) = f_s(\boldsymbol{x}) - \lfloor f_s(\boldsymbol{x}) \rfloor, \qquad (2.2)$$

where $\lfloor \cdot \rfloor$ is the floor operation. In the backward pass, the derivative is replaced with the derivative of the expectation, which is equal to 1, and hence, the derivative passes through the quantization layer unchanged.

A smooth approximation of vector quantization that was annealed toward hard quantization during training is used in [29]. Another approach in [30] adds uniform noise during training as an approximation to rounding at test time. In this scheme, the loss in Eq. (2.1) becomes

$$\mathcal{L} = - \underbrace{\log_2 p(f_s(\boldsymbol{x}) + \boldsymbol{u})}_{\text{Number of bits}} + \lambda \underbrace{d(\boldsymbol{x}, g_s(f_s(\boldsymbol{x}) + \boldsymbol{u}))}_{\text{Distortion}}, \qquad (2.3)$$

where $p$ is now a probability density function, and $\boldsymbol{u}$ is a random vector with independent and identically distributed (i.i.d.) elements drawn from a uniform distribution $u_i \sim U(-0.5, 0.5)$. If the distortion measure is the mean-squared error, then this approach is equivalent to a variational autoencoder [31] with a uniform encoder. Later in [32–34], the spatial dependencies that might exist in the source (e.g., spacial correlations between pixels in an image) were further exploited by transmitting such information as side information to the decoder, which is in parallel to what is being used by conventional compression methods to improve performance. These DNN-based methods surpass conventional compression standards in various performance metrics such as peak SNR (PSNR) and also in terms of subjective perceptual quality [35].

## 2.3.2    Channel Coding Using DL

Channel codes can also be designed using autoencoders or variational autoencoders. In this approach, the output of the encoder neural network is passed on to a modulator and transmitted over the channel. The channel output is then passed to a decoder, which estimates the input bits, *b*, from the noisy channel output. The network can be trained end-to-end to learn new channel codes, but to allow joint training of the encoder and decoder neural networks, the modulation scheme used and the stochastic channel both must be modeled or approximated as neural network layers. Using this approach, the system learns a reversible transformation from the input data to a latent space, and then from noisy observations of the latent space back to the estimate of the original input data. This process is illustrated in Fig. 2.2(b).

The design of the encoder and decoders using autoencoders was first proposed in [36, 37] for short blocklength codes. For such codes it was shown that the DL-based channel code design can achieve the same or better performance compared to Hamming codes when used over AWGN channels. These results were then extended to OFDM channels in [38], convolutional codes over AWGN channels in the short blocklength regime in [39], and turbo codes over AWGN and additive T-distribution noise (ATN) channels in [40]. Autoencoders were used in [41, 42] for novel code design for the feedback channel, for which no practical coding scheme exists even under known channel statistics. Despite the promising performance achieved by DL-aided channel code design over short to moderate blocklengths, extending this result to longer codes has been a challenge due to the exponentially growing codebook size with the blocklength.

Due to this difficulty, some of the prior works have explored using DNNs to decode existing channel codes, such as linear [24], convolutional [43], turbo [44], or polar [45] codes. Since the decoder of any known channel code can be treated as a classifier on the noisy channel output, this approach has provided promising results in either improving the error probability or reducing the computational complexity of the decoder. For example, in [46] a multiple-in multiple-out (MIMO) channel decoder using DNN was proposed that achieves a performance close to approximate message passing and semidefinite relaxation at a much reduced computational complexity. Even more impressive results have been achieved through DNN-based designs in settings where current codes fall short of the fundamental theoretical limits. For example, this is the case for channels that are harder to model, such as optical [47] and molecular [48] communications, or even blind channel equalization [49].

Autoencoder-based channel codes do not yet provide significant improvements over existing conventional codes, especially at longer blocklengths. However, state-of-the-art channel codes are the result of decades-long intense efforts and expertise, whereas data-driven DNN techniques that were introduced only a few years ago have already achieved impressive results, especially for short blocklengths. Moreover, channel codes designed using DL tend to be more resilient than conventional codes, such as when the channel conditions change with respect to the underlying model they were designed for. This effect is also observed when DNNs are used for decoding of conventional channel codes.

## 2.3.3    JSCC with DL

As we have mentioned earlier, there is a long history of research on JSCC, and there are numerous studies both on information theoretical limits and on practical code design. However, existing results have either been for specific designs exploiting the properties of a particular source signal or were too complex to be used in practice. Moreover, such code designs have not provided significant improvements in practice to justify the introduced complexity in their design. The lack of low-complexity, high-performance JSCC solutions together with the recent advances in DL-aided coding schemes for source compression and channel coding have motivated the application of DL to design novel JSCC schemes.

In DL-aided JSCC, the input to the DNN encoder is the source signal, while its outputs are the symbols transmitted over the channel. The channel is modeled or approximated as another layer in the DNN architecture, where the output of the channel is passed to the DNN decoder. The encoder and the decoder for JSCC can be trained end-to-end using a data-driven approach. This process is illustrated in Fig. 2.2(c).

One of the first works that proposed JSCC using neural networks is [50], where simple neural network architectures were used as encoder and decoder for Gauss-Markov sources over the additive white Gaussian noise channel. More recently in [51–56], autoencoder-based solutions for end-to-end design and optimization of JSCC were proposed.

Specifically, [51, 57] focus on text as the information source that is communicated over discrete channels, [52] considers JSCC for lossy data storage, and image transmission over an AWGN wireless channel is studied in [54–56, 58, 59]. In [54], the authors propose a fully convolutional autoencoder architecture, which maps the input images directly to channel symbols, without going through any digital interface. The authors show that the proposed DeepJSCC architecture not only improves upon the concatenation of state-of-the-art compression and channel coding schemes in a separate architecture [55], but it also provides graceful degradation as the channel SNR degrades. This latter property, which is common to analog transmission schemes, provides significant benefits compared to digital schemes, which exhibit catastrophic error when the channel SNR significantly deviates from an expected value determined on system's design. This is particularly common when broadcasting to multiple receivers or when transmitting over a time-varying channel. DeepJSCC is also shown in [56, 58] to be almost *successively refinable*; that is, an image can be transmitted in stages, where each stage refines the quality of the previous stages at almost no additional cost. Finally, in [59] JSCC of images transmitted over binary symmetric and over binary erasure channels are considered. To overcome the challenges imposed by the nondifferentiability of discrete latent random variables (i.e., the channel inputs), unbiased low-variance gradient estimation is used, and the model is trained using a lower bound on the mutual information between the images and their binary representations.

One of the other benefits of JSCC using DL is the ability to jointly optimize the encoder and the decoder for the downstream DL task. For example, if the receiver is interested in object detection using DL on an image received from the transmitter, the JSCC encoder and decoder can be trained together with the image-detection network in an end-to-end manner, thereby optimizing the encoder and decoder for the downstream task rather than just for image reconstruction.

In the rest of this chapter, we separately focus on JSCC for *discrete* and *continuous* channels and provide specific examples on how the general approach proposed in this section can be applied for JSCC design in different domains. Specifically, we first consider a specific JSCC design for text transmission over discrete input channels in Section 2.4 and then a specific JSCC design for image transmission over continuous input channels in Section 2.5.

## 2.4          DL-Aided JSCC for Text

In this section, we focus on the JSCC of text over discrete channels using a DL approach. In several applications, system performance is not measured by fidelity to the transmitted data but by the performance of downstream systems that use the received data. That is, the receiver is less interested in the exact recovery of the transmitted data than in the relevant information of interest or a facsimile of the data that would be used in downstream applications. In the case of text data, the receiver would be interested in recovering the semantic content of a sentence including facts, relations, topics, or keywords as opposed to the exact sentence that can include noninformative carrier phrases. Downstream natural language processing tasks of summarization, topic classification, sentiment detection, intent, and named-entity extraction would use these artifacts. We declare decoded sentences error free if they convey the equivalent information as the original sentence, even if they are paraphrased (e.g., "the car stopped" and "the automobile came to a halt"). The neural network architecture for JSCC of text we develop is inspired by recent state-of-the-art results of DL in natural language processing tasks such as machine translation, summarization, and semantic understanding.

Our model is composed of a recurrent neural network (RNN) encoder, a stochastic binarization layer, the channel layer, and a decoder based on RNNs. We use this architecture to train a JSCC encoder-decoder pair and show that it is possible to obtain different but equivalent sentences that preserves the semantic content of the transmitted sentence. We introduce schemes where a fixed-length binary encoding is produced for an input sentence as well as one in which encodings of variable lengths are produced for sentences that vary in length and complexity. The latter variable-length architecture would improve on the performance of the fixed-length encoding scheme with a less strict average sentence encoding length constraint expending more bits to encode longer sentences than frequently occurring shorter ones.

The performance of our DL encoder and decoder is contrasted with separate source and channel coding design. In the separate design, channel coding is done using Reed-Solomon codes. For compression or source coding, we consider three different methods: a universal source coding scheme, Huffman coding, and a 5-bit character ASCII encoding. We show that the proposed DL encoder and decoder does better than the separate design on the metric of word error rate (WER) or edit distance, when each sentence is encoded using fewer bits. In several cases, the DL decoder may insert, replace, or substitute words that preserve the semantic content of the sentence in a

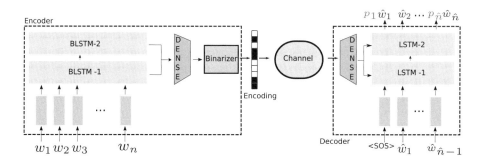

**Figure 2.3** The DL encoder-decoder architecture for JSCC of text.

qualitative sense, but this will not be reflected in the edit distance that penalizes these transformations. In order to capture the impact of replacing words with synonyms, a new metric is also proposed that scores word substitutions with a similarity score between them.

## 2.4.1    System Model

The system model in this particular application is defined as follows: Let $\mathcal{V}$ be the entire vocabulary, indexing the set of all the words in the language. Then the source in our model is $x \in \mathcal{V}^n$, where $n$ is the length of the sentence and $x = [w_1, w_2, \ldots, w_n]$ is a vector representing the sequences of words in the sentence. Note that, although the source is not a real number in this setup, $\mathcal{V} \subset \mathcal{R}$. As the channel, we consider the binary erasure channel (BEC), the binary symmetric channel (BSC), and the AWGN channel. Let $\hat{x} = [\hat{w}_1, \hat{w}_2, \ldots, \hat{w}_{\hat{n}}]$ be the output of the JSCC decoder (i.e., the recovered sentence). With this framework, the number of words in the decoded sentence can differ, or equivalently, we can have $n \neq \hat{n}$. Specifically, we design the JSCC encoder and decoder (i.e., $f_\theta$ and $g_\phi$) such that the meaning between the transmitted sentence $x$ and the recovered sentence $\hat{x}$ is preserved. Therefore, the transmitted and recovered sentences may have different words and different lengths. We now further describe the component modules in the system.

**Neural JSCC Architecture for Text**

The DL architecture we implement is inspired from the sequence-to-sequence learning framework [60]. The end-to-end neural JSCC architecture is shown in Fig. 2.3. It has primarily three components: the encoder, the channel, and the decoder. The encoder $f_\theta$ takes a sentence $x$ as input and produces a binary encoding $z \in \mathcal{Z}^k$, where $\mathcal{Z} = \{0, 1\}$. The channel transforms this bit vector $z$ to realize an output vector $\hat{z}$ at the receiver. This module is stochastic. We will consider different cases, where the channel output alphabet is either binary, ternary, or continuous. The channel output vector $\hat{z}$ is the input to the decoder $g_\phi$, and the output of the decoder is the estimated sentence $\hat{x}$. We now describe each of these modules in detail.

**Encoder**

The encoder first applies an embedding layer to generate a continuous vector representation for each word of the input sentence. We make use of pretrained Glove word vectors [61] to initialize the embedding layer. Glove word vectors have been obtained using joint co-occurence statistics of words from a large text corpora; they have been shown to capture the semantic meaning of words as demonstrated by performance in the word analogy task. The embedding is represented by $E = \theta_e(s)$, where $E = [e_1, e_2, \ldots, e_n, e_{eos}]$ is the $n + 1$ embeddings of words in the sentence. We have $n + 1$ words in the sentence as an additional end of sentence symbol is affixed in the data preparation process.

In the next step of the encoder, the word embeddings are inputs to a stacked bidirectional long short term memory (BLSTM) network [62]. LSTM cells with peepholes have been used in this work similar to that used in [63]. We can represent the BLSTM layers by

$$r = \theta_{\text{BLSTM}}(E), \tag{2.4}$$

where $r$ is the output state of the BLSTM stack. Each individual layer of the stack comprises two states from the forward network and backward network. These states from all the layers are concatenated to form $r$.

The output state $r$ is then fed to a feed-forward dense layers with tanh activation or a multilayer perceptron (MLP). This layer is used to modulate the dimension of the binary encoding of the sentence. Or, the MLP is used to increase or decrease the dimension of $r$ to $\ell_{\max}$, the maximum number of bits used to encode the sentence. This is represented by

$$v = \theta_{\text{MLP}}(r), \tag{2.5}$$

where $v \in [-1, 1]^{\ell_{\max}}$.

The final step in the encoder is to binarize $v$ from the interval $[-1, 1]$ to binary values $\{-1, 1\}$. We define a scalar stochastic binarization function as

$$\theta_\beta^{\text{sto}}(x) = x + Z_x, \tag{2.6}$$

where $Z_x$ is a random variable with distribution

$$Z_x \sim \begin{cases} 1 - x & \text{w.p. } \frac{1+x}{2} \\ -x - 1 & \text{w.p. } \frac{1-x}{2} \end{cases}. \tag{2.7}$$

This final binarization step is

$$z = \theta_\beta^{\text{sto}}(v) \tag{2.8}$$

in the forward pass. A custom operation is defined to back-propagate gradients through the binarization layer during training. The gradients are defined to pass straight through or unchanged through $\theta_\beta^{\text{sto}}$. This is obtained by using the derivative with respect to the expectation $[\theta_\beta^{\text{sto}}(v)] = v$ to calculate the gradient [64]. When the trained

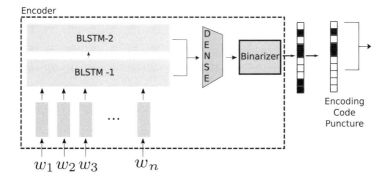

**Figure 2.4** The variable length encoder that can produce binary codes with lengths proportional to the sentence length.

model is deployed for inference, the deterministic binarizer $\theta_\beta^{\text{det}}(v) = 2u(v) - 1$ is used, where $u(x)$ is the unit step function in place of the stochastic $\theta_\beta^{\text{sto}}(v)$.

**Variable Length Encoding**

In the described encoder, all sentences are mapped to a binary codeword of constant length regardless of differences in sentence lengths and complexities. If we are constrained by the average length of encoding of each sentence, it is optimal to encode shorter sentences with encodings of a shorter length and expend more bits to encode longer sentences. This would also occur if we are to transmit a block of text or collection of sentences using as few bits as possible.

Figure 2.4 describes the architecture of the variable length encoder. This is accomplished by transmitting the first $\ell_n = L(n)$ encoded bits, where $L$ is a function that maps the sentence length $n$ to the length of binary encoding $\ell_n$. We implement this code puncturing in the neural network model by zeroing out the last $\ell_{\max} - \ell_n$ bits in $x$. As bits are represented using -1 and 1, symbol 0 is equivalent to not transmitting the bit.

**Channel**

The next module in the system models the communication channel. The transform function of the channel that maps the input to the output at the receiver must be differentiable to facilitate joint training of the encoder and decoder using variants of stochastic gradient descent. In this chapter, we consider three different channels: the BEC, the BSC, and the AWGN channels. The framework can be extended to models with memory such as inter-symbol interference (ISI) channels or channels with nonlinearities.

The BEC can be implemented in training via a dropout layer [65],

$$\hat{z} = \eta_{\text{bec}}(z, p_d), \tag{2.9}$$

where $\hat{z}$ is the received vector, and $p_d$ is the probability of erasing a bit. The elements of $\hat{z}$ are in the ternary set $\{-1, 0, 1\}$, where 0 indicates an erasure. Each bit in $z$ may or may not be dropped independent of other bits.

The BSC can be modeled as

$$\hat{z} = \eta_{\text{bsc}}(\boldsymbol{x}, p_e) = \boldsymbol{n}_{p_e} \odot \boldsymbol{z}, \tag{2.10}$$

where $\boldsymbol{n}_{p_e}$ is the noise introduced by the channel with an element of the vector equal to $-1$ with probability $p_e$ and 1 otherwise, and $\odot$ denotes element-wise multiplication. In other words, bits may be inverted with probability $p_e$.

Finally, the real-valued AWGN channel is represented by

$$\hat{z} = \eta_{\text{awgn}}(\boldsymbol{z}, \sigma^2) = \boldsymbol{z} + \boldsymbol{n}_{\sigma^2}, \tag{2.11}$$

where $\boldsymbol{n}_{\sigma^2}$ is an additive noise term with elements arising from i.i.d. Gaussian random variables with zero mean and variance $\sigma^2$. Note that the output $\boldsymbol{y}$ consists of a vector of real values here.

**Decoder**

At the receiver, the first step in the decoding process is to change the dimension of the observation vector $\hat{z}$ using an MLP or a feed-forward network:

$$\boldsymbol{c} = \phi_{\text{MLP}}(\hat{z}). \tag{2.12}$$

The MLP is an adapter module that transforms the received vector so that it can serve as the initial state of the next submodule in the decoder, a stacked LSTM decoder. The stacked unidirectional decoder with initial state $\boldsymbol{c}$ emits each word of the decoded sentence auto-regressively. This is given by

$$\hat{x} = \phi_{\text{LSTM}}(\boldsymbol{c}, \langle \text{sos} \rangle), \tag{2.13}$$

where $\hat{s}$ represents the decoded sentence. The embedding vector for the special start of the sentence symbol $\langle \text{sos} \rangle$ serves as the first input to the LSTM stack. At each decoding step, the word is sampled from the distribution over words in the vocabulary represented by the logits of the output layer and this embedding of this word serves as the input for the next decoding step. During training, we set up a schedule for annealing the probability of using the previously decoded word as the decoder input or using the ground truth data. During model deployment and inference, we have a choice of greedy decoding that uses the previously decoded word or a beam search algorithm that maintains a fixed number of likely beams. Beam decoding ensures that a high probability decoding error early on does not cascade to the entire sentence that results in overall lower beam probability.

## 2.4.2   Experimental Setup

**Dataset**

The sentences that are transmitted are drawn from the News Crawl 2015 dataset [66]. This dataset was constructed by crawling through articles of various online English publications. Our vocabulary is selected using the 20,000 most popular words in the dataset. We filter sentences from lengths 4 to 30 with less than 20% of the vocabulary words. We used 18.5 million sentences, or 90%, for model training and validation, and the remaining 2 million sentences form the test set.

## Model Training Details

The loss function used to train the model is categorical cross-entropy function that minimizes the Kullback-Leibler (KL) divergence between the probabilities of the decoded words, represented by the logits emitted by the decoder final layer, and the one-hot distribution of the ground truth sequence.

To initialize the encoder word embedding layer, 200-dimensional pretrained Glove embeddings [61] are used. The special symbols out-of-vocabulary, padding, start of sentence, and end of sentence are randomly initialized using the uniform distribution scaled inversely be embedding length. The data processing pipeline prepares batches of 512 sentences bucketed by their sequence lengths as input to the JSCC encoder. The encoder BLSTM has two layers of 256 units with peephole connections.

To obtain a constant length encoding, the encoder MLP transforms the concatenated end states of the BLSTM layers to a vector of dimension $\ell$, the number of transmission bits. To obtain variable length embeddings, we log the histogram of sentence lengths. To map the sequence length to binary encoding length, the smallest bucket (of length 4–7) are allotted 250 bits. Subsequent buckets of width 4 are allotted 50 more bits linearly. The average number of bits per sentence amounts to 400. The encoder MLP maps the BLSTM states to the maximum bit length, which is then punctured. The decoder LSTM stack has 2 512-unit layers with peephole connections. An Adam optimizer with initial learning rate of $10^{-3}$ is used for 6 epochs on the training dataset. At inference time, a beam decoder with width 10 is used.

## Baselines

Separate source and channel coding schemes serve as baselines for comparison. As previously noted, the separate design is optimal for arbitrarily large blocklengths with no constraints on delays for the channels we consider. We consider the following source coding approaches:

1. Universal compressors: Lempel-Ziv universal compression [67] implemented in gzip is the first method. Universal compression asymptotically reaches the entropic compression limit for any arbitrary source of data. Empirically, large blocks of sentences are required for good compression performance, and we use blocks of 30 sentences for evaluation. This method is unsuitable for the transmission of single sentences unlike the other baselines and the proposed DL JSCC encoder.
2. Huffman coding: The character frequencies obtained using the validation set are used in a Huffman coding scheme to encode characters of single sentences.
3. Fixed-length character encoding: This is the computationally simple baseline that assigns a fixed 5-bit ASCII coding for the characters including the lower case alphabet and the special symbols. We implement this baseline as the receiver can retrieve the sentence partially when the channel introduces noise to the source code. This partial retrieval capability is not available in the other baselines.

In the separate design baselines, channel coding is done via Reed-Solomon codes [68] after source coding. With $x$ bits of added redundancy, this can correct up to $x$ erasures in the binary erasure channel or $\lfloor x/2 \rfloor$ inversion errors of the binary symmetric channel or the maximum likelihood decoded binary vector in the AWGN channel. The message cannot be decoded at all with gzip or Huffman codes if the channel introduces more erasures or errors than the channel code can correct for. We tune the number of redundant bits added to maximize word throughput based on channel statistics via the following trade-off: too few parity bits would result in a high probability of decoding failures of the entire sentence and too many parity bits will require words to be dropped so that the sentence can be transmitted with the required bit budget.

Performance is measured via the normalized edit or Levenshtein distance. This metric is obtained by a dynamic programming algorithm that finds the shortest sequence of insert, delete, or substitute operations to map the reference sentence to the hypothesis. To capture the semantic similarity more accurately, we underweight the substitution cost in a new metric by using the Wu-Palmer score for relatedness of words [69].

## 2.4.3    Results

In Fig. 2.5(a), we first observe the effect of the bit erasure rate of the BEC on the word error rate (WER). Gzip is the best performing baseline with large batches of sentences. In the regime where the binary erasure probability is large, the DL JSCC model proposed outperforms the baselines, suggesting that the former is robust to channel errors and exhibits graceful degradation in performance. In Fig. 2.5(b) we use the BSC and observe an increase in WER with sentence lengths across the board with the performance of the proposed method comparable to baselines in the limited bit regime for longer sentences. We reiterate that word errors for the baselines imply that a word was lost at the receiver as it either was not transmitted or incorrectly decoded, whereas the DL JSCC may preserve the semantic content of the sentence. Figure 2.5(c) reinforces the conclusions of the earlier plots with the AWGN channel model.

Figure 2.6(a) first compares the fixed-length encoding architecture to the variable length one. Longer sentences are allotted more bits in the latter scheme, leading to significantly fewer errors without impacting the performance for shorter sentences encoded with fewer bits. As can be seen, the increased bit allotment for longer sentences results in fewer errors without much loss in performance for shorter sentences. Finally, in Fig. 2.6(b) we investigate the impact of source and channel coding in JSCC by contrasting a DL JSCC network with 450 bits/sentence for the binary erasure channel with a scheme where the neural network is used only for source coding that generates a 400 bit representation after which Reed-Solomon coding with 50 added parity bits is used. The plots suggest that the latter scheme performs fractionally better, implying that source coding accounts for most performance gains for DL JSCC. As noted in Section 2.3.2, DNN based channel coders outperform conventional

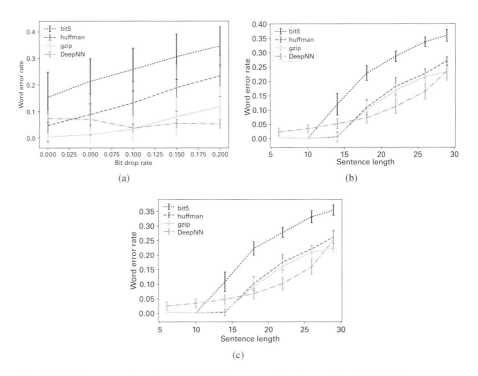

**Figure 2.5** Performance plots of DL JSCC with constant length encoding: (a) Word error rate (WER) increases with erasure probability in BEC for 400 bit encoding, (b) WER for sentences of different lengths for a binary symmetric channel with error rate 0.5, and (c) WER for sentences of different length with an AWGN channel with a standard deviation of 0.6.

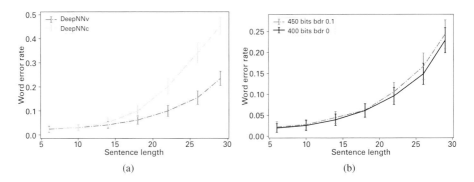

**Figure 2.6** (a) Contrasting WER with sentence length for fixed and variable length encoding with the BEC channel with erasure probability 0.1 and 400 bit average sentence encoding length, and (b) comparing DL JSCC with neural network source coding and Reed-Solomon coding.

techniques for channels that are hard to model such as optical or molecular channels, and we similarly expect the gap between DL JSCC and source coding with Reed-Solomon coding to reduce or invert with such channels.

In the next section, we describe JSCC for images over continuous input channels.

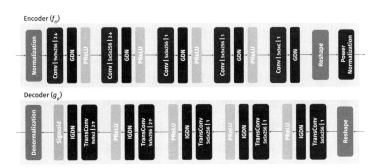

**Figure 2.7** Encoder and decoder architectures used in experiments.

## 2.5  DL-Aided JSCC for Images

An increasing number of applications involve transmission of images over wireless channels. This is not only for the traditional human-to-human (e.g., social networks, messaging, online content) communications, but also for human-to-machine (e.g., artificial or virtual reality, telepresence) and machine-to-machine (e.g., surveillance, pattern recognition) applications that are increasingly connected through wireless links. This exponentially growing demand for high quality image communication under strict latency constraints present new challenges on the wireless infrastructure.

This section reviews recent developments on *DeepJSCC*, an autoencoder-based solution for generating robust and compact codes directly from images pixels. The channel input symbols produced by DeepJSCC are not constrained to a specific constellation, so it is able to operate with continuous values (instead of discrete, as in the previous section). This property ensures that DeepJSCC can present analog behavior such as graceful degradation. We demonstrate with experimental results that Deep-JSCC achieves superior performance over the state-of-the-art digital communication schemes (BPG/ JPEG2000 compression followed by LDPC+QAM for transmission) on static channels, presenting graceful degradation as the channel quality degrades, and can successfully adapt to time-varying channels. Finally, we present a hardware implementation of the scheme, showing the model's application to real physical channels.

## 2.5.1  System Model

Consider an image source with height $H$, width $W$, and $C$ color channels, represented as a vector of pixel intensities $x \in \mathcal{R}^n$; $n = H \times W \times C$. An encoder $f_{\theta_i}: \mathcal{R}^n \to \mathcal{C}^{k_i}$ maps $x$ into a block of channel input symbols $z_i \in \mathcal{C}_i^k$. The transmission of $k$ symbols is split over $L$ layers, so the total bandwidth $k$ is achieved by accumulating $L$ transmissions $\left( \sum_{i=1}^{L} k_i = k \right)$. Unlike in the previous section, the encoder outputs continuous complex symbols, that is, $\mathcal{Z} = \mathcal{C}$. These symbols are transmitted over a noisy channel, characterized by a random transformation $\eta: \mathcal{C}^{k_i} \to \mathcal{C}^{k_i}$, resulting in the corrupted channel output $\hat{z}_i = \eta(z_i)$. We consider $L$ distinct decoders, where the channel outputs for the first $i$ layers are decoded using $g_{\phi_i} : \mathcal{C}^{k_l} \rightarrow \mathcal{R}^n$ (where $l = \sum_{j=0}^{i} k_j$), creating reconstructions $\hat{x}_i = g_{\phi_i}(\hat{z}_1, \ldots, \hat{z}_i) \in \mathcal{R}^n$, for $i \in 1, \ldots, L$.

Figure 2.7 shows the component blocks of the chosen architecture and its hyper-parameters. The encoder and decoder are composed of a series of trainable convolutional neural network (CNN) blocks, using generalized normalization transformations (GDN/IGDN) [70], followed by a parametric rectified linear unit (PReLU) [71] activation function (or a sigmoid, in the last decoder block). This architecture was inspired by [72] and improved with ablation studies. Intuitively, convolutional layers extract image features, GDN apply local divisive normalization, and the nonlinear activation allows the learning of nonlinear mapping from the source signal space to the coded signal space. The communication channel is incorporated into the model as a nontrainable layer. Although different channel models are considered, as described in the experimental section, all of them are differentiable transfer functions, allowing their inclusion in the general architecture and enabling gradient computation and error back propagation.

Before transmission, the latent vector $z_i'$ generated at the encoder's last convolutional layer is normalized to enforce an average power constraint so that $\frac{1}{k_i} \mathbb{E}[z_i^* z_i] \leq P$, by setting

$$z_i = \sqrt{k_i P} \frac{z_i'}{\sqrt{z_i'^* z_i'}}. \tag{2.14}$$

where $\tilde{z}_i^*$ is the conjugate transpose of $\tilde{z}_i$.

The model can be optimized to minimize the average distortion between input $x$ and its reconstructions $\hat{x}_i$ at each layer $i$:

$$(\theta_i^*, \phi_i^*) = \arg\min_{\theta_i, \phi_i} \mathbb{E}_{p(x, \hat{x})}[d(x, \hat{x}_i)], \tag{2.15}$$

where $d(x, \hat{x}_i)$ is a specified distortion measure, usually the mean squared error (MSE), although other metrics are also considered. Since the true distribution of $p(x)$ is unknown, an analytical form of the expected distortion in Eq. (2.15) is also unknown, so we estimate the expected distortion by sampling from a dataset.

When $L > 1$, we have a multiobjective problem. However, we simplify it so that the optimization of multiple layers is done either jointly, by considering a weighted combination of distortions, or greedily, by optimizing $(\theta_i, \phi_i)$ successively. Please see [56, 58, 73] for more details.

## 2.5.2    Evaluation Metrics

In order to measure the performance of the proposed JSCC algorithm and alternative schemes, we use the peak signal-to-noise ratio (PSNR), given by

$$\text{PSNR} = 10 \log_{10} \frac{\text{MAX}^2}{||x - \hat{x}||^2} \quad (dB), \tag{2.16}$$

where $MAX^2$ is the maximum power our input signal can have. When evaluating 24-bits RGB images, $MAX = 255$ is given by the maximum value a pixel can represent.

The quality of the channel is measured by the average signal-to-noise ratio (SNR) given by

$$SNR = 10 \log_{10} \frac{P}{\sigma^2} \quad (dB), \tag{2.17}$$

representing the ratio of the average power of the channel input signal to the average noise power. $P$ is set to 1 in all experiments.

To compare the performance of DeepJSCC to traditional separation-based digital schemes, we consider different well established source codes followed by LDPC codes for error correction. For the problem of image transmission, we use as source codes JPEG, WebP, JPEG2000, and BPG, and we discount the header information for BPG and JPEG2000 when computing bit rates and transmission sizes for fair comparison. For the channel code, we consider all possible combinations of (4096, 8192), (4096, 6144), and (2048, 6144) LDPC codes (which correspond to 1/2, 2/3, and 1/3 rate codes) with BPSK, 4-QAM, 16-QAM, and 64-QAM modulation schemes.

For each channel code configuration, we can define the maximum rate $R_{max}$ (bits per pixel) at which we can transmit an image (using the channel code rate) and empirically evaluate the frame error rate $\epsilon$ for each channel model and condition we consider. Then, we compress the images (using the different codecs) at the largest rate $R$ that satisfies $R \leq R_{max}$. We consider that the transmission can either be successful or fail, with probability of failure $\epsilon$. When the transmission fails, we consider that the reconstruction at the receiver is set to the mean value for all the pixels. When the transmission is successful the distortion is dictated by $R$ and the compression scheme used. We then measure the average performance over the evaluation dataset.

### 2.5.3    Experimental Setup

Using the model described in the previous section we perform a series of experiments using RGB source images. We train and evaluate the model with distinct datasets and all plotted performance values are averaged from 10 realizations of the channel for every image on the evaluation dataset. All results presented use the same compression ratio of $k/n = 1/6$, although similar results apply to other ratios.

The model was implemented in Tensorflow [74] and optimized using the Adam algorithm [75]. We used a learning rate of $10^{-4}$ and a batch size of 16. Models were trained until convergence, when the loss does not decrease after new iterations. The loss function used for training of the model is the average mean squared error (MSE) over $N$ samples:

$$\mathcal{L} = \frac{1}{N} \sum \|x - \hat{x}\|^2. \tag{2.18}$$

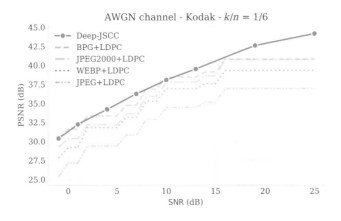

**Figure 2.8** DeepJSCC performance compared to digital schemes.

### 2.5.4    Results

**Baseline**

Our first set of results demonstrates the base case in which an image $x$ is encoded by a single encoder and a single decoder and thus $L = 1$. As channel model we consider a complex AWGN channel, which is a common model for static wireless links. Its transfer function is given by

$$\eta_{\text{cawgn}}(z) = z + n, \tag{2.19}$$

where $n \in \mathbb{C}^k$ is a vector with i.i.d. elements sampled from a circularly symmetric complex Gaussian distribution $n \sim \mathcal{CN}(0, \sigma^2 \mathbb{I}_k)$, where $\sigma^2$ is the average noise power.

In Fig. 2.8, we compare DeepJSCC at different channel SNRs. For comparison, we also plot the performance of well established separation-based schemes with JPEG, JPEG2000, WebP, and BPG codecs followed by LDPC channel coding. We see that the performance of DeepJSCC is either above or comparable to the performance of these separate source and channel coding schemes for a wide range of channel SNRs. These results show that we can obtain significant performance gain by a joint design.

In Fig. 2.9 we plot a visual comparison between the reconstructed output of DeepJSCC and a separation-based scheme using BPG+LDPC, for transmission over an AWGN channel, with channel SNR equal to 1 dB and $k/n = 1/24$. DeepJSCC achieves considerably higher PSNR and multiscale strucutural similarity (MS-SSIM) [76]. The performance difference can be seen clearly in the images, in which the DeepJSCC output presents more richness in details and sharpness, especially in high frequency components, such as background trees and leaves, while the BPG+LDPC output exhibits blurry artifacts over the whole image.

**Graceful Degradation**

These results are obtained by training a different encoder/decoder model for each SNR value evaluated in the case of DeepJSCC, and considering the best performance achieved by the separation-based scheme at each SNR.

(a) Original image

(b) **DeepJSCC**: PSNR = 24.40 dB / MS-SSIM = 0.907

(c) **BPG+LDPC**: PSNR = 22.27 dB / MS-SSIM = 0.779

**Figure 2.9** Visual comparison of reconstructed output for transmission under AWGN channel, SNR = 1 dB, and $k/n = 1/24$.

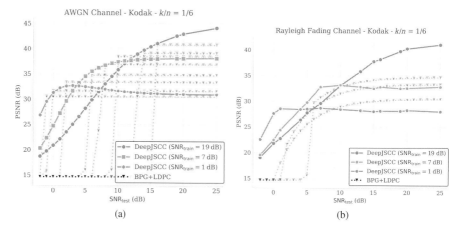

**Figure 2.10** (a) Effects of graceful degradation for DeepJSCC compared to cliff effect in separation-based scheme, and (b) performance of DeepJSCC and graceful degradation on Rayleigh fading channel.

Here, we consider the situation where there is a mismatch between the channel conditions during design and deployment. In practice this might occur for several reasons. First of all, it is impractical to assume a different model can be stored and employed at the transceivers for each SNR value. Instead, the same model needs to be used at least for a range of SNR values. There may also be mismatch between the SNR estimate at the encoder and the real SNR value during transmission, either due to time variations in the channel or due to imperfect channel estimation and feedback.

For this, we consider different DeepJSCC models trained for specific SNRs and show the evaluation of the test dataset for a range of SNRs, lower and higher than that is used for training. In Fig. 2.10(a) we show a separate curve for each model trained at a specific channel SNR ($\text{SNR}_{\text{train}}$), and the performance of each model is plotted against the test SNR ($\text{SNR}_{\text{test}}$). We also plot the performance of the separation-based scheme with BPG, showing the best performing LDPC code at each SNR.

It can be clearly seen that DeepJSCC presents *graceful degradation*; that is, the performance gradually decreases as the channel SNR deteriorates, while the digital scheme presents a *cliff-effect* when the quality of the channel goes below the capacity for which the code was designed, resulting in indistinguishable transmission output. Thus, we can see that DeepJSCC not only produces high quality transmissions (when compared to digital schemes), but also *analog behavior*, being more robust to nonergodic channels.

## Channel Versatility

A big advantage of DeepJSCC being data-driven is the possibility of training for different channel models, objective functions, or specific domains.

To better illustrate the advantage of graceful degradation, we consider transmission over the more challenging Rayleigh fading channel, which models variations in channel quality over time due to physical changes in the environment. The channel is modelled by a random channel gain:

$$\eta_{\text{fading}}(z) = hz + n, \tag{2.20}$$

where $h \sim \mathcal{CN}(0, H_c)$ is a complex normal random variable.

Here we consider a slow Rayleigh fading channel and assume that the channel gain $h$ remains constant during the transmission of a single image and changes to an independent value for the next image following the Rayleigh distribution. We do not assume channel state information (CSI) either at the receiver or the transmitter, but we consider that the phase shift introduced by the channel is known at the receiver, making the model equivalent to a real fading channel with bandwidth $2k$. In order to emulate and measure the average channel SNR, we define $H_c = 1$ and vary the noise variance $\sigma^2$ over transmissions.

In Fig. 2.10(b) we present results for models trained at different average channel SNRs and tested at different average SNR values. We also compare our performance with different LDPC configurations, using BPG as compression. We do not consider any explicit channel estimation or feedback in the case of DeepJSCC, whereas we assume that the channel gain is known by the receiver for the digital transmission scheme, providing a clear advantage for the latter. Yet, DeepJSCC still outperforms the digital scheme significantly for the whole range of average SNRs considered. This is mainly due to the graceful degradation property of DeepJSCC, which means that accurate CSI knowledge at the transmitter is not required. On the other hand, we stipulate that the network architecture learns to employ channel estimation sufficient for the decoder network to correctly reconstruct the image over varying channel gains.

We can also consider a channel model with bursty noise, which can model a scenario in which individual symbols of the transmitted signal, apart from being perturbed by AWGN noise $n$, can also be perturbed by a high variance noise with probability $p$. Formally, this is a Bernouille-Gaussian noise channel with the transfer function

$$\eta_{\text{bursty}}(z) = z + n + B(k, p)w, \tag{2.21}$$

where $B(k, p)$ is the binomial distribution, and $w \sim \mathcal{CN}(0, \sigma_b^2 \mathbb{I})$ the high variance noise component with $\sigma_b^2 \gg 0$. In practice, this models an occasional random interference from a nearby transmitter. This channel model is also used in Kim et al. [43] to illustrate the robustness of DNN-based channel encoders with channel output feedback.

Figure 2.11(a) shows the effect of the probability $p$ on the model's performance for a bursty channel with AWGN component's SNR equal to 10 dB. We consider both a low-power ($\sigma_b = 0.5$) and a high-power ($\sigma_b = 3.5$) burst and compare the performance with a digital scheme with BPG+LDPC. As expected, the performance degrades as $p$ increases, but the DeepJSCC scheme is much more robust against the increasing power of the burst noise. This improved robustness of DeepJSCC is particularly obvious for $\sigma_b = 3.5$. A high-power burst degrades the performance of the digital scheme very quickly, even if the burst probability is very low, completely

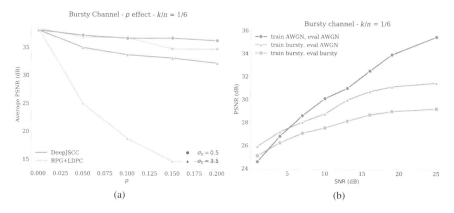

**Figure 2.11** (a) Performance of DeepJSCC on a bursty interference channel (b) compared to AWGN performance.

destroying the signal when $p > 0.15$. In this region, all combinations of compression rates and codes are considered to have failed. In contrast, DeepJSCC exhibits a graceful degradation even in the presence of bursty channel noise, showing its advantages in practical scenarios, particularly for communications over unlicensed bands, where occasional burst noise is common.

Figure 2.11(b) shows the effect in performance of the bursty channel at different channel conditions, considering fixed values of $p = 0.10$ and $\sigma_b = 3.5$. The SNR displayed is with regards to the fixed AWGN component of the channel. We also plot the AWGN performance without any noise bursts, for comparison. The results show that, although impacted by the bursty noise, DeepJSCC still achieves a reasonable performance in the whole range of SNRs considered.

Interestingly the gap between the AWGN performance and the bursty channel increases with the SNR; quite surprisingly, we notice that in the low SNR regime (SNR $\leq 2.5$) the model trained and evaluated on the bursty channel achieves higher performance than the one trained and evaluated with only AWGN. This could be explained due to the fact that the network trained on the bursty channel better generalizes its latent vector representation given its probability of erasure – an effect similar to the use of dropout layers [77] as a regularization technique. For comparison, we also plot the performance of the model trained on the bursty channel and evaluated with fixed AWGN, where we can see that this generalization effect is even stronger at higher SNRs (SNR $\leq 5$).

## Domain Specific Communication

Traditional image compression schemes are independent of the type of images being compressed. However, in principle, statistical properties of different types of images can be exploited to obtain a more efficient compression algorithm. A trivial example of this would be compressing black-and-white images instead of color images. But, a learning based image transmission strategy can exploit even less obvious statistical properties of the dataset that may not even be visible to human eye. To test the

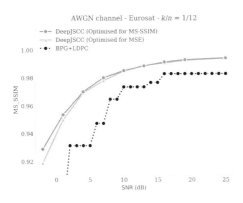

**Figure 2.12** Performance of DeepJSCC trained with MS-SSIM as an objective function.

capability of the DeepJSCC approach to adapt to a more specific class of images, we experimented by training our model with satellite image data [78], a plausible application of our model. Here we use the distortion measure of MS-SSIM [76] – a widely accepted image quality measure that better represents human visual perception than pixel-wise differences. The results, presented in Fig. 2.12 show that, when more specific image domains are considered, DeepJSCC can better adapt to it, significantly increasing the performance gap to conventional separation-based techniques.

**Successive Refinement**

Yet another advantage of DeepJSCC is its flexibility to adapt the transmission to different paths or stages. We now consider a model with $L > 1$, in which the same image is transmitted progressively in blocks of size $k_i$, $i = 1, \ldots L$ and $\sum_{i=1}^{L} k_i = k$. We aim to be able to reconstruct the complete image after each transmission, with increasing quality, thus performing *successive refinement* [79, 80]. Progressive transmission can be applied to scenarios in which communication is either expensive or urgent. For example, in surveillance applications, it may be beneficial to quickly send a low-resolution image to detect a potential threat as soon as possible, while a higher resolution description can be later received for further evaluation or archival purposes. Or, in a multiuser communication setting, one could send a different number of components for different users, depending on the available bandwidth.

We therefore expand our system by creating $L$ encoder and decoder pairs, each responsible for a partial transmission $z_i$ and trained jointly (see [56] for implementation details and alternative architectures). Figure 2.13(a) presents results for the case $L = 2$, for $k_1/n = k_2/n = 1/12$ and shows the performance of each layer for different channel SNRs, for the AWGN channel. Results show that the loss of dividing the transmission into multiple stages is not significant; when compared to a single transmission with $k/n = 1/6$ (dotted black curve in Fig. 2.13(a)), the model performs with approximately the same quality for most channel conditions. Moreover, we observe that every layer of the layered transmission scheme preserves all features of the single transmission, such as graceful degradation and adaptability to different channel models.

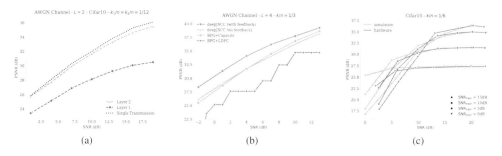

**Figure 2.13** (a) Successive refinement with $L = 2$; (b) layered transmission with channel output feedback, for $L = 4$; and (c) comparison between simulated and hardware performance.

## Channel Output Feedback

Another interesting direction to be explored by DeepJSCC is the use of channel output feedback, when it is available. Suppose that alongside the *forward* communication channel considered so far, there is also a *feedback* channel able to send back to the transmitter an estimation of the channel output $\tilde{z}_i$ after its realization. In a multi-layered transmission, this information can be used to inform subsequent layers and enhance the reconstruction at the receiver. Thus, a transmission of a source $x$ is done sequentially in $L$ steps, in which each step $i$ a channel input $z_i$ is generated from input $x$ and feedback $\tilde{z}_{i-1}$ (for $i > 1$), transmitted and decoded to generate successively refined representations $\hat{x}_i$ (see [73] for specific architecture and implementation details). There has also been recent advances in the use of channel output feedback to improve the performance of channel coding [41]; however, the design is for a specific blocklength and code rate, whereas the proposed DeepJSCC scheme can transmit large content, such as images.

Figure 2.13(b) shows the results for a scenario considering noiseless feedback (i.e., $\tilde{z}_i = \hat{z}_i$) and three uses of the feedback channel ($L = 4$), for channel inputs with size $k_i/n = 1/12$, $i = 1, \dots, 4$. We see that by exploiting the feedback information, DeepJSCC can further increase its performance, establishing its superiority over other schemes. Note that we compare DeepJSCC with feedback with a theoretical capacity achieving channel code and can still outperform the separation-based scheme.

This architecture enables other communication strategies, such as variable length coding, in which a minimum number of layers $z_i$ are transmitted and the quality of the reconstruction is estimated through feedback, until a target quality is achieved and the further transmission is interrupted. This scheme can provide gains of over 50 percent in bandwidth, when compared to separation-based approaches [73]. Further experiments also demonstrate that our model successfully operates under noisy channel feedback and even present graceful degradation when the feedback channel changes between training and evaluation.

### Hardware Implementation

Finally, to validate the real world performance of the proposed architecture, we implemented our DeepJSCC scheme on a real communication channel, enabled by a software defined radio platform. We used models trained on the AWGN model, with different SNRs. Results can be seen in Fig. 2.13(c) and show that the simulated performance closely matches the hardware performance, especially in higher SNRs.

The JSCC approach to image transmission also provides significant gains in terms of end-to-end encoding and decoding delays. We observed that the average encoding and decoding time per image with DeepJSCC is 6.40 ms on a GPU, or 15.4 ms on a CPU, while a scheme with JPEG2000+LDPC and BPG+LDPC takes on average 4.53 ms and 69.9 ms on a CPU, respectively, using the same hardware as the used for DeepJSCC evaluation. This shows the competitive processing times of DeepJSCC together with its outstanding performance. Note that, as with all CNNs, both the encoder and decoder components of DeepJSCC are parallelized and benefit from multicore hardware such as GPUs; this is not true for the standard implementations of source and channel codes considered, hence only CPU times are shown. Moreover, the DeepJSCC implementation uses standard and generic Tensorflow libraries and could still be further optimized for speed, if necessary.

## 2.6     Conclusion

Optimality of separation between compression and channel coding promised by Shannon's theorem assumes no constraint on the complexity of the system or the associated delay. However, in practice, large blocklength source and channel codes may not be feasible due to the computational complexity and delay constraints. Therefore, many practical communication systems can benefit from designing the source/channel codes jointly. However, despite the suboptimality of separate design, the lack of low-complexity JSCC schemes that can notably outperform the alternative modular design approach has prevented the adoption of JSCC in practice. Recent successes of machine learning, in particular, DNNs, for a variety of complex tasks in image processing and natural language processing have motivated their use for JSCC design.

In this chapter, we reviewed DL-aided JSCC design over both continuous and discrete input channels for the transmission of different information sources, particularly focusing on images and text. The results show that for both types of channels, DL-aided JSCC design achieves a performance comparable or superior to the state-of-the-art separation-based schemes. Moreover, when the channel conditions deteriorate, DL-based JSCC achieves graceful degradation in performance, in contrast to the cliff effect observed in separation-based schemes. This data-driven approach is versatile and can be applied to different channels and domains. In the case of text data, this technique can also be used to recover sentences that preserve semantic meaning at the receiver relevant for downstream tasks rather than the exact sentence. Moreover, the JSCC design based on DL can be successively refined, and can exploit channel output feedback in order to improve the communication. It is also possible to build support for variable length encoding for the JSCC using DL to further optimize performance with average encoding length constraints. These properties, coupled with its lower computational complexity, make deep JSCC code design an attractive approach for signal transmission in wireless communications, particularly for low-latency and power-limited constraints.

## References

[1] C. E. Shannon and W. Weaver, *The Mathematical Theory of Communication*. University of Illinois Press, 1998.

[2] S. Vembu, S. Verdu, and Y. Steinberg, "The source-channel separation theorem revisited," *IEEE Trans. on Information Theory*, vol. 41, no. 1, pp. 44–54, 1995.

[3] A. E. Gamal and Y. H. Kim, *Network Information Theory*. Cambridge University Press, 2011.

[4] D. Gündüz et al., "Source and channel coding for correlated sources over multiuser channels," *IEEE Trans. on Information Theory*, vol. 55, no. 9, pp. 3927–3944, Sep. 2009.

[5] A. Goldsmith, "Joint source/channel coding for wireless channels," in *IEEE Vehicular Technology Conf. Countdown to the Wireless Twenty-First Century*, vol. 2, pp. 614–618, July 1995.

[6] G. Davis and J. Danskin, "Joint source and channel coding for image transmission over lossy packet networks," in *Conf. Wavelet Applications to Digital Image Processing*, pp. 376–387, 1996.

[7] O. Y. Bursalioglu, G. Caire, and D. Divsalar, "Joint source-channel coding for deepspace image transmission using rateless codes," *IEEE Trans. on Communications*, vol. 61, no. 8, pp. 3448–3461, 2013.

[8] M. Gastpar, B. Rimoldi, and M. Vetterli, "To code, or not to code: Lossy source-channel communication revisited," *IEEE Trans. on Information Theory*, vol. 49, no. 5, pp. 1147–1158, May 2003.

[9] V. Kostina, Y. Polyanskiy, and S. Verdu, "Joint source-channel coding with feedback," *IEEE Trans. on Information Theory*, vol. 63, no. 6, pp. 3502–3515, June 2017.

[10] M. Skoglund, N. Phamdo, and F. Alajaji, "Design and performance of VQ-based hybrid digital-analog joint source-channel codes," *IEEE Trans. on Information Theory*, vol. 48, no. 3, pp. 708–720, March 2002.

[11] U. Mittal and N. Phamdo, "Hybrid digital–analog (HDA) joint source-channel codes for broadcasting and robust communications," *IEEE Trans. on Information Theory*, vol. 48, no. 5, pp. 1082–1102, May 2002.

[12] F. Hekland, P. A. Floor, and T. A. Ramstad, "Shannon-Kotel'nikov mappings in joint source-channel coding," *IEEE Trans. on Communications*, vol. 57, no. 1, pp. 94–105, Jan. 2009.

[13] I. Estella Aguerri and D. Gündüz, "Joint source-channel coding with time-varying channel and side-information," *IEEE Trans. on Information Theory*, vol. 62, no. 2, pp. 736–753, 2016.

[14] G. Cheung and A. Zakhor, "Bit allocation for joint source/channel coding of scalable video," *IEEE Trans. on Image Processing*, vol. 9, no. 3, pp. 340–356, March 2000.

[15] N. Farvardin, "A study of vector quantization for noisy channels," *IEEE Trans. on Information Theory*, vol. 36, no. 4, pp. 799–809, July 1990.

[16] N. Farvardin and V. Vaishampayan, "On the performance and complexity of channel-optimized vector quantizers," *IEEE Trans. on Information Theory*, vol. 37, no. 1, pp. 155–160, Jan. 1991.

[17] M. Skoglund, "Soft decoding for vector quantization over noisy channels with memory," *IEEE Trans. on Information Theory*, vol. 45, no. 4, pp. 1293–1307, May 1999.

[18] I. Kozintsev and K. Ramchandran, "Robust image transmission over energy-constrained time-varying channels using multiresolution joint source-channel coding," *IEEE Trans. on Signal Processing*, vol. 46, no. 4, pp. 1012–1026, April 1998.

[19] M. Skoglund, "On channel-constrained vector quantization and index assignment for discrete memoryless channels," *IEEE Trans. on Information Theory*, vol. 45, no. 7, pp. 2615–2622, Nov. 1999.

[20] H. Y. Shutoy et al., "Cooperative source and channel coding for wireless multimedia communications," *IEEE Journal of Selected Topics in Signal Processing*, vol. 1, no. 2, pp. 295–307, 2007.

[21] C. T. K. Ng et al., "Distortion minimization in Gaussian layered broadcast coding with successive refinement," *IEEE Trans. on Information Theory*, vol. 55, no. 11, pp. 5074–5086, 2009.

[22] T. M. Cover and J. A. Thomas, *Elements of Information Theory*. Wiley-Interscience, 1991.

[23] G. E. Hinton and R. R. Salakhutdinov, "Reducing the dimensionality of data with neural networks," *Science*, vol. 313, no. 5786, pp. 504–507, 2006.

[24] E. Nachmani et al. "Deep learning methods for improved decoding of linear codes," *IEEE Journal of Selected Topics in Signal Processing*, vol. 12, no. 1, pp. 119–131, Feb. 2018.

[25] T. Gruber et al., "On deep learning-based channel decoding," in *2017 51st Annual Conf. on Information Sciences and Systems (CISS)*, pp. 1–6, 2017.

[26] G. Toderici et al., "Variable rate image compression with recurrent neural networks," in *Proc. Int. Conf. on Learning Representations (ICLR)*, 2016.

[27] N. Johnston et al., "Improved lossy image compression with priming and spatially adaptive bit rates for recurrent networks," in *IEEE Conf. on Computer Vision and Pattern Recognition (CVPR)*, June 2018.

[28] L. Theis et al., "Lossy image compression with compressive autoencoders," in *Proc. Int. Conf. on Learning Representations (ICLR)*, 2017.

[29] E. Agustsson et al., "Soft-to-hard vector quantization for end-to-end learning compressible representations," in *Advances in Neural Information Processing Systems*, vol. 30, pp. 1141–1151, 2017.

[30] J. Balle, V. Laparra, and E. P. Simoncelli, "End-to-end optimized image compression," in *Proc. Int. Conf. on Learning Representations (ICLR)*, pp. 1–27, Apr. 2017.

[31] D. P. Kingma and M. Welling, "Auto-encoding variational Bayes," in *Int. Conf. on Learning Representations (ICLR)*, 2014.

[32] J. Ballé et al., "Variational image compression with a scale hyperprior," in *Proc. Int. Conf. on Learning Representations (ICLR)*, 2018.

[33] D. Minnen, J. Ballé, and G. D. Toderici, "Joint autoregressive and hierarchical priors for learned image compression," in *Advances in Neural Information Processing Systems 31*, pp. 10771–10780, 2018.

[34] J. Lee, S. Cho, and S.-K. Beack, "Context-adaptive entropy model for end-to-end optimized image compression," in *Int. Conf. on Learning Representations*, 2019.

[35] Z. Cheng et al., "Perceptual Quality Study on Deep Learning based Image Compression," *arXiv e-print*, p. arXiv:1905.03951, May 2019.

[36] T. J. O'Shea, K. Karra, and T. C. Clancy, "Learning to communicate: Channel auto-encoders, domain specific regularizers, and attention," in *Proc. IEEE Int. Symp. on Signal Processing and Information Technology (ISSPIT)*, pp. 223–228, Dec. 2016.

[37] T. O'Shea and J. Hoydis, "An introduction to deep learning for the physical layer," *IEEE Trans. on Cognitive Communications and Networking*, vol. 3, no. 4, pp. 563–575, Dec. 2017.

[38] A. Felix et al., "OFDM autoencoder for end-to-end learning of communications systems," in *Proc. IEEE Int. Workshop Signal Processing Advances in Wireless Communications (SPAWC)*, Jun. 2018.

[39] Y. Jiang et al., "Learn codes: Inventing low-latency codes via recurrent neural networks," in *ICC 2019 - 2019 IEEE Int. Conf. on Communications (ICC)*, pp. 1–7, May 2019.

[40] Y. Jiang et al., "Turbo autoencoder: Deep learning based channel codes for point-to-point communication channels," in *Advances in Neural Information Processing Systems*, pp. 2758–2768, 2019.

[41] H. Kim et al., "Deepcode: Feedback codes via deep learning," in *Advances in Neural Information Processing Systems 31*, pp. 9436–9446, 2018.

[42] Y. Jiang et al., "Feedback turbo autoencoder," in *IEEE Int. Conf. on Acoustics, Speech and Signal Processing (ICASSP)*, pp. 8559–8563, 2020.

[43] H. Kim et al., "Communication algorithms via deep learning," in *Proc. Int. Conf. on Learning Representations (ICLR)*, 2018.

[44] Y. Jiang et al., "DeepTurbo: Deep Turbo Decoder," *arXiv e-print*, p. arXiv:1903.02295, March 2019.

[45] M. Ebada et al., "Deep Learning-based Polar Code Design," *arXiv e-print*, p. arXiv:1909.12035, Sep. 2019.

[46] N. Samuel, T. Diskin, and A. Wiesel, "Deep MIMO detection," in *IEEE 18th Int. Workshop on Signal Processing Advances in Wireless Communications (SPAWC)*, pp. 1–5, July 2017.

[47] B. Karanov et al., "End-to-end deep learning of optical fiber communications," *Journal of Lightwave Technology*, vol. 36, no. 20, pp. 4843–4855, Oct. 2018.

[48] N. Farsad and A. Goldsmith, "Neural network detection of data sequences in communication systems," *IEEE Trans. on Signal Processing*, vol. 66, no. 21, pp. 5663–5678, Nov. 2018.

[49] A. Caciularu and D. Burshtein, "Blind channel equalization using variational autoencoders," in *Proc. IEEE Int. Conf. on Comms. Workshops, Kansas City, MO*, pp. 1–6, May 2018.

[50] L. Rongwei, W. Lenan, and G. Dongliang, "Joint source/channel coding modulation based on bp neural networks," in *Proc. Int. Conf. on Neural Networks and Signal Processing*, vol. 1. pp. 156–159, 2003.

[51] N. Farsad, M. Rao, and A. Goldsmith, "Deep learning for joint source-channel coding of text," in *Proc. IEEE Int. Conf. on Acoustics, Speech and Signal Processing (ICASSP)*, April 2018.

[52] R. Zarconee et al., "Joint source-channel coding with neural networks for analog data compression and storage," in *2018 Data Compression Conference*, pp. 147–156, March 2018.

[53] E. Bourtsoulatze, D. B. Kurka, and D. Gündüz, "Deep joint source-channel coding for wireless image transmission," in *IEEE Int. Conf. on Acoustics, Speech and Signal Processing (ICASSP)*, pp. 4774–4778, May 2019.

[54] E. Bourtsoulatze, D. Burth Kurka, and D. Gündüz, "Deep joint source-channel coding for wireless image transmission," *IEEE Trans. on Cognitive Communications and Networking*, vol. 5, no. 3, pp. 567–579, Sep. 2019.

[55] D. Burth Kurka and D. Gündüz, "Joint source-channel coding of images with (not very) deep learning," in *Proc of Int. Zurich Seminar on Information and Communication (IZS 2020)*, pp. 90–94, 2020.

[56] D. B. Kurka and D. Gündüz, "Bandwidth-Agile Image Transmission With Deep Joint Source-Channel Coding," in *IEEE Transactions on Wireless Communications*, vol. 20, no. 12, pp. 8081–8095, Dec. 2021.

[57] M. Rao, N. Farsad, and A. Goldsmith, "Variable length joint source-channel coding of text using deep neural networks," in *IEEE International Workshop on Signal Processing Advances in Wireless Communications (SPAWC)*, pp. 1–5, 2018.

[58] D. B. Kurka and D. Gündüz, "Successive refinement of images with deep joint source-channel coding," in *IEEE 20th Int. Workshop on Signal Processing Advances in Wireless Communications (SPAWC)*, pp. 1–5, July 2019.

[59] K. Choi et al., "Neural joint source-channel coding," in *Int. Conf. on Machine Learning*, pp. 1182–1192, 2019.

[60] I. Sutskever, O. Vinyals, and Q. V. Le, "Sequence to sequence learning with neural networks," in *Advances in Neural Information Processing Systems*, pp. 3104–3112, 2014.

[61] J. Pennington, R. Socher, and C. Manning, "Glove: Global vectors for word representation," in *Proc. Conf. on Empirical Methods in Natural Language Processing (EMNLP)*, pp. 1532–1543, 2014.

[62] A. Graves and J. Schmidhuber, "Framewise phoneme classification with bidirectional lstm and other neural network architectures," *Neural Networks*, vol. 18, no. 5, pp. 602–610, 2005.

[63] H. Sak, A. Senior, and F. Beaufays, "Long short-term memory recurrent neural network architectures for large scale acoustic modeling," in *15th Annual Conf. of the Int. Speech Communication Association*, 2014.

[64] T. Raiko et al., "Techniques for learning binary stochastic feedforward neural networks," *stat*, vol. 1050, p. 11, 2014.

[65] N. Srivastava et al., "Dropout: a simple way to prevent neural networks from overfitting," *Journal of Machine Learning Research*, vol. 15, no. 1, pp. 1929–1958, 2014.

[66] "Shared Task: Machine Translation of News," *2nd Conf. on Machine Translation (WMT17)*, 2017.

[67] J. Ziv and A. Lempel, "A universal algorithm for sequential data compression," *IEEE Trans. on Information Theory*, vol. 23, no. 3, pp. 337–343, 1977.

[68] I. S. Reed and G. Solomon, "Polynomial codes over certain finite fields," *Journal of the Society for Industrial and Applied Mathematics*, vol. 8, no. 2, pp. 300–304, 1960.

[69] Z. Wu and M. Palmer, "Verbs semantics and lexical selection," in *Proc. 32nd Annual Meeting on Association for Computational Linguistics*. pp. 133–138, 1994.

[70] J. Ballé, V. Laparra, and E. P. Simoncelli, "Density modeling of images using a generalized normalization transformation," *arXiv preprint*, arXiv:1511.06281, 2015.

[71] K. He et al., "Delving deep into rectifiers: Surpassing human-level performance on imagenet classification," in *IEEE Int. Conf. on Computer Vision (ICCV)*, Dec. 2015.

[72] J. Ballé et al., "Variational image compression with a scale hyperprior," in *Proc. 6th Int. Conf. on Learning Representations (ICLR)*, 2018.

[73] D. B. Kurka and D. Gündüz, "Deepjscc-f: Deep joint source-channel coding of images with feedback," *IEEE Journal on Selected Areas in Information Theory*, vol. 1, no. 1, pp. 178–193, 2020.

[74] M. Abadi et al., "Tensorflow: a system for large-scale machine learning," in *12th USENIX Symp. on Operating Systems Design and Implementation (OSDI)*, pp. 265–283, 2016.

[75] D. P. Kingma and J. Ba, "Adam: a method for stochastic optimization," *arXiv preprint*, arXiv:1412.6980 [cs.LG], 2014.

[76] Z. Wang, E. P. Simoncelli, and A. C. Bovik, "Multiscale structural similarity for image quality assessment," in *37th Asilomar Conf. on Signals, Systems & Computers*, vol. 2,pp. 1398–1402, 2003.

[77] N. Srivastava et al., "Dropout: a simple way to prevent neural networks from overfitting," *Journal of Machine Learning Research*, vol. 15, no. 1, pp. 1929–1958, 2014.

[78] P. Helber et al., "Eurosat: a novel dataset and deep learning benchmark for land use and land cover classification," *IEEE Journal of Selected Topics in Applied Earth Observations and Remote Sensing*, 2019.

[79] Y. Steinberg and N. Merhav, "On hierarchical joint source-channel coding," in *Proc. Int. Symp. on Information Theory (ISIT)*, pp. 365–365, Jun. 2004.

[80] W. H. R. Equitz and T. M. Cover, "Successive refinement of information," *IEEE Trans. on Information Theory*, vol. 37, no. 2, pp. 269–275, March 1991.

# 3 Neural Network Coding

Litian Liu, Amit Solomon, Salman Salamatian, Derya Malak,
and Muriel Medard

## 3.1 Introduction

The problem of communicating sources over a network has a decades-long history [1]. In various examples such as the Internet of Things (IoT) and autonomous driving, we see a merging of new scenarios of the classical communication problem with novel features in power budget, latency requirements, etc. For example, Fig. 3.1(a) demonstrates such a scenario, where the signals generated by distributed measurements are transmitted over a communication network and reconstructed at both the destination nodes. As shown in the example, the signals in IoT and autonomous driving are often task-specific: they are generated by specific measurements and therefore they have a nontrivial underlying distribution. Moreover, the signals are often of small size, because of the limited buffer size of edge devices and the stringent latency constraints in IoT and autonomous driving. The energy budget for such scenarios is low as edge devices typically have limited energy. Furthermore, the end goal of communication can be versatile. Figure 3.1(a) demonstrates a multicast problem, where both destination nodes seek to reconstruct all the input signals. In other cases, a destination node might be interested in a subset of the input signals or a function of the input signals rather than the signals themselves. For example, in smart wearable body sensors, one central processor could be solely interested in whether the person needs emergency care. The exact value of the heart rate or blood pressure could matter less. The features in merging scenarios both impose challenges on the classical communication models based on long-block length source and channel coding techniques and open up new opportunities for designing application-specific schemes.

Communicating over a network is an intrinsically hard problem. For reconstructing the input signals, Ramamoorthy et al. [2] demonstrated the failure of the separation between source and network coding for multicast networks. Effros et al. [3] demonstrated the same failure for nonmulticast networks. The requirement for joint coding indicates the insufficiency of using efficient source codes, followed by efficient network codes. A joint source and network coding scheme, namely random linear network coding, was shown in [4] to be capacity-achieving for multicasting sources over a network, with the error exponent generalizing the linear Slepian-Wolf coding [5]. However, for sources that are neither independent nor linearly correlated, minimum

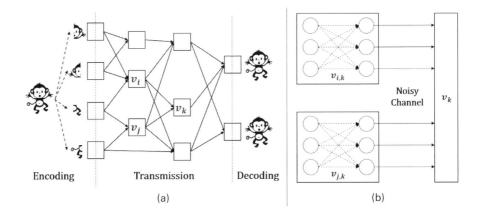

**Figure 3.1** NNC overview. (a) Overview with $N = 4$ source nodes and $M = 2$ destination nodes. (b) Nodes $v_i$, $v_j$ to node $v_k$. Within each box (in (b)) is a standard notation of a neural network, where each circle is a perception and each arrow indicates the information flow.

entropy (ME) decoders or maximum a posteriori (MAP) decoders are required for random linear network coding. Despite the efforts toward reducing the complexity of ME or MAP decoders [6, 7], it is in general NP-hard to find a ME or MAP solution. As a result, joint source and network coding schemes with low-complexity decoders have been proposed, but those apply only to restricted settings. For example, in [8] sources are restricted to binary ones and in [9] the number of sources is restricted to two. To the best of our knowledge, no practical joint source and network coding schemes exist for multicasting arbitrarily correlated real-life signals over a point-to-point network of arbitrary topology. The general nonmulticast communication problem where each destination seeks to reconstruct an arbitrary subset of the input signals is considerably harder than the multicast problem. Sufficient conditions for solving a nonmulticast problem using random linear network coding are derived in [10]. A subset of nonmulticast problems that can be solved using arbitrary nonlinear coding strategies is derived in [11]. To the best of our knowledge, there is no existing solution for nonmulticast problems in general. For the case where destinations are interested in the functions of the input signals, distributed functional computation adds another layer of complexity as computation can happen within the network itself. A modularized coding scheme for distributed functional computing is derived in [12] for tree networks with independent sources. A flow-based analysis for functional compression in general network topologies is conducted in [13]. However, this analysis might not be implemented in practice because it is based on NP-hard compression concepts, such as compression of characteristic graphs. To the best of our knowledge, there is no existing practical solution for functional reconstruction problems in general.

   The difficulty of the tasks motivates us to take an application-specific approach. In recent years, there has been an increased effort toward data-driven code constructions, for both signal reconstruction problems [14–18] and functional reconstruction problems [19]. As opposed to the traditional methods, data-driven approaches make no assumptions on the source statistics, which leads to a significant improvement when looking at complex sources of data such as images. Instead, such methods aim at discovering efficient codes by making use of a (potentially large) pool of data, in conjunction with a learning algorithm. A successful candidate for the latter is a neural network (NN), which has tremendous potential in many application domains. All works that we know have focused on the point-to-point communication problem under various channels. For example, learning the physical layer representation is studied for the single-input and single-output (SISO) system in [14], the multiple input and multiple output (MIMO) system in [15], and the orthogonal frequency-division multiplexing (OFDM) system in [16]. An NN-based joint source channel coding (JSCC) scheme is proposed for images in [17] and text in [18]. Though traditional techniques are optimal in the asymptotic regimes, in practical scenarios, it was shown in [14–18] that NN-based methods were competitive and even outperformed the state-of-the-art methods in information theory in some signal-to-noise ratio (SNR) regimes. In this chapter, we shift the scope to the network perspective and make a first step in broadening our understanding of the benefits and challenges of data-driven approaches for a networked communication scheme.

To this end, we propose an *application-specific* novel code construction based on NNs, which we refer to as *neural network coding* [20], or NNC for short. The NNC scheme has the following main benefits: (a) it makes no assumptions on the statistics of the source, but rather makes use of a seed dataset of examples; (b) it is an end-to-end communication scheme, or equivalently, a joint source and network coding scheme; (c) its decoding is of low complexity; (d) it can be applied to any network topology; (e) it can adapt to specific tasks; and (f) it can be used with various power constraints at each of the source and intermediate nodes. Figure 3.1(a) demonstrates an NNC applicable scenario, where arbitrarily correlated signals are multicast over a communication network. The network has four source nodes and two destination nodes. In NNC, the encoders at the source nodes and intermediate nodes, as well as the decoders at the destination nodes, are NNs as shown in Fig. 3.1(b). The resulting network code construction is jointly designed with the encoding phase and decoding phase of the transmission, where real-valued input signals are mapped into channel inputs, and channel outputs are reconstructed into real-valued signals. The end-to-end NNC scheme can be optimized through training and testing offline over a large dataset and can be readily implemented. Of particular interest for these codes is the expected distortion level they achieve for a given topology, power constraints, and distortion measure specified for the communication tasks. NNC is reminiscent of the autoencoder [21] structure. An autoencoder is a NN trained to minimize the distortion, such as mean square error (MSE), between its output and input. The end-to-end NN structure that results from NNC scheme is similar to the auto-encoder mentioned earlier, with additional constraints imposed by the topological structure of the physical communication network. Our experimental results showed the benefit of having a nonlinear rather than linear code construction in multicast problems. We also illustrate through experiments on images that NNC achieves better performance for multicasting problems compared to a separation-based scheme that relies on a compression scheme (JPEG [22]), followed by capacity-achieving network coding. Furthermore, we demonstrate through experiments that NNC can adapt to a variety of network topologies and tasks, including theoretically-hard nonmulticasting problems and functional reconstruction problems. We numerically illustrate the benefit of adopting a task-specific scheme like NNC, especially in a low power regime. While still in its infancy, we believe that NNC and its variants may pave the way to an efficient way to exploit nonlinearity in codes, which appears to be an important component to more complex networked settings.

The rest of this chapter is organized as follows. Section 3.2 describes our system model where we model the network as a graph: Nodes are devices and edges are (potentially multiple) connections between devices. Section 3.3 presents our design principle where each node has an NN for network code construction and decoding and we jointly optimize the scheme. Section 3.4 first studies the performance of NNC in a multicasting task over the butterfly network with additive white Gaussian noise (AWGN) links. We then build upon the baseline experiment and study the performance of NNC with one property of the baseline changed. The properties we consider in this section include network topologies, channel statistics, and the communication task. Section 3.5 explores various applications of NNC. Section 3.6 summarizes the paper and discusses possible extensions.

## 3.2        System Model

Throughout this chapter, we use $x, \vec{x}, X, \vec{X}$ to denote a scalar, a vector, a random variable, and a random vector, respectively. We model the communication network as an acyclic directed graph $G = (\mathcal{V}, \mathcal{E})$ [23]. Elements $v_i$ of $\mathcal{V}$ are called nodes and elements $(i, j)$ of $\mathcal{E}$ are called links. Each link $(i, j)$ is assigned an energy constraint $p_{i,j} \geq 0$, which specifies the maximum signal energy that can be sent from node $v_i$ to node $v_j$. We consider two disjoint subsets $\mathcal{S}, \mathcal{D}$ of $\mathcal{V}$, where each element in $\mathcal{S}$ is called a source node and each element in $\mathcal{D}$ is called a destination node. Let $N$ and $M$ denote $|\mathcal{S}|$ and $|\mathcal{D}|$, respectively. We consider $n$ virtual sources $\{s_i\}_{i=1}^n$ located at $N$ source nodes. Each $s_i$ generates a random variable $X_i \in \mathbb{R}$, $i = 1, \ldots, n$, according to the joint probability density function $f_{X_1, \ldots, X_n}$. The resulting random vector is denoted by $\vec{X} \in \mathbb{R}^n$. Observe that $n$ may not be equal to $N$. This setup encompasses the case in which some of the sources are colocated, or some physical sources generate random variables of higher dimension, by grouping some of the sources into a source node in $\mathcal{S}$. Thus, when appropriate we refer to a source node $s \in \mathcal{S}$ to represent the

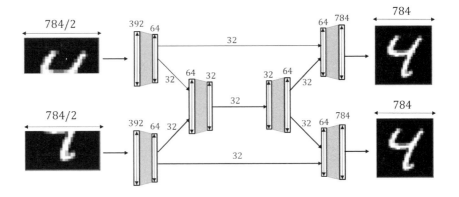

**Figure 3.2** An end-to-end NN example of NNC. There are $n = 784$ virtual sources located at $N = 2$ source nodes (at the left), and there are $M = 2$ destination nodes (at the right). Each link consists of 32 channels, as indicated by the number above each arrow. Each node has an inner NN to construct and decode network codes. Each inner NN is represented by a quadrilateral where the left-hand side is the input dimension and the right side is the output dimension of all the outgoing links. For example, the inner NN at top source node has input dimension 392 and output dimension 64. Half of the output goes to the upper destination node and the other half goes to the middle node. Each inner NN is implemented to be trainable. Each link is implemented with a nontrainable NN layer of width 32.

collection of virtual sources which are colocated at $s$ (see experiment Sections 3.4 and 3.5).

We model each link in the network as a set of parallel channels. More precisely, two nodes $v_i$ and $v_j$ are connected via $k_{i,j}$ parallel noisy channels. On each link $(i, j)$, $v_i$ may transmit any signal $\vec{W}_{i,j}$ in $\mathbb{R}^{k_{i,j}}$, subject to the average power constraint on the link (i.e., $\mathbb{E}[\vec{W}_{i,j}^2] \le p_{i,j}$). The signal $\vec{W}_{i,j}$ on each link is corrupted by noise, fading, and/or interference from other signals. The received signal at node $v_j$ is $\vec{Y}_{i,j} \in \mathbb{R}^{k_{i,j}}$. In the special case of independent zero-mean AWGN channels, the node $v_j$ receives $\vec{Y}_{i,j} = \vec{W}_{i,j} + \vec{N}_{i,j}$, where each element of the $k_{i,j}$-dimensional vector $\vec{N}_{i,j}$ is a zero-mean Gaussian random variable with variance $\sigma_{i,j}^2$. Note that this setup can model either band-limited channels in wireless communication or the parallel channels in the Ethernet connection.

We study the general communication problem where each destination node $t \in \mathcal{D}$ is interested in a target function $f_t$ of the sources. For example in a multicast scenario, the target function $f_t$ for $\forall t \in \mathcal{D}$ is the identity function of all the sources.[1] The performance of the scheme can be evaluated by a tuple of distortion measures $\delta_t$, $t \in \mathcal{D}$, where each measure is defined between the target function over the source $f_t(\vec{X})$ at $t$ and the functional reconstruction $g_t(\vec{X}^t)$ at $t$. For example for hand-written digits in Fig. 3.2, the target function $f_t$ can be digit and $\delta_t$ could be the accuracy.

[1] Therefore, the reconstruction of the input signals is a special case of the functional reconstruction. We refer to both as functional reconstruction from now on.

## 3.3      Neural Network Coding

In NNC, we jointly design the channel inputs at the source nodes and at the intermediate nodes – this makes NNC a joint source and network coding scheme. Existing joint source and network coding schemes (e.g., [4, 8–10]) assume error-free point-to-point transmission on each link and focus on the network layer operations. The physical layer then relies on a separate channel coding scheme with potentially a high latency, as it is assumed that each link employs an error correction coding with a large block length. In contrast, in NNC the signal inputs are directly transmitted over the noisy channels; that is, there are no underlying physical layer codes. As such, the communication problem described in Section 3.2 can be decomposed into three phases as shown in Fig. 3.1(a): the encoding phase, the transmission phase, and the decoding phase. NNC operates in a *one-shot* manner over all three phases. In the encoding phase, real-valued signals at the source nodes are directly mapped into network codes. The length of a network code $\vec{W}_{i,j}$ is designed to match the number of independent channels $k_{i,j}$ contained in link $(i, j)$. Therefore, $\vec{W}_{i,j}$ can concurrently be transmitted through the noisy channels over link $(i, j)$. In the transmission phase, network codes $\vec{W}_{i,j}$s are directly constructed at node $v_i$ from the incoming noise-corrupted network codes $\{\vec{Y}_{l,i} : l \in c(i)\}$, where $c(i)$ is the set of direct predecessors of $v_i$. In the decoding phase, each destination node reconstructs the target function of the transmitted signals directly from the noise-corrupted network codes it receives. NNC does not involve any long block-length source or channel coding techniques and therefore is free of their associated latency penalty and computational overhead.

Note that by picking a nonlinear activation, the resulting joint source and network code is nonlinear by design. As mentioned in Section 3.1, the nonlinearity in codes may be crucial in constructing efficient codes for the problem at hand. We design the network code from node $v_i$ to node $v_j$ by constructing a NN with input dimension $d_{in}^i$ and output dimension $k_{i,j}$. The last layer of the NN is a normalization layer to enforce the power constraints. When $v_i \notin S$, $d_{in}^i$ is the number of incoming channels. When $v_i \in S$, $d_{in}^i$ is the dimension of signal generated at $v_i$. During a transmission, noise-distorted network codes received at $v_i$, $\{\vec{Y}_{l,i} : l \in c(i)\}$ are fed into the NN if $v_i \notin S$. Or the generated signal is fed into the NN if $v_i \in S$. The NN output is the network code $\vec{W}_{i,j}$ to be transmitted over link $(i, j)$.

Similarly, for functional reconstruction of the inputs signal, we decode the received noise-distorted network codes with a NN at each destination node $t$. Note that NNs at destination nodes are low-complexity decoders since each layer of a NN is an affine transformation followed by an element-wise nonlinear function. We say that the set of functions for constructing and decoding network codes at each node specifies a *NNC policy* for the communication system if each of them can be represented as a NN. Under a NNC policy, the resulting end-to-end encoding-decoding can be seen as a series of NNs connected via communication links, as given by the physical network topology. We simulated the channel statistics of communication links by NN-layers as well, with the main difference being that those layers have fixed (nontrainable) parameters that correspond to the channel statistics. Thus, under a NNC policy, we construct an end-to-end NN, where some of the layers have nontrainable parameters. The end-to-end NN has the physical topology of the communication system embedded and has NNs that are used for constructing and decoding network code as its subgraphs. We refer to the NNs for constructing and decoding network code as *inner NNs* henceforth. Overall, there are $N$ input layers and $M$ output layers in the end-to-end NN. Each input layer has a width equal to the dimension of the source generated at the node. Each output layer has a width equal to the output dimension of the target function at the node. An illustration of an end-to-end NN is given in Fig. 3.2.

With $\vec{X}$ partitioned and fed into the input layers, the outputs of the end-to-end NN simulate the functional reconstruction at the destination nodes under the current NNC policy. Recall that $\delta_t$ is the distortion measure between the target function over the source $f_t(\vec{X})$ and the functional reconstruction $g_t(\vec{X})$ at a destination node $t$, as defined in Section 3.2. Parameters $\{\theta_l\}_l$ of the NNC policy are initialized randomly and are trained to minimize

$$\sum_{t \in \mathcal{D}} \delta_t \tag{3.1}$$

over a large dataset sampled from $f_{X_1,...,X_n}$. The parameters of the NN policy can be trained and tested offline[2] using a variety of available tools (e.g., [24] [25]). Note that for a point-to-point link, NNC reduces to deep JSCC in [17].

## 3.4 Performance Evaluation

### 3.4.1 Baseline

We start with an experiment of multicasting an MNIST image [26] over a butterfly network, as shown in Fig. 3.2. Technical details of implementation can be found in Section 3.7.

In this setup, there are two source nodes ($N = 2$) and two destination nodes ($M = 2$). A normalized MNIST image, with pixel values between 0 and 1, is split between the two source nodes, such that each source node observes only one half of the image. In other words, 392 out of $n = 28 \times 28$ virtual sources (pixels) are colocated at each source node, where the top 392 pixels are located at the first source node, and the rest at the second. Note that in this setup, sources are correlated. Each link in the butterfly network consists of 32 independent parallel AWGN channels ($k_{i,j} = 32$, $\forall i, j$), with zero-mean noise of variance $10^{-4}$. Performance is evaluated by the peak signal to noise ratio (PSNR) at each destination node, defined as

$$PSNR_t = 10 \log_{10} \left( \frac{\max\{X_i\}^2}{\mathbb{E}[(\vec{X}^t - \vec{X})^2]} \right), \tag{3.2}$$

---

[2] For the best performance, efforts are in general required to optimize over the choice of hyper-parameters as well, as is the case in other applications of NNs. Hyper-parameters, such as the number of layers and the activation functions in every inner NN, can also be tuned and tested off-line before implementation.

where $\max\{X_i\}$ is the largest value of the input pixels. The PSNR is essentially a normalized version of the MSE and can be used for performance comparison between inputs with different pixel ranges. The choice of PSNR as a distortion measure is natural for images [27].

A NNC policy is trained to minimize

$$\min_{\{\theta_l\}_l} \sum_{t=1}^{2} H_b(\vec{X}, \vec{X}^t) \ s.t. \ \mathbb{E}[\vec{W}_{i,j}^2] \leq P \ \forall i, j, \qquad (3.3)$$

where $H_b(\vec{X}, \vec{X}^t)$ is the binary cross entropy between the original image and the reconstructed image at destination node $t$. Note the dependence of $H_b$ and $\vec{W}_{i,j}$ on $\{\theta_l\}_l$ is omitted in the expression for simplicity. We use binary cross entropy $H_b(\vec{X}, \vec{X}^t)$, defined as

$$H_b(\vec{X}, \vec{X}^t) = -\frac{1}{n}\sum_i X_i \log(X_i^t) + (1 - X_i) \log(1 - X_i^t), \qquad (3.4)$$

in the objective function instead of PSNR as an engineering tweak to speed up the training process. Note that the power constraints on all links are the same and this setup demonstrates a homogeneous butterfly network.

We studied the performance of NNC in a homogeneous butterfly network under different power budgets by varying $P$ in Eq. (3.3). Note that since the noise statistic on each link is fixed, the power-distortion trade-off we demonstrated is equivalent to the SNR-distortion trade-off. As expected, the quality of the reconstruction improves as the transmission power increases, from Fig. 3.3(a) (low power at $1.75 \times 10^{-2}$ per

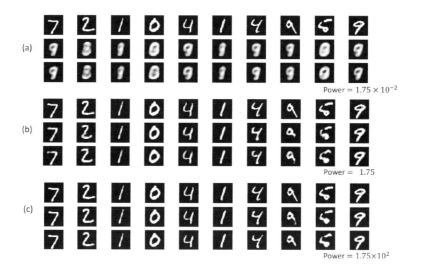

**Figure 3.3** NNC transmission quality with different power: (a) low power, (b) medium power, (c) high power. The first row of each subfigure is the original image, and the following row is a reconstruction at one of the two destination nodes of a butterfly network. The two reconstructed images at both destination nodes were similar.

**Table 3.1.** Minimum transmission power per image for achieving the corresponding mean PSNR values under NNC, ANC, and JPEG schemes.

| $pSNR$ | NNC | ANC | JPEG |
|--------|-----|-----|------|
| 11.7 | $7.0 \times 10^{-6}$ | 2.10 | N/A |
| 12.0 | $7.0 \times 10^{-3}$ | 2.23 | N/A |
| 13.8 | $7.0 \times 10^{-2}$ | 3.31 | N/A |
| 17.4 | $7.0 \times 10^{-1}$ | 7.12 | $4.6 \times 10^{51}$ |
| 20.6 | $7.0 \times 10^{0}$ | 14.6 | $2.6 \times 10^{53}$ |
| 22.0 | $7.0 \times 10^{1}$ | 20.9 | $4.7 \times 10^{54}$ |

image transmission) to Fig. 3.3(b) (medium power at 1.75 per image transmission), and finally to Fig. 3.3(c) (high power at $1.75 \times 10^2$ per image transmission). Note that when the transmission power is forced to be nearly zero, as shown in Fig. 3.3(a), the reconstruction at both destination nodes is close to the average of all the training data, which is essentially the best possible reconstruction when no information flows from the sources to destinations.

We compared the performance of NNC with two baseline methods. The first competitor is a *linear* analog of NNC, the analog network coding (ANC) scheme [28]. In ANC, transmission power ANC is controlled by the amplification factor at the source nodes. Each intermediate node forwards the sum of its inputs. At both destination nodes, the amplification factor, as well as the network topology, are known and used to reconstruct the source signals. Note that a $28 \times 28$ MNIST image can be sent by NNC in a one-shot, but it has to be sent over the network in 13 transmissions by ANC, as there is no compression scheme in the ANC baseline and thus at most 64 pixels can be sent in a single transmission. All distortion in the reconstruction comes from the noise in the channel rather than compression under the ANC baseline.

We compared the minimum required transmission powers for achieving certain PSNR values under NNC and ANC scheme in Table 3.1. For NNC, we set the power constraints $P$ on all links to be the same and recorded the performance of both ends with 10 different values of $P$ evenly spaced in log-scale from $10^{-6}$ to $10^{-2}$ For ANC, transmission power is controlled by amplification factor at the source nodes. Note that performances of both schemes are "symmetric" at destination nodes, reconstructing images with similar PSNR. Such "symmetric" performance can be expected since the network is isotropic. Overall, NNC outperforms the ANC baseline when the transmission power is low, and the ANC baseline outperforms NNC when the transmission power is high. This is consistent with [28], which shows that ANC is in fact capacity-achieving in the high-SNR regime. We see the benefit of allowing nonlinearity in the construction of application-specific short codes.

The second competitor, the JPEG baseline, is a scheme that separates source coding from network coding: images are compressed through the JPEG compression algorithm [22] at the source node and are then transmitted by capacity-achieving network codes through error-free channel codes over the network. Distortion in the reconstruction under the JPEG baseline only comes from compression in source coding, as the transmission is assumed to be error-free. Notice that the JPEG baseline has potentially a high latency due to error-free channel coding, which operates on a large block length.

In our experiments, the JPEG baseline reconstructs high-quality images with impractically high power. With the JPEG baseline, the average PSNR between a reconstructed image and the original image ranges from 17 to 46. In Table 3.1, some representative PSNR values along with the required transmission powers for the JPEG baseline are given. However, the minimum power threshold for using the JPEG baseline is tremendously high as $10^{51}$ per image. The need for such high transmission power is explained by the JPEG algorithm hardly compressing MNIST images. Before compression, each half MNIST image is 392 bytes. After compression, the average file size of half images ranges from 350 bytes to 580 bytes for different qualities. For small images like MNIST, the overhead of the JPEG algorithm is significant. The same problem exists for other compression schemes like JPEG2000 [29]. Ineffective compression by JPEG is a representative example of how traditional schemes may lack the ability to adapt their rates to different communication scenarios.

In each of the following subsections, we exhibit a different experiment building on the baseline experiment we just described. In each experiment, we change a single property of the baseline simulation. Unless specified, the source distribution, the network topology, the number of channels per link, and channel statistics are the same as the baseline experiment.

## 3.4.2    Topology

Our second set of experiments studies the performance of NNC under different network topologies. Figure 3.4(a) shows an experiment where both destination nodes aim to infer the whole image while only the lower half of the image can be sent through the network. In the instance visualized in Fig. 3.4(a), the power constraints $P$ on all links are set to be 25. Both destination nodes reconstruct a mixture of "5" and "3" from the lower half of digit "5". The experiment illustrates that NNC can partially infer the missing signals from the available signals that are correlated to the missing part: This is an example of NNC learning and utilizing the correlation in data. Since no existing transmission scheme that we know of could recover the missing part of the image from the rest, any baseline competitor can only reconstruct the lower half of the image in our setting. As the baseline competitors are worse than NNC at low SNR in the multicasting setting presented in Section 3.4.1, we expect that the NNC would do even better in this case.

We demonstrate in Fig. 3.4(b) the multicast of MNIST images in a modified butterfly network where there are three destinations. The link connecting to the middle consists of 32 parallel AWGN channels with zero-mean noise of variance $10^{-4}$, which is the same as all other links in the network. In the instance visualized in Fig. 3.4(b), with power constraints set to be 25 on each link, the source image can be reconstructed at the middle destination node with a slightly worse reconstruction quality compared to the top and bottom destination nodes. We have the same observation on the PSNRs of the reconstructed images as well. Over the testing set, the average PSNR 3.2 is

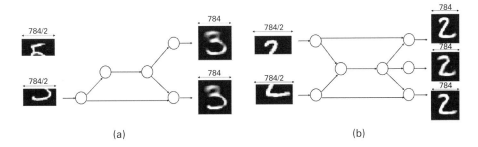

**Figure 3.4** NNC performance under different topologies. (a) Butterfly network where the top source node has zero transmission power. (b) Butterfly network with an additional destination.

21.4, 18.9, 21.5 at the top, middle, and bottom destination node, respectively. This is because, as shown in Fig. 3.4(b), the middle destination node has less information available compared to the top and bottom destination nodes: the middle node only has access to the information on its one incoming link, whereas the top and bottom destination nodes each has an additional incoming link and therefore has additional information available. This experiment further demonstrates the benefit of NNC over linear network coding schemes: Under a linear scheme, a linear combination of the top and bottom source is sent on the middle link, from which reconstructing at the middle destination node is impossible. Therefore, we expect ANC to perform poorly in this scenario. As for the comparison with the JPEG competitor, we expect NNC to outperform the baseline competitor when SNR is low, as shown in Section 3.4.1.

### 3.4.3    Channel Statistics

In wireless communication, channel impairment is often not limited to noise corruption [30]. In this section, we consider a wireless setup where links are subject to fading and simultaneous transmissions interfere. We evaluate the performance of NNC under this setup.

**Interference**

Consider the multicasting of MNIST images over a butterfly network with interfering channels. As shown in Fig. 3.5(a), for each node with multiple preceding nodes, signals on each link interfere with each other. The received signal for such a node is $\vec{W}_u + \vec{W}_l + \vec{N}$, where $\vec{W}_u, \vec{W}_l, \vec{N} \in \mathbb{R}^{32}$ are the signal sent by the upper preceding nodes, the signal sent by the lower preceding nodes, and the AWGN noise with zero mean and variance $10^{-4}$ on the channels, respectively. Our setup models the scenario where preceding nodes use the same set of subchannels in a frequency-division multiple access (FDMA) scheme simultaneously. With transmission power $P = 25$ on all links, images are reconstructed well with an average PSNR of 19.9 and 19.7 over the test set for the top and bottom destination node, respectively. Some of the reconstructed images are visualized in Fig. 3.5(b). We compared the NNC performance with

**Table 3.2.** Minimum transmission power per image for achieving the corresponding mean PSNR values under NNC and ANC with channel interference.

| $pSNR$ | NNC | ANC |
|---|---|---|
| 11.71 | $7.0 \times 10^{-3}$ | 2.10 |
| 11.73 | $7.0 \times 10^{-2}$ | 2.12 |
| 13.73 | $7.0 \times 10^{-1}$ | 3.31 |
| 19.62 | $7.0 \times 10^{0}$ | 11.76 |

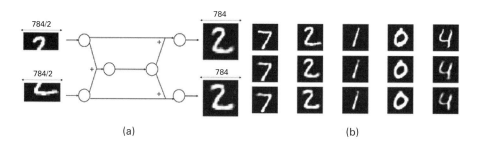

(a)　　　　　　　　　　　　　(b)

**Figure 3.5** NNC performance with channel interference. (a) Butterfly network with inference. Links where interference occurs are marked with plus signs. (b) Transmission quality. The first row of is the original image, and the following rows are reconstructions at one of the two destination nodes of a butterfly network. The two reconstructed images at both destination nodes were similar.

ANC in Table 3.2. As shown in the table, NNC outperforms ANC when the power budget is low. As for the JPEG competitor, which separates the source code from network code, we expect to encounter the same problem as in the baseline experiment in Section 3.4.1. The JPEG scheme cannot sufficiently compress the MNIST image and requires tremendous transmission power.

### Fading Channel

Consider the multicasting of MNIST images over a butterfly network with Rayleigh fading channels [30]. Instead of having parallel channels transmitting real numbers in each link as in Section 3.4.1, we instead consider parallel channels transmitting complex numbers. Channel gains $h$ are drawn in an independent and identically distributed (i.i.d.) fashion from a complex Gaussian distribution for each channel, namely $h \sim \mathcal{CN}(0, H_c)$. The channel coefficients remain unchanged for the entire duration of the transmission and only change between different images. It is worth mentioning that the capacity of the channel is zero, as the state of the channel is unknown to the communicating parties, hence information cannot be sent reliably [17].

We simulated the fading channel with nontrainable layers. Figure 3.6 shows the reconstruction of MNIST image over a butterfly network with $H_c = 1$, power constraints on each link $P = 25$, and 150 channels per link. We trained a NNC policy with a batch size of 256. Observe that although the capacity of the channel is zero,

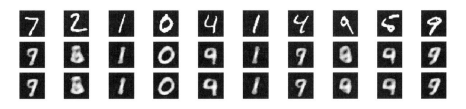

**Figure 3.6** NNC Performance over fading channels. The first row of is the original image, and the following rows are reconstructions at one of the two destination nodes of a butterfly network.

NNC can still reconstruct some of the images. As the channel capacity in Shannon's sense is zero in this scenario, our JPEG competitor is ill-defined here. As the channel gains are unknown to the destination nodes, we expect ANC to perform poorly in this scenario. This is because the channel gains are i.i.d sampled and therefore the signals are unproportionally scaled. As a result, a plain addition and subtraction cannot recover the signal without knowing the channel gains.

### 3.4.4 Nonmulticasting

Our next experiment shows the versatility of NNC by demonstrating its adaption into a nonmulticasting task. In this experiment, MNIST images are transmitted through a modified butterfly network with three destinations, and each destination node aims to reconstruct a different pixel range, as shown in Fig. 3.7. This is an example of a nonmulticasting task as a different subset of the source signals is reconstructed at each destination node. The link connecting to the middle consists of 32 parallel AWGN channels with zero-mean noise of variance $10^{-4}$. Note that this setup is not solvable by linear network coding as it does not satisfy the generalized Min-Cut Max-Flow condition derived in [10]. To the best of our knowledge, there is no existing solution to the nonmulticasting task. However, we can learn a NNC policy for the given task by setting the target function $g_t$ at each destination node $t$ to be the identity function of the corresponding subset of the source. With transmission power on each link constrained to be 25, the mean PSNR 3.2 over the test set is 21.2, 16.4, and 21.7 for the top, middle, and bottom destination node, respectively. An instance of the test image reconstruction is sketched in Fig. 3.7. Observe that nonlinear network codes constructed by the NNC policy can solve the nonmulticasting communication problem with tolerable distortion. Note that there is no existing solution to the nonmulticasting task and any baseline competitor can only multicast to all the destinations in our setting. As the baseline competitors are worse than NNC at low SNR in the multicasting setting presented in Section 3.4.1, we expect that the NNC would perform even better here.

### 3.4.5 Functional Reconstruction

Our next set of experiments reemphasizes the versatility of NNC and further shows how NNC adapts to tasks other than reconstruction. Consider the network shown

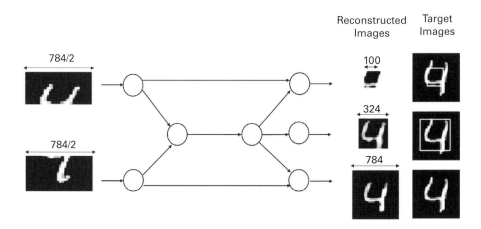

**Figure 3.7** Nonmulticasting task. The column of reconstructed images show the images reconstructed at corresponding destination nodes. The column of target images shows the source subset to be constructed in white rectangles.

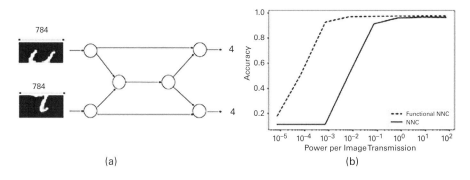

**Figure 3.8** Functional reconstruction task. (a) A butterfly network where both destination nodes aim to recover the digit in the source image. (b) Performance of functional NNC and NNC followed by a classification function under different transmission power per image. The PSNR at the two destination nodes are equal under both schemes.

in Fig. 3.8(a). In this network, both destination nodes are interested in determining which digit is sent rather than reconstructing the original MNIST image. This is an example of functional reconstruction of the source signals at the destination nodes. NNC can adapt into functional reconstruction by selecting a corresponding target function $f_t$ and distortion measure $\delta_t$ introduced in Session 3.2. In the example of digit recovery, the target function $f_t$ is a digit recognition function and distortion measure $\delta_t$ is the digit accuracy. Thus we learn a functional NNC policy that recovers the digit in a distributed manner while forwarding the information from the source nodes to the destination nodes. We compared the performance of functional NNC with a centralized computational scheme in Fig. 3.8(b). In the centralized scheme, MNIST images are first transmitted by an NNC policy through the butterfly network as in Section 3.4.1. Each destination then learns a digit recognition function from the image it reconstructs. We set the power constraints $P$ on all links to be the same and recorded the performance of both ends with 10 different values of $P$ evenly spaced in log-scale from $10^{-5}$ to $10^{-4}$. As shown in Fig. 3.8, in the high SNR regime destinations both schemes successfully recover the sent digit with functional NNC slightly better than NNC; in the low SNR regime, functional NNC significantly outperforms the centralized scheme. This is because functional NNC optimizes the network for the tasks of digit recovering, whereas NNC optimizes for the PSNR of the reconstructed images, which is not specialized for the task. In the high SNR regime, PSNR is very high, so digit recognition should be an easy task. However, in the low SNR regime, lowering image quality might focus energy on irrelevant parts of the image. The result demonstrates the benefit of adopting a task-specific scheme when the transmission power is limited. Since no existing transmission scheme that we know of could decompose the digit recognition function within the network, any baseline competitor could only multicast the source and determine the digit at the destinations under our setting. As the baseline competitors are worse than NNC at low SNR in the multicasting setting presented in Section 3.4.1, we postulate that functional the performance of NNC would improve in this scenario.

So far each of our experiments modifies a single property of the baseline experiment. What we present next in Section 3.5 are experiments in which multiple properties are different from the baseline experiment.

## 3.5     Experiments

In this section, we exploit the various use of NNC. Overall, NNC adapts to different tasks by selecting the corresponding target function $f_t$ and distortion measure $\delta_t$ introduced in Section 3.2. Unless specified, the source distribution, the network topology, the number of channels per link, and the channels statistics are the same as the baseline experiment in Section 3.4.1. Technical details of implementation can be found in Section 3.7.

### 3.5.1     Tasks of Different Complexity

As shown in Section 3.4.4, NNC can adapt to the different reconstruction goals at different destinations. In the following experiment, we demonstrate the analogy in functional reconstruction. As illustrated in Fig. 3.9(a), in this experiment, one destination node of the butterfly network aims to exactly recover the digit in the source image while the other destination is only interested in if the digit is smaller than five. We learned a NNC policy for the network by selecting a different target function $f_t$ for different destination nodes $t$: $f_t$ at the top destination node is the digit recognition function as in Section 3.4.5, whereas $f_t$ at the bottom is a binary classification function of the range. The distortion measure $\delta_t$ at both destinations is the accuracy. We set the power constraints $P$ on all links to be the same and recorded the performance of both ends with 10 different values of $P$ evenly spaced in log-scale from $10^{-5}$

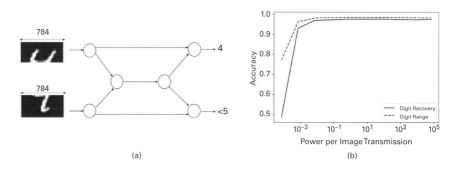

(a)              (b)

**Figure 3.9** Butterfly network where two tasks of different difficulty coexist. (a) A butterfly network where the top destination node aims to recover the digit in source image and the bottom to recover the range. (b) Performance at two destination nodes over different transmission power per image, where the solid line shows the accuracy of the digit recovery task at the top destination node and the dashed line shows the accuracy of the digit range task.

to $10^{-4}$ in Fig. 3.9(b). As shown in the figure, the digit recovery task always has a lower accuracy compared to the digit range task. We therefore quantified that the digit range task is easier. When the power budget is high, the tasks at both the destination nodes succeed with an accuracy higher than 97 percent. As power decreases, the performance at both destinations drops simultaneously. Compared with Fig. 3.8, we observe that the performance of the digit recovery task at the top destination is not affected as the lower destination node changes to an easier task. This gives an example of NNC not sacrificing performance under heterogeneous demands. Since no existing transmission scheme that we know of could decompose the task within the network, any baseline competitor could only multicast the source and compute the functions at the destinations under our setting. As the baseline competitors are worse than NNC at a low SNR in the multicasting setting presented in Section 3.4.1, we expect that functional NNC would do even better here.

### 3.5.2    Multiresolution Reconstruction

In this experiment, we consider the case where two ends of the butterfly network are interested in reconstructing the same image with different resolutions. As shown in Fig. 3.10, the upper destination is interested in reconstructing the original image with the same resolution and the lower destination is interested in a lower-resolution version. With the power constraint $P = 25$ on all links, the PSNR at the lower destination node is 24.7, which is higher than PSNR at the upper destination node 21.8. This is a result of the lower destination node requiring less information to complete its task. Since no existing transmission scheme that we know of could decompose the multiresolution reconstruction task within the network, any baseline competitor could only multicast the source and adjust the resolution at the destinations under our setting. As the baseline competitors are worse than NNC at a low SNR in the multicasting setting presented in Section 3.4.1, we expect that functional NNC's performance

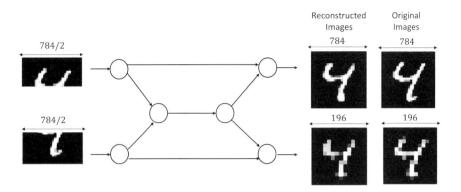

**Figure 3.10** Multiresolutional reconstruction task. The source image is reconstructed with a different resolution at each destination node. The column of reconstructed images shows the images reconstructed at corresponding destination nodes. The column of original images shows the source image with desirable resolution.

would improve in this case. The example of multiresolution reconstruction models the scenarios where different destination nodes have different reconstruction requirements and resource constraints. A practical scenario closely linked to our experiment is video streaming at multiple rates, where the video rate is customized into user requirement, device resolution, available caching size, bandwidth, etc. [31]. Our example shows the potential of NNC in solving such problems.

### 3.5.3    Same Source for Distinct Tasks

In the experiment in Fig. 3.11, we consider a butterfly network where two uncorrelated MNIST images are located at the top and bottom source nodes, respectively. The top destination node is interested in the addition of the two digits in the source images and the bottom is interested in the subtraction. This experiment presents a difficult communication task, as both destination nodes have to recognize which digits are sent as well as perform the required operations. We learned an NNC policy for the network by selecting a different target function $f_t$ for different destination nodes $t$: $f_t$ at the top destination node has input dimension $784 \times 2$, output dimension 1, mapping two MNIST images into the addition of numerical digits; $f_t$ at the bottom has the same dimension but maps to the subtraction. The distortion measure $\delta_t$ at both destinations is the accuracy. Despite the difficulty of the task, with the power constraints on all links set to be $P = 25$, our NNC policy succeeds over the test set with an average probability of 89.7 percent and 90.0 percent at the top and bottom destination nodes, respectively. This is an example of how the same source data can be used to complete different tasks in NNC. This approach can model the practical example of wearable health devices, where blood pressure and body temperature collected from sensors can be used to diagnose multiple diseases, say COVID-19 and the flu, simultaneously. Since no existing transmission scheme that we know of could decompose the task within the network, any baseline competitor could only

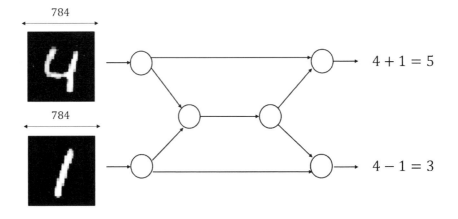

**Figure 3.11** Using the same source for distinct tasks. The butterfly network has two uncorrelated MNIST images located at two source nodes. The top destination node is interested in the addition of two digits, and the bottom is interested in the subtraction.

multicast the source and compute the functions at the destinations under our setting. As the baseline competitors are worse than NNC at a low SNR in the multicasting setting presented in Section 3.4.1, we expect that functional NNC would do even better here.

### 3.5.4    Functional Reconstruction of Fashion-MNIST Images

In this experiment, we extend the experiments in Section 3.4.5 to the Fashion-MNIST dataset [32]. The Fashion-MNIST dataset consists of images of fashion products from 10 classes, and the task of this experiment is to identify the class label of the source image, as illustrated in Fig. 3.12(a). Our setup is essentially the same as the experiments in Section 3.4.5, so details are omitted here. Similarly, we compared the performance of functional NNC with a centralized computational scheme in Fig. 3.12(b), where Fashion-MNIST images are first transmitted by an NNC policy through the butterfly network and then classified at each destination node. We observed the same pattern as in Fig. 3.8. As shown in Fig. 3.12(b), in the high SNR regime destinations under both schemes successfully recover the label with functional NNC slightly better than NNC; in the low SNR regime, functional NNC significantly outperforms the centralized scheme. Note that overall the accuracy of classifying Fashion-MNIST images is lower than classifying MNIST images, which is as expected since Fashion-MNIST images are more complicated and the task of classifying Fashion-MNIST images is harder. Also, the gap at the high SNR regime between the performance of functional NNC and NNC for Fashion-MNIST images is larger than that for MNIST images. The reason is that the compression in transmitting the Fashion-MNIST images becomes the bottleneck of the centralized scheme, which is limited by the bandwidth (the number

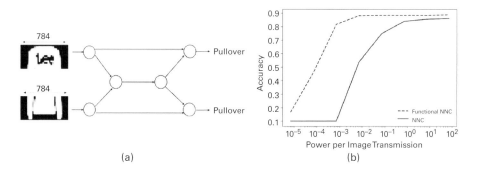

(a)                                       (b)

**Figure 3.12** Functional reconstruction of Fashion-MNIST images. (a) A butterfly network where both destination nodes aim to identify the class label of the source image. (b) Performance of functional NNC and NNC followed by a classification function under different transmission power per image. The PSNR at the two destination nodes are equal under both schemes.

of parallel channels) of the network. As we move to more complicated datasets, such as CIFAR [33] and Imagenet [34], we expect the phenomenon to worsen: the exact reconstruction performance of the source will be limited by the bandwidth of the network. Furthermore, we expect the training to be harder and require more engineering effort in hyper-parameter tuning. When the end goal is to reconstruct functions of the original source, we expect the performance gap between NNC and functional NNC to increase with the complexity of the dataset. The gap might be massive such that functional NNC becomes the only practical approach out of the two.

## 3.6     Conclusion and Open Questions

In this chapter, we proposed a novel way of constructing network codes using NN. Our scheme, NNC, provides a practical solution for communicating sources over networks of arbitrary topologies in an application-specific manner. NNC massively lowers the energy used compared to existing schemes. Furthermore, NNC can be easily learned and tested offline and implemented in a distributed fashion. We examined the performance of NNC under a variety of network topologies, channel statistics, and communication tasks through experiments.

A natural extension of NNC arises where multiple tasks exist asynchronously in a system. A separate NNC policy can be learned for each task. Thanks to the simplicity of NNC implementation, multiple functions can be implemented at one node. Signals can be designed with a flag piggybacked to trigger task-specific functions for constructing and decoding network codes at each node.

As we only present the initial study of NNC in this chapter, many open questions remain unanswered. One such question is how NNC would perform when data are distributed not only across the source nodes but also over the network. In our experiments, all the data are located at the source node. In other communication systems, however, data can be partially available at the intermediate nodes or even the destination nodes. Such a system is in general more robust as a single point of failure is avoided. Another question is how NNC would perform with constraints imposed on computational power. In our experiments, the communication power per link is fixed and the computation power depends on the NN structure. In our experiment, we did not put an upper bound on the computational power. This is in general a reasonable assumption, as computation with NN is of low complexity. However, in cases where edges devices are extremely constrained in power, things might be different if we link the problem to computation constraints: the size of the NN at the node is limited and the performance might be undermined as a result. In addition, in all our experiments, the physical topology of the communication system is fixed. Further study is needed to understand if and how NNC can adapt into a communication system with an uncertainty of the topology. Furthermore, we only studied the trade-off between the energy available and NNC performance in this chapter. It remains to study if a similar trade-off exists between the bandwidth available (i.e., the number of parallel channels), and NNC performance. Last but not the least, we took an analytical point of view in this chapter and studied NNC performance. It would be worthwhile to explore the design aspect: how we can design a wireless system such that it naturally exploits the findings provided by the NNC scheme. Some of our initial observations indicate promising directions in the design aspect. For example, we observed that the signal magnitude of NNC network codes is Gaussian distributed when the channel is AWGN, which is theoretically optimal. Another example is that for channels with interference, we observed that NNC tends to overlap the signals instead of allocating signals to distinct subchannels when SNR is low.

## 3.7     Technical Details

Unless specified otherwise, each NNC policy in Sections 3.4 and 3.5 is learned through training over 60,000 MNIST training images for 50 epochs and is tested on 10,000 MNIST testing images. Corresponding to each MNIST image, there is a label ranging from 0 to 9, representing the digit in the image. The target functions $f_t$'s in functional reconstructions are built from the labels. Note that the training set and the test set are disjoint.

We implemented the NN architecture in Keras [24] with TensorFlow [25] backend. We used Adadelta optimization framework [35] with a learning rate of 1.0 and a mini-batch size of 32 samples. In each of our experiment, we learned a NNC policy with every inner NN set to be two-layer fully connected with activation function ReLU: $f(x) = x^+ = \max(0, x)$. In the centralized computational scheme of Section 3.4.5, each of the digit recognition functions is also learned with a two-layer fully connected with activation function ReLU. Note that the hyper-parameters here may not be optimal, but the results still serve as a proof of concept.

## Acknowledgment

The authors would like to thank Yushan Su, Rui Wang, and Alejandro Cohen for their technical help and constructive comments.

## References

[1] A. El Gamal and Y.-H. Kim, *Network Information Theory*. Cambridge University Press, 2011.

[2] A. Ramamoorthy et al., "Separating distributed source coding from network coding," *IEEE/ACM Trans. on Networking*, vol. 14, no. SI, pp. 2785–2795, 2006.

[3] M. Effros et al., "Linear network codes: A unified framework for source, channel, and network coding," *Advances in Network Information Theory*, vol. 3, pp. 197–216, 2003.

[4] T. Ho et al., "A random linear network coding approach to multicast," *IEEE Trans. on Information Theory*, vol. 52, no. 10, pp. 4413–4430, 2006.

[5] I. Csiszar, "Linear codes for sources and source networks: Error exponents, universal coding," *IEEE Trans. on Information Theory*, vol. 28, no. 4, pp. 585–592, 1982.

[6] G. Maierbacher, J. Barros, and M. Médard, "Practical source-network decoding," in *6th Int. Symp. on Wireless Communication Systems*, pp. 283–287, 2009.

[7] T. P. Coleman, M. Médard, and M. Effros, "Towards practical minimum-entropy universal decoding," in *Data Compression Conf.*, pp. 33–42, 2005.

[8] Y. Wu et al., "On practical design for joint distributed source and network coding," *IEEE Trans. on Information Theory*, vol. 55, no. 4, pp. 1709–1720, 2009.

[9] A. Lee et al., "Minimum-cost subgraphs for joint distributed source and network coding," in *Proc. NETCOD*, 2007.

[10] R. Koetter and M. Médard, "An algebraic approach to network coding," *IEEE/ACM Trans. on Networking (TON)*, vol. 11, no. 5, pp. 782–795, 2003.

[11] R. W. Yeung and Z. Zhang, "Distributed source coding for satellite communications," *IEEE Trans. on Information Theory*, vol. 45, no. 4, pp. 1111–1120, 1999.

[12] S. Feizi and M. Médard, "On network functional compression," *IEEE Trans. on Information Theory*, vol. 60, no. 9, pp. 5387–5401, 2014.

[13] D. Malak, A. Cohen, and M. Médard, "How to distribute computation in networks," in *IEEE Conf. on Computer Communications (INFOCOM)*, pp. 327–336, 2020.

[14] T. J. O'Shea, K. Karra, and T. C. Clancy, "Learning to communicate: Channel autoencoders, domain specific regularizers, and attention," in *Proc. IEEE Int. Symp. on Signal Processing and Information Technology (ISSPIT)*, pp. 223–228, 2016.

[15] T. J. O'Shea and J. Hoydis, "An introduction to deep learning for the physical layer," *IEEE Trans. on Cognitive Communications and Networking*, vol. 3, no. 4, pp. 563–575, 2017.

[16] A. Felix et al., "OFDM-autoencoder for end-to-end learning of communications systems," in *Proc. IEEE Int. Workshop Signal Proc. Adv. Wireless Commun. (SPAWC)*, 2018.

[17] E. Bourtsoulatze, D. B. Kurka, and D. Gündüz, "Deep joint source-channel coding for wireless image transmission," *IEEE Trans. on Cognitive Communications and Networking*, vol. 5, no. 3, pp. 567–579, Sept. 2019.

[18] N. Farsad, M. Rao, and A. Goldsmith, "Deep learning for joint source-channel coding of text," in *Proc. IEEE Int. Conf. on Acoustics, Speech and Signal Processing (ICASSP)*, pp. 2326–2330, 2018.

[19] M. Jankowski, D. Gunduz, and K. Mikolajczyk, "Joint device-edge inference over wireless links with pruning," *IEEE Int'l Workshop on Signal Processing Advances in Wireless Communications (SPAWC)*, Jul. 2020.

[20] L. Liu et al., "Neural network coding," in *Proc. IEEE Int. Conf. on Communications (ICC)*, 2020.

[21] I. Goodfellow, Y. Bengio, and A. Courville, *Deep Learning*. MIT Press, 2016.

[22] G. K. Wallace, "The JPEG still picture compression standard," *IEEE Trans. on Consumer Electronics*, vol. 38, no. 1, pp. xviii–xxxiv, 1992.

[23] T. H. Cormen et al., *Introduction to Algorithms*. MIT Press, 2009.

[24] F. Chollet et al., "Keras," software, 2015; https://keras.io.

[25] M. Abadi et al., "TensorFlow: Large-scale machine learning on heterogeneous systems," software, 2015; http://tensorow.org.

[26] Y. LeCun, "The MNIST database of handwritten digits," database, 1998; http://yann.lecun.com/exdb/mnist/.

[27] S. T. Welstead, *Fractal and Wavelet Image Compression Techniques*. SPIE Optical Engineering Press, 1999.

[28] I. Maric, A. Goldsmith, and M. Médard, "Analog network coding in the high-SNR regime," in *3rd IEEE Int. Workshop on Wireless Network Coding*, pp. 1–6, 2010.

[29] A. Skodras, C. Christopoulos, and T. Ebrahimi, "The JPEG 2000 still image compression standard," *IEEE Signal Processing Magazine*, vol. 18, no. 5, pp. 36–58, 2001.

[30] D. Tse and P. Viswanath, *Fundamentals of Wireless Communication*. Cambridge University Press, 2005.

[31] B. Li and J. Liu, "Multirate video multicast over the internet: An overview," *IEEE Network*, vol. 17, no. 1, pp. 24–29, 2003.

[32] H. Xiao, K. Rasul, and R. Vollgraf, "Fashion-MNIST: A novel image dataset for benchmarking machine learning algorithms," *arXiv preprint*, arXiv:1708.07747, 2017.

[33] A. Krizhevsky and G. Hinton, "Learning multiple layers of features from tiny images," tech. report, University of Toronto, 2009.

[34] J. Deng et al., "Imagenet: A largescale hierarchical image database," in *IEEE Conf. on Computer Vision and Pattern Recognition*, pp. 248–255, 2009.

[35] M. D. Zeiler, "Adadelta: An adaptive learning rate method," *arXiv preprint*, arXiv:1212.5701, 2012.

# 4 Channel Coding via Machine Learning

Hyeji Kim

## 4.1 Introduction

Channel coding is one of the key elements in the physical layer communication system. The role of channel coding is to introduce redundancy in a controlled manner so that the receiver can reliably and efficiently recover a message from a *corrupted* received signal. Over the past 70 years, the design of codes has been one of the central areas of study in communication theory. Several novel codes have been invented, including convolutional codes, Turbo codes, low-density parity-check (LDPC) codes, and Polar codes. The practical impact of these codes is enormous. For example, global cellular phone standards, ranging from the second generation to the fifth generation models, rely on these codes.

The canonical setting is point-to-point additive white Gaussian noise (AWGN) channels, which closely fit the wireline and wireless communications, although the front end of the receiver may have to be specifically designed. Examples of this include inter-symbol equalization in cable modems and beamforming and sphere decoding in multiple antenna wireless systems. Accordingly, the performance of a code over the AWGN channels has been the gold standard. Several codes, such as LDPC, Turbo, and Polar codes, operate close to the (infinite block length) information theoretic "Shannon limit" for AWGN channels. A recent development also includes practical implementation of random codes with efficient guessing random additive noise decoding, which operates close to the (finite block length) theoretical limit.

In communication theory, there are two long-term goals. The first goal is to improve the state of the art over the point-to-point channels by designing new, computationally efficient codes. Since the current codes operate close to the information theoretic limit for AWGN channels, the emphasis is on robustness and adaptability to deviations from the AWGN settings, such as in urban, pedestrian, and vehicular settings. The second goal of communication theory involves the design of new codes for multiterminal and bidirectional communication settings. Examples of this include the feedback channel, the relay channel, and the interference channel.

Progress on these long-term goals has been sporadic, with several long-standing open problems. Deep learning is an emerging and powerful tool that has demonstrated huge success in a variety of domains, including computer vision and natural language processing. Motivated by this success, deep learning is now applied to solve open problems in coding theory and has demonstrated initial successes and great potential.

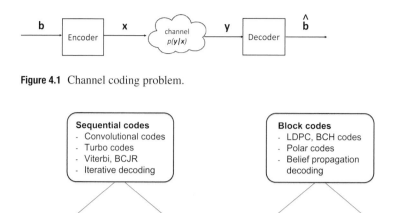

**Figure 4.1** Channel coding problem.

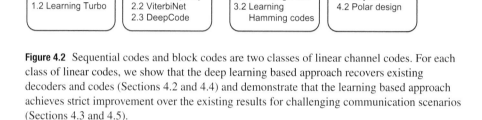

**Figure 4.2** Sequential codes and block codes are two classes of linear channel codes. For each class of linear codes, we show that the deep learning based approach recovers existing decoders and codes (Sections 4.2 and 4.4) and demonstrate that the learning based approach achieves strict improvement over the existing results for challenging communication scenarios (Sections 4.3 and 4.5).

Two aspects where we can benefit from deep learning are in dealing with (a) algorithm deficiency and (b) model deficiency. Algorithm deficiency refers to the challenge in designing algorithms when the space of an algorithm is huge – for example, in the game of Go, where the possible number of moves grows exponentially over time. Model deficiency refers to the scenario where the underlying dynamics of the problem are hard to analyze or unknown. For example, it is hard to model the distribution of the natural image mathematically. Recent advances have demonstrated that we can leverage both aspects of deep learning.

In this chapter, we introduce several recent breakthroughs on applying deep learning to solve open problems in coding theory, focusing on the unique challenges posed and how they are addressed. We consider a simplified channel coding problem, illustrated in Fig. 4.1, where the channel encoder maps a message (e.g., a random bit sequence $\mathbf{b} \in [0,1]^K$) to a codeword $\mathbf{x} \in \mathbb{R}^n$. The channel decoder receives $\mathbf{y} \sim P(\mathbf{y}|\mathbf{x})$, where $P(\mathbf{y}|\mathbf{x})$ denotes the channel, based on which it then estimates $\hat{\mathbf{b}}$.

The traditional approach to the design of codes involves imposing some structure on encoders (example: linearity of codes) and decoders (example: iterative decoding), thus paring down the complexity of space of encoders and decoders. The design of codes and decoders involves tremendous human effort and creativity. The two broad classes of linear codes, sequential codes and block codes (as illustrated in Fig. 4.2) are important as they facilitate computationally efficient decoding.

The deep learning based approach to the design of codes also involves imposing certain structures on the encoder and the decoder. Unlike in the traditional approach, however, the imposed structures are parametric functions, the parameters of which can be *learned* from data as opposed to being manually derived. Hence, the deep learning based approach allows for a broader family of structures, so called neural networks, to be considered. One example is the use of nonlinear codes for the encoders and decoders, with the challenges lying in both choosing the right neural networks, and using data to learn the best parameters within the neural network.

In addressing these challenges, several works have successfully enlarged and strengthened existing codes, including both families of block codes and the sequential codes, with their corresponding decoders. Within each family, we will first investigate how deep learning can recover existing algorithms in a data-driven manner. We will then consider communication scenarios where existing algorithms are far from optimal, and finally, we will examine the improvement driven by a deep learning based approach. We will see a variety of canonical contexts in which the enlarged family strictly improves upon the state of the art. In Sections 4.2 and 4.4, we will show that the deep learning based approach can recover existing decoders and codes for the family of sequential codes and block codes, respectively. In Sections 4.3 and 4.5, we will review new results demonstrating the success of deep learning in enlarging and improving the encoding and decoding of existing sequential codes and block codes, respectively.

## 4.2    Recovering Sequential Codes

The very first question one can ask is whether we can recover existing decoding algorithms via deep learning. We show that well-known sequential codes, such as convolutional codes and Turbo codes, and their corresponding decoders, such as Bahle-Cocke-Jelinek-Raviv (BCJR) decoders, can be learned, but only when neural architectures are carefully chosen and trained with the right training examples. We focus on sequential codes, such as convolutional codes and Turbo codes, for which there are well-known decoders for AWGN channels. For convolutional codes, the Viterbi and BCJR algorithms are optimal in terms of block error rate and bit error rate, respectively. In Section 4.2.1, we show that we can learn *decoders* for convolutional codes via deep learning that match the reliability of optimal decoders. We then study the capability of deep learning to match the performance of existing codes. In Section 4.2.2, we show that we can learn a code that achieves the reliability of Turbo codes. We start with an overview of sequential codes in the following section.

### Convolutional Codes and Recurrent Neural Networks

A *rate-1/2 recursive systematic convolutional (RSC) code* is a canonical example of sequential codes. As shown in Fig. 4.3, the encoder generates a sequence of binary

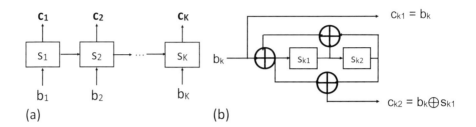

**Figure 4.3** (a) A sequential encoder of a convolutional code is a recurrent network. (b) An example of one cell for a rate-1/2 RSC code.

**Figure 4.4** The family of convolutional codes is a canonical sequential code, with a family of efficient decoders that is also optimal. We subsequently replace the decoder (Fig. 4.7) and also the encoder (Figure 4.11) with trained neural networks, while maintaining the sequential structure for efficient training and decoding.

vector *states* $(s_1, \ldots, s_K) \in \{0,1\}^{2 \times K}$ and a sequence of *coded bits* (or a codeword) $(c_1, \ldots, c_K) \in \{0,1\}^{2 \times K}$, recursively and causally based on a sequence of *message bits* (or *information bits*), $\mathbf{b} = (b_1, \ldots, b_K) \in \{0,1\}^{K}$. At each time $k$, the dynamic system is associated with a *state* represented by a two-dimensional binary vector $s_k = (s_{k1}, s_{k2})$, which is updated based on $s_{k-1}$ and $b_k$, as $s_k := (b_k \oplus s_{(k-1)1} \oplus s_{(k-1)2}, s_{(k-1)1})$, with the initial state $s_1 = (0,0)$. The corresponding output at time $k$ is a two-dimensional binary vector $c_k = (c_{k1}, c_{k2}) = (b_k, b_k \oplus s_{k1}) \in \{0,1\}^2$, where $x \oplus y = |x - y|$.

As illustrated in Fig. 4.4, the coded bits $\mathbf{c} \in \{0,1\}^{2K}$ are mapped to *transmitted symbols* $\mathbf{x} = 2\mathbf{c} - 1 \in \{-1,1\}^{2 \times K}$ via binary phase shift keying (BPSK) modulation. Throughout this section, we omit the explicit description of the modulation when we use the BPSK modulation (as seen in Fig. 4.4). The receiver receives a noisy version of the transmitted symbols; we let $\mathbf{y} = (y_1, \ldots, y_K) \in \mathbb{R}^{2 \times K}$ denote the *received symbols*. For AWGN channels, the received symbols are $y_{ki} = x_{ki} + z_{ki}$ for all $k \in \{1, \ldots, K\}$ and $i \in \{1,2\}$, where $z_{ki}$'s are independent and identically distributed (i.i.d. ) Gaussian variables with a mean of zero and a variance of $\sigma^2$.

Based on received symbols $\mathbf{y}$, a decoder attempts to find the maximum a posteriori (MAP) estimate of the message bits $\mathbf{b}$, with the goal of minimizing a target error metric. Two commonly used error metrics in evaluating the reliability of a decoder are bit error rate (BER) or block error rate (BLER). The BER evaluates the average fraction of message bits that are incorrect, i.e., $\frac{1}{K}\sum_{k=1}^{K} \mathbb{P}(\hat{b}_k \neq b_k)$. The BLER evaluates the average fraction of the blocks that are incorrect, i.e., $\mathbb{P}(\hat{\mathbf{b}} \neq \mathbf{b})$. There are two MAP decoders, each optimized for BER or BLER. Due to the recurrent structure of the encoder, the evolution of received symbols can be represented as a hidden Markov model (HMM); the MAP estimates can be computed efficiently and iteratively by MAP decoders [1, 2].

The *Viterbi algorithm*, which is a dynamic programming, finds a MAP estimate in time linear in the block length $K$ (and exponential in the size of the state, which is two in the earlier example of rate-1/2 RSC code), with the resulting estimate $\hat{\mathbf{b}} = \arg\max_{\mathbf{b} \in \{0,1\}^K} \mathbb{P}(\mathbf{b}|\mathbf{y})$. The *BCJR algorithm*, which is a forward-backward algorithm, finds a MAP estimate in time linear in block length $K$ as well, with the $k$th message bit of the resulting estimate being $\hat{b}_k = \arg\max_{b_k} \mathbb{P}(b_k|\mathbf{y})$, for all $k \in \{1, \ldots, K\}$. In both cases, the optimal (ML) decoder is computationally efficient (linear complexity) and crucially depends on the Markov structure of the encoder. Both decoders are special cases of a general family of efficient methods to solve inference problems on HMMs using the principle of dynamic programming.

A natural first step in evaluating whether sequential neural networks are suitable for new codes is to verify that the optimal decoders, the Viterbi and BCJR algorithms, can be recovered via neural network training. In Section 4.2.1, we demonstrate that one can recover the accuracy of the optimal decoders by training a sequential neural network from samples (the message sequence $\mathbf{b}$ and the noisy received sequence $\mathbf{y}$), without explicitly specifying the underlying probabilistic model. At a high level, this constructive example shows that highly engineered dynamic programming (Viterbi algorithm) can be matched by a neural network that only has access to the samples from a dynamic system (pairs of messages and the noisy received sequences). The challenge lies in finding both the appropriate neural network architecture and the training examples that are effective in training the decoder.

## Turbo Codes

Turbo codes concatenate codewords generated from two convolutional encoders, where one of the convolutional encoders take a permuted message bit stream as an input. Turbo codes are the first practical capacity-approaching codes [3]. *The rate-1/3 Turbo code*, which concatenates rate-1/2 RSC codes in parallel, is illustrated in Fig. 4.5. A rate-1/2 RSC encoder takes $\mathbf{b}$ as an input and generates coded bit streams $\mathbf{c}_1, \mathbf{c}_2$, where $\mathbf{c}_1 = \mathbf{b} \in \{0,1\}^K$ denotes the systematic bits (information bits) and $\mathbf{c}_2$ denotes the (remainder of the) coded bits. A rate-1/2 RSC encoder also takes an interleaved (permuted) bit sequence $\pi(\mathbf{b}) \in \{0,1\}^K$ as an input and generates coded bit streams $\pi(\mathbf{b}), \mathbf{c}_3$. Turbo code selects $(\mathbf{c}_1, \mathbf{c}_2, \mathbf{c}_3) = (\mathbf{b}, \mathbf{c}_2, \mathbf{c}_3) \in \{0,1\}^{3 \times K}$ as the codeword for the input sequence $\mathbf{b} \in \{0,1\}^K$. (Systematic bit stream $\pi(\mathbf{b})$ is not included in the codeword as it is redundant.) Due to the interleaver, long-range correlations exist in the codewords, which contributes to the reliability of Turbo codes.

The coded bit streams are transmitted through a noisy channel via the BPSK modulation, where $\mathbf{x}_i = 2\mathbf{c}_i - 1$ for $i \in \{1, 2, 3\}$. Let $\mathbf{y} = (\mathbf{y}_1, \mathbf{y}_2, \mathbf{y}_3) \in \mathbb{R}^{3 \times K}$ denote the received vectors. For AWGN channels, the received noisy signal is $\mathbf{y} = (\mathbf{y}_1, \mathbf{y}_2, \mathbf{y}_3) \in \mathbb{R}^{3 \times K}$, where $\mathbf{y}_i = \mathbf{x}_i + \mathbf{n}_i = (2\mathbf{c}_i - 1) + \mathbf{n}_i \in \mathbb{R}^K$ for $i \in \{1, 2, 3\}$.

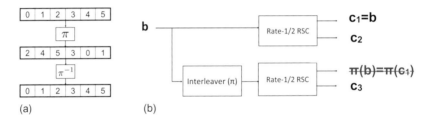

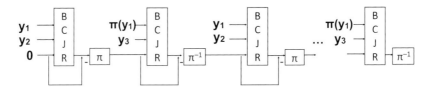

**Figure 4.5** (a) Interleaver ($\pi$) and de-interleaver ($\pi^{-1}$). (b) A rate-1/3 Turbo code that concatenates rate-1/2 RSC codewords in parallel.

**Figure 4.6** A Turbo decoder is an alternating recursion of two BCJR algorithms with interleaver ($\pi$) and de-interleaver ($\pi^{-1}$) in between.

Turbo codes are accompanied with an efficient and reliable decoder, which we call the Turbo decoder. The Turbo decoder successively refines the posterior distribution via the BCJR algorithm with interleavers and de-interleavers, as shown in Fig. 4.6. The Turbo decoder first computes the likelihood $\mathbf{L} = (L_1, \ldots, L_K)$ for $L_k = \log(\mathbb{P}(b_k = 1 | (\mathbf{y}_1, \mathbf{y}_2)) / \mathbb{P}(b_k = 0 | (\mathbf{y}_1, \mathbf{y}_2)))$ based on the noisy rate-1/2 RSC codeword $(\mathbf{y}_1, \mathbf{y}_2)$ via the BCJR algorithm. The Turbo decoder then estimates $\mathbb{P}(b_k | (\pi(\mathbf{y}_1), \mathbf{y}_3, \mathbf{L}))$ based on the noisy rate-1/2 RSC codeword $(\pi(\mathbf{y}_1), \mathbf{y}_3)$ with a prior given by $\mathbf{L}$ from the previous execution of a BCJR algorithm. The Turbo decoder continues to update the likelihood for a predefined number of iterations or until the likelihood converges.

In Section 4.2.1, we show that the reliability of a Turbo decoder for AWGN channels can be matched by a sequential neural network trained with samples. In Section 4.2.2, we investigate if one can train a sequential neural network based encoder together with a sequential neural network based decoder that achieves the reliability of Turbo codes for AWGN channels. At a high level, we aim to (a) demonstrate by a constructive example that a code with the reliability of modern codes can be learned in an end-to-end manner and (b) acquire the know-how to train both the encoder and decoder jointly. The challenge lies in finding both the right neural network architecture and the correct training procedures.

## 4.2.1    Learning the BCJR Algorithm

As a first step toward inventing novel codes via deep learning, we verify the capability of deep learning in matching the known performance of existing decoders. We fix the encoder as a convolutional code and recover the performance of the optimal decoder,

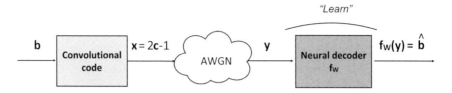

**Figure 4.7** The first step in designing new codes is to showcase that efficient and reliable decoders can be trained from samples. This is achieved by replacing the optimal decoder with a neural network, training the parameters **W** using training samples, and matching the optimal performance.

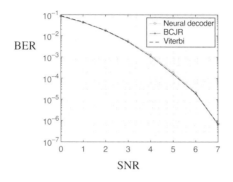

**Figure 4.8** The neural decoder trained on $K = 100$ sequences generalizes to match the BCJR decoder when tested on block length $K = 10,000$ sequences.

by training a neural network decoder from samples (Fig. 4.7). This is a necessary precursor to the more challenging tasks of training both encoders and decoders to obtain a new code in Section 4.2.2 and shown in Fig. 4.11.

The BCJR algorithm is a forward-backward algorithm, and one would claim that the BCJR algorithm is recovered if a neural decoder trained at a small block length can be applied to larger block lengths. Figure 4.8 shows that a strong generalization in block length is achieved by the proposed approach of learning a neural network decoder (for a rate-1/2 RSC code as shown in Fig. 4.3).

**Setup.** We first consider a simpler task of training a neural network decoder from $N$ labeled samples of the form $\{(\mathbf{y}^{(i)}, \mathbb{P}(b_k|\mathbf{y}^{(i)}))\}_{i=1}^{N}$ for each $k \in \{1, \ldots, K_{tr}\}$, and testing the trained neural decoder on $M$ labeled samples of the form $\{(\mathbf{y}^{(i)}, b_k)\}_{i=1}^{M}$ for each $k \in \{1, \ldots, K_{test}\}$. We let $K_{tr} \leq K_{test}$ to evaluate the generalization capability. For training, we fix $K_{tr} = 100$. For testing, we vary $K_{test}$ from 100 to 10,000. To generate training samples, we randomly generate codewords and simulate the channel to obtain $\mathbf{y}^{(i)}$, the input to the neural network, and run the BCJR algorithm to generate the label $\mathbb{P}(b_k|\mathbf{y}^{(i)})$. A more challenging task of training with samples of the form $\{(\mathbf{y}^{(i)}, b_k)\}_{i=1}^{N}$, which does not rely on the BCJR output $\mathbb{P}(b_k|\mathbf{y}^{(i)})$, is a difficult but necessary step, and is addressed at the end of this section.

One of the key challenges in learning a dynamic programming is that we want the end product to be an *algorithm*. In other words, we require a strong *generalization*

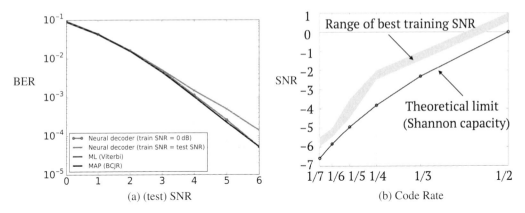

**Figure 4.9** Neural decoder performance for the rate 1/2-RSC code in Fig. 4.3. (a) Neural decoder trained at a mismatched 0 dB SNR improves upon a standard matched training, when trained and tested with block length $K = 100$. (b) The region of optimal training SNR is shown in grey as a function of the rate of the code. It is close to an information theoretic prediction shown in the dotted line.

*beyond the examples shown in the training.* It has been shown in [4] that neural network based solutions such as those in [5] do not generalize, especially over the size of the problem. This is especially concerning as training over a large dimensional examples is prohibitibly slow. A key idea to overcome this challenge is to impose stronger structures on the neural network decoder, such as sequential neural networks, as demonstrated in [4]. The recurrent structure allows achieving generalization to larger problems. We utilize this strength of sequential neural networks, on top of the fact that they match the structure of the encoder (i.e., convolutional codes), as well as the structure of the optimal decoder (i.e., BCJR decoder), to achieve strong generalization over the size of the problem (i.e., block length).

Another challenge in learning a dynamic programming is in choosing the *right training examples* for efficient training and faster convergence. As we generate training samples via mathematically well-defined models, we have the freedom to generate (a possibly infinite number of) training examples. However, selecting the right training examples is critical in efficient training and allows us to achieve the optimal reliability as shown in Fig. 4.9(a). We propose a novel scheme for choosing such training examples by fully utilizing the mathematical description of our problem. Further, this principle turns out to be crucial for extending applications beyond communications [4].

**Architecture.** To achieve strong generalization, we use recurrent neural networks, specifically, bidirectional gated recurrent units (bi-GRU), as the architecture for decoding convolutional codes. The detailed architecture is shown in Fig. 4.10. This is also a natural choice as the encoding process is sequential. It is shown in [6, 7] that optimal decoders can be represented as a sequential neural network with specific choices of parameters. We aim to show that the parameters can be learned from data as

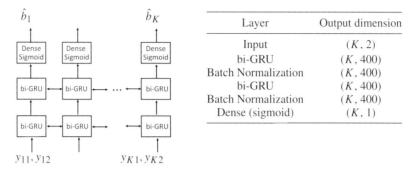

| Layer | Output dimension |
|---|---|
| Input | $(K, 2)$ |
| bi-GRU | $(K, 400)$ |
| Batch Normalization | $(K, 400)$ |
| bi-GRU | $(K, 400)$ |
| Batch Normalization | $(K, 400)$ |
| Dense (sigmoid) | $(K, 1)$ |

**Figure 4.10** Neural decoder architecture in [8] for the rate 1/2-RSC code in Fig. 4.4.

opposed to being mathematically driven. Let $\mathbf{y} = ((y_{11}, y_{12}), \ldots, (y_{K1}, y_{K2})) \in \mathbb{R}^{2 \times K}$ denote the received symbols (as defined in Section 4.2), and let $f_{\mathbf{W}}(\mathbf{y}) = \hat{\mathbf{b}} \in \mathbb{R}^K$ denote the output of the neural network with a parameter $\mathbf{W}$. Given training samples generated with an oracle access to the optimal BCJR algorithm, we solve for

$$\text{minimize}_{\mathbf{W}} \quad \mathbb{E}\left[ \sum_{k=1}^{K} \left( \hat{b}_k - \mathbb{P}(b_k = 1 | \mathbf{y}) \right)^2 \right], \tag{4.1}$$

where the expectation is over the samples used in training, and the conditional probability is evaluated using the BCJR algorithm. This is a mean square loss commonly used for a regression problem.

**Results.** We show that the trained neural network decoder recovers the reliability of the optimal decoder at various signal-to-noise ratios (SNRs) at various block lengths $K$. While the training framework is relatively straightforward, we see that depending on the choice of training examples, the end result can be very different. A natural choice, which is to sample the training data and test data from the same distribution, results in a suboptimal reliability, as shown in Fig. 4.9(a). To be precise, consider decoding convolutional codes under the AWGN channel with a specific SNR defined as $-10 \log_{10} \sigma^2$, where $\sigma^2$ is the variance of the Gaussian noise in the channel (assuming the transmitted sequence $\mathbf{x}$ satisfies $\mathbb{E}[\mathbf{x}^2] = 1$). A natural choice would be to train the decoder on examples generated on the AWGN channel with the target SNR. However, this fails to achieve the BCJR performance as shown in Fig. 4.9(a). Surprisingly, there exists an empirically optimal training SNR (0 dB in the rate-1/2 example as shown in Fig. 4.9) that primarily depends on the rate of the code and does not depend on the testing SNRs. The optimal training SNR increases as the rate of code increases. We conjecture that this is because as the rate of code increases, the distance between codewords becomes smaller, and in turn, examples lying at the decoding boundary are realizations of AWGN channels at a higher SNR.

We propose a rule for choosing such training examples, borrowing from information theoretic insight. Intuitively, we aim to select training examples that are just difficult enough (close to the decision boundaries) that no training time is wasted (on easy examples) or we avoid misguiding the training (on hard examples such as outliers). The decoding boundary is typically unknown, for example, in the image classification. However, for the channel decoding, where underlying dynamics are mathematically well defined, we can analyze the decoding boundary and the distance from a codeword to its decoding boundary. The average distance from a codeword to its decision boundary depends on the encoding schemes. Hence, to derive a universal rule of thumb, we consider the ideal codewords with a Gaussian codebook, with a maximum distance between neighboring codes. For a given code rate $r$, this ideal distance to the boundary can be translated into

$$\mathrm{SNR}_{\mathrm{train}} = 10 \log_{10}(2^{2r} - 1), \tag{4.2}$$

and is shown and named the "theoretical limit" in Fig. 4.9(b). This is derived from the sphere packing bound (and also Shannon capacity of the Gaussian channel). As practical codes have smaller inter-codeword distances than the ideal code, the SNR that corresponds to the decoding boundary will be higher than $\mathrm{SNR}_{\mathrm{train}}$. We also see that the empirical best choice of training SNR for convolutional codes is slightly higher (but close to the theoretical prediction), as shown in grey.[1] Similar efforts toward understanding optimal choices of data for training decoders have been taken in [9], and active learning approaches for training a decoder were explored in [10].

**Related work.** Building upon the above neural decoder, [8] has shown that one can learn to decode Turbo codes by replacing the BCJR blocks in the Turbo decoder as previously shown in Fig. 4.6 using recurrent neural networks. However, this requires access to the BCJR algorithm. It was then shown in [11] that the access to the BCJR algorithm is not necessary. They showed that a neural network trained on the data of the form $\{(\mathbf{b}^{(i)}, \mathbf{y}^{(i)})\}_{i=1}^{N}$ can recover the reliability of Turbo decoders, without an oracle access to the BCJR algorithm. It uses a standard cross-entropy loss of $\ell(\hat{b}_k, b_k) = -b_k \log(\hat{b}_k) - (1 - b_k) \log(1 - \hat{b}_k)$, a standard loss used in training binary classifiers. In [11], multidimensional vectors are used to encode the posterior beliefs from one iteration to the next. It is empirically shown that this technique is critical in achieving the target reliability. The standard Turbo decoder encodes its posterior belief as a scalar valued log-likelihood and updates it over iterations. Using multidimensional posterior beliefs allows the neural network to search over a larger class of functions that can potentially encode more information between iterations and potentially make the training process easier by the over-parametrization, thus achieving the performance of a Turbo decoder without the help of BCJR codes.

The learnability of optimal decoding algorithms for various convolutional codes is explored in [12], especially as the memory length of convolutional codes increases. The MAP decoders' complexity scales exponentially in the memory length. Accordingly, training neural network decoders for convolutional codes takes longer and requires larger neural networks as the memory length increases. The work [12] systematically compares the capability of various neural networks in achieving Viterbi performance on convolutional codes and proposes a new training method for learning Viterbi decoders for convolutional codes with a long memory.

---

[1] The codes from [8] are available at https://github.com/deepcomm/RNNViterbi.

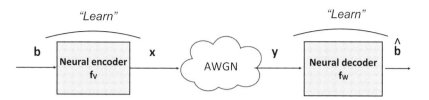

**Figure 4.11** Learning both the encoder and decoder via deep learning

## 4.2.2 Learning Turbo Codes (Turbo Autoencoder)

Motivated by the success of learning a decoder, we turn our attention to a harder task: Can we learn both the encoder and decoder for AWGN channels and recover the reliability of modern codes without a mathematical analysis? For AWGN channels, the current codes operate close to the theoretical limit. Hence, our focus is to recover the reliability of the existing codes rather than to outperform the existing codes. We will shown in Section 4.3.3 that deep learning based codes outperform existing codes for channels beyond the point-to-point AWGN channels. To recover the reliability of modern codes via deep learning, we model both the encoder and the decoder as sequential neural networks parametrized by $\mathbf{V}$ and $\mathbf{W}$, respectively, as illustrated in Fig. 4.11. We then train both of them using training samples for AWGN channels. While the setup might look like a simple extension of the decoder learning framework, learning a code (both the encoder and decoder) is fundamentally different from and much harder than learning a decoder. We introduce the associated challenges and techniques to mitigate them in the following.

We investigate if one can learn a sequential neural network based code (a pair of encoder and decoder) from samples, without explicitly specifying the underlying probabilistic model, and also recover the accuracy of Turbo codes. At a high level, we want to prove, by a constructive example, that a code driven by human ingenuity can be matched by a neural network that only has access to the samples from a dynamic system (the message sequence $\mathbf{b}$ and the noisy received sequence $\mathbf{y}$). Additionally, we aim to build techniques that are used to strengthen the deep learning based approach to design codes that outperform the state of the art. The challenge lies in finding the right architecture and training with the right training examples.

**Setup.** For concreteness, we consider learning a *rate-1/3* code for $K = 100$ information bits. The encoder maps $\mathbf{b} \in \{0, 1\}^{100}$ to a codeword $\mathbf{x} = (\mathbf{x}_1, \mathbf{x}_2, \mathbf{x}_3) \in \mathbb{R}^{3 \times 100}$, and the decoder maps $\mathbf{y} = (\mathbf{y}_1, \mathbf{y}_2, \mathbf{y}_3) \in \mathbb{R}^{3 \times 100}$ to $\hat{\mathbf{b}} \in \{0, 1\}^{100}$. We consider AWGN channels: $\mathbf{y}_i = \mathbf{x}_i + \mathbf{z}_i$ for i.i.d. Gaussian noise sequence $\mathbf{z}_i$ for $i = 1, 2, 3$. We compare our results against existing codes, including Turbo codes [3], Polar codes [13], and LDPC codes [14]. Our proposed approach can be applied to other code rates and block lengths.

**Architecture.** Turbo Autoencoder (TurboAE), introduced in [15], is inspired by a rate-1/3 Turbo code and consists of a channel encoder and a decoder, both modeled as neural networks. As shown in Fig. 4.12, the encoder of TurboAE combines the

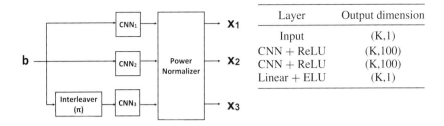

| Layer | Output dimension |
|---|---|
| Input | (K,1) |
| CNN + ReLU | (K,100) |
| CNN + ReLU | (K,100) |
| Linear + ELU | (K,1) |

**Figure 4.12** The encoder architecture in Turbo Autoencoder in [15] for a rate-1/3 code.

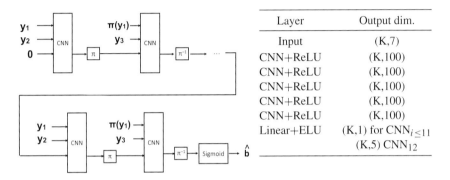

| Layer | Output dim. |
|---|---|
| Input | (K,7) |
| CNN+ReLU | (K,100) |
| CNN+ReLU | (K,100) |
| CNN+ReLU | (K,100) |
| CNN+ReLU | (K,100) |
| CNN+ReLU | (K,100) |
| Linear+ELU | (K,1) for $CNN_{i \leq 11}$ |
| | (K,5) $CNN_{12}$ |

**Figure 4.13** The decoder architecture in Turbo Autoecoder for a rate-1/3 code.

interleaver and convolutional neural networks (CNNs). It consists of three trainable CNN encoding blocks followed by a power normalization layer. Input to the first two CNN encoding blocks is the original bit sequence, while input to the last CNN block is an interleaved bit sequence. Each CNN block is a two-layered CNN, with the filter size of five, as shown in the table in Fig. 4.12. The three CNN block outputs are concatenated and passed through a power normalization layer. TurboAE has two versions of power normalizations. TurboAE-continuous, studied mainly in this chapter, allows real-valued codewords. TurboAE-binary, on the other hand, binarizes the codewords. The channel decoder of TurboAE also combines the interleaver and CNNs, as illustrated in Fig. 4.13.

**Training.** We introduce three key techniques that result in exploiting the full potential of the long-range correlations provided by the interleavers in the TurboAE. While the TurboAE can in principle mimic the Turbo code and its decoder, learning them from data alone requires a careful training methodology.

1. *Alternating the training of the encoder and the decoder.* The encoder parameters **V** and the decoder parameters **W** are updated iteratively and asymmetrically. Stochastic gradient descent (SGD) is used to train **V** and **W**, which solves

$$\text{minimize}_{\mathbf{V},\mathbf{W}} \quad \mathbb{E}\left[\sum_{k=1}^{K} \ell(b_k, \text{sigmoid}(q_k))\right],$$

where we use a cross-entropy loss $\ell$ and the expectation is over the samples used in the training. We observe that updating $\mathbf{V}$ and $\mathbf{W}$ simultaneously results in a poor reliability, and alternating the training of $\mathbf{V}$ and $\mathbf{W}$ prevents converging to a poor local optimum [16, 17]. By alternating the training, we can also train them in an asymmetric manner. For example, the decoder $\mathbf{W}$ is updated for a larger number of iterations than the encoder $\mathbf{V}$ is (e.g., $\mathbf{W}$ is updated with 500 examples, $\mathbf{V}$ with 100 examples). This turns out to also be critical in achieving an improved accuracy.

2. *Different training noise levels for the encoder and decoder.* As the updates of encoder and decoder parameters are separated, we can use different noise levels for the SGD update of the encoder and the decoder. We train the encoder at the target SNR and train the decoder at a mixture of SNRs. It has been shown empirically that it is best to train the decoder at a lower training SNR than the SNR it will be tested on [8, 17]. (We refer to Section 4.2.1 for the insight on how to choose the training SNRs for decoder learning only). However, when both the encoder and decoder are trained from scratch, the encoder is evolving over the training process, and corresponding decision boundaries between codewords shift over time (and are arbitrary in the beginning). Hence, [15] trains the decoder at various SNRs (with a random selection from $-1.5$ to 2 dB).

3. *Large batch size.* It is crucial to choose a batch size that is large enough. TurboAE is trained with a mini-batch size of 500. We observe that a TurboAE trained with a batch size less than 500 results in a dramatic drop in accuracy.

**Result.** The BER of the rate-1/3 TurboAE for the block length 100 as a function of the SNR is shown in Fig. 4.14. The BERs of widely used baseline codes including Turbo, Polar, LDPC, and tail-biting convolutional codes (TBCCs), generated via Vienna 5G simulator [18] are also shown for comparison. We also plot the BER of a CNN-based encoder-decoder architecture that does not use an interleaver as another baseline, labeled as CNN-AE. The reliability of CNN-AE is much worse than that of TurboAE, which shows that simply replacing the encoder and decoder with general-purpose neural networks does not achieve a high level of accuracy. We conjecture that this is because learning a code with a long-range memory with general-purpose neural networks is challenging. Both versions of TurboAE (TurboAE-continuous and TurboAE-binary) are shown to perform comparable to Turbo codes at a low SNR, while at a high SNR (over 2 dB with BER $< 10^{-5}$), performance is worse than both LDPC and Polar codes.

Modern codes that have a long-range memory, such as Turbo codes, achieve improved reliability as the block length increases. We refer to this as the *block length gain*. A long-range memory is necessary for the block length gain; convolutional codes have a limited memory and do not achieve a block length gain. We test the reliability of TurboAE as the block length increases as a way to verify if TurboAE exhibits a long-range memory. The BERs of TurboAE, Turbo code, and CNN-AE versus the block length are shown in Fig. 4.15(a), where the SNR of the AWGN channel is fixed at 2 dB. As shown in the figure, TurboAE achieves the block length gain, implying

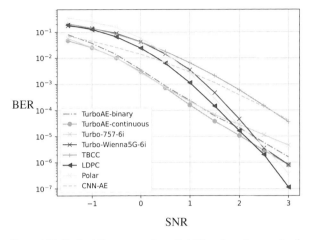

**Figure 4.14** TurboAE recovers the reliability of modern codes for moderate block length (100 information bits, rate-1/3) in low and mid-range of SNRs.

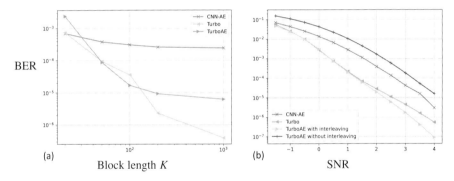

**Figure 4.15** (a) The error probability of TurboAE (at SNR 2 dB) decreases as block length $K$ increases. (b) The interleaver is critical in achieving the reliability of TurboAE.

that it exploits a long-term memory by the embedded interleaver, while well-known neural architectures (e.g., CNN-AE) do not achieve the block length gain as they tend to learn a code with a short-range memory. The key difference between the TurboAE and CNN-AE is that TurboAE includes an interleaver. To further understand the effect of interleavers, in Fig. 4.15(b), we plot the reliabilities of TurboAE with and without a random interleaver. The large gap between two BERs empirically demonstrates that the random interleaver is critical in achieving the reliability comparable to Turbo codes.

## 4.3 New Results on Sequential Codes

While optimal (and close-to-optimal) codes and matched decoders are analytically designed for AWGN channels, on several channels that are not AWGN, reliable and

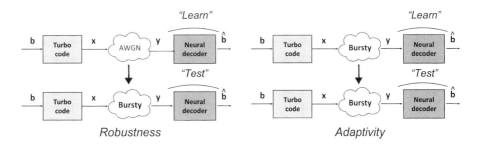

**Figure 4.16** Trained neural decoders exhibit superior robustness and adaptivity.

efficient codes and decoders are lacking. We demonstrate that deep learning serves as a promising new tool in designing decoders and codes for those scenarios. In Section 4.3.1, we show that deep learning based decoders are more robust to mismatched or unknown channel models and adapt better to the underlying channel models than the existing decoders. In Section 4.3.2, we show that deep learning can be integrated into the existing Viterbi algorithm for learning a decoder that is capable of tracking time-varying channels without the instantaneous channel knowledge. We next consider in Section 4.3.3 the AWGN channels with noisy output feedback, whose study was initiated by Claude E. Shannon. Feedback is known theoretically to improve reliability of communication, but linear coding methods to incorporate the feedback have proven inferior compared to nonlinear codes. We demonstrate that nonlinear deep learning based codes achieve significant improvement in reliability over the state of the art.

## 4.3.1 Robust and Adaptive Decoding for Non-AWGN Channels

In practical wireless communications, the channels do not exactly match the mathematical models or are not precisely known at the decoder. We show that deep learning based decoders are more reliable than traditional decoders for those scenarios. As illustrated in Figure 4.16, *robustness* refers to the ability of a decoder trained for a particular channel (e.g., AWGN) to work well, without retraining, when the test channel deviates from the training channel. *Adaptivity* refers to the decoder's ability to adapt to various non-AWGN channel models, by training with examples from those channels. We demonstrate both the robustness and the adaptivity of deep learning based decoders.

To evaluate the robustness of deep learning based decoders, we take the neural decoder trained on an AWGN channel for Turbo codes, introduced in Section 4.2.1, and evaluate its reliability for non-AWN channels. We consider bursty noise channels that model the $k$th received output as $y_k = x_k + z_k + n_k$, where $x_k$ denotes the $k$th transmitted symbol, $z_k \sim \mathcal{N}(0, \sigma^2)$ denotes the background Gaussian noise, and $n_k$ denotes the bursty noise $n_k \sim \mathcal{N}(0, \sigma_b^2)$ with probability $\rho$ and $n_k = 0$ with probability $1 - \rho$. Bursty channel models are widely used to model the interference or jamming environment. In Figure 4.17, we show the results for bursty channels with

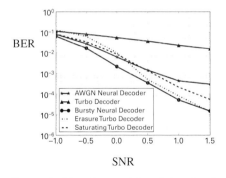

**Figure 4.17** Neural network based decoders for Turbo codes are more robust when tested on bursty channels and adapt to bursty noise channels to outperform heuristics used in practice.

$\sigma_b = 3.5$; without any retraining, neural decoders trained on an AWGN channel, labeled as "AWGN neural decoder" is more robust than the Turbo decoder designed for AWGN channels, labeled as "Turbo decoder" for bursty channels. Further, investigations reveal that the BCJR blocks in a Turbo decoder are often wrongly over-confident about some received symbols, and the wrong beliefs are propagated over the iteration, which degrades the reliability significantly. On the other hand, neural decoders are inherently conservative, making them robust against changes in the channel.

To evaluate the adaptivity of deep learning based decoders, we take the neural decoder trained on an AWGN channel for Turbo codes and further train the decoder using data generated from non-AWGN channels. In Figure 4.17, we demonstrate adaptivity results on bursty noise channels. We compare the neural decoder trained for bursty channels, labeled as "bursty neural decoder" against well-known heuristic baselines. To mitigate the bursty noise, heuristics are used in practice; the received symbols with high magnitudes are first thresholded before passing through the Turbo decoder. Two such heuristics are the "saturating Turbo decoder" (which shrinks large magnitude signals) and the "erasure Turbo decoder" (which sets large magnitude signals to zero). The performance of two such heuristics are shown in the figure. The figure shows that the bursty neural decoder trained on the bursty channel adapts to the new channel and achieves the highest reliability.

## 4.3.2      Symbol Detectors for Finite-Memory Channels (ViterbiNet)

As opposed to learning the decoder solely from data, an alternative approach in designing a decoder via deep learning is to start with the existing decoder and then introduce a neural network that replaces some operations of the existing decoder. Considering the problem of symbol detection for finite-memory channels without the channel knowledge, [19] proposes ViterbiNet which maintains the structure of the Viterbi algorithm but replaces the component of the Viterbi algorithm that depends on the channel knowledge with a trainable neural network. We consider channels with memory of length $l$:

$$p(y_i | s^t, y^{i-1}, y_{i+1}^t) = p(y_i | s_{i-l+1}^i). \tag{4.3}$$

**Setup.** We consider the intersymbol interference (ISI) channel modeled as

$$y_i = \sqrt{\rho} \sum_{\tau=1}^{l} (\mathbf{h}(\gamma))_\tau s_{i-\tau+1} + z_i, \tag{4.4}$$

where $\rho > 0$ denotes the SNR of the channel, $z_i \sim \mathcal{N}(0,1)$, the length of memory $l = 4$, $\mathbf{h}(\gamma) \in \mathcal{R}^l$ denotes the channel vector representing an exponentially decaying profile (i.e., $(\mathbf{h}(\gamma))_\tau = e^{-\gamma(\tau-1)}$ for $\gamma > 0$). We also consider a block fading ISI channel, for which the channel vector for the $j$th block is modeled as $(\mathbf{h}^{(j)})_\tau = e^{-0.2(\tau-1)}(0.8 + 0.2 \cos\left(\frac{2\pi j}{\mathbf{p}_\tau}\right))$, with $\mathbf{p} = [59, 39, 33, 21]$.

**Architecture.** ViterbiNet maintains the structure of the Viterbi algorithm while replacing the channel cost computation ($c_k^{NN}$ in Algorithm 4.1) with a neural network that is trained with samples. The cost computation block, as illustrated in Fig. 4.18, consists of two modules. One is the finite mixture model PDF estimator used to estimate the marginal probability for the observation $p(y_i)$, and the other is a fully connected network which approximates $\hat{p}(\mathbf{s}|y_i)$, the conditional probability of state given the observation. These two estimator outputs are combined via Bayes rule to generate the log-likelihood estimate of the observation $y_i$ given states $\mathbf{s}$, which corresponds to the cost function used in the Viterbi algorithm.

**Training.** For the time-invariant ISI channel, ViterbiNet without channel state information (CSI) is trained with samples generated from various realizations of noisy channels $\mathbf{h}(\gamma)$. Specifically, each entry of $\mathbf{h}(\gamma)$ is corrupted by i.i.d. zero-mean Gaussian noise with a variance of 0.1. The test result is compared to the Viterbi algorithm

---

**Algorithm 4.1** ViterbiNet [19]

**Input:** Block of channel outputs $\mathbf{y}^t$, where the memory length $l < t$.
**Initialization:** path cost $\tilde{c}_0(\tilde{\mathbf{s}}) = 0$ for each state $\tilde{\mathbf{s}} \in \mathcal{S}^l$.
**while** $k = 1, \ldots, t$ **do**
> For each sate $\tilde{\mathbf{s}} \in \mathcal{S}^l$, compute
> $\tilde{c}_k(\tilde{\mathbf{s}}) = \min_{\mathbf{u} \in \mathcal{S}^l : \mathbf{u}^{l-1} = \tilde{\mathbf{s}}_2^l} (\tilde{c}_{k-1}(\mathbf{u}) + c_k^{NN}(\tilde{\mathbf{s}}))$.
> **if** $k \geq l$ **then**
> > set $\hat{\mathbf{s}}_{k-l+1} := (\arg\min_{\tilde{\mathbf{s}} \in \mathcal{S}^l} \tilde{c}_k(\tilde{\mathbf{s}}))_1$.
> **end**

**end**
**Output:** decoded output $\hat{\mathbf{s}}^t$, where $\hat{\mathbf{s}}_{t-l+1}^t := \arg\min_{\tilde{\mathbf{s}} \in \mathcal{S}^l} \tilde{c}_t(\tilde{\mathbf{s}})$.

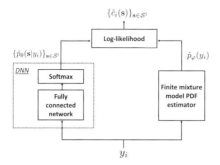

**Figure 4.18** DNN-based cost computation for ViterbiNet in [19].

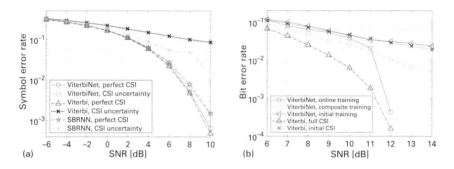

**Figure 4.19** The accuracy of ViterbiNet without CSI is significantly higher than the accuracy of Viterbi algorithm without CSI for (a) time-invariant ISI channels and (b) time-varying ISI channels [19].

without CSI, which computes the log-likelihoods using the noisy channel estimate under the same corruption model for $\mathbf{h}(\gamma)$.

For the time-varying channel, ViterbiNet is trained both offline and online. Before the transmission starts, ViterbiNet is trained using 5 000 training samples generated via initial channel vector $\mathbf{h}^{(1)}$ offline. During the transmission of blocks, when a new block $\mathbf{y}^{(j)}$ arrives, it first estimates the transmitted symbols $\hat{\mathbf{s}}$. When forward error correcting codes (e.g., convolutional codes) are used, which is very often the case, as long as the number of symbol errors is smaller than the minimum distance between codewords, the encoded message can be still recovered [20]. Hence, the estimated correct symbol $\tilde{\mathbf{s}}^{(j)}$ can be generated by encoding the estimated message, after which the pair $(\mathbf{y}^{(j)}, \tilde{\mathbf{s}}^{j})$ is used to retrain ViterbiNet online if the estimated number of symbol errors is below a certain threshold.

**Results.** The accuracy of ViterbiNet for time-invariant and time-varying ISI channels are shown in Fig. 4.19. In both cases, ViterbiNet without the CSI achieves a significantly higher accuracy compared to the Viterbi algorithm without the CSI. For time-invariant scenario in Fig. 4.19(a), ViterbiNet without CSI closely achieves the accuracy of the optimal Viterbi algorithm with the CSI. The performance of a sliding bidirectional recurrent neural network (SBRNN) based sequence detector by [21] is

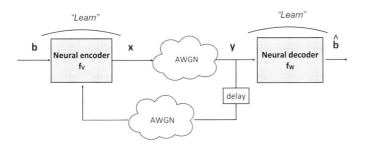

**Figure 4.20** An end-to-end deep learning based communication system for channels with output feedback: both the encoder and decoder are modeled as neural networks and the parameters **V** and **W** are trained from samples.

also shown as a reference, which achieves a higher error rate compared to ViterbiNet when the channel state information is not available. For the time-varying scenario in Fig. 4.19(b), ViterbiNet trained online based on recent decoding decisions achieves a significant reliability improvement upon the initial training at a high SNR and closely achieves the accuracy of the optimal Viterbi algorithm that knows the channel state information at a high SNR. As a reference, ViterbiNet trained with a mixture of channel vectors offline (labeled composite trainining) is shown to achieve higher error rates compared to ViterbiNet at a high SNR.

**Related work.** In addition to ISI channels, [19] also demonstrates that the ViterbiNet without CSI is significantly more reliable than the Viterbi algorithm without CSI for time-invariant and time-varying Poisson channels with a finite memory. The hybrid approach of integrating deep learning with existing decoding algorithms has been applied to several other sequential decoding algorithms, such as the BCJR algorithm [22] and iterative Turbo decoders [23], and also to MIMO channels [24, 25].

## 4.3.3    Designing Codes for Channels with Output Feedback (DeepCode)

The advantages of deep learning based approach for coding is that it can be applied to learn new codes for channels for which analytically designing reliable codes is challenging. One such example is the channel with output feedback, proposed by Shannon [26], shown in Fig. 4.20, where the received symbols are sent back to the transmitter with some delay. In [26] it is shown that the presence of output feedback does not improve the capacity. However, accuracy in the finite block length regime can in theory increase significantly. In the ideal output feedback model (the output symbol is fed back noiselessly with unit delay), the block error rate decreases doubly exponentially with block length with feedback (the celebrated result of Schalkwijk and Kailath [27]). In practice, however, the Schalkwijk and Kailath scheme (S-K scheme) is sensitive to the finite machine precision and noise in the feedback [27, 28]. It is shown that any linear code, which includes the S-K scheme, that incorporates noisy output feedback, cannot achieve a positive rate of communication [29]. If the

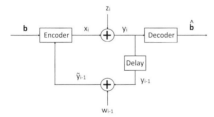

**Figure 4.21** An AWGN channel with noisy output feedback

feedback channel is noisy, however small, the communication with noisy feedback is fundamentally different from the ideal output feedback scenario.

We show that *nonlinear* codes learned via deep learning significantly outperform the state of the art on the AWGN channel with noisy output feedback. Sequential neural networks are used to model the encoder and the decoder since the nature of the feedback channel is sequential.[2]

**Setup.** We consider the channel with output feedback proposed by Shannon. As illustrated in Fig. 4.21, both forward channel and feedback channel are modeled as AWGN channels. At time $i$, the decoder receives $y_i = x_i + z_i$, where $z_i \sim \mathcal{N}(0, \sigma^2)$, and transmits $y_i$ back to the encoder with a unit-step delay. At time $i + 1$, the encoder receives $\tilde{y}_i = y_i + w_i$ for $w_i \sim \mathcal{N}(0, \sigma_F^2)$, which denotes the feedback noise. An *encoder* sequentially maps the information bit sequence $\mathbf{b} \in \{0, 1\}^K$ and the feedback signal available at time $i$, $\tilde{y}_1^{i-1} = (\tilde{y}_1, \ldots, \tilde{y}_{i-1})$, to the next transmission symbol $x_i$. Formally, a rate-$K/n$ encoder consists of mappings $f_i \colon (\mathbf{b}, \tilde{y}_1^{i-1}) \mapsto x_i \in \mathbb{R}$, $i \in \{1, \ldots, n\}$. Without loss of generality, we constrain the average power of $\mathbf{x}$ to one; that is, $(1/n)\mathbb{E}[\|\mathbf{x}\|^2] \leq 1$, where $\mathbf{x} = (x_1, \ldots, x_n)$. The randomness of $\mathbf{x}$ is due to the randomness of the information bits $\mathbf{b}$ and noisy feedback signals $(\tilde{y}_1, \ldots, \tilde{y}_n)$. A *decoder* maps the received sequence $\mathbf{y} = (y_1, \ldots, y_n)$ into an estimated information bit sequence $g \colon \mathbf{y} \mapsto \hat{\mathbf{b}} \in \{0, 1\}^K$.

We demonstrate that a trained pair of an encoder and a decoder outperforms the S-K scheme and the state-of-the-art linear codes in accuracy for the channels with noisy output feedback. For concreteness, we focus on rate-1/3 codes with $K = 50$ and a unit-step delayed feedback.

**Architecture.** We model both the encoder and the decoder as recurrent neural networks, as illustrated in Figs. 4.22 and 4.23, since (a) communication with feedback is naturally a sequential process, and (b) we can exploit the sequential structure for efficient decoding. We refer to this framework and the corresponding newly discovered code as *DeepCode*.

A rate-1/3 encoder operates in two phases. In the first phase, the encoder generates an uncoded BPSK transmission sequence of length $K$ in $\mathbb{R}^K$, with an appropriate power control. In the second phase, as shown in Fig. 4.22, the encoder generates a

---

[2]  Implementation of [30] is available at `https://github.com/hyejikim1/Deepcode`, `https://github.com/yihanjiang/feedback_code`.

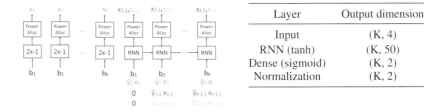

| Layer | Output dimension |
|---|---|
| Input | (K, 4) |
| RNN (tanh) | (K, 50) |
| Dense (sigmoid) | (K, 2) |
| Normalization | (K, 2) |

**Figure 4.22** The encoder architecture of the rate-1/3 DeepCode in [30].

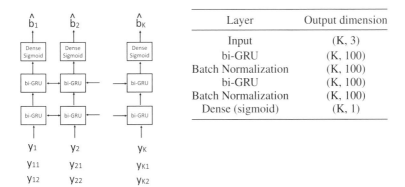

| Layer | Output dimension |
|---|---|
| Input | (K, 3) |
| bi-GRU | (K, 100) |
| Batch Normalization | (K, 100) |
| bi-GRU | (K, 100) |
| Batch Normalization | (K, 100) |
| Dense (sigmoid) | (K, 1) |

**Figure 4.23** The decoder architecture of the rate-1/3 DeepCode in [30].

coded transmission sequence of length $2K$ via an RNN followed by a learnable power allocation block. For $j \in \{1, \ldots, K\}$, each $j$th RNN cell updates the hidden state based on the (a) previous state, (b) input information bit $b_j$, (c) $\tilde{y}_j - x_j$ (the estimated noise added to the $j$th transmission in phase 1), and (d) $(\tilde{y}_{j-1,1} - x_{j-1,1}, \tilde{y}_{j-1,2} - x_{j-1,2})$ (the estimated noise added to the $(j - 1)$th transmission in phase 2). The $j$th RNN cell then maps the updated hidden state to a pair of coded symbols $(x_{j,1}, x_{j,2}) \in \mathbb{R}^2$.

The decoder is modeled as gated recurrent units (GRU), which fits naturally with the sequential codes, as shown in Fig. 4.23. Let $\mathbf{y} = (y_1, \ldots, y_K, y_{1,1}, y_{1,2}, \ldots, y_{K,1}, y_{K,2}) \in \mathbb{R}^{3K}$ denote the received sequence (i.e., $y_j = x_j + z_j$, $y_{j,1} = x_{j,1} + z_{j,1}$, and $y_{j,2} = x_{j,2} + z_{j,2}$, where $z_j, z_{j,1}, z_{j,2}$ denote the Gaussian noise added in the AWGN channel for $j = 1, \ldots, K$). The decoder waits until it receives the complete received sequence $\mathbf{y} \in \mathbb{R}^{3K}$ and then maps $\mathbf{y}$ to $\hat{\mathbf{b}} \in \{0, 1\}^K$ via two-layered bidirectional GRUs. The inputs to the $k$th first-layer GRU cell are three symbols, $y_k$ and $(y_{k,1}, y_{k,2})$, which correspond to the $k$th symbol of the first phase and the second phase, respectively. The encoder and the decoder are trained jointly via back propagation through time (on the entire input message bit sequence) to minimize the binary cross-entropy loss function. We train DeepCode for both noiseless and noisy settings. We always set the training SNR to be matched to the test SNR.

**Result.** We evaluate DeepCode for both noiseless and noisy feedback settings. Block length $K = 50$ and code rate-1/3 are assumed in all experiments. In Fig. 4.24(a),

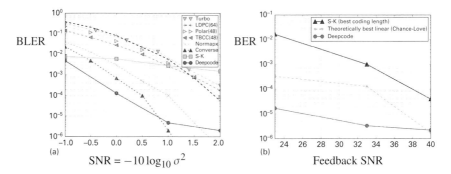

**Figure 4.24** (a) DeepCode outperforms the baseline of S-K and all state-of-the art codes (that do not use feedback) by several orders of magnitude in BLER, on block-length 50, and on noiseless feedback in BER. (b) DeepCode also outperforms S-K with the empirically best coding length and the theoretical reliability (assuming an infinite precision) of linear codes [33] for noisy output feedback channels. Forward SNR is fixed as 0 dB.

we plot the BLER of DeepCode and several baselines for channels with noiseless feedback (i.e., $\sigma_F^2 = 0$) with a finite machine precision. The S-K scheme is implemented on MATLAB with a precision of 64 bits to represent floating-point numbers. Since the scheme is sensitive to finite numerical precision, numerical errors dominate the performance of the S-K scheme, as shown in Fig. 4.24. DeepCode outperforms the S-K scheme by several orders of magnitude in BLER on a wide range of SNR. Notably, DeepCode significantly improves over all state-of-the-art codes of similar block lengths and identical rates that do not utilize feedback. In addition, an approximate achievable BLER (labeled Normapx) and a converse to the BLER (labeled Converse) from [31] and [32] are shown as a reference. The theoretical estimate of the best code (with no efficient decoding schemes) for channels without feedback lies between these two.

In Fig. 4.24(b), we show the BER of DeepCode for channels with noisy feedback as a function of the feedback SNR ($-\log_{10} \sigma_F^2$). The S-K scheme is shown as a reference. Unlike in Fig. 4.24(a), we plot the empirically optimized performance of the S-K scheme over the encoding block length to possibly remove the effect of the precision issue in evaluating the sensitivity of the S-K scheme to the noise in the feedback. We also plot the BER of the theoretically expected optimal performance of linear coding schemes by [33] with infinite precision implementation and optimized encoding block length. The results in Fig. 4.24(b) demonstrate that the nonlinearity of DeepCode is playing a crucial role in achieving the improved reliability and that nonlinear codes can be successfully learned via deep learning.

**Interpretation of DeepCode.** Can we understand the nonlinear behavior of Deep-Code (i.e., how coded bits are generated via RNN in Phase 2)? We run systematic experiments to learn an insight from DeepCode that we have learned. We observe that in the second phase of DeepCode (a) the encoder focuses on refining information bits

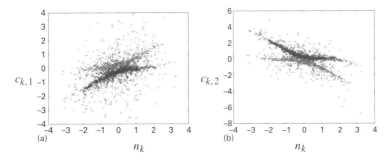

**Figure 4.25** (a) Noise in first phase $n_k$ versus first parity bit $c_{k,1}$ and (b) second parity bit $c_{k,2}$ under noiseless feedback channel and forward AWGN channel of SNR 0 dB. The x data points correspond to those samples conditioned on $b_k = 1$, and the o points correspond to those samples conditioned on $b_k = 0$.

that were corrupted by large noise in the first phase and (b) the coded bit depends on past as well as current information bits (i.e., coupling in the coding process occurs). *Correcting noise from the previous phase.* The main motivation behind the proposed two-phase encoding scheme is to use the Phase 2 to correct symbols that were highly corrupted in Phase 1. The encoder at Phase 2, exactly or approximately, knows how much noise was added in Phase 1, depending on the amount of noise in the feedback. The encoder can then focus on correcting the symbols with a large and adversarial noise. Interpreting the parity bits confirms this conjecture as shown in Fig. 4.25. We show as a scatterplot multiple instances of the pairs of random variables, $(n_k, c_{k,1})$ Fig. 4.25(a) and $(n_k, c_{k,2})$ Fig. 4.25(b), where $n_k$ denotes the noise added to the transmission of $b_k$ in the first phase. We are plotting 1,000 sample points: 20 samples for each $k$ and for $k \in \{1, \ldots, 50\}$.

We observe that the encoder first checks if the noise was favorable or adversarial. If a positive noise is added to a positive transmitted symbol, the noise is counted as favorable (e.g., $n_k > 0$, $c_k = 1$). If a negative noise is added to a positive transmitted symbol, the noise is counted as adversarial (e.g., $n_k < 0$, $c_k = 1$). If the noise at the $k$th symbol in the first phase is favorable, the encoder does not transmit anything in the $k$th transmission in the second phase. Otherwise, the encoder transmits the magnitude of the noise in Phase 2, so that the decoder can refine the corrupted information bits sent in Phase 1. This illustrates how the encoder has learned to send rectified linear unit ($\text{ReLU}(x) = \max\{0, x\}$), functional of the noise $n_k$, to send the noise information while efficiently using the power. Precisely, the dominant term in the parity bit can be closely approximated by $c_{k,1} \simeq -(2b_k - 1) \times \text{ReLU}(-n_k(2b_k - 1))$, and $c_{k,2} \simeq (2b_k - 1) \times \text{ReLU}(-n_k(2b_k - 1))$. By generating coded bits close to zero (i.e., does not further refine $b_k$) if $n_k$ is favorable and generating coded bits proportional to the noise $n_k$, the encoder can use the power more efficiently.

*Coupling.* One of the key questions for DeepCode is whether it jointly encodes information bits. To see if DeepCode exploits the memory of RNN and coding information bits jointly, we measure the correlation between the information bits and the coded bits. We find that $\mathbb{E}[c_{k,1}b_k] = -0.42, \mathbb{E}[c_{k,1}b_{k-1}] = -0.24, \mathbb{E}[c_{k,1}b_{k-2}] = -0.1,$ $\mathbb{E}[c_{k,1}b_{k-3}] = -0.05,$ and $\mathbb{E}[c_{k,2}b_k] = 0.57, \mathbb{E}[c_{k,2}b_{k-1}] = -0.11, \mathbb{E}[c_{k,2}b_{k-2}] = -0.05, \mathbb{E}[c_{k,2}b_{k-3}] = -0.02$ (for the encoder for forward SNR 0 dB and noiseless feedback). This result implies that the RNN encoder utilizes the memory of length at least two to three. Since the correlation captures linear associations, having a small correlation does not necessarily imply a small association. The actual memory length might be larger than two or three.

**Related work.** Taking a step further, exploiting the advantage of end-to-end trainability, several works have proposed recurrent neural network based joint source channel coding schemes for point-to-point channels both without feedback [34] and with feedback [35].

## 4.4      Recovering Linear Block Codes

We turn our attention to linear block codes that do not have a sequential nature of encoding and decoding. The architectures and training methodologies used to recover and extend linear block codes are different from the ones used to recover and extend sequential codes. In this section, we demonstrate that neural network based codes and decoders can recover the existing linear block codes and their optimal decoders. In Section 4.4.1, we show that the reliability of the optimal (MAP) decoder for Polar codes can be matched by a feedforward neural network that is trained only on data. In Section 4.4.2, we show that the reliability of Hamming codes can be matched by a pair of feedforward neural networks, representing the encoder and decoder, trained from data.

### 4.4.1      Learning the MAP Decoder for Polar Codes

A linear block code that maps a $K$-bit message $\mathbf{b}$ to a length-$N$ codeword $\mathbf{c}$ is represented as $\mathbf{c} = \mathbf{bG}$, where $\mathbf{G}$ denotes a $K \times N$ generator matrix. For binary codes, the generator matrix is $\mathbf{G} \in \{0,1\}^{K \times N}$ and additions are calculated using modulo 2. Polar codes are linear codes; the coded bit sequence $\mathbf{c} = \mathbf{Fu}$ for a $N \times N$ polarization matrix $\mathbf{F} = [1,0;1,1]^{\times N}$ (the $N$th Kronecker power), where $\mathbf{u} \in \{0,1\}^N$ includes $K$ information bits $\mathbf{b}$ and $N - K$ frozen (known) bits.

**Setup.** Learning a decoder for a rate-1/2 Polar code for $K = 16$ information bits is considered in [36]. We demonstrate that feedforward neural networks, trained from samples, can match the reliability of the optimal (MAP) decoder, which can be efficiently implemented for short block lengths (e.g., $K \leq 32$). As the decoder is modeled as a feedforward neural network that does not exhibit strong structure, one can conjecture that training with abundant data is important. We show empirically that the number of training data required for achieving the MAP reliability scales exponentially as the message length $K$ increases ($\approx 2^K$). This is different from learning a recurrent neural network based decoder for sequential codes, which exhibits strong

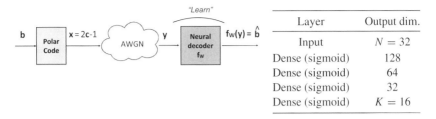

| Layer | Output dim. |
| --- | --- |
| Input | $N = 32$ |
| Dense (sigmoid) | 128 |
| Dense (sigmoid) | 64 |
| Dense (sigmoid) | 32 |
| Dense (sigmoid) | $K = 16$ |

**Figure 4.26** Neural decoder architecture in [36] for the rate-1/2 Polar code.

recurrent structures and can achieve the generalization toward unseen codewords as well as longer block lengths.

**Architecture.** The decoder for Polar codes is modeled as a four-layer fully connected neural network, as shown in Fig. 4.26, and is trained on a loss of the form Eq. (4.1). The $\ell_2$ loss and the cross-entropy loss both achieve similar performances.

**Result.** In training, each *epoch* consists of showing the entire codebook of $2^{16}$ codewords. After $2^{18}$ epochs, the neural network decoder approaches the performance of the optimal MAP decoder for the AWGN channel with SNR in the range of 0–8 dB. At SNR 8 dB, a BER of $10^{-5}$ is achieved for both the MAP decoder and the trained Neural decoder. It is observed that the optimal training SNR for the rate-1/2 Polar code is 1 dB. This matches the prediction from Fig. 4.8.

This trained network has no hope of generalizing to larger block lengths, as no structure of the Polar code is exploited. In fact, it is impossible to train a neural decoder in this manner for a moderate length Polar code, as the required training time (and samples) increase exponentially in $k$ as reported in [36]. It is shown that the performance significantly degrades unless the entire codebook (of size $2^K$) is shown in training.

**Related work.** An important remaining question is whether a carefully designed neural network architecture can exploit the structure of Polar codes and achieve generalization in block lengths. There are two popular efficient decoders that exploit the structure of Polar codes: successive cancellation-based decoding [13] and the belief propagation (BP) algorithm. The work [37] builds upon the BP algorithm and replaces the lower-level BP update with a small-length neural network decoder. It is shown that the performance of the BP decoder is unharmed when the lower level is switched by a neural decoder, which is separately trained on small block lengths. A similar approach of replacing lower-level successive cancellation operation by trained neural networks is proposed in [38], where again the performance of the successive cancellation decoder is retained. The work [39] starts from the BP algorithm and replaces each BP update function with a six-layer fully connected neural network. As the structure of Polar code is exploited, the experiments can handle up to rate-1/2 Polar code with $K = 256$. This is further made more efficient by replacing the fully connected layers with recurrent neural networks as seen in [40].

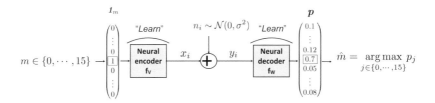

**Figure 4.27** End-to-end joint training of the encoder and decoder in [41].

| Encoder Layer | Output dimension |
|---|---|
| Input | $2^k$ |
| Dense (ReLU) | $2^k$ |
| Dense (Linear) | n |
| Normalization | n |

| Decoder Layer | Output dimension |
|---|---|
| Dense (ReLU) | $2^k$ |
| Dense (softmax) | $2^k$ |

**Figure 4.28** Neural encoder and decoder architectures in [41] for the rate-4/7 code in Fig. 4.27.

### 4.4.2   Learning Hamming Codes

The framework for end-to-end joint training of the encoder and the decoder are introduced in [41], which shows that the jointly trained encoder and decoder can recover the reliability of Hamming code for a short information block length of $K = 4$.

**Setup.** Consider the communication system where the transmitter communicates an information bit sequence $\mathbf{b} \in \{0,1\}^k$ via AWGN channels. The encoder maps the information bit sequence $\mathbf{b} \in \{0,1\}^K$ to a codeword $\mathbf{x} \in \mathbb{R}^N$. Without loss of generality, a unit power constraint is imposed on the codewords; that is, $\mathbb{E}(1/N)[\|\mathbf{x}\|^2] \leq 1$, where $\mathbf{x} = (x_1, \ldots, x_N)$. The decoder receives $\mathbf{y} = \mathbf{x} + \mathbf{n}$, where $n_i \sim \mathcal{N}(0, \sigma^2)$ for $i = 1, \ldots, N$. The work in [41] aims to learn a code for a fixed size and rate ($K = 4$ and $N = 7$) and recover the reliability of a rate-4/7 Hamming code [42] combined with maximum likelihood decoding.

**Architecture.** The end-to-end neural network based communication system is illustrated in Fig. 4.27. The encoder first maps a message bit sequence $\mathbf{b} = (b_0, b_1, b_2, b_3)$ to a real-valued message $m = \sum_{j=0}^{3} 2^j b_j \in \{0, \ldots, 15\}$, which is then mapped to a one-hot representation $\mathbf{1}_m \in \{0,1\}^{16}$, where $\mathbf{1}_m(j) = 1$ if $j = m$ and $\mathbf{1}_m(j) = 0$ otherwise. The one-hot representation $\mathbf{1}_m$ is given as an input to the encoder neural network, which consists of two consecutive layers of feedforward neural networks followed by a power normalization layer, as illustrated in Fig. 4.28. The power normalization layer maps an input sequence $\mathbf{x}_{in}$ to an output sequence $\mathbf{x}_{out} = \mathbf{x}_{in} / \|\mathbf{x}_{in}\|_2$ so that $\|\mathbf{x}_{out}\|_2 = 1$. Similarly, the decoder is also modeled as two layers of feedforward neural networks. The last layer has a softmax activation function (i.e., the output $\mathbf{p}$ is a probability distribution over $2^k$ possible messages). Finally, $\hat{m} = \arg\max_{j\in\{0,\ldots,15\}} p_j$ and $\hat{\mathbf{b}}$ is a binary representation of $\hat{m}$.

Let $\mathbf{x} = f_{\mathbf{V}}(\mathbf{1}_m)$ denote the the encoder neural network parameterized by $\mathbf{V}$, and let $\mathbf{p} = f_{\mathbf{W}}(\mathbf{y})$ denote the decoder neural network, parameterized by $\mathbf{W}$. The parameters $\mathbf{V}$ and $\mathbf{W}$ are found via stochastic gradient descent (backpropagation) that solves minimize $_{\mathbf{V},\mathbf{W}}$  $\mathbb{E}\big[\ell(\mathbf{1}_m, \mathbf{p})\big]$, where the expectation is over samples used in training, and a canonical cross-entropy loss function for multiclass classification $\ell(\mathbf{1}_m, \mathbf{p}) = -\sum_{j=1}^{2^K} 1_m(j)\log(p(j))$ is used.

**Result.** The end-to-end joint training of encoder and decoder achieves reliability comparable to a rate-4/7 Hamming code with a BPSK modulation with with maximum likelihood decoding for AWGN channels of a wide range of SNRs (from $-4$ dB to 8 dB) that corresponds to block error rates 0.5 to $10^{-5}$. Here, the information block length is $K = 4$, and [41] notes that generalizing the performance of an end-to-end trained code to longer block lengths is challenging due to the increasing size of the model and the required training samples. This demonstrates how structural choices of the neural architecture are critical in efficient training. Without further structure (such as recurrent structure of convolutional codes), end-to-end training requires all codewords to be shown at training and does not generalize to larger block lengths.

## 4.5     New Results on Linear Block Codes

The efficient BP decoders for linear block codes often achieve reliability far from optimal (MAP) reliability, even for AWGN channels, and are often highly complex. Deep learning serves as a promising tool for improving the reliability and efficiency of decoders for linear block codes, such as BCH, LDPC, and Polar codes. Since learning a decoder for moderate to long linear block codes solely from data is infeasible (requires training time exponential to the information block length), focus has been on integrating deep learning into the existing decoding algorithm. In Section 4.5.1, we show that the BP can be improved by integrating trainable weights into the BP algorithm for BCH codes. In Section 4.5.2, we show that deep learning can also serve as a tool for designing linear block codes (e.g., Polar codes), especially when the decoder is restricted to a certain form (e.g., belief propagation decoders).

## 4.5.1     Decoding Linear Block Codes Beyond Belief Propagation

An iterative BP decoder for a linear code is built on the Tanner graph that characterizes the linear code. The BP algorithm can be represented as a trellis where the nodes in the hidden layers correspond to edges in the Tanner graph and the messages are passed over the the the trellis. The works [43, 44] introduce learnable weights into the BP decoder and train those parameters via supervised training using examples of pairs of noisy codewords and the true message bit sequence.

**Setup.** Decoders for several high density parity code (HDPC) codes (e.g., BCH(127, 106), BCH(63,36), and BCH(63,45)) are considered, for which the BP decoder and ML decoder have a large gap in the decoding error probability.

**Architecture.** The work [43] proposes a neural BP decoder that is represented as

$$x_{i,e=(v,c)} = \tanh\left(\frac{1}{2}\left(w_{i,v}l_v + \sum_{e'=(v,c'),c'\neq c} w_{i,e,e'}x_{i-1,e'}\right)\right) \tag{4.5}$$

for odd $i \in [1:2L]$,

$$x_{i,e=(v,c)} = 2\tanh^{-1}\left(\prod_{e'=(v',c),v'\neq v} x_{i-1,e'}\right) \tag{4.6}$$

for even $i \in [1:2L]$, and

$$o_v = \sigma\left(w_{2L+1,v}l_v + \sum_{e'=(v,c')} w_{2L+1,v,e'}x_{wL,e'}\right), \tag{4.7}$$

where $\sigma(\cdot)$ denotes a sigmoid function. Variable nodes, check nodes, and the edges are represented as $v, c, e = (v, c)$. The number of iterations, log-likelihood ratio (LLR), and output messages are denoted as $L$, $l_v$ and $x_{i,e}$, respectively. The weights $w_{i,v}, w_{i,e,e'}$, and $w_{i,v,e'}$ denote the trainable weights. The neural BP has a strong symmetric structure. It satisfies the message passing symmetry conditions [45], which implies that the error rate for the all-zero codeword coincides with the error rate for an arbitrary codeword. As a result, it is sufficient to train the decoder with only all-zero codewords because generalization across unseen codewords occurs naturally due to the strong structure in the neural BP decoder.

**Training.** The learnable weights introduced to the BP decoder are initialized as ones so that, initially, the neural BP decoder mimics the traditional BP decoder. In learning the parameters introduced in the BP decoder, [43, 44] uses stochastic gradient descent via supervised training (with all-zero codewords). The parameters that minimize the decoding loss (binary cross-entropy loss) is found via stochastic gradient descent. It is shown that the error rate of the neural BP decoder achieves error rates that are independent of the transmitted codeword when communicating over a binary memoryless symmetric channels; hence, noisy examples of one codeword (an all-zero codeword) are used for training.

**Result.** In decoding HDPC codes for AWGN channels of various code (e.g., BCH(127,106), BCH(63,36), and BCH(63,45)), [43, 44] shows an improvement in BER up to 0.75 dB in the high SNR region, when compared to the BP decoders.

**Related work.** The work [46] shows that a neural min-sum algorithm, which introduces the trainable parameters to a min-sum algorithm instead of a sum-product algorithm, achieves the reliability of a neural BP with less parameters and reduced complexity. The work [47] further integrates graph neural networks into the BP decoding and demonstrates the reliability improvement over the vanilla BP for a large family of linear block codes (namely, the Polar, BCH, and LDPC codes).

## 4.5.2    Designing New Polar Codes

Polar code construction methods tailored to the BP decoder are designed in [48]. BP decoding is inferior in reliability to successive cancellation decoding, which achieves reliability close to the maximum likelihood decoding for a sufficiently large list size; however, it has advantages in complexity and latency. The work in [48] applies deep learning to design Polar codes that are tailored to work well under BP decoding.

**Setup.** Polar codes are linear codes. The coded bit sequence $\mathbf{x} = \mathbf{Fu}$, for an $n \times n$ polarization matrix $\mathbf{F} = [1,0;1,1]^{\times n}$ (the $n$th Kronecker power), where $\mathbf{u} \in \{0,1\}^n$ includes $k$ information bits and $n - k$ frozen (known) bits. Let $\mathbb{A} \in \{0,1\}^n$ denote the positions where $k$ information bits are located ($\mathbb{A}_i = 1$ if $u_i$ denotes an information bit and $\mathbb{A}_i = 0$ otherwise). A Polar code and its BP decoder are precisely characterized by $\mathbb{A}$, and the error probability of the Polar code depends on $\mathbb{A}$. A deep learning methodology to learn $\mathbb{A}$ (for a fixed block length and rate) is demonstrated in [48] that results in a high reliability for BP decoding.

**Architecture.** Stochastic gradient descent is used to find $\mathbb{A}$ that minimizes the BP decoding error. As SGD cannot be directly applied to a discrete sequence $\mathbb{A}$, [48] introduces a learnable parameter $\mathbb{A}_{soft} \in \mathbb{R}^n$, which denotes $\text{Prob}(\mathbb{A} = 1) = \text{sigmoid}(\mathbb{A}_{soft}(i))$, and a binarization layer that maps $\mathbb{A}_{soft}$ to $\mathbb{A}$ by letting $\mathbb{A}_i = 1$, with a probability of $\text{sigmoid}(\mathbb{A}_{soft}(i))$ (via random generation) for $i = 1, \dots, n$. The BP decoder then maps $(\mathbb{A}, \mathbf{y})$ to $\hat{\mathbf{b}}$. SGD is used to find $\mathbb{A}_{soft}$; backpropagation is implemented from the output through the predefined BP decoder and the straight-through-estimator for the binarizer layer.

In implementing the SGD, [48] introduces three loss functions, each of which quantifies the error performance, how far the code rate is from the target code rate, and the degree to which $\mathbb{A}_{soft}$ is deterministic. In the beginning, the training is focused on the error loss, and toward the end, the training is focused more on matching the rate and then learning a deterministic $\mathbb{A}_{soft}$.

**Result.** The performance of the learned Polar code combined with BP decoding is compared to the 5G Polar code combined with BP decoding for AWGN and Rayleigh fading channels, where the number of BP iterations are fixed at 5 and 20. The input block length and code rates are fixed at 128 and 1/2, respectively. The gap is most noticeable for Rayleigh channels with a small (5) number of iterations. The learned Polar code outperforms the 5G Polar code (with both under BP decoding) by 0.5 dB in achieving the BER of $10^{-2}$ (at SNR 0.6 dB). In other scenarios, the learned Polar code outperforms the 5G polar under BP decoding by 0.1–0.2 dB.

## 4.6    Discussion and Open Problems

The recent advances on channel coding via deep learning demonstrate that deep learning is a promising tool for designing the decoders and codes that improve upon the state of the art. From the deep learning perspective, the challenges and opportunities in learning decoders and codes are unique and notably different from the challenges encountered in other applications, such as computer vision and natural language processing. We can generate potentially infinite training samples and often model the underlying communication channels mathematically; however, the codebook size is exponentially large, and we desire generalization in block lengths for efficient training. These challenges need to be handled via the careful design of highly structured neural architectures; the appropriate choice of training block length and SNR; and finally, the precise choice of the loss function, the regularizers, and the hyperparameters for stochastic gradient descent or its variant. Looking forward, there are several remaining challenges and opportunities in applications of deep learning for channel coding.

**Inventing novel codes with a dynamic memory.** We showed that a recurrent neural network based channel code (DeepCode) outperforms the state-of-the-art for channels with noisy output feedback. Specifically, a vanilla RNN is used to model the encoder of DeepCode. In deep learning literature, there are important variants of vanilla RNNs, called GRU and long short-term memory (LSTM), designed to overcome the limited memory behavior of vanilla RNNs. In terms of representation-capability, GRU and LSTM can represent a channel coding with dynamically varying memory, and this variation can be designed to functionally depend on an aspect of the communication environment. We can make an analogy to switched linear systems, where it is known that state-dependent switching allows for both long-term dependence of the state and output on the original input. How to harness this added capability of GRUs in channel coding (to further enhance the performance of sequential encoding and decoding schemes) is a promising and open direction for further research.

**Channel impairments and computational efficiency.** A promising next direction is to apply the deep learning driven decoders and codes to practical wireless communications. There are several challenges to overcome. First is to demonstrate a robust performance of deep learning based algorithms for simulated channel environments and to verify robustness under channel impairments via over-the-air experiments. Meta-learning is a viable solution that allows deep learning based algorithms to adapt quickly to time-varying channel environments. Second is to improve the computational efficiency. Deep learning based algorithms, especially ones obtained from training samples, often require more computations than the traditional algorithms. Improving the computational efficiency and latency of the deep learning based models is necessary from the practical point of view.

**Insights for design of channel codes.** Whether one can derive an intuition about the design of channel codes from deep learning is an interesting open question. For example, the first-order behavior of DeepCode can be understood; designing an analytical code based on this understanding is a challenging but interesting direction of research. The ability to understand and interpret the behavior of deep learning based algorithms is also helpful for the practical deployment of deep learning based codes and decoders.

**Multiuser communications.** Practical coding schemes are lacking for multiuser communications, such as interference and relaycite channels, as the mathematical analysis for these channels is highly challenging. Deep learning can be a highly promising solution for these scenarios due to its ability to learn from data without the need for analytical models.

# References

[1] A. Viterbi, "Error bounds for convolutional codes and an asymptotically optimum decoding algorithm," *IEEE Trans. on Information Theory*, vol. 13, no. 2, pp. 260–269, 1967.

[2] L. Bahl et al., "Optimal decoding of linear codes for minimizing symbol error rate (corresp.)," *Trans. on Information Theory*, vol. 20, no. 2, pp. 284–287, 1974.

[3] C. Berrou, A. Glavieux, and P. Thitimajshima, "Near Shannon limit error-correcting coding and decoding: Turbo-codes. 1," in *Proc. IEEE Int. Conf. on Communications (ICC)*, vol. 2, pp. 1064–1070, 1993.

[4] S. B. Venkatakrishnan, M. Alizadeh, and P. Viswanath, "Graph2seq: Scalable learning dynamics for graphs," *arXiv preprint*, arXiv:1802.04948, 2018.

[5] E. Khalil et al., "Learning combinatorial optimization algorithms over graphs," in *Advances in Neural Information Processing Systems*, pp. 6348–6358, 2017.

[6] X.-A. Wang and S. B. Wicker, "An artificial neural net viterbi decoder," *IEEE Trans. on Communications*, vol. 44, no. 2, pp. 165–171, Feb. 1996.

[7] M. H. Sazli and C. Isik, "Neural network implementation of the BSJR algorithm," *Digital Signal Processing*, vol. 17, no. 1, pp. 353–359, 2007.

[8] H. Kim et al., "Communication algorithms via deep learning," in *Int. Conf. on Learning Representations*, 2018.

[9] M. Benammar and P. Piantanida, "Optimal training channel statistics for neural-based decoders," in *52nd Asilomar Conf. on Signals, Systems, and Computers*, pp. 2157–2161, 2018.

[10] I. Be'ery et al., "Active deep decoding of linear codes," *arXiv preprint*, arXiv:1906.02778, 2019.

[11] Y. Jiang et al., "Deepturbo: Deep turbo decoder," in *IEEE 20th Int. Workshop on Signal Processing Advances in Wireless Communications (SPAWC)*, pp. 1–5, 2019.

[12] D. Tandler et al., "On recurrent neural networks for sequence-based processing in communications," in *53rd Asilomar Conf. on Signals, Systems, and Computers*, pp. 537–543, 2019.

[13] E. Arikan, "Channel polarization: A method for constructing capacity-achieving codes," in *IEEE Int. Symp. on Information Theory*, pp. 1173–1177, 2008.

[14] D. J. MacKay and R. M. Neal, "Near Shannon limit performance of low density parity check codes," *Electronics Letters*, vol. 33, no. 6, pp. 457–458, 1997.

[15] Y. Jiang et al., "Turbo autoencoder: Deep learning based channel codes for point-to-point communication channels," in *Advances in Neural Information Processing Systems*, pp. 2754–2764, 2019.

[16] F. A. Aoudia and J. Hoydis, "End-to-end learning of communications systems without a channel model," in *52nd Asilomar Conf. on Signals, Systems, and Computers*, pp. 298–303, 2018.

[17] Y. Jiang et al., "Learn codes: Inventing low-latency codes via recurrent neural networks," *arXiv preprint*, arXiv:1811.12707, 2018.

[18] M. K. Muller et al., "Flexible multi-node simulation of cellular mobile communications: the Vienna 5G System Level Simulator," *EURASIP Journal on Wireless Communications and Networking*, vol. 2018, no. 1, p. 17, Sep. 2018.

[19] N. Shlezinger et al., "ViterbiNet: A deep learning based viterbi algorithm for symbol detection," *IEEE Trans. on Wireless Communications*, 2020.

[20] A. Goldsmith, *Wireless Communications*. Cambridge University Press, 2005.

[21] N. Farsad and A. Goldsmith, "Neural network detection of data sequences in communication systems," *IEEE Trans. on Signal Processing*, vol. 66, no. 21, pp. 5663–5678, Jan. 2018.

[22] N. Shlezinger et al., "Data-driven factor graphs for deep symbol detection," *arXiv preprint*, arXiv:2002.00758, 2020.

[23] Y. He et al., "Model-driven DNN decoder for turbo codes: Design, simulation and experimental results," *IEEE Trans. on Communications*, vol. 68, no. 10, pp. 6127–6140, 2020.

[24] N. Shlezinger, R. Fu, and Y. C. Eldar, "Deepsic: Deep soft interference cancellation for multiuser mimo detection," *arXiv preprint*, arXiv:2002.03214, 2020.

[25] N. Farsad et al., "Data-driven symbol detection via model-based machine learning," *arXiv preprint*, arXiv:2002.07806, 2020.

[26] C. Shannon, "The zero error capacity of a noisy channel," *IRE Trans. on Information Theory*, vol. 2, no. 3, pp. 8–19, 1956.

[27] J. Schalkwijk and T. Kailath, "A coding scheme for additive noise channels with feedback–i: No bandwidth constraint," *IEEE Trans. on Information Theory*, vol. 12, no. 2, pp. 172–182, 1966.

[28] R. G. Gallager and B. Nakiboglu, "Variations on a theme by schalkwijk and kailath," *IEEE Trans. on Information Theory*, vol. 56, no. 1, pp. 6–17, Jan. 2010.

[29] Y. H. Kim, A. Lapidoth, and T. Weissman, "The gaussian channel with noisy feedback," in *IEEE Int. Symp. on Information Theory*, pp. 1416–1420, 2007.

[30] H. Kim et al., "Deepcode: Feedback codes via deep learning," in *Advances in Neural Information Processing Systems*, pp. 9436–9446, 2018.

[31] Y. Polyanskiy, H. V. Poor, and S. Verdú, "Channel coding rate in the finite blocklength regime," *IEEE Trans. on Information Theory*, vol. 56, no. 5, pp. 2307–2359, 2010.

[32] T. Erseghe, "On the evaluation of the polyanskiy-poor-verdu converse bound for finite block-length coding in AWGN," *IEEE Trans. on Information Theory*, vol. 61, no. 12, pp. 6578–6590, 2014.

[33] Z. Chance and D. J. Love, "Concatenated coding for the AWGN channel with noisy feedback," *IEEE Trans. on Information Theory*, vol. 57, no. 10, pp. 6633–6649, Oct. 2011.

[34] N. Farsad, M. Rao, and A. Goldsmith, "Deep learning for joint source-channel coding of text," in *IEEE Int. Conf. on Acoustics, Speech and Signal Processing (ICASSP)*, pp. 2326–2330, 2018.

[35] E. Bourtsoulatze, D. B. Kurka, and D. GÃijndÃijz, "Deep joint source-channel coding for wireless image transmission," in *IEEE Int. Conf. on Acoustics, Speech and Signal Processing (ICASSP)*, pp. 4774–4778, 2019.

[36] T. Gruber et al., "On deep learning-based channel decoding," in *51st Annual Conf. on Information Sciences and Systems (CISS)*, pp. 1–6, 2017.

[37] S. Cammerer et al., "Scaling deep learning based decoding of polar codes via partitioning," in *IEEE Global Communications Conf (GLOBECOM)*, pp. 1–6, 2017.

[38] N. Doan, S. A. Hashemi, and W. J. Gross, "Neural successive cancellation decoding of polar codes," in *IEEE 19th Int. Workshop on Signal Processing Advances in Wireless Communications (SPAWC)*, pp. 1–5, 2018.

[39] W. Xu et al. "Improved polar decoder based on deep learning," in *IEEE Int. Workshop on Signal Processing Systems (SiPS)*, pp. 1–6, 2017.

[40] C.-F. Teng et al., "Low-complexity recurrent neural network-based polar decoder with weight quantization mechanism," in *IEEE Int. Conf. on Acoustics, Speech and Signal Processing (ICASSP)*, pp. 1413–1417, 2019.

[41] T. O'Shea and J. Hoydis, "An introduction to deep learning for the physical layer," *IEEE Trans. on Cognitive Communications and Networking*, vol. 3, no. 4, pp. 563–575, 2017.

[42] R. W. Hamming, "Error detecting and error correcting codes," *The Bell System Technical Journal*, vol. 29, no. 2, pp. 147–160, April 1950.

[43] E. Nachmani, Y. Be'ery, and D. Burshtein, "Learning to decode linear codes using deep learning," in *54th Annual Allerton Conf. on Communication, Control, and Computing*, pp. 341–346, 2016.

[44] E. Nachmani et al., "Deep learning methods for improved decoding of linear codes," *IEEE Journal of Selected Topics in Signal Processing*, vol. 12, no. 1, pp. 119–131, 2018.

[45] T. Richardson and R. Urbanke, *Modern Coding Theory*. Cambridge University Press, 2008.

[46] L. Lugosch and W. J. Gross, "Neural offset min-sum decoding," in *IEEE Int. Symp. on Information Theory (ISIT)*, pp. 1361–1365, 2017.

[47] E. Nachmani and L. Wolf, "Hyper-graph-network decoders for block codes," in *Advances in Neural Information Processing Systems 32*, pp. 2329–2339, 2019.

[48] M. Ebada et al., "Deep learning-based polar code design," in *57th Annual Allerton Conf. on Communication, Control, and Computing*, pp. 177–183, 2019.

# 5 Channel Estimation, Feedback, and Signal Detection

Hengtao He, Hao Ye, Shi Jin, and Geoffrey Y. Li

## Abbreviations

| | |
|---|---|
| AI | Artificial intelligence |
| DL | Deep learning |
| mmWave | Millimeter-wave |
| OFDM | Orthogonal frequency division multiplexing |
| CSI | Channel state information |
| MIMO | Multiple-input multiple-output |
| RF | Radio-frequency |
| BS | Base station |
| LDAMP | Learnable denoising-based approximate message passing |
| DnCNN | Denoising convolutional neural network |
| ReLU | Rectified linear unit |
| SE | State evolution |
| MSE | Mean-squared error |
| NMSE | Normalized MSE |
| CS | Compressed sensing |
| THz | Terahertz |
| DFT | discrete Fourier transform |
| LDGEC | Learned denoising-based generalized expectation consistent |
| GEC-SR | Generalized expectation consistent signal recovery |
| SURE | Stein's unbiased risk estimator |
| IDFT | Inverse discrete Fourier transform |
| CP | Cyclic-prefix |
| LS | Least-square |
| SNR | Signal-to-noise ratio |
| DNN | Deep neural network |
| CNNs | Convolutional neural networks |
| SRCNN | Super-resolution CNN |
| UE | User equipment |
| FDD | Frequency-division duplexing |
| TDD | Time-division duplexing |
| LSTM | Long short-term memory |
| FCN | Fully connected network |

DetNet    Detection network
OAMP    Orthogonal approximate message passing
JCESD    Joint channel estimation and signal detection
PAPR    Peak-to-average power ratios
ADC    Analog-to-digital converter

## Notation

| | |
|---|---|
| $K$ | The number of subcarriers |
| $N_t$ | The number of transmitting antennas |
| $N_r$ | The number of receiving antennas |
| $\mathbf{H}$ | Channel matrix |
| $\hat{\mathbf{H}}$ | Estimated channel matrix |
| $T$ | The number of layers |
| $N_{RF}$ | The number of radio-frequency chains |
| $P$ | The number of paths |
| $\mathbf{x}$ | Transmitted signal vector |
| $\mathbf{y}$ | Received signal vector |

## 5.1 Introduction

Future wireless communications need to handle a large amount of wireless data, recognize and dynamically adapt to complex environments, and satisfy various requirements, such as high speed and low latency. However, conventional algorithms and solutions are no longer sufficient, and more advanced technologies are desired. Intelligent communication, which applies artificial intelligence (AI) techniques to communications, is promising to meet the aforementioned requirements. In physical layer communications, transceiver design is an important research topic, which includes channel estimation, feedback, and signal detection. Therefore, this chapter focuses on deep learning (DL) based channel estimation, feedback, and signal detection. We first discuss DL-based channel estimation for millimeter-wave (mmWave) and orthogonal frequency division multiplexing (OFDM) systems. Then, we briefly investigate DL-based channel state information (CSI) feedback. Next, DL-based signal detection approaches, including multiple-input multiple-output (MIMO) and OFDM detection, are presented. The concluding remarks are provided at the end.

## 5.2 Channel Estimation

The availability of CSI is critical to achieve potential advantages of future wireless communications. This section describes how channel estimation is carried out by the DL technology. The mmWave channel estimation is investigated in Section 5.2.1 for narrowband and wideband channels. Then, several DL-based OFDM channel estimation networks are briefly discussed in Section 5.2.2.

## 5.2.1     MmWave/Terahertz Channel Estimation

Almost all mmWave communication systems use massive MIMO to offer high data rates because of huge bandwidth and large antenna arrays, which is regarded as an important technique in future wireless communications. However, the high costs of hardware and power consumption become unaffordable when a dedicated radio-frequency (RF) chain is used for each antenna. Therefore, the beamspace channel model and the lens antenna array-based architecture have been proposed to reduce the number of RF chains. However, channel estimation for this beamspace mmWave massive MIMO system is extremely challenging, especially when the antenna array is large and the number of RF chains is limited. Furthermore, the beam squint effect [1] in wideband terahertz systems makes channel estimation even more challenging. To solve the problem, this section investigates DL-based mmWave/terahertz channel estimation.

### Narrowband mmWave Channel Estimation

Figure 5.1 illustrates a three-dimensional (3D) lens-based typical mmWave massive MIMO system where the base station (BS) is equipped with an $M_1 \times M_2$ antenna array. The $M_1 M_2$ antennas are connected to $N_{RF}$ RF chains through the $N_{RF} \times M_1 M_2$ selection network. In Fig. 5.1, the selection network is denoted by matrix $\mathbf{W} \in \mathbb{R}^{N_{RF} \times M_1 M_2}$, with each entry being $\pm 1$. That is, fully connected one-bit phase shifters are used and normalized by dividing $\sqrt{M_1 M_2}$.

The beamspace channel matrix can be expressed as

$$\mathbf{H} = \sqrt{\frac{M_1 M_2}{P+1}} \sum_{i=0}^{P} \alpha^{(i)} \mathbf{A}(\phi^{(i)}, \theta^{(i)}), \tag{5.1}$$

where $\mathbf{H} \in \mathbb{R}^{M_1 \times M_2}$ denotes beamspace channel matrix; $P+1$ is the number of paths; $\alpha^{(i)}$ denotes the gain of the $i$th path; $\phi^{(i)}$ and $\theta^{(i)}$ represent the azimuth and elevation angles of arrival (AoAs) of the incident plane wave, respectively; and $\mathbf{A}(\phi^{(i)}, \theta^{(i)})$ refers to the antenna array response matrix, which is determined by its geometry.

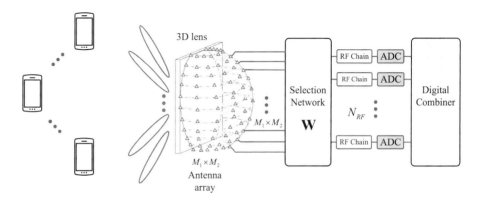

**Figure 5.1**  Millimeter-Wave receiver with a lens-based antenna array.

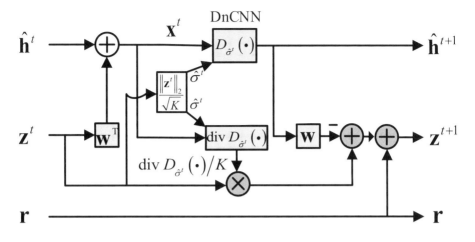

**Figure 5.2** The $t$th layer architecture of the LDAMP network.

The beamspace massive MIMO channel estimation can be regarded as a compressed signal recovery problem. In the uplink training phase, the user sends the training symbol, $s$, to the BS and the received signal vector $\mathbf{y} \in \mathbb{R}^{M_1 M_2 \times 1}$ at the BS is given by

$$\mathbf{y} = \mathbf{h}s + \mathbf{n}, \tag{5.2}$$

where $\mathbf{n} \sim \mathcal{N}(\mathbf{0}, \sigma_n^2 \mathbf{I})$ represents an additive white Gaussian noise (AWGN) vector. The beamspace channel vector $\mathbf{h} \in \mathbb{R}^{M_1 M_2 \times 1}$ is obtained by vectorizing $\mathbf{H}$. Given a selection network $\mathbf{W}$ as in Fig. 5.1, the received signal from the RF chain can be expressed as

$$\mathbf{r} = \mathbf{W}\mathbf{y} = \mathbf{W}\mathbf{h} + \bar{\mathbf{n}}, \tag{5.3}$$

where $\bar{\mathbf{n}} = \mathbf{W}\mathbf{n}$ is the equivalent noise after the selection network at the receiver, which follows $\mathcal{N}(\mathbf{0}, \sigma_n^2 \mathbf{I})$ as each entry in matrix $\mathbf{W}$ is normalized by dividing $\sqrt{M_1 M_2}$. As illustrated in Fig. 5.2 from [2], channel estimation can exploit learnable denoising-based approximate message passing (LDAMP), where the beamspace channel vector, $\mathbf{h}$, is estimated from the received signal, $\mathbf{r}$, for the given selection network, $\mathbf{W}$, in Eq. (5.3). The LDAMP neural network consists of $L$ layers connected in cascade way. In Fig. 5.2, each layer has the same structure and contains denoiser $D_{\hat{\sigma}^t}(\cdot)$, a divergence estimator $\mathrm{div}D_{\hat{\sigma}^t}(\cdot)$. Denoiser $D_{\hat{\sigma}^t}(\cdot)$ is performed by the advanced denoising convolutional neural network (DnCNN) and is used to update $\mathbf{h}$. Channel vector is estimated iteratively by

$$\mathbf{z}^{t+1} = \mathbf{r} - \mathbf{W}\hat{\mathbf{h}}^{t+1} + \frac{1}{K}\mathbf{z}^t \mathrm{div}D_{\hat{\sigma}^t}(\hat{\mathbf{h}}^t + \mathbf{W}^T \mathbf{z}^t), \tag{5.4}$$

$$\hat{\mathbf{h}}^{t+1} = D_{\hat{\sigma}^t}(\hat{\mathbf{h}}^t + \mathbf{W}^T \mathbf{z}^t). \tag{5.5}$$

Figure 5.3 illustrates the network architecture of the DnCNN denoiser. It consists of 20 convolutional layers. The first convolutional layer uses 64 different $3 \times 3 \times 1$

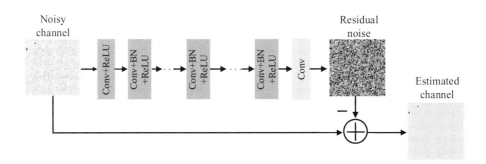

**Figure 5.3** Network architecture of the DnCNN denoiser.

filters and followed by a rectified linear unit (ReLU). Each of the succeeding 18 convolutional layers uses 64 different $3 \times 3 \times 64$ filters, each followed by batch-normalization and a ReLU. The final convolutional layer uses one separate $3 \times 3 \times 64$ filters to reconstruct the signal. Furthermore, residual learning technology is also considered.

The LDAMP neural network consists of a series of state evolution (SE) equations that predict the performance of the network over each layer with a large-system limit $(M_1, M_2 \to \infty)$. The SE equations are given by

$$\theta^{t+1}(\mathbf{h}_o, \delta, \sigma_n^2) = \frac{1}{M_1 M_2} \mathbb{E} \| D_{\sigma^t}(\mathbf{h}_o + \sigma_e^t \epsilon) - \mathbf{h}_o \|_2^2, \tag{5.6}$$

$$(\sigma_e^t)^2 = \frac{1}{\delta} \theta^t(\mathbf{h}_o, \delta, \sigma_n^2) + \sigma_n^2, \tag{5.7}$$

where $\mathbf{h}_o$ is a deterministic realization of channel $\mathbf{h}$, $\delta = N_{RF}/M_1 M_2$ represents the measurement ratio, $\theta^t(\mathbf{h}_o, \delta, \sigma_n^2)$ is the average mean-squared error (MSE) of the denoiser output in the $l$th layer network, and $\sigma_n^2$ denotes the noise variance.

Figure 5.4 compares the normalized MSE (NMSE) performance of different channel estimation methods. From the figure, the LDAMP network outperforms other compressed sensing (CS) based channel estimation methods.

## Wideband Terahertz Channel Estimation

In the last subsection, we considered the narrowband mmWave massive MIMO systems. However, for wideband terahertz (THz) massive MIMO systems, the physical propagation delays of electromagnetic waves traveling across the whole array will become large and comparable to the time-domain sample period. Therefore, different antenna elements will receive different time-domain symbols. This phenomenon is known as the spatial-wideband effect [1] and thus makes channel estimation challenging.

For an uplink wideband beamspace THz MIMO-OFDM system, the spatial channel $\mathbf{h}_k \in \mathbb{C}^{N_r \times 1}$ over subcarrier $k$ is given by

$$\mathbf{h}_k = \sqrt{\frac{N_r}{P}} \sum_{p=1}^{P} \alpha_p e^{-j2\pi \tau_p f_k} \mathbf{a}(\phi_{p,k}), \tag{5.8}$$

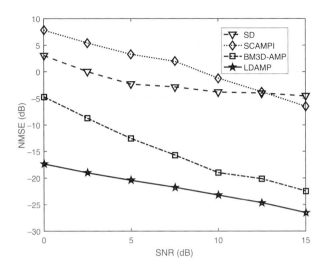

**Figure 5.4** Comparison of NMSE performance between the LDAMP network and other methods.

for $k = 1, 2, \ldots, K$, where $P$ is the number of resolvable paths, $\alpha_p$ and $\tau_p$ are the complex gain and the time delay of the $p$th path, respectively. Furthermore, $\mathbf{a}(\phi_{p,k})$ is the array response vector and $\phi_{p,k}$ is the spatial direction at subcarrier $k$ defined as $\phi_{p,k} = \frac{f_k}{c} d \sin \theta_p$, $f_k = f_c + \frac{f_s}{K}(k - 1 - \frac{K-1}{2})$ is the frequency of subcarrier $k$ with $f_c$ and $f_s$ representing the carrier frequency and bandwidth, respectively, $c$ is the speed of light, $\theta_p$ is the physical direction, and $d$ is the antenna spacing. Usually, $d = c/2f_c$. For a uniform linear lens array in the BS, the array response vector $\mathbf{a}(\phi_{p,k})$ is given by,

$$\mathbf{a}(\phi_{p,k}) = \frac{1}{\sqrt{N_r}}[e^{-j2\pi\phi_{p,k}(-\frac{N_r-1}{2})}, e^{-j2\pi\phi_{p,k}(-\frac{N_r+1}{2})}, \ldots . e^{-j2\pi\phi_{p,k}(\frac{N_r-1}{2})}]^T. \quad (5.9)$$

Accordingly, the wideband beamspace channel $\tilde{\mathbf{h}}_k$ at subcarrier $k$ can be expressed as

$$\tilde{\mathbf{h}}_k = \mathbf{F}^H \mathbf{h}_k, \quad (5.10)$$

and $\mathbf{F}$ is $N_r$-element spatial discrete Fourier transform (DFT) matrix $\mathbf{F}$. The DFT matrix is given by

$$\mathbf{F} = [\mathbf{a}(\bar{\phi}_1), \mathbf{a}(\bar{\phi}_2), \ldots , \mathbf{a}(\bar{\phi}_{N_r})], \quad (5.11)$$

where $\bar{\phi}_{n_r} = \frac{1}{N_r}(n - \frac{N_r+1}{2})$ for $n_r = 1, 2, \ldots , N_r$ are the spatial directions predefined by the lens antenna array.

The received signal vector $\mathbf{y}_{k,q}$ in $q$th instant can be expressed as

$$\mathbf{y}_{k,q} = \mathbf{W}_q \tilde{\mathbf{h}}_k + \mathbf{n}_{k,q}, \quad (5.12)$$

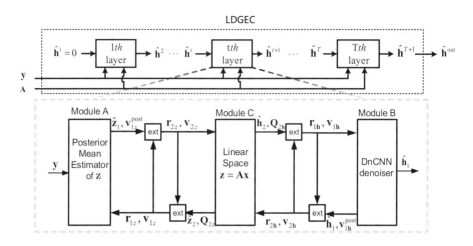

**Figure 5.5** The network structure of LDGEC-based channel estimator.

for $k = 1, 2, \ldots, K$ and $q = 1, 2, \ldots, Q$, where $\mathbf{n}_{k,q}$ is the Gaussian noise. Thus, the received signal $\mathbf{y}_k$ in $Q$ instants is given by

$$\mathbf{y}_k = \bar{\mathbf{W}}\tilde{\mathbf{h}}_k + \mathbf{n}_k, \tag{5.13}$$

where $\bar{\mathbf{W}} = [\mathbf{W}_1^T, \mathbf{W}_2^T, \ldots, \mathbf{W}_Q^T]^T$, $\mathbf{y}_k = [\mathbf{y}_{k,1}^T, \mathbf{y}_{k,2}^T, \ldots, \mathbf{y}_{k,Q}^T]^T$, and $\mathbf{n}_k = [(\mathbf{W}_1\mathbf{n}_{k,1})^T, \ldots, (\mathbf{W}_Q\mathbf{n}_{k,Q})^T]^T$. A compact system model can be obtained by stacking $\mathbf{y}_k$, $\mathbf{h}_k$ and $\mathbf{n}_k$ into

$$\mathbf{y} = \mathbf{A}\mathbf{h} + \mathbf{n}, \tag{5.14}$$

where $\mathbf{y} = [\mathbf{y}_1^T, \mathbf{y}_2^T, \ldots, \mathbf{y}_K^T]^T$, $\mathbf{h} = [\tilde{\mathbf{h}}_1^T, \tilde{\mathbf{h}}_2^T, \ldots, \tilde{\mathbf{h}}_K^T]^T$, $\mathbf{n} = [\mathbf{n}_1^T, \mathbf{n}_2^T, \ldots, \mathbf{n}_K^T]^T$, and $\mathbf{A} = (\mathbf{I} \otimes \bar{\mathbf{W}})$. We denote the matrix Kronecker product by $\otimes$.

A model-driven unsupervised DL network for wideband beamspace channel estimation, named learned denoising-based generalized expectation consistent (LDGEC) based channel estimator, from [3], is shown in Fig. 5.5. The network architecture is obtained by deep unfolding of the generalized expectation consistent signal recovery (GEC-SR) algorithm [4].

In Fig. 5.5, the LDGEC network consists of $T$ layers connected in cascade, and each iteration of the GEC-SR algorithm can be interpreted as a layer of the LDGEC network, but with different denoisers. Each layer of the LDGEC network has three modules, where module A computes the posterior mean and variance of $\mathbf{z} = \mathbf{A}\mathbf{h}$, module B performs denoising from the noisy signal, $\mathbf{r}_{1\mathbf{h}}$, by using the advanced DnCNN denoiser, and module C provides the framework that constrains the estimation problem into the linear space $\mathbf{z} = \mathbf{A}\mathbf{h}$. Modules A, B, and C are executed iteratively as in the figure. In addition, each module uses the turbo principle as in iterative decoding; that is, each module passes the extrinsic messages to its next module. The three modules are executed iteratively until convergence or terminated by a fixed number of layers. By using the Stein's unbiased risk estimator (SURE) [5], the LDGEC network can be

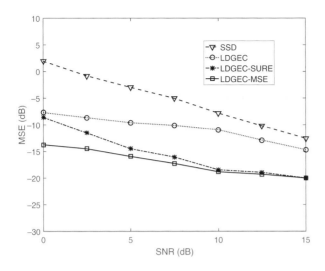

**Figure 5.6** MSEs performance comparisons of the LDGEC network with other channel estimation algorithms.

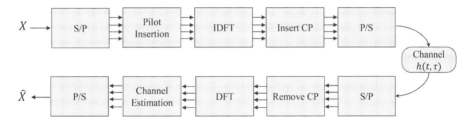

**Figure 5.7** OFDM architecture.

trained without ground truth data. Furthermore, the layer-by-layer training can further improve the network performance.

Figure 5.6 compares the performance of the LDGEC network with other channel estimation algorithms. From the figure, the LDGEC-based channel estimator outperforms the traditional CS-based algorithms with different training methods.

## 5.2.2   OFDM Channel Estimation

OFDM has been investigated for a long time and is now a dominant technology for wireless broadband communications. The OFDM architecture is depicted in Fig. 5.7. In the OFDM systems, the data blocks are transmitted in parallel on a large number of orthogonal subcarriers. The channel frequency response of each subcarrier is estimated for coherent signal detection.

To be specific, we consider a time-varying frequency-selective channel with the channel impulse response

$$h(t, \tau) = \sum_p \alpha_p(t)\delta(\tau - \tau_p), \tag{5.15}$$

where $\tau_p$ and $\alpha_p(t)$ denote the delay and the complex amplitude of $p$th path, respectively. With channel impulse response, the frequency response of the channel at time $t$ can be expressed as

$$H(t, f) = \int_{-\infty}^{\infty} h(t, \tau)e^{-j2\pi f\tau} = \sum_p \alpha_p(t)e^{-j2\pi f\tau_p}. \tag{5.16}$$

As shown in Fig. 5.7, given the $n$th data block $X[n,k]$, inverse discrete Fourier transform (IDFT) is performed to modulate the data in the OFDM transmitter and a cyclic-prefix (CP) is inserted to convert the linear convolution of a frequency-selective multipath channel to a circular convolution. Under an ideal condition, where CP is sufficient and the intercarrier interference can be omitted, the received signal after DFT can be represented as

$$Y[n,k] = H[n,k]X[n,k] + W[n,k], \tag{5.17}$$

where $W[n,k]$ denotes the additive Gaussian noise and $H[n,k]$ is channel frequency response at the $k$th subcarrier. The channel frequency response at the $k$th subcarrier of the $n$th block can be expressed as

$$H[n,k] = H(nT_s, k/T_d) = \sum_p \alpha_p(nT_s)e^{-j2\pi k\tau_p/T_d} \tag{5.18}$$

where $T_s$ and $T_d$ represent the durations of a OFDM symbol with and without CP.

In order to obtain $H[n,k]$, pilots are inserted in both the time and frequency dimensions. Based on the received signal of the pilots, the frequency response on pilot positions can be estimated. The channel frequency responses at nonpilot positions value are usually estimated via interpolation or decision feedback [6]. In particular, two basic estimation algorithms are commonly utilized, namely least-square (LS) and minimum MSE (MMSE) estimators. If the statistics of the channel response is unavailable, LS estimation is often applied, where the estimation is obtained by solving a problem,

$$\hat{H}_{LS} = \arg\min_H \|Y - XH\|^2, \tag{5.19}$$

and the solution is given by $\hat{H}_{LS}[n,k] = Y[n,k]/X[n,k]$.

In general, LS estimation performance is not good enough due to exploiting not using statistical information. On the other hand, with the statistics of the channel response, the MMSE estimator, which is designed to minimize $\mathbb{E}\|\hat{H} - H\|^2$, can be expressed as

$$\hat{H}_{MMSE} = R_H \left(R_H + \frac{1}{\gamma}I\right)^{-1} \hat{H}_{LS} \tag{5.20}$$

where $R_H = \mathbb{E}(H^H H)$ is the channel correlation matrix and $\gamma$ is the signal-to-noise ratio (SNR). Although the MMSE estimator improves the performance of LS

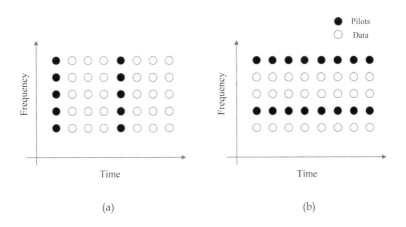

**Figure 5.8** Examples of pilot placements. (a) Block-type. (b) Comb-type.

significantly with the usage of the second-order statistics, the computational complexity also increases due to a matrix inversion.

The pilots can be inserted in the OFDM blocks with different patterns [7] and the widely used examples are shown in Fig. 5.8. With the estimated channel on the pilot positions by LS or MMSE, the channel frequency responses at nonpilot positions are usually obtained by interpolation, such as using linear or cubic interpolation.

### Image Super-Resolution and Channel Estimation

The DL-based channel estimation in OFDM system is mainly motivated by the success of DL-based image super-resolution restoration. A deep neural network (DNN) can learn to estimate the whole channel response, which is considered as a high-resolution image, from the received pilot data, which is regarded as a low-resolution image. The conventional estimation approach can also be used as a prepossessing step to get a coarse estimation.

Image super-resolution is a class of techniques used for restoring high-resolution images from low-resolution ones. In general, image super-resolution is a challenging and ill-posed problem since a specific low-resolution image may correspond to a set of possible high-resolution images. DL algorithms have been a standard solution for this ill-posed problem since it can restore the high-resolution image by learning the statistical relationship between a low-resolution image and its corresponding high-resolution counterpart from tremendous training samples. Among these various DL algorithms, DNN, especially convolutional neural networks (CNNs) have become the state-of-art approaches due to the strong capability of learning effective representations in an end-to-end manner. The DNN based image super-resolution can be expressed as

$$I_y = \mathcal{F}(I_x; \theta),\tag{5.21}$$

where $I_x$ and $I_y$ denote the low- and high-resolution images, respectively, and $\mathcal{F}$ denotes the DL model with parameters $\theta$.

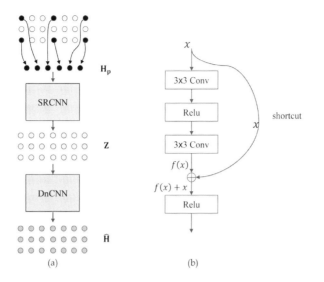

**Figure 5.9**  (a) Architecture of ChannelNet [8]. (b) Architecture of residual block [11].

The channel estimation with pilots can be recast as the image super-resolution problem, where the estimated channel at pilot positions are considered as a low-resolution image and the true channel time-frequency response is considered as a high-resolution image. Inspired by the triumph of the CNN-based super-resolution approaches, the strong expression capability of CNN is leveraged in approximating the complex mapping between the low-resolution and high-resolution channel images. Therefore, CNN based channel estimation can be expressed as

$$\hat{H} = \mathcal{F}(\hat{H}_p; \theta), \tag{5.22}$$

where $\hat{H}$ represents the high-resolution channel images and $\hat{H}_p$ is the initial estimation based on pilots as in Eqs. (5.19) or (5.20).

### CNN-Based Channel Estimation

ChannelNet, a DL-based OFDM channel estimation approach, has been proposed in [8]. The channel time-frequency response matrix in the OFDM system is viewed as a two-dimensional (2D) image. To estimate the channel matrix, $H$, two CNNs are designed, where an image super-resolution CNN (SRCNN) [9] is first utilized to enhance the resolution of the low-resolution input and the DnCNN [10] is followed to mitigate the noise effects. The structure of ChannelNet is shown in Fig. 5.9(a).

In particular, SRCNN takes initial channel estimation by the LS algorithm at the pilot locations as input and predicts all unknown values in the channel response matrix. The low-resolution input is first processed by an interpolation framework to get a coarse high resolution channel image, which is then improved by a three-layer convolutional network. SRCNN is trained to minimize the MSE between its output $Z$ and the real channel responses, $H$, that is

$$L_1 = ||Z - H||^2. \tag{5.23}$$

The output of SRCNN, $Z$, is then employed as the input of DnCNN, which further refines the estimation results by denoising. The DnCNN is a residual network with 20 convolutional layers. The structure of a residual block is shown in Fig. 5.9(b), which uses shortcut connections to the ease vanishing gradient problem in training a very deep model [11]. During training of DnCNN, the parameters of the SRCNN remain unchanged, and those of the DnCNN are learned to minimize the MSE between the output of DnCNN, $\hat{H}$, and the real channel response, $H$, as

$$L_2 = ||\hat{H} - H||^2, \tag{5.24}$$

where $\hat{H}$ is the output of DnCNN. In the experiment, ChannelNet outperforms several approximated MMSE but the SNR has a notable impact on the performance. Based on the simulation results, at low SNRs, ChannelNet algorithm has comparable performance with ideal MMSE and better than an approximation to linear MMSE. But the performance degrades if SNR is above 20 dB.

### Further Improvement

Despite good performance, ChannelNet has several drawbacks and can be further improved from the following aspects.

#### Computational Efficient Channel Estimation CNN

The network size of ChannelNet is huge, where there are 23 convolutional layers in SRCNN and DnCNN with about 670,000 parameters. A more efficient DL based channel estimation approach, ReEsNet, has been proposed in [12], where a compact model with 11 convolutional layers and only 53,000 parameters is employed. Instead of using two CNNs that are separately trained, ReEsNet only consists one CNN and thus can be trained in an end-to-end manner. In addition, the interpolation in ChannelNet is conducted at the beginning of the network, which increases the input size and the computational complexity significantly. On the contrary, the up-sampling function is performed at the end of the ReEsNet with the deconvolution layer, which can upscale the image height and width by different factors and therefore can work with any pilot pattern. From the experimental results, ReEsNet outperforms ChannelNet by 2–3 dB in low SNR range and 4–5 dB in high SNR range.

#### Joint Channel Estimation and Signal Detection:

Instead of estimating channel based on pilots alone, the channel estimation can be jointly performed with signal detection since the received signal at the receiver carries the channel information implicitly. We can build a joint channel estimation and signal detection framework with DL [13], where the transmitted data can be recovered directly without explicitly estimating the channel. This part will be further discussed in Section 5.4.2.

## 5.3     Channel Feedback

MIMO can greatly increase the link capacity and energy efficiency of a communication system by exploiting a large number of antennas at the BS. However, these benefits can be achieved only when the BS is with downlink CSI and the accuracy of available CSI will significantly affect the performance gain [14].

The BS can estimate the uplink CSI via the pilots sent by the user equipment (UE) easily. However, compared with the uplink CSI, the downlink CSI is difficult to be obtained at the BS, especially in frequency-division duplexing (FDD) systems. In time-division duplexing (TDD) systems, the BS can easily infer the downlink CSI exploiting the reciprocity between the uplink and downlink channels, which does not exist in FDD systems. Therefore, in FDD systems, the UE has to feed the downlink CSI back to the BS, which consumes precious bandwidth resource. In massive MIMO, a huge number of antennas at the BS significantly increase the CSI dimension, making the feedback overhead unaffordable.

To reduce the feedback overhead, many algorithms have been applied to compress downlink CSI. One of the most promising technologies is CS, which exploits the sparsity of massive MIMO CSI in certain domain to compress and reconstruct the downlink CSI [15, 16]. However, the sparsity is the only prior information used during the reconstruction in the CS-based methods, and the reconstruction problem is underdetermined and often solved by iterative algorithms, thereby consuming much computational and time resources. As a result, the CS-based feedback is difficult to be implemented to practical communication systems.

Inspired by the success of DL in image compression [17] and physical layer communication [18, 19], DL-based feedback scheme has been proposed, which provides a new way for solving the CSI feedback problem in FDD systems.

### 5.3.1     Massive MIMO-OFDM CSI Feedback

We consider a single-cell FDD massive MIMO-OFDM system over $K$ subcarriers, where the BS is equipped with $N_t \gg 1$ transmit antennas and the UE is equipped with a single receiver antenna. The received signal at the $k$th subcarrier can be expressed as follows:

$$y_k = \mathbf{h}_k^H \mathbf{v}_k s_k + n_k, \tag{5.25}$$

where $\mathbf{h}_k \in \mathbb{C}^{N_t \times 1}$, $s_k \in \mathbb{C}$, $n_k \in \mathbb{C}$ denote the channel vector, the transmit data symbol, and the additive noise of the $k$th subcarrier, respectively. $\mathbf{v}_k \in \mathbb{C}^{N_t \times 1}$ represents the precoding vector designed by the BS based on the received CSI. Therefore, the downlink CSI matrix in the spatial-frequency domain can be expressed as

$$\tilde{\mathbf{H}} = \begin{bmatrix} \mathbf{h}_1 & \mathbf{h}_2 & \cdots & \mathbf{h}_K \end{bmatrix}^H = \begin{bmatrix} h_{11} & h_{12} & \cdots & h_{1N_t} \\ h_{21} & h_{22} & \cdots & h_{2N_t} \\ \vdots & \vdots & \ddots & \vdots \\ h_{K1} & h_{K2} & \cdots & h_{KN_t} \end{bmatrix}_{K \times N_t}, \tag{5.26}$$

where each column is a frequency domain channel vector and each row is a space domain channel vector.

Due to limited delay spread of wireless channels, the time delay among multiple paths is within a limited period. Therefore, the CSI matrix shows sparsity in the delay domain. Meanwhile, if the number of the transmit antennas is very large, that is, $N_t \to \infty$, the CSI matrix in the angular domain can be derived from the spatial domain by performing DFT and also exploit sparsity [20]. Combining the two aspects, we can convert the CSI matrix in the spatial-frequency domain into the angular-delay domain by 2D DFT as follows:

$$\mathbf{H} = \mathbf{F}_d \tilde{\mathbf{H}} \mathbf{F}_a, \tag{5.27}$$

where $\mathbf{F}_d$ is a $K \times K$ DFT matrix and $\mathbf{F}_a$ is a $N_t \times N_t$ matrix. Then, we retain the first $K'$ nonzero rows and remove the remaining to get an $K' \times N_t$ truncated matrix $\mathbf{H}$. The number of the real parameters of $\mathbf{H}$ is $N = 2K'N_t$, which will be further reduced using DL.

By utilizing the sparse characteristics of the channel matrix, the CS algorithms compress the above $N$-dimensional vector into $M$-dimensional measurements to further reduce the feedback overhead. However, the traditional CS algorithms rely heavily on the prior assumption and cost much time, which promotes the rapid development of the DL-based feedback methods.

DL has been first introduced to massive MIMO CSI feedback in [21], where a novel CSI feedback neural network, namely, CsiNet, is developed. The architecture of CsiNet is similar to that of the autoencoder [22], which includes an encoder and a decoder, as shown in Fig. 5.10. The encoder at the UE is used for CSI compression, which converts the $N$-dimensional channel matrix into $M$-dimensional codeword by utilizing the sparse characteristic of the channel matrix. The process can be mathematically described as

$$\mathbf{s} = f_{en}(\mathbf{H}), \tag{5.28}$$

where $\mathbf{H}$ and $\mathbf{s}$ represent the channel matrix and the corresponding codeword, respectively. The compression ratio (CR) is $M/N$, where $M < N$. The decoder at the BS is used for CSI reconstruction, which tries to recover the original channel matrix from the received codeword. It can be expressed as

$$\hat{\mathbf{H}} = f_{de}(\mathbf{s}), \tag{5.29}$$

where $\hat{\mathbf{H}}$ is the recovered channel.

As in Fig. 5.10, the first layer of the encoder is a convolutional layer, which uses two $3 \times 3$ filters to extract CSI features and generates two feature maps. Then, a fully connected layer with $M$ neurons is used to compress the feature maps into a lower dimension. The decoder consists of a fully connected layer, two RefineNet units, and a convolutional layer. The first fully connected layer decomposes the received codeword into two matrices with the same dimension as the encoder input, which is served as the initial estimation of the real and imaginary parts of $\mathbf{H}$. The reconstructed channel matrix is further refined by the two RefineNet units. Each consists of four

**Table 5.1.** NMSE (dB) performance of CsiNet and CS-based methods.

| Scenario | CR | LASSO | TVAL3 | BM3D-AMP | CsiNet |
|---|---|---|---|---|---|
| Indoor | 1/4 | −7.59 | −14.87 | −4.33 | **−17.36** |
| | 1/16 | −2.72 | −2.61 | 0.26 | **−8.65** |
| | 1/32 | −1.03 | −0.27 | 24.72 | **−6.24** |
| | 1/64 | −0.14 | 0.63 | 26.22 | **−5.84** |
| Outdoor | 1/4 | −5.08 | −6.90 | −1.33 | **−8.75** |
| | 1/16 | −1.01 | −0.43 | 0.55 | **−4.51** |
| | 1/32 | −0.24 | 0.46 | 22.66 | **−2.81** |
| | 1/64 | −0.06 | 0.76 | 25.45 | **−1.93** |

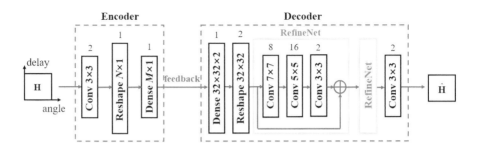

**Figure 5.10** The architecture of CsiNet. The encoder at the UE is used for compressing CSI, and the decoder at the BS is used for reconstructing CSI.

convolutional layers, which use $3 \times 3$ filters and generate 2, 8, 16, and 2 feature maps. The final convolutional layer normalizes the output and generates the final reconstructed channel matrix.

To generate the dataset, we use the uniform linear array with $N_t = 32$ antennas at the BS and $K = 1024$ subcarriers under the COST 2100 channel model [23]. Two scenarios are selected as channel environments, which are the indoor picocellular scenario at the 5.3 GHz band and the outdoor rural scenario at the 300 MHz band. We utilize NMSE to measure the difference between the recovered channel, $\hat{\mathbf{H}}$, and the original channel, $\mathbf{H}$, which is calculated as follows:

$$\text{NMSE} = \text{E} \{\|\mathbf{H} - \hat{\mathbf{H}}\|_2^2 / \|\mathbf{H}\|_2^2\}. \tag{5.30}$$

where $\| \cdot \|_2$ is the Euclidean norm.

We compare the reconstruction performance of CsiNet with three CS-based methods, namely, LASSO [24], TVAL3 [25], and BM3D-AMP [26]. The corresponding NMSE of all the methods mentioned is shown in Table 5.1, where the best results are presented in bold font. CsiNet outperforms all the CS-based methods at different CRs for both scenarios. Since the data-driven model can be trained in the end-to-end manner, CsiNet can learn how to effectively utilize the channel structure characteristics from the training data, which is independent of the prior knowledge of channel distribution. Meanwhile, the DL-based method is several orders of magnitude faster than the CS-based algorithm. Even at the low CR where the traditional CS-based methods fails, CsiNet still has a good reconstruction performance, which suggests that CsiNet is a novel channel feedback mechanism with a great potential.

## 5.3.2    Time-Varying Channel Feedback

Inspired by the real-time architecture for high rate video compression [27], we can address channel feedback for time-varying channels by setting the $T_s$ adjacent channel matrices in the angular-delay domain within the coherent time as a channel group. Since the correlation between the channel matrices is similar to the interframe correlation in video signals, the CsiNet architecture has been extended with long short-term memory (LSTM) in [28] to improve the recovery quality of the network. The architecture of CsiNet-LSTM is shown in Fig. 5.11, whose encoder and decoder modules are same as CsiNet. Two CRs are used when extracting the features in the angular-delay domain. The first encoder is with a high CR, thus retaining enough structural information of the first channel matrix for the subsequent high-resolution recovery. Due to the correlation within the channel group, the remaining channels contain less extra information, so they can be compressed with a low CR. Before reconstruction, the first high-CR codeword is concatenated to the front of all low-CR codewords to make full use of channel correlation information for decoding. The outputs of the decoders form a sequence of length $T_s$ and then are sent into a three-layer LSTM, which can implicitly learn the time correlation through the input at the previous moment and then merge with the input at the current moment to improve the reconstruction quality at a low CR.

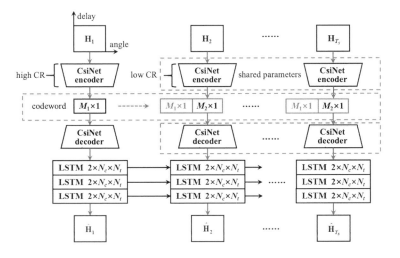

**Figure 5.11**  The architecture of CsiNet-LSTM. The encoder and decoder modules are the same as CsiNet. The outputs of the decoders are fed into LSTMs to extract time correlation.

**Table 5.2.** NMSE (dB) performance and runtime(s) of CsiNet-LSTM.

| Scenario | | CR | LASSO | TVAL3 | BM3D-AMP | CsiNet | CsiNet-LSTM |
|---|---|---|---|---|---|---|---|
| Indoor | NMSE | 1/16 | −2.96 | −3.20 | 0.25 | −10.59 | **−23.06** |
| | | 1/32 | −1.18 | −0.64 | 20.85 | −7.35 | **−22.23** |
| | | 1/64 | −0.18 | 0.60 | 26.66 | −6.09 | **−21.24** |
| | runtime | 1/16 | 0.2471 | 0.3148 | 0.3454 | **0.0001** | 0.0003 |
| | | 1/32 | 0.2137 | 0.3148 | 0.5556 | **0.0001** | 0.0003 |
| | | 1/64 | 0.2479 | 0.2860 | 0.6047 | **0.0001** | 0.0003 |
| Outdoor | NMSE | 1/16 | −1.09 | −0.53 | 0.40 | −3.60 | **−9.86** |
| | | 1/32 | −0.27 | 0.42 | 18.99 | −2.14 | **−9.18** |
| | | 1/64 | −0.06 | 0.74 | 24.42 | −1.65 | **−8.83** |
| | runtime | 1/16 | 0.2122 | 0.3145 | 0.4210 | **0.0001** | 0.0003 |
| | | 1/32 | 0.2409 | 0.2985 | 0.6031 | **0.0001** | 0.0003 |
| | | 1/64 | 0.2166 | 0.2850 | 0.5980 | **0.0001** | 0.0003 |

We generate training samples using $N_t = 32$ antennas at the BS and $K = 256$ subcarriers under the COST 2100 channel model. Some parameters of the training network are initialized by loading from CsiNet in advance. The reconstruction performance and runtime of CsiNet-LSTM, three CS-based methods and CsiNet have been tested with different CRs. The comparison results are summarized in Table 5.2, and the best ones are shown in bold font. The NMSE of CsiNet-LSTM is lower than that of the CS-based methods and CsiNet, indicating that it has higher reconstruction accuracy. Moreover, with the reduction of CR, the NMSE performance of CsiNet-LSTM does not decline significantly, thus solving the problem of poor reconstruction quality of CsiNet at a low CR. Due to the introduction of LSTM, the running time of CsiNet-LSTM is slightly longer than that of CsiNet, but it is still several orders of magnitude faster than the CS-based methods.

LSTM on the basis of CsiNet has also been introduced in [29], where a new architecture, namely, RecCsiNet, has been proposed as in Fig. 5.12. By comparison, RecCsiNet focuses on optimizing feature compression and decompression modules, whereas CsiNet-LSTM focuses on optimizing channel recovery modules. The encoder of RecCsiNet includes two modules for feature extraction and feature compression, respectively, and the decoder includes two modules for feature decompression and channel recovery, respectively. The feature extraction and channel recovery modules are the same as those of CsiNet, whereas the LSTM network is added to feature compression and decompression modules to extract and exploit the time correlation. The input of the feature compression module is divided into two parallel streams, namely LSTM network and fully connected network (FCN). LSTM extracts the time correlation of time-varying channels, and FCN serves as a jump connection to accelerate convergence and reduce vanishing gradient problem. Accordingly, the decompression module also includes a LSTM network and a linear FCN,

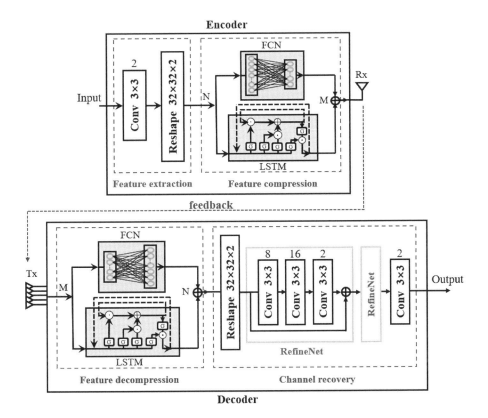

**Figure 5.12** The architecture of RecCsiNet. The input of the feature compression module is divided into two parallel streams, and LSTM is used for extract time correlation.

and the input and output sizes of the compression and decompression modules are symmetric. Similar to the residual network, the LSTM network can learn the residual characteristics with the connection of the linear FCN, making the network more robust. To reduce the training parameters, the structure of RecCsiNet has been changed in [29], where a variant network, called PR-RecCsiNet, has been proposed. In PR-RecCsiNet, FCN and LSTM are connected in serial rather than in parallel. The compression and decompression modules in PR-RecCsiNet are illustrated in Fig. 5.13. FCN reduces the input dimension of LSTM and the training parameters of the whole network.

Table 5.3 and Fig. 5.14 compare the parameter complexity and reconstruction performance of RecCsiNet with other architectures. From Fig. 5.14, RecCsiNet obviously outperforms CsiNet, indicating that the optimized compression and decompression modules can improve the reconstruction accuracy by utilizing the time correlation of CSI. Combined with Table 5.3, the parameter size of CsiNet-LSTM is almost five times that of RecCsiNet, and the parameter complexity of PR-RecCsiNet is significantly reduced although the performance of CsiNet-LSTM is better than RecCsiNet at some CRs.

**Table 5.3.** Number of the parameters.

| CR | 1/16 | 1/32 | 1/64 |
|---|---|---|---|
| CsiNet | 530,656 | 268,448 | 137,344 |
| CsiNet-LSTM | 102,009,892 | 101,354,404 | 101,026,660 |
| RecCsiNet | 19,478,584 | 18,118,392 | 17,450,584 |
| PR-RecCsiNet | 793,144 | 333,816 | 153,304 |

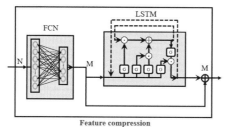

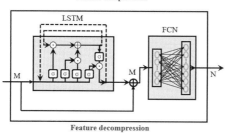

**Figure 5.13** The compression and decompression modules in PR-RecCsiNet.

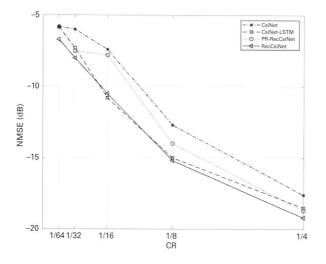

**Figure 5.14** The NMSE performance of different methods in the indoor scenario.

### 5.3.3    Other Feedback Methods

In addition to the methods introduced in Sections 5.3.1 and 5.3.2, there are many other feedback methods. Here, we discuss a couple of examples.

**Channel Reciprocity-Based Feedback**

In TDD MIMO systems, the BS can take advantage of strong channel reciprocity to obtain the downlink CSI while the FDD systems need the UE to feedback the downlink CSI to the BS because the channel reciprocity is not obvious. The correlation between uplink and downlink CSI in FDD systems has been investigated in [30]. It has been found that there still exists some hiding reciprocity even if there is no direct reciprocity between uplink and downlink instantaneously CSI. Therefore, an architecture, namely, DualNet-MAG or DualNet-ABS, has been developed in [30] to use the uplink CSI to improve the reconstruction accuracy of the downlink CSI.

From [30], the distribution of correlation coefficient between uplink CSI and downlink CSI in FDD systems is at different confidence intervals, which is obviously unstable. However, if transforming the CSI in the delay domain into polar coordinates, the CSI magnitudes in uplink and downlink show a strong correlation while there is no obvious correlation between phases. Similarly, by separating the signs of the real and imaginary parts of CSI, the absolute values of CSI have a strong correlation between the uplink and downlink while the signs exhibit little correlation.

Therefore, when the downlink CSI is compressed by the UE, more resources can be used to feed back the phases while the magnitudes or absolute values can be highly compressed to reduce the feedback overhead. Accordingly, when the BS receives the magnitudes or absolute values of downlink CSI, it can combine the feedback magnitudes or absolute values of downlink CSI with the self-estimated uplink CSI magnitudes or absolute values to make full use of their correlation to improve the recovery accuracy of downlink CSI.

Figure 5.15 shows the general architecture of DualNet-MAG, which separates the magnitudes and phases of CSI. We test CsiNet, DualNet-MAG, and DualNet-ABS at different CRs. The simulation results show that the two methods using the correlation between the uplink and downlink CSI have better reconstruction performance than the CsiNet in both COST 2100 scenarios. In addition, under most test conditions, DualNet-MAG outperforms DualNet-ABS, which also reflects that the correlation of magnitudes between uplink and downlink CSI is stronger than that of absolute values.

**Bit-Level Channel Feedback**

The output of the encoder needs to be converted into bitstream for feedback in many practical applications. However, in the previous studies, the UE is assumed to directly send the floating-point compression code back to the BS, which ignores quantization errors in the actual wireless communication systems.

Quantization errors usually have a great impact on the reconstruction accuracy of the network. Therefore, in [31], the structure of CsiNet is optimized, and a new architecture, CsiNet+, as shown in Fig. 5.16, is robust to quantization errors, where

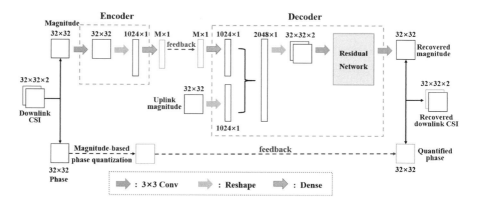

**Figure 5.15** The architecture of DualNet-MAG. The magnitudes and phases of downlink CSI are separately compressed, and the magnitudes were reconstructed with the help of uplink CSI.

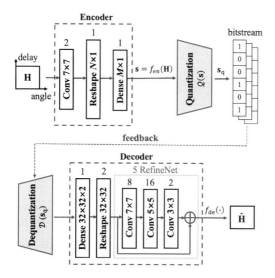

**Figure 5.16** The architecture of bit-level CsiNet+. The output of the encoder is quantized with finite bits, further reducing the feedback overhead.

quantization module and dequantization module are added to the existing CSI feedback architecture to realize bitstream transformation. When the channel matrix is compressed by the encoder into a codeword at the UE, it is quantized with finite bits and then it is encoded by entropy into bitstream for feedback. Since entropy encoding is lossless, this step can be ignored. The processing at the UE can be expressed as

$$\mathbf{s}_{\mathrm{q}} = \mathcal{Q}(f_{\mathrm{en}}(\mathbf{H})), \tag{5.31}$$

where $f_{\mathrm{en}}(\cdot)$ represents the compression process and $\mathcal{Q}(\cdot)$ denotes the quantization process. To make full use of the distribution of the CSI amplitudes, a $\mu$-law nonuniform quantizer is adopted in [31]. Specifically, the compressed codeword, $\mathbf{s} = f_{\mathrm{en}}(\mathbf{H})$, is first passed through a companding function $\Phi(\cdot)$, which is nondecreasing and smooth, and then quantized using a uniform quantizer and passed through an expander function $\Phi^{-1}(\cdot)$ to generate the final quantized codeword $\mathbf{s}_q$. The companding function $\Phi(\cdot)$ in the $\mu$-law nonuniform quantizer can be described as

$$\Phi(\mathbf{s}) = \pm \frac{\ln(1 + \mu|\mathbf{s}|)}{1 + \mu}, \qquad (5.32)$$

where $\mathbf{s} \in [-1, 1]$ is the output of the encoder and $\mu$ is a constant that determines the companding amplitude. The parameter $\mu$ can be adjusted according to the distribution of the CSI amplitudes.

The BS then dequantizes, decodes, and decompresses the quantized codeword to reconstruct CSI, which can be expressed as

$$\hat{\mathbf{H}} = f_{de}(\mathcal{D}(\mathbf{s}_q)), \qquad (5.33)$$

where $\mathcal{D}(\cdot)$ and $f_{de}(\cdot)$ denotes the dequantization and decompression process, respectively.

Since quantization is taken into account during the training of this architecture, the trained network is robust to quantization errors. As the quantization bits increase, the reconstruction becomes increasingly accurate, as shown in Table 5.4. The experimental results show that the reconstruction performance of this architecture is equivalent to that of the architecture without quantization when the quantization bits are six, but the corresponding overhead of feedback is greatly reduced.

The DL-based CSI feedback schemes have made up the shortcomings of the traditional feedback methods, which not only significantly reduce the feedback overhead, but also ensure reconstruction accuracy and thus have a broad application prospect in future mobile communication systems. However, the application of DL technology to CSI feedback is still in the early stage of exploration, and many problems and challenges remain to be further investigated. Here are some initial research results in the topic.

**Table 5.4.** NMSE (dB) performance of bit-level CsiNet+.

| | Indoor | | | | | | | |
|---|---|---|---|---|---|---|---|---|
| CR | 4 | | | | 16 | | | |
| Quantization bits | 3 | 4 | 5 | 6 | 3 | 4 | 5 | 6 |
| CsiNet+ | | −27.37 | | | | −14.14 | | |
| bit-level CsiNet+ | −18.55 | −22.00 | −24.95 | −12.45 | −12.56 | −13.64 | −14.04 | −7.82 |

| | Outdoor | | | | | | | |
|---|---|---|---|---|---|---|---|---|
| CR | 4 | | | | 16 | | | |
| Quantization bits | 3 | 4 | 5 | 6 | 3 | 4 | 5 | 6 |
| CsiNet+ | | −12.40 | | | | −5.73 | | |
| bit-level CsiNet+ | −9.45 | −11.17 | −11.96 | −12.28 | −4.65 | −5.33 | −5.59 | −5.68 |

1. Model-driven feedback: The existing DL-based feedback architecture all are based on an autoencoder, in which the traditional algorithms are completely replaced by standard neural networks. The networks autonomously learn the channel structure characteristics in a data-driven way, which can be regarded as a black box and is unable to explain and hard to control even if achieving an excellent performance. To solve these problems, a model-driven feedback scheme, combined with the traditional algorithms, has been developed. The CSI compression and reconstruction process is addressed in the same way as in sparse signal processing [32]. A sparse antoencoder is employed to learn the sparse transformation and inverse transformation. By unfolding the Iterative Shrinkage-Thresholding Algorithm and mapping the algorithm into a network to reconstruct CSI, a model-driven CSI feedback architecture is proposed. Joint optimization of the sensing matrix, sparse transformation, and reconstruction algorithm makes its performance better than the traditional CS-based methods and the existing DL-based methods.

2. Multiple-rate compression: The sparsity characteristics of CSI are different in various channel environments while the network with a fixed CR cannot adapt well to the change of characteristic granularity. As a direct way to adjust the CR according to the environments, the UE needs to store several network architectures and corresponding parameters, which is obviously not desirable for the UE with limited storage space. Therefore, it is of great practical significance to design and implement a CSI feedback architecture with multirate compression but with a reasonable number of parameters. Two different multiple-rate compression frameworks, namely, SM-CsiNet+ and PM-CsiNet+, have been proposed in [31]. If CSI compression is performed by the fully connection layer, it will take up 99.9 percent of the total parameters. By recompressing the compressed codeword with a high CR to generate one with a low CR, the input dimension of the fully connected layer for the low CR is reduced, thereby reducing the number of parameters. SM-CsiNet+, which compresses CSI matrix from a high CR to a low CR in serial, decreases the parameter number by approximately 38.0 percent, compared to saving the frameworks with four CRs separately. The other framework, PM-CsiNet+, first compresses CSI matrix by fourfold and then the compressed codewords with low CRs are selected from the four-fold compressed codeword. In this case, only a fully connected layer used for four-fold compression needs to be stored in the framework, which decreases the parameter number by approximately 46.7 percent.

3. Lightweight structure: The existing DL-based CSI feedback schemes have relatively a large number of network parameters, which are not conducive to actual deployment. According to the research on compression and acceleration of neural networks in [33], a lighter network can be designed to reduce the network parameters. Since the convolutional layers of ConvCsiNet in [34] have the highest proportion of the parameter number, a lightweight structure, called ShuffleCsiNet, has been proposed, which replaces the convolutional layers and the average pooling in ConvCsiNet with three shuffle network units and a mean pooling. A depthwise separable convolutional structure, composed of a depthwise convolutional layer with a $3 \times 3$ filter and a convolutional layer with a $1 \times 1$ filter, is used in the shuffle network unit, which can ensure the performance of network compared with the traditional convolution

structure, but the corresponding parameters are considerably reduced. The simulation results show that the parameter number and the complexity of ShuffleCsiNet are much lower than those of ConvCsiNet at the cost of a small NMSE performance loss. The lightweight network promotes the actual deployment of DL-based CSI feedback in the future.

## 5.4        Signal Detection

Signal detection determines the performance of wireless communications directly. An efficient signal detection algorithm should also balance performance and complexity. This section discusses DL-based signal detection algorithms for MIMO and OFDM systems.

### 5.4.1      MIMO Detection

MIMO is the mainstream technology in 4G cellular networks owing to its ability to increase spectral efficiency and link reliability. For a conventional uplink MIMO system, the receiver signal vector, $\mathbf{y}$, is given by

$$\mathbf{y} = \mathbf{Hx} + \mathbf{n}, \tag{5.34}$$

where $\mathbf{H} \in \mathbb{C}^{N_r \times N_t}$ is the channel matrix, $\mathbf{x}$ is a random transmit signal vector with each element from some finite constellation $\mathcal{A}$ (e.g., QPSK or QAM), $\mathbf{n}$ is the noise vector with elements being independent, zero mean Gaussian variables of variance $\sigma^2$. MIMO detection is to estimate the transmit signal vector $\mathbf{x}$ from the received signal vector, $\mathbf{y}$, where channel matrix $\mathbf{H}$ is usually assumed to be known or can be obtained by channel estimation.

**MIMO Detection with Perfect CSI**
Recently, many DL-based MIMO detectors have been developed under the assumption that channel $\mathbf{H}$ and noise variance $\sigma^2$ are perfectly known to the receiver. First, a model-driven DL approach, called detection network (DetNet), has been designed in [35] by applying a projected gradient descent method for maximum likelihood detection in a neural network. DetNet recovers the transmitted signal by treating received signal $\mathbf{y}$ and perfect CSI $\mathbf{H}$ as inputs. This approach outperforms the traditional iterative detection and is comparable with the $K$-best sphere decoder [35]. The robustness of DetNet to some complex channels, such as the deterministic ill-conditioned channel and the varying channel model with a known distribution, is also demonstrated. DetNet achieves the performance equal to or better than those of the AMP [36] and semi-definite relaxation algorithms [37]. DetNet has $1 \sim 10$ million trainable parameters. The parameters of DL-based MIMO detection can be significantly reduced if some domain knowledge is embedded.

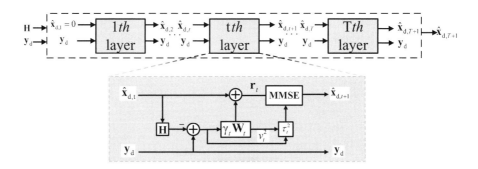

**Figure 5.17** The network structure of OAMP detector.

To further reduce the number of the trainable parameters and incorporate more domain knowledge, a model-driven DL network, called orthogonal approximate message passing net (OAMP-Net), has been developed in [38] for MIMO detection. As in Fig. 5.17, it is obtained by unfolding the iterative OAMP detector. For the $t$th layer of the OAMP-Net, the input is the estimated signal $\hat{\mathbf{x}}_t$ from the $(t-1)$th layer and the received signal, $\mathbf{y}$, and signal detection is performed as follows:

$$\mathbf{r}_t = \hat{\mathbf{x}}_t + \gamma_t \mathbf{W}_t (\mathbf{y} - \mathbf{H}\hat{\mathbf{x}}_t), \tag{5.35}$$

$$\hat{\mathbf{x}}_{t+1} = \mathbb{E}\left\{\mathbf{x}|\mathbf{r}_t, \tau_t\right\}, \tag{5.36}$$

$$v_t^2 = \frac{\|\mathbf{y} - \mathbf{H}\hat{\mathbf{x}}_t\|_2^2 - N_r \sigma^2}{\mathrm{tr}(\mathbf{H}^T \mathbf{H})}, \tag{5.37}$$

$$\tau_t^2 = \frac{1}{2N_t}\mathrm{tr}(\mathbf{C}_t \mathbf{C}_t^T)v_t^2 + \frac{\theta_t^2 \sigma^2}{4N_t}\mathrm{tr}(\mathbf{W}_t \mathbf{W}_t^T), \tag{5.38}$$

where $\mathbf{W}_t = \frac{2N_t}{\mathrm{tr}(\hat{\mathbf{W}}_t \mathbf{H})}\hat{\mathbf{W}}_t$, and $\hat{\mathbf{W}}_t$ is the linear MMSE matrix as follows:

$$\hat{\mathbf{W}}_t = v_t^2 \mathbf{H}^T \left(v_t^2 \mathbf{H}\mathbf{H}^T + \frac{\sigma^2}{2}\mathbf{I}\right)^{-1}. \tag{5.39}$$

Furthermore, matrix $\mathbf{B}_t$ in the OAMP-Net is given by $\mathbf{C}_t = \mathbf{I} - \mathbf{W}_t \mathbf{H}$. Compared with the OAMP algorithm, OAMP-Net introduces only a few trainable parameters to incorporate the side information from the data. The number of trainable variables is independent of the number of antennas and is only determined by the number of layers, thereby giving OAMP-Net an advantage in addressing large-scale MIMO. With only few trainable variables, the stability and speed of convergence can be significantly improved in the training process. The network can also handle a time-varying channel with only a single training and improve the performance of the OAMP detector in Rayleigh and correlated MIMO channels.

By considering practical 3GPP 3D MIMO channels, a DL-based MIMO detector, named MMNet, has been developed with an unfolded NN based on the AMP algorithm and adding a large number of trainable parameters to track different channel realizations. The neural network models are different for i.i.d. Gaussian channels and practical MIMO channels. In the i.i.d. Gaussian case, the MMNet-iid is performed as~follows:

$$\mathbf{z}_t = \hat{\mathbf{x}}_t + \theta_t^{(1)} \mathbf{H}^H (\mathbf{y} - \mathbf{H}\hat{\mathbf{x}}_t)$$
$$\hat{\mathbf{x}}_{t+1} = \eta_t (\mathbf{z}_t; \sigma_t^2).$$

(5.40)

where the equivalent variance is given by

$$\sigma_t^2 = \frac{\theta_t^{(2)}}{N_t} \left( \frac{\|\mathbf{I} - \mathbf{A}_t \mathbf{H}\|_F^2}{\|\mathbf{H}\|_F^2} \left[ \|\mathbf{y} - \mathbf{H}\hat{\mathbf{x}}_t\|_2^2 - N_r \sigma^2 \right]_+ + \frac{\|\mathbf{A}_t\|_F^2}{\|\mathbf{H}\|_F^2} \sigma^2 \right). \qquad (5.41)$$

The principle behind this is that the noise at the input of the denoiser is comprised of two parts: (a) the residual error caused by distortion of $\mathbf{x}_t$ from the true value of $\mathbf{x}$ and (b) the contribution of the channel noise $\mathbf{n}$. The first component is amplified by the linear transformation $(\mathbf{I} - \mathbf{A}_t \mathbf{H})$ and the second component is amplified by $\mathbf{A}_t$. This model has only two parameters per layer: $\theta_t^{(1)}$ and $\theta_t^{(2)}$. In this example, for the i.i.d. Gaussian channel matrix, a simple model that adds a small number of trainable variables to the existing algorithms like AMP can perform very well.

The MMNet neural network for an arbitrary channel matrix is as follows:

$$\mathbf{z}_t = \mathbf{x}_t + \mathbf{\Theta}_t^{(1)} (\mathbf{y} - \mathbf{H}\mathbf{x}_t)$$
$$\mathbf{x}_{t+1} = \eta_t (\mathbf{z}_t; \sigma_t^2)$$

(5.42)

where $\mathbf{\Theta}_t^{(1)}$ is an $N_t \times N_r$ complex-valued trainable matrix. To handle different levels of noise, we add an extra degree of freedom to the estimation of noise per transmitter, which is given by

$$\sigma_t^2 = \frac{\theta_t^{(2)}}{N_t} \left( \frac{\|\mathbf{I} - \mathbf{A}_t \mathbf{H}\|_F^2}{\|\mathbf{H}\|_F^2} \left[ \|\mathbf{y} - \mathbf{H}\mathbf{x}_t\|_2^2 - N_r \sigma^2 \right]_+ + \frac{\|\mathbf{A}_t\|_F^2}{\|\mathbf{H}\|_F^2} \sigma^2 \right) \qquad (5.43)$$

where $N_t \times 1$ parameter vector $\theta_t^{(2)}$ scales the noise variance by different amounts for each symbol. This approach distinguishes MMNet from both the OAMP-Net and DetNet. In particular, MMNet uses a flexible linear transformation to construct intermediate signal $\mathbf{z}_t$, but it applies the standard optimal denoiser for Gaussian noise. Furthermore, MMNet requires no matrix inverse operation, and therefore has lower complexity.

The MMNet significantly outperforms existing DL-based detectors on realistic channels. Furthermore, a trained model for one channel realization can serve as a strong initialization for training adjacent channel realizations to accelerate online training in MIMO-OFDM systems because of the temporal and frequency correlation in practical MIMO channels. By leveraging the emerging idea of hypernetworks, a HyperMIMO detector has been proposed for MIMO detection. It consists of a secondary NN, referred to as the hypernetwork, that generates an optimized set of weights for an NN-based detector. Combined with the MMNet detector from [39], HyperMIMO replaces the training procedure that would be required for each channel

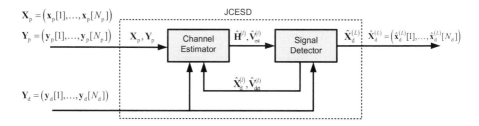

**Figure 5.18** The diagram of the turbo-like JCESD architecture. The channel estimator and signal detector exchange information iteratively until convergence.

realization by a single inference of the hypernetwork. The HyperMIMO achieves a performance close to that of MMNet trained for each channel realization and outperforms the recently proposed OAMP-Net. Furthermore, the HyperMIMO is robust to user mobility up to a certain point, which is encouraging for practical use.

## Joint Channel Estimation and Signal Detection

In practical systems, channel estimation error always exists and cannot be ignored. Therefore, investigating DL-based MIMO detection with imperfect CSI is very meaningful. In [40], an OAMPNet2-based joint channel estimation and signal detection architecture has been developed, where the DL-based detection considers the characteristics of channel estimation error and channel statistics. Furthermore, a data-aided channel estimation method is utilized to further improve channel estimation performance. A turbo-like joint channel estimation and signal detection (JCESD) architecture for MIMO systems is demonstrated in Fig. 5.18, which shares the same spirit as iterative decoding. In JCESD, channel estimator and signal detector exchange information iteratively until convergence. In the first iteration, channel estimation is only based on pilots. In the subsequent iterations, data-aided channel estimation is employed.

As in Fig. 5.18, the input of the JCESD architecture is the pilot signal matrix, $\mathbf{X}_p$, received signal matrix corresponding to the pilot matrix, $\mathbf{Y}_p$, received signal matrix corresponding to the transmit data, $\mathbf{Y}_d$, in each time slot. In the figure, $\hat{\mathbf{H}}^{(l)}$ and $\hat{\mathbf{X}}_d^{(l)}$ are the estimated channel matrix and data matrix in the $l$th iteration, respectively. $\hat{\mathbf{V}}_{est}^{(l)}$ and $\hat{\mathbf{V}}_{det}^{(l)}$ are used to compute the covariance matrix for equivalent noise in signal detector and channel estimator, respectively. The final output of the signal detector is the detected data matrix at the $l$th iteration, $\hat{\mathbf{X}}_d^{(L)}$, where $L$ is the total number of turbo iterations.

Compared with the conventional receiver design where the channel estimator and signal detector work separately, this architecture can improve the performance of the receiver by jointly considering the characteristics of channel estimation error and channel statistics.

## Joint Signal Detection and Channel Decoding

If the MIMO technique is used with channel coding, as in almost all communication systems, a significant issue is how the MIMO detector exploits the soft information from the decoder and produces the soft-output. In [41], a joint signal detection and decoding framework has been investigated, where signal detection is realized by representing the expectation propagation algorithm as multilayer deep feed-forward networks. The joint channel estimation, signal detection, and channel decoding framework in [42] significantly outperforms the existing algorithms and is also effective for the channel estimation with insufficient pilots.

## 5.4.2    OFDM Detection

From Section 5.2.2, the received OFDM signal at the $k$th subcarrier can be expressed as

$$Y[k] = H[k]X[k] + W[k], \tag{5.44}$$

where $W[k]$, $H[k]$, $X[k]$, $Y[k]^1$ represent the additive Gaussian noise, channel frequency response, the transmit symbol, and the received signal at the $k$th subcarrier, respectively. Therefore, the signal detection becomes much simpler as it can be conducted in each subcarrier, which can be performed by

$$\hat{X}[k] = \frac{Y[k]}{\hat{H}[k]}, \tag{5.45}$$

where $\hat{H}$ is the estimation of channel frequency response. As discussed in Section 5.2.2, training pilots are inserted into OFDM symbols in both the time and frequency dimensions. Conventionally, the frequency responses on the pilot positions are first estimated based on the received signal corresponding to the pilot data. Then the channel frequency response at non-pilot positions can be estimated via interpolations or some DL approaches.

### DNN for OFDM Detection

Although the detection has been significantly simplified in the ideal OFDM system, many imperfection in the real systems, including the imperfect CSI and hardware constraints, will degrade the performance.

Recently, the data-driven DL approach has been exploited in OFDM detection to deal with the non-ideal situations in real system [43]. Rather than building analytical models for the distortions and imperfectness, the DNN based OFDM detection system directly approximates the relationship between the received signals and the transmitted data. The parameters of the DNN are learned with massive training samples to minimize an empirical loss function, which often measures the distance between the DNN output and the transmitted data. The robustness to the imperfection can be obtained automatically via training with data generated from the OFDM system with distortion or imperfection. In particular, two types of examples will be discussed in the following, including DNN based joint channel estimation and signal detection and OFDM with hardware constraints.

---

$^1$ The index $n$ is omitted for simplicity.

## Joint Signal Detection with Channel Estimation

The performance of OFDM detection heavily relies on CSI while perfect CSI is usually unavailable in real systems. In the conventional communication systems, the channel estimator and signal detector are designed separately. The estimated CSI is usually obtained by means of the LS or the MMSE channel estimator and is then leveraged in the signal detector. In fact, the estimated CSI is only the intermediate result used for acquiring the detection results, similar to the latent variables used in DL approaches. From Section 5.2.2, DL can be used for channel estimation in the OFDM systems. Therefore, it is natural to move forward to build a joint channel estimation and signal detection framework with DL, where the transmitted data can be recovered directly without explicitly estimating the channel. In this way, characteristics of the channel estimation error can be taken into consideration in a data-driven manner during the training, which improves the robustness to many undesired circumstances.

A DNN-based joint channel estimation and signal detection scheme in an OFDM system has been proposed in [13]. In particular, an OFDM frame is assumed to consist of several OFDM blocks, where the first block contains the pilot data and the other blocks are the transmitted data. The channel is considered as constant over the pilot block and the data blocks. As shown in Fig. 5.19, in order to directly recover the transmit data without explicitly estimating the current channel, the input of the DNN consists of the received signal of the pilot block and a data block while the output of the DNN is the estimated transmitted data. Since the received signal is usually in a complex form while the commonly used DNN is with real inputs, the real part and the imaginary part of the received signals are concatenated together as the input of the DNN. The DNN is designed to consist of five layers, with 256, 500, 250, 120, and 16 neurons in each layer. The ReLU activation function is used in all layers except the output layer, where the Sigmoid activation function is used in the output layer in order to map the output to the interval [0, 1].

The training data is obtained offline, which can be simulated data from predefined channel models or real data collected in real environment. The parameters of the DNN are learned with stochastic gradient descent to minimize an empirical loss on the training data, which measures the difference between the output of DNN and real transmitted data. A commonly used loss function is the $l_2$-loss, which can be expressed as

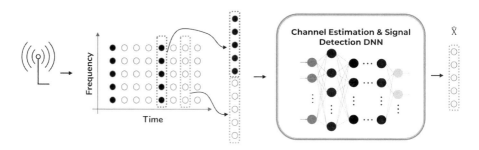

**Figure 5.19** DNN-based joint channel estimation and signal detection in OFDM receiver.

$$L_{loss} = \sum_{k=1}^{K} (X[k] - \hat{X}[k])^2, \tag{5.46}$$

where $\hat{X}[k]$ and $X[k]$ denote the output of the DNN and the original transmitted data, respectively.

Experiments are conducted for the WINNER II channel model. The DNN based approach is compared with the traditional methods in terms of bit-error rates (BERs) under different SNRs. Figure 5.20 demonstrates that the DNN-based joint channel estimation and detection method outperforms the MMSE when no adequate pilots or CP are available or there exists nonlinear distortion. From the figure, the DL method can improve the performance via remembering and leveraging the complicated characteristics of the channels with serious distortion and interference.

The pure data-driven DNN approach relies on the extraordinary expressive power of DL and the massive training data to approximate the relationship between the received signal and the transmitted data without any prior information about OFDM modulation and the channel. On the other hand, the high sample complexity and huge number of parameters of data-driven DNN approaches can be alleviated with prior expert information. This motives the development of model-driven DNNs based approaches to combine the expert knowledge and DL in the OFDM receiver.

An example of the model-driven DNN based OFDM receiver, named ComNet, has been investigated in [44], as shown in Fig. 5.21. Instead of using a single DNN to output the detection results directly, two DNN subnets are designed for channel estimation and signal detection, respectively. Both the channel estimation subnet and the signal detection subnet use the traditional communication solutions as coarse input that are further refined DL. The LS channel estimation result is refined in the channel

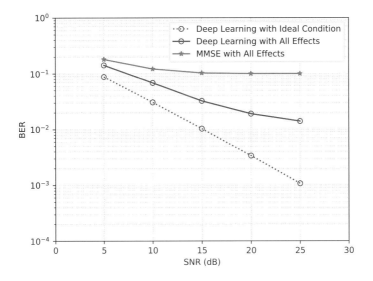

**Figure 5.20** Performance DNN-based and MMSE-based joint channel estimation and signal detection in OFDM systems [13].

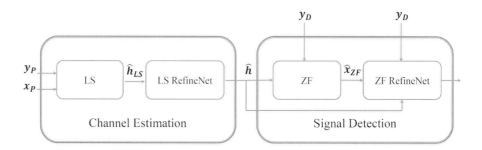

**Figure 5.21** Architecture of ComNet [44].

estimation subnet, which is an one-layer linear DNN. The refined channel estimation result is then leveraged by the zero-forcing method to obtain the coase signal detection result. The zeros-forcing detection result is taken as the input of the signal detection subnet and refined by a LSTM [45]. Both of the two subnets are trained with $l_2$ loss and the signal detection subnet is trained with the channel estimation subnet being fixed. From the experimental results, the model-driven approach can accelerate the training process and reduce the computational complexity compared with the purely data-driven DNN.

## DNN-Based OFDM Detector with Hardware Constraints

It is well known that the OFDM suffers from high peak-to-average power ratios (PAPR) [46], which not only generates nonlinear distortion at RF amplifier but also requires high-resolutions ADC at the receiver. OFDM systems with low-resolution ADC can dramatically reduce the power consumption, but the detection algorithms become much more complicated [47, 48]. Recently, the DL provides effective alternate methods to deal with the hardware constraints [49, 50].

### PAPR Reduction

Given a OFDM modulated signal $x[n]$, the PAPR of $x[n]$ is defined as the ratio of the maximum instantaneous power to its average power, as

$$PAPR[x[n]] = \frac{\max_{0 \le k \le N-1} x[n]^2}{\mathbb{E}[|x[n]^2|]}. \tag{5.47}$$

Due to the addition of the large number of independently modulated subcarriers, high amplitude peaks exist and lead to high PAPR. A deep autoencoder [51] based PAPR reduction OFDM system has been developed in [49]. The autoencoder and the autodecoder are exploited at the transmitter and the receiver, respectively, to learn constellation mapping and demapping of symbols on each subcarrier. Both the autoencoder and autodecoder are composed of fully connected connected layers, batchnorm layers, and ReLU activation functions. The loss function used for training the parameters is a summation of two components:

$$L_{loss} = (X - \hat{X})^2 + \lambda PAPR(x), \tag{5.48}$$

where $\lambda$ is used to tradeoff two components. The loss function considers the loss for recovering and PAPR penalty. The simulation results in [49] indicate that the DNN-based PAPR reduction method can effectively reduce PAPR while preserving the detection performance.

*Low-Resolution ADC*
High-resolution ADCs cost a notable portion of the power in the OFDM systems. Systems with low-resolution ADCs have practical benefits but they also introduce strong nonlinear distortion and drive the system performance away from the optimal. DL-based one-bit OFDM receiver has been proposed in [50] to address nonlinear distortion, where the quantized signal with one-bit ADC can be expressed as

$$r = \frac{1}{\sqrt{2}}\mathrm{sign}(\Re(y)) + \frac{j}{\sqrt{2}}\mathrm{sign}(\Im(y)) \tag{5.49}$$

Similar to the PAPR reduction approach earlier, the deep autoencoder and autodecoder are utilized in the transmitter and the receiver, respectively. The encoder and the decoder networks consist of fully connected layers and are trained with $l_2$ loss. Since quantization prevents end-to-end training, a two-step sequential training strategy is used, where the decoder network is first trained and remains fixed when training the encoder network. From [50], DL can achieve even better performance than the unquantized OFDM system under certain conditions.

## 5.5 Conclusion

This chapter focuses on DL-based channel estimation, feedback, and signal detection. Specifically, we have investigated DL-based channel estimation and used mmWave and OFDM channel estimation as two examples. Then, DL-based CSI feedback is introduced in detail. In this direction, with excellent expressive capacity, the CNN can learn channel features and estimate channel from a large number of training data. Furthermore, deep unfolding is also a promising technology by inheriting the superiority of iterative signal recovery algorithms and DL technology and thus presents excellent performance. We can improve the existing signal detectors by unfolding the iterative detector and adding several trainable parameters. With only few trainable variables, the stability and speed of convergence can be improved in the training process.

DL for channel estimation, feedback, and signal detection is still in its infancy. First, the recently developed DL-based communication algorithms lack unified theoretical foundation and framework, thereby preventing their widespread usage. Hence, performing a theoretical analysis for DL-based communication algorithms is meaningful. Then, the aforementioned DL network is trained offline and deployed online, which requires all possible effects of the practical system to be considered. Therefore, online training can be considered to handle all possible distortions for channel estimation, feedback, and signal detection. Finally, designing specialized DL architectures for channel estimation, feedback, and signal detection is another promising direction.

## References

[1] B. Wang et al., "Spatial- and frequency-wideband effects in millimeter-wave massive MIMO systems," *IEEE Trans. on Signal Processing*, vol. 66, no. 13, pp. 3393–3406, July 2018.

[2] H. He et al., "Deep learning-based channel estimation for beamspace mmwave massive MIMO systems," *IEEE Wireless Communication Letters*, vol. 7, no. 5, pp. 852–855, Oct. 2018.

[3] H. He et al., "Beamspace channel estimation in terahertz communications: A model-driven unsupervised learning approach," *arXiv preprint*, arXiv: 2006.16628, 2020.

[4] H. He, C. Wen, and S. Jin, "Generalized expectation consistent signal recovery for nonlinear measurements," *Proc. IEEE Int. Symp. Information Theory (ISIT)*, pp. 2333–2337, June 2017.

[5] C. M. Stein, "Estimation of the mean of a multivariate normal distribution," *Annals of Statistics*, vol. 9, no. 6, pp. 1135–1151, Nov. 1981.

[6] Y. Liu et al., "Channel estimation for OFDM," *IEEE Communications Surveys and Tutorials*, vol. 16, no. 4, pp. 1891–1908, 2014.

[7] S. Coleri et al., "Channel estimation techniques based on pilot arrangement in OFDM systems," *IEEE Trans. on Broadcasting*, vol. 48, no. 3, pp. 223–229, Nov. 2002.

[8] M. Soltani et al., "Deep learning-based channel estimation," *IEEE Communication Letters*, vol. 23, no. 4, pp. 652–655, April 2019.

[9] C. Dong et al., "Image super-resolution using deep convolutional networks," *IEEE Trans. on Pattern Analysis and Machine Intelligence,* vol. 38, no. 2, pp. 295–307, Feb. 2016.

[10] K. Zhang et al., "Beyond a Gaussian denoiser: residual learning of deep CNN for image denoising," *IEEE Trans. on Image Processing*, vol. 26, no. 7, pp. 3142–3155, July 2017.

[11] K. He et al., "Deep residual learning for image recognition," in *Proc. IEEE Conf. on Computer Vision and Pattern Recognition (CVPR)*, June 2016.

[12] L. Li et al., "Deep residual learning meets OFDM channel estimation," *IEEE Wireless Communications Letters*, vol. 9, no. 5, pp. 615–618, May 2020.

[13] H. Ye, G. Y. Li, and B. H. Juang, "Power of deep learning for channel estimation and signal detection in OFDM systems," *IEEE Wireless Communication Letters*, vol. 7, no. 1, pp. 114–117, Feb. 2018.

[14] E. G. Larsson et al., "Massive MIMO for next generation wireless systems," *IEEE Communications Magazine*, vol. 52, no. 2, pp. 186–195, Feb. 2014.

[15] P.-H. Kuo, H. T. Kung, and P.-A. Ting, "Compressive sensing based channel feedback protocols for spatially-correlated massive antenna arrays," in *Proc. IEEE Wireless Communications Networking Conf. (WCNC)*, pp. 492–497, 2012.

[16] X. Rao and V. K. N. Lau, "Distributed compressive CSIT estimation and feedback for FDD multi-user massive MIMO systems," *IEEE Trans. on Signal Processing*, vol. 64, no. 12, pp. 3261–3271, June 2014.

[17] S. Lohit et al., "Convolutional neural networks for non-iterative reconstruction of compressively sensed images," *IEEE Trans. Computer Imaging*, vol. 4, no. 3, pp. 326–340, June 2018.

[18] J. Zhang et al., "An overview of wireless transmission technology utilizing artificial intelligence," *Telecommunications Science*, vol. 34, no. 8, pp. 46–55, Aug. 2018.

[19] T. Wang et al., "Deep learning for wireless physical layer: opportunities and challenges," *China Communications*, vol. 14, no. 11, pp. 92–111, Oct. 2017.

[20] C. Wen et al., "Channel estimation for massive MIMO using Gaussian-mixture bayesian learning," *IEEE Trans. on Wireless Communications*, vol. 14, no. 3, pp. 1356–1368, Mar. 2015.

[21] C. Wen, W. Shih, and S. Jin, "Deep learning for massive MIMO CSI feedback," *IEEE Wireless Communications Letters*, vol. 7, no. 5, pp. 748–751, Oct. 2018.

[22] I. Goodfellow, Y. Bengio, and A. Courville, *Deep Learning*. MIT Press, 2016.

[23] L. Liu et al., "The COST 2100 MIMO channel model," *IEEE on Wireless Communications*, vol. 19, no. 6, pp. 92–991, Dec. 2012.

[24] I. Daubechies, M. Defrise, and C. D. Mol, "An iterative thresholding algorithm for linear inverse problems with a sparsity constraint," *Communications on Pure and Applied Mathematics*, vol. 57, no. 11, pp. 1413–1457, Nov. 2004.

[25] C. Li, W. Yin, and Y. Zhang, "User's guide for TVAL3: TV minimization by augmented Lagrangian and alternating direction algorithms," Rice University, 2010; www.caam.rice.edu/~optimization/L1/TVAL3/v.1/User_Guide_v1.0b.pdf.

[26] C. A. Metzler, A. Maleki, and R. G. Baraniuk, "From denoising to compressed sensing," *IEEE Trans. on Information Theory*, vol. 62, no. 9, pp. 5117–5144, Sep. 2016.

[27] K. Xu and F. Ren, "Csvideonet: A real-time end-to-end learning framework for high-frame-rate video compressive sensing," in *Proc. IEEE Winter Conf. on Applications of Computer Vision (WACV)*, pp. 1680–1688, 2018.

[28] T. Wang et al., "Deep learning-based CSI feedback approach for time-varying massive MIMO channels," *IEEE Wireless Communications Letters*, vol. 8, no. 2, pp. 416–419, April 2019.

[29] C. Lu et al., "MIMO channel information feedback using deep recurrent network," *IEEE Communications Letters*, vol. 23, no. 1, pp. 188–191, Jan. 2019.

[30] Z. Liu, L. Zhang, and Z. Ding, "Exploiting bi-directional channel reciprocity in deep learning for low rate massive MIMO CSI feedback," *IEEE Wireless Communications Letters*, vol. 8, no. 3, pp. 889–892, June 2019.

[31] J. Guo et al., "Convolutional neural network based multiple-rate compressive sensing for massive MIMO CSI feedback: design, simulation, and analysis," *IEEE Trans. on Wireless Communications*, vol. 19, no. 4, pp. 2827–2840, Jan. 2020.

[32] Y. Wang et al., "Learnable sparse transformation-based massive MIMO CSI recovery network," *IEEE Communications Letters*, vol. 24, no. 7, pp. 1468–1471, July 2020.

[33] J. Guo et al., "Compression and acceleration of neural networks for communications," *IEEE Wireless Communications*, pp. 1–8, July 2020.

[34] Z. Cao et al., "Lightweight convolutional neural networks for CSI feedback in massive MIMO," *arXiv preprint*, arXiv: 2005.00438, 2020.

[35] N. Samuel, T. Diskin, and A. Wiesel, "Deep MIMO detection," in *Proc. 18th Int. Workshop Signal Processing Advances in Wireless Communications (SPAWC)*, pp. 1–5, 2017.

[36] C. Jeon et al., "Optimality of large MIMO detection via approximate message passing," in *Proc. IEEE Int. Symp. Information Theory (ISIT)*, pp. 1227–1231, 2015.

[37] Z. Luo et al., "Semidefinite relaxation of quadratic optimization problems," *IEEE Signal Processing Magazine*, vol. 27, no. 3, pp. 20–34, May 2010.

[38] H. He et al., "A model-driven deep learning network for MIMO detection," in *Proc. IEEE Global Conf. Signal Inf. Process. (GlobalSIP)*, pp. 584–588, Nov. 2018.

[39] M. Khani et al., "Adaptive neural signal detection for massive MIMO," *IEEE Trans. on Wireless Communications*, vol. 19, no. 8, pp. 5635–5648, Aug. 2020.

[40] H. He et al., "Model-driven deep learning for MIMO detection," *IEEE Trans. on Signal Processing*, vol. 68, pp. 1702–1715, 2020.

[41] S. Sahin et al., "Doubly iterative turbo equalization: Optimization through deep unfolding," in *Proc. IEEE 30th Annu. Int. Symp. Personal, Indoor and Mobile Radio Communications (PIMRC)*, pp. 1–6, 2019.

[42] J. Zhang et al., "Model-driven deep learning based turbo-MIMO receiver," *Proc. 21st Int. Workshop Signal Processing Advances in Wireless Communications (SPAWC)*, pp. 1–5, May 2020.

[43] Z. Qin et al., "Deep learning in physical layer communications," *IEEE Wireless Communications*, vol. 26, no. 2, pp. 93–99, April 2019.

[44] X. Gao et al., "ComNet: Combination of deep learning and expert knowledge in OFDM receivers," *IEEE Communication Letters*, vol. 22, no. 12, pp. 2627–2630, Dec. 2018.

[45] A. Graves and J. Schmidhuber, "Framewise phoneme classification with bidirectional LSTM networks," in *Proc. IEEE Int. Joint Conf. Neural Networking*, vol. 4, pp. 2047–2052, 2005.

[46] T. Jiang and Y. Wu, "An overview: Peak-to-average power ratio reduction techniques for OFDM signals," *IEEE Trans. on Broadcasting*, vol. 54, no. 2, pp. 257–268, June 2008.

[47] C. Studer and G. Durisi, "Quantized massive MU-MIMO-OFDM uplink," *IEEE Trans. on Communications*, vol. 64, no. 6, pp. 2387–2399, June 2016.

[48] H. Wang, C. Wen, and S. Jin, "Bayesian optimal data detector for mmwave OFDM system with low-resolution ADC," *IEEE Journal on Selected Areas in Communications*, vol. 35, no. 9, pp. 1962–1979, Sep. 2017.

[49] M. Kim, W. Lee, and D. Cho, "A novel papr reduction scheme for OFDM system based on deep learning," *IEEE Communication Letters*, vol. 22, no. 3, pp. 510–513, Mar. 2018.

[50] E. Balevi and J. G. Andrews, "One-bit OFDM receivers via deep learning," *IEEE Trans. on Communications*, vol. 67, no. 6, pp. 4326–4336, June 2019.

[51] P. Vincent et al., "Stacked denoising autoencoders: Learning useful representations in a deep network with a local denoising criterion," *Journal on Machine Learning Research*, vol. 11, no. 12, pp. 3371–3408, 2010.

# 6 Model-Based Machine Learning for Communications

Nir Shlezinger, Nariman Farsad, Yonina C. Eldar, and Andrea Goldsmith

## 6.1 Introduction

Traditional communication systems design is dominated by methods that are based on statistical models. These statistical-model-based algorithms, which we refer to henceforth as *model-based methods*, rely on mathematical models that describe the transmission process, signal propagation, receiver noise, interference, and many other components of the system that affect the end-to-end signal transmission and reception. Such mathematical models use parameters that vary over time as the channel conditions, the environment, network traffic, or network topology change. Therefore, for optimal operation, many of the algorithms used in communication systems rely on the underlying mathematical models as well as the estimation of the model parameters. However, there are cases where this approach fails, in particular when the mathematical models for one or more of the system components are highly complex, hard to estimate, poorly understood, do not well-capture the underlying physics of the system, or do not lend themselves to computationally-efficient algorithms. In some other cases, although mathematical models are known, accurate parameter estimation may not be possible. Finally, common hardware limitations, such as the restriction to utilize low-resolution quantizers or nonlinear power amplifiers, can significantly increase the complexity of the underlying channel model.

An alternative data-driven approach is based on machine learning (ML). ML techniques, and in particular, deep learning, have been the focus of extensive research in recent years due to their empirical success in various applications, including computer vision and speech processing [1, 2]. The benefits of ML-driven methods over traditional model-based approaches are threefold: First, ML methods are independent of the underlying stochastic model and thus can operate efficiently in scenarios where this model is unknown or its parameters cannot be accurately estimated. Second, when the underlying model is extremely complex, ML algorithms have demonstrated the ability to extract and disentangle the meaningful semantic information from the observed data [3], a task which is difficult to carry out using traditional model-based approaches, even when the model is perfectly known. Finally, the main complexity in utilizing ML methods is in the training stage, which is typically carried out offline. Once trained, they tend to implement inference at a lower computational burden and delay compared to their analytical model-based counterparts [4].

Although ML has been the focus of significant research attention over the last decade, it has yet to significantly contribute to practical designs in one of the most important technologies of the modern era: digital communication. The fact that ML-based algorithms, which have revolutionized the fields of computer vision and natural language processing, do not yet play a fundamental role in the design of physical layer communication systems, and particularly digital receivers, may be due to one or more of the following reasons:

1. The large amount of possible outputs impose a major challenge in efficiently applying ML algorithms. In particular, the constellation size of the modulation and the blocklength of the channel code, combined with the time-varying nature of communication channels, leads to an exponentially large number of possible channel outputs that an ML receiver algorithm must be trained on.
2. Traditional deep learning techniques require high computational resources, while communication devices, such as wearable devices and mobile phones, are typically limited in hardware and power.
3. To date, conventional communication schemes, which assume a simplified channel model with parameters that are dynamically estimated, have been very successful.

The third reason is likely to become less relevant as the spectrum congestion of existing cellular standards forces future communication systems to explore new frequency ranges and share spectrum with other application such as radar [5]. As these new frequency bands and spectrum sharing techniques become widespread, the simplified channel, interference, and noise models used in current communication receiver techniques may no longer work well. Moreover, the strict cost, power, and memory constraints imposed on communicating devices lead to the usage of low-resolution analog-to-digital convertors (ADCs) and power amplifiers with dominant nonlinearities [6]. This makes the successful application of model-based techniques significantly more complex. Thus, conventional model-based approaches may no longer be able to meet the performance and throughput demands of future wireless devices, motivating their combination with data-driven approaches based on ML such as deep learning. Such techniques must still overcome the challenges we have identified with respect to the large computational resources and datasets needed for training.

Despite its unprecedented success, deep learning is subject to several challenges that limit its applicability in some important communication scenarios. In particular, deep neural networks (DNNs) consist of highly parameterized systems that can represent a broad range of mappings. As such, massive data sets are typically required to learn a desirable mapping, and the computational burden of training and utilizing these networks may constitute a major drawback. For example, consider the two receivers illustrated in Fig. 6.1, which carry out symbol detection using model-based algorithms and model-agnostic DNNs, respectively. The dynamic nature of wireless channels implies that the receivers should track channel variations in order to reliably detect the transmitted messages over long periods of time. To do so, the model-based receiver in Fig. 6.1(a) typically estimates the model parameters imposed on the underlying

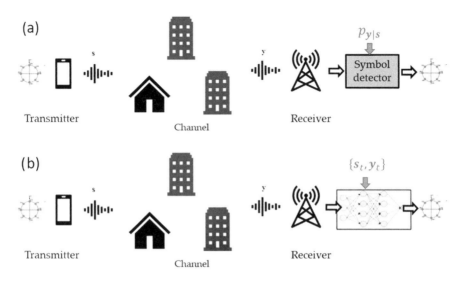

**Figure 6.1** Model-based methods versus deep learning for symbol detection: (a) a receiver uses its knowledge of the underlying statistical channel model, denoted $p_{Y|S}$, to detect the transmitted symbols for the channel output in a model-based manner, and (b) a receiver uses a DNN trained using the data set $\{s_t, y_t\}$ for recovering the symbols.

statistics using periodic pilots. For the same purpose, the DNN-based receiver in Fig. 6.1(b) should periodically retrain its DNN to track channel variations. The fact that doing so requires a large dataset leads to a significant decrease of spectral efficiency and increase in computational complexity associated with this training. Furthermore, DNNs are commonly utilized as black boxes and thus do not offer the interpretability, flexibility, versatility, and reliability of model-based techniques.

The limitations associated with model-based methods and black-box deep learning systems gave rise to a set of techniques on the interface of traditional model-based communication and ML, attempting to benefit from the best of both worlds [7]. Such *model-based ML* systems can be divided into two main categories. The first of the two utilizes model-based methods as a form of domain knowledge in designing a DNN architecture, which is then trained and used for inference. The most common example of this strategy is the family of *deep unfolded networks* [8, 9], which design the layers of a DNN to imitate the iterations of a model-based optimization algorithm, and has been utilized in various communication-related tasks [10]. The second strategy, which we call *DNN-aided hybrid algorithms*, uses model-based methods with integrated DNNs for inference by incorporating ML in a manner that makes the system more robust and model-agnostic. Communication receivers designed using DNN-aided inference include the data-driven implementations of the Viterbi algorithm [11] and the Bahle-Cocke-Jelinek-Raviv (BCJR) detector [12].

In this chapter we present an introduction to model-based ML for communication systems. We begin by reviewing existing strategies for combining model-based algorithms and ML from a high-level perspective in Section 6.2, and compare them to the conventional deep learning approach that utilizes established DNN architectures trained in an end-to-end manner. Then, in Section 6.3 we focus on symbol detection, which is one of the fundamental tasks of communication receivers. We show how each strategy (i.e., conventional DNN architectures, deep unfolding, and DNN-aided hybrid algorithms), can be applied to this problem. The last two approaches constitute a middle ground between the purely model-based and the DNN-based receivers illustrated in Fig. 6.1. By focusing on this specific task, we highlight the advantages and drawbacks of each strategy, and present guidelines to facilitate the design of future model-based deep learning systems for communications. We conclude this chapter with a summary provided in Section 6.4.

## 6.2      Model-Based Machine Learning

We begin by reviewing the leading approaches for combining ML, and particularly deep learning, with model-based algorithms. The neural networks in this hybrid approach are trained in a supervised manner and then used during inference. Then, in Section 6.3 we focus on symbol detection and provide concrete examples on how this approach can be used to design data-driven detectors.

In a broad family of problems, a system is required to map an input variable $Y$ into a prediction of a label variable $S$. ML systems learn such a mapping from a training set consisting of $n_t$ pairs of inputs and their corresponding labels, denoted $\{s_t, y_t\}_{t=1}^{n_t}$. Model-based methods carry out such inference based on prior knowledge of the statistical model relating $Y$ and $S$, denoted $p_{Y|S}$. Model-based ML systems reviewed in this chapter combine model-based methods with learning techniques; namely, they tune their mapping of the input $Y$ based on both a labeled training set $\{s_t, y_t\}_{t=1}^{n_t}$ as well as some knowledge of the underlying distribution. Such hybrid data-driven model-aware systems can typically learn their mappings from smaller training sets compared to purely model-agnostic DNNs and commonly operate without full and accurate knowledge of $p_{Y|S}$, upon which model-based methods are based. We next elaborate on the main strategies of combining ML and model-based techniques, beginning with extreme cases of DNNs that rely solely on data and purely model-based inference algorithms.

## 6.2.1      Conventional Deep Learning

The conventional application of deep learning is to carry out inference using some standard DNN architecture. This DNN uses the training data to learn how to map a realization of the input $Y = y$ into a prediction $\hat{s}$. Such highly parameterized networks can effectively approximate any Borel measurable mapping, as it follows from the universal approximation theorem [13, Ch. 6.4.1]. Therefore, by properly tuning their parameters using a *sufficiently large* training set, typically using optimization based on some variant of stochastic gradient descent (SGD), one should be able to obtain the desirable inference rule.

While standard DNNs structures are highly model-agnostic and are commonly treated as black boxes, one can still incorporate some level of domain knowledge in the selection of the specific network architecture. For instance, when the input is known to exhibit temporal correlation, architectures based on recurrent neural networks (RNNs) or transformers are known to be preferable. Alternatively, in the presence of spatial patterns, one may prefer to utilize convolutional layers. An additional method to incorporate domain knowledge into a black-box DNN is by preprocessing of the input via, for example, feature extraction.

Conventional deep learning based on established black-box DNNs is data-driven; that is, it requires data representing the problem at hand, possibly combined with a basic level of domain knowledge to select the specific architecture. A major drawback of using such networks, which is particularly relevant in the context of communication systems, is that learning a large number of parameters requires a massive dataset to train. In dynamic environments, even when a sufficiently large dataset is available, it is difficult to train a model that performs optimally over the whole range of the dynamically changing system. Moreover, online training as the system dynamics change tends to be computationally expensive because of the large number of parameters.

## 6.2.2    Model-Based Methods

Model-based algorithms carry out inference based on prior knowledge of the underlying statistics relating the input $Y$ and the label $S$, i.e., $p_{Y|S}$. A common family of model-based methods is based on iterative algorithms, which allow us to infer with provable performance and controllable complexity in an iterative fashion, as illustrated in Fig. 6.2(a). This generic family of iterative algorithms consists of some input and output processing stages, with an intermediate iterative procedure. The latter can in turn be divided into a model-based computation, namely, a procedure that is determined by $p_{Y|S}$, and a set of generic mathematical manipulations.

These algorithms vary significantly between different statistical models. For instance, for the symbol detection task, model-based methods such as the Viterbi detector [14] or the BCJR algorithm [15] are valid for finite-memory channels, while for multiple-input multiple-output (MIMO) detection one may utilize the family of interference cancellation methods [16]. Each such algorithm may be specifically tailored to a given scenario, as opposed to black-box DNNs, in which the parameterized inference rule is generic, and the unique characteristics of the scenario at hand are encapsulated in the parameters learned during training.

Model-based techniques do not rely on data to learn their mapping, though data is often used to estimate unknown model parameters. In practice, accurate knowledge of the statistical model relating the observations and the desired information is typically unavailable, and thus applying such techniques commonly requires imposing some assumptions on the underlying statistics, which in some cases reflects the actual behavior, but often do not. In the presence of inaccurate knowledge of $p_{Y|S}$ due to estimation errors or due to enforcing a model that does not fully capture the environment, the performance of model-based methods tends to degrade considerably. This limits the

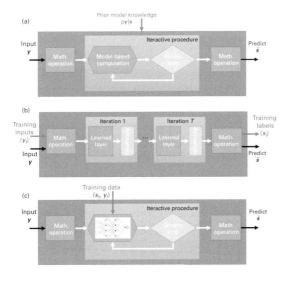

**Figure 6.2** (a) Model-based iterative algorithm, (b) unfolding the algorithm into a DNN, and (c) converting the scheme into a DNN-aided method.

applicability of model-based algorithms in scenarios where $p_{Y|S}$ is unknown, costly to estimate accurately, or too complex to express analytically.

### 6.2.3    Model-Based Deep Learning by Deep Unfolding

Deep unfolding [8, 9], also referred to as *deep unrolling*, is a common strategy to combine deep learning with model-based algorithms. Here, model-based methods are utilized as a form of domain knowledge in designing a DNN architecture, which is trained end-to-end and then used for inference. Unlike the application of conventional black-box DNNs discussed in Section 6.2.1, deep unfolding utilizes a unique DNN structure designed specifically for the task at hand.

The main rationale in deep unfolding is to design the network to imitate the operation of a model-based iterative optimization algorithm corresponding to the considered problem. In particular, each iteration of the model-based algorithm is replaced with a dedicated layer with trainable parameters in which the structure is based on the operations carried out during that iteration. An illustration of a neural network obtained by unfolding the model-based iterative model of Fig. 6.2(a) is depicted in Fig. 6.2(b), where a network with $T$ layers is designed to imitate $T$ iterations of the optimization method. Once the architecture is fixed, the resulting network is trained in an end-to-end manner as in conventional deep learning.

Deep unfolded networks, in which the iterations consist of trainable parameters, are typically capable of inferring with a smaller number of layers compared to the amount of iterations required by the model-based algorithm. Consequently, even when the model-based algorithm is feasible, processing $Y$ through a trained unfolded DNN is typically faster than applying the iterative algorithm [4]. Furthermore, converting a model-based algorithm into an unfolded deep network can also improve its performance. For example, iterative algorithms based on some relaxed optimization commonly achieve improved accuracy when unfolded into a DNN, due to the ability to learn to overcome the error induced by relaxation in the training stage. The main benefits of deep unfolding over using end-to-end networks stem from its incorporation of domain knowledge in the network architecture. As such, unfolded networks can achieve improved performance at reduced complexity (i.e., when operating with fewer parameters) compared to conventional end-to-end networks [10]. Nonetheless, deep unfolded networks are highly parameterized DNNs, which often require large datasets for training, though usually not as much as the generic DNNs. Furthermore, deep unfolding typically requires a high level of domain knowledge, such as explicit knowledge of the statistical model $p_{Y|S}$ up to possibly some missing parameters, in order to formulate the optimization algorithm in a manner that can be unfolded.

## 6.2.4   Model-Based Deep Learning by DNN-Aided Algorithms

The second strategy for combining model-based methods and deep learning, which we refer to as *DNN-aided hybrid algorithms*, aims at integrating ML into model-based techniques. Such DNN-aided systems mainly utilize conventional model-based methods for inference, while incorporating DNNs to make the resultant system more robust and model-agnostic. This approach builds upon the insight that model-based algorithms typically consist of a set of generic manipulations that are determined by the *structure of the statistics* (e.g., whether it obeys a Markovian structure). Beside these generic manipulations, there are also computations that require actual knowledge of $p_{Y|S}$, as illustrated in Fig. 6.2(a). Consequently, when one has prior knowledge on the structure of the underlying distribution but not of its actual distribution, ML can be utilized to fill in the missing components required to carry out the algorithm.

In particular, DNN-aided hybrid algorithms start with a model-based algorithm that is suitable for inference when the statistics of $p_{Y|S}$ are available. For instance, symbol detection over finite-memory channels can be carried out accurately and with affordable complexity using either the Viterbi algorithm [14] or the BCJR method [15], assuming $p_{Y|S}$ is known. Then, ML-based techniques, such as dedicated DNNs, are used to estimate only $p_{Y|S}$ from data. These dedicated DNNs can be trained individually, separately from the inference task, or in an end-to-end manner along with the overall algorithm that maps $Y$ into an estimate of $S$. An illustration of a DNN-aided algorithm obtained by integrating ML into the iterative methods illustrated in Fig. 6.2(a) is depicted in Fig. 6.2(c).

DNN-aided hybrid algorithms have several advantages: First, they use DNNs for specific intermediate tasks, such as computing a conditional probability measure, which are much simpler compared to end-to-end inference. Consequently, relatively simple networks that are trainable using small training sets can be used. Furthermore, once trained the system effectively implements the model-based algorithm in a data-driven manner without imposing a model on the underlying distribution and estimating its parameters. Concrete examples of DNN-aided symbol detection algorithms are detailed in Section 6.3.4.

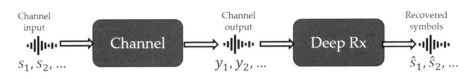

**Figure 6.3** Symbol detection.

## 6.3    Deep Symbol Detection

In digital communication systems, the receiver is required to reliably recover the transmitted symbols from the observed channel output. This task is commonly referred to as *symbol detection*. In this section, we present how the strategies for combining ML and model-based algorithms detailed in the previous section can be applied for data-driven symbol detection. We first formulate the symbol detection in Section 6.3.1, after which we discuss the applications of data-driven receivers based on conventional DNN architectures, deep unfolding, and DNN-aided algorithms, in Sections 6.3.2 through 6.3.4, respectively. For each strategy we begin with the main rationale behind this approach, present at least one concrete example, and discuss its pros and cons. Finally, we numerically compare the data-driven receivers to their model-based counterparts in Section 6.3.5.

### 6.3.1    Symbol Detection Problem

To formulate the symbol detection problem, we let $S_i \in \mathcal{S}^K$ be the symbol transmitted at time index $i \in \{1, 2, \ldots, T\} := \mathcal{T}$. Here, $T$ represents the blocklength and $K$ denotes the number of symbols transmitted at each time instance (e.g., the number of users transmitting simultaneously in the uplink communications channel). Each symbol is uniformly distributed over a set of $M$ constellation points, thus $|\mathcal{S}| = M$. We use $Y_i \in \mathcal{Y}^N$ to denote the channel output at time index $i$, where $N$ represents the number of receive antennas. When both $N$ and $K$ are larger than one, the resulting setup corresponds to MIMO communications. Symbol detection refers to the recovery of $S^T := \{S_i\}_{i \in \mathcal{T}}$ from the observed $Y^T := \{Y_i\}_{i \in \mathcal{T}}$. An illustration of the symbol detection problem using a DNN-aided receiver is depicted in Fig. 6.3.

We focus on finite-memory stationary causal channels, where each $Y_i$ is given by a stochastic mapping of $\{S_l\}_{l=i-L+1}^{i}$, and $L$ is the memory of the channel, assumed to be smaller than the blocklength $T$. The special case in which $L = 1$ is referred to as *flat* or memoryless channel conditions. The conditional probability density function (PDF) of the channel output given its input thus satisfies

$$p_{Y^T|S^T}\left(y^T|s^T\right) = \prod_{i=1}^{T} p_{Y_i|\{S_l\}_{l=i-L+1}^{i}}\left(y_i|\{s_l\}_{l=i-L+1}^{i}\right), \tag{6.1}$$

where the lower-case $y_i$ and $s_i$ represent the realizations of the random variables $Y_i$ and $S_i$, respectively. The fact that the channel is stationary implies that the conditional PDF $p_{Y_i|\{S_l\}_{l=i-L+1}^{i}}$ does not depend on the index $i$.

The symbol detection mapping that minimizes the error rate is the maximum a posteriori probability (MAP) rule, given by

$$\hat{s}_i = \arg\max_{s \in \mathcal{S}^K} p_{S_i|Y^T}(s|y^T), \qquad i \in \mathcal{T}. \tag{6.2}$$

For the memoryless case $L = 1$, solving Eq. (6.2) reduces to maximizing $p_{S|Y}(s|y_i)$ over $s \in \mathcal{S}^K$. However, when $K$ is large, as is commonly the case in uplink MIMO systems, solving Eq. (6.2) may be computationally infeasible, even when the PDF $p_{S|Y}$ is perfectly known. The application of deep learning for symbol detection thus has two main motivations: the first is to allow symbol detection to operate in a model-agnostic manner (i.e., without requiring knowledge of $p_{S|Y}$), and the second is to facilitate inference when the computational complexity of Eq. (6.2) renders solving it infeasible.

## 6.3.2    Symbol Detection via Established Deep Networks

The first approach to designing data-driven symbol detectors treats the channel as a black box and relies on well-known deep learning architectures used in computer vision, speech, and language processing for detection. We now describe how conventional deep neural architectures can be used for symbol detection.

### Overview of Design Process

Different DNN architectures have shown promising results for detection and estimation in applications such as image processing [17–19], speech recognition [20–22], machine translation [23–25], and bioinformatics [26, 27]. Some of these neural network architectures can be used to design a symbol detector for channels with unknown models using supervised learning. This process typically consists of the following steps:

1. Identify the conventional neural network architectures that are suitable for the channel under consideration, and use these networks as building blocks for designing the detection algorithm. For example, RNNs are more suitable for sequential detection in channels with memory, while convolutional and fully connected networks are more suitable for memoryless channels.
2. Next, use channel input-output pairs to train the network. Two approaches can be used for training: In the first approach, a model is trained for each channel condition (e.g., each SNR). In the second approach, a large training dataset consisting of various channel conditions is used to train a single neural network detector for detection over a wide range of channel conditions. The training data can be generated by randomizing the transmitted symbols and generating the corresponding received signal using mathematical models, simulations, experiments, or field measurements.
3. Train the overall resulting network in an end-to-end fashion.

We next demonstrate how this rationale is translated into a concrete data-driven symbol detector architecture for finite-memory channels.

**Example: SBRNN for Finite-Memory Channels**

The sliding bidirectional RNN (SBRNN) is a sequence detection algorithm for finite-memory channels proposed in [28]. Generally, sequence detection can be performed using RNNs [1], which are well established for sequence estimation in different problems such as neural machine translation [23], speech recognition [2], or bioinformatics [26]. For simplicity, we assume in our description that the input cardinality $K$ is $K = 1$. The estimated symbol in this case is given by

$$\hat{s}_i = \arg\max_{s_i \in \mathcal{S}} P_{\text{RNN}}(s_i | \mathbf{y}^i), \qquad i \in \mathcal{T}, \tag{6.3}$$

where $P_{\text{RNN}}$ is the probability of estimating each symbol based on the DNN model used. One of the main benefits of this detector is that after training, it can perform detection on any data stream as it arrives at the receiver. This is because the observations from previous symbols are summarized as the state of the RNN, which is represented by a vector. Note that the observed signal during the $j$th transmission, $\mathbf{Y}_j$ where $j > k$, may carry information about the $k$th symbol $\mathbf{S}_k$ due to the memory of the channel. However, since RNNs are feed-forward only, during the estimation of $\hat{\mathbf{S}}_k$, the observation signal $\mathbf{Y}_j$ is not considered.

One way to overcome this limitation is by using bidirectional RNNs (BRNNs). In such networks, a sequence of $B$ received signals are fed once in the forward direction into one RNN cell and fed once in the backward direction into another RNN cell [29], for some fixed $B$ representing the BRNN length. The two outputs are then concatenated and may be passed to more bidirectional layers. A signal whose blocklength $T$ is larger than the BRNN length $B$ is divided into multiple distinct subsequences of length $B$. Ideally, the $B$ must be at least the same size as the memory length $L$. However, if this is not known in advance, the BRNN length can be treated as a hyperparameter to be tuned during training. At time instance $i$ belonging to the $k$th subsequence, the estimated symbol for BRNN is given by

$$\hat{s}_i = \arg\max_{s_i \in \mathcal{S}} P_{\text{BRNN}}(s_i | \mathbf{y}_k^{k+B-1}), \qquad k - B + 1 \leq i \leq k. \tag{6.4}$$

To simplify the notation, we use $\hat{\mathbf{p}}_i^{(k)}$ to denote the $M \times 1$ matrix whose entries are the probability mass function (PMF) $P_{\text{BRNN}}(s_i | \mathbf{y}_k^{k+B-1})$ for each $s_i \in \mathcal{S}$.

The BRNN architecture ensures that in the estimation of a symbol, future signal observations are taken into account. During training, blocks of $B$ consecutive transmissions are used for training. Once the network is trained, BRNNs detect the stream of incoming data in fixed blocks of length $B$, as shown in the top portion of Fig. 6.4. The main drawback here is that the symbols at the end of each block may affect the symbols in the next block, and since each block is treated independently, this relation is not captured in this scheme. Another issue is that the block of $B$ symbols must be received before detection can be performed. The top portion of Fig. 6.4 shows this scheme for $B = 3$.

To overcome these limitations, in the SBRNN proposed in [28], the first $B$ symbols are detected using the BRNN. Then, as each new symbol arrives at the receiver,

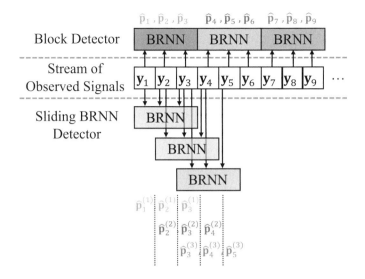

**Figure 6.4** The SBRNN detector versus using a BRNN as a block detector.

the subsequence processed by the BRNN slides ahead by one symbol. Let the set $\mathcal{K}_i = \{k \mid k \leq i \wedge k + B > i\}$ be the set of all valid starting positions for a BRNN detector of length $B$, such that the detector overlaps with the $i$th symbol. For example, if $B = 3$ and $i = 4$, then $k = 1$ is not in the set $\mathcal{K}_i$ since the BRNN detector overlaps with symbol positions 1, 2, and 3, and not the symbol position 4. The estimated PMF corresponding to the $i$th symbol is given by the weighted sum of the estimated PMFs for each of the relevant windows:

$$\hat{\boldsymbol{p}}_i = \sum_{k \in \mathcal{K}_i} \alpha_k \hat{\boldsymbol{p}}_i^{(k)}, \qquad \alpha_k \geq 0 \text{ and } \sum \alpha_k = 1. \tag{6.5}$$

The weighted sum coefficients can be set to $\alpha_k = \frac{1}{|\mathcal{K}_i|}$, as we do in the numerical evaluations in Section 6.3.5. An illustration of the operation of the SBRNN detector based on Eq. (6.5) is depicted in the bottom portion of Fig. 6.4.

## Summary

Well-known DNN architectures can be trained in an end-to-end manner to perform symbol detection. This approach builds upon the success of existing model-agnostic DNN structures, resulting in symbol detectors operating without any knowledge about the underlying channel models. Furthermore, this strategy allows combining some basic level of domain knowledge in the selection of the architecture as well as the preparation of its input. For example, the SBRNN detector detailed earlier identifies the BRNN architecture as one that is capable of handling temporal correlation in finite-memory channels, while using a sliding subsequence to overcome some of the limitations of BRNN architectures when applied to different blocks independently. Also, DNNs can well-capture the nonlinearities that may exist in the channel. For example, the SBRNN was used as an autoencoder in [30, 31] to achieve state-of-the-art performance over optical channels, outperforming model-based nonlinear equalizers such as Voltera. Finally, these networks can also be computationally more efficient than the optimal maximum-likelihood sequence detector over finite-memory channels, specifically for channels with long memory.

The main drawback in using established deep networks for end-to-end symbol detection is that such architectures typically have a very large number of parameters and thus require massive datasets for training. This renders online training using pilot sequences impractical. Moreover, when they are trained using data from a large set of channel conditions, the resulting network will not perform optimally for each of those channel conditions individually. Furthermore, even when the dataset is extremely large and diverse, it is not likely to capture all expected channel conditions. Finally, conventional DNN architectures are treated as black boxes, and are in general not interpretable, making it difficult to come up with performance guarantees.

## 6.3.3    Symbol Detection via Deep Unfolding

Unlike conventional DNNs, which utilize established architectures, in deep unfolding the network structure is designed following a model-based algorithm. We next describe how this model-based ML technique can be applied for symbol detection and detail a concrete example for flat Gaussian MIMO channels.

### Overview of Design Process

Deep unfolding is a method for converting an iterative algorithm into a DNN by designing each layer of the network to resemble a single iteration. As such, the rationale in applying deep unfolding consists of the following steps:

1. Identify an iterative optimization algorithm that is useful for the problem at hand. For instance, recovering the "MAP" symbol detector for flat MIMO channels can be tackled using various iterative optimization algorithms, such as projected gradient descent, unfolded into DetNet [32], as described in the sequel.
2. Fix a number of iterations in the optimization algorithm.
3. Design the layers to initiate the operation of each iteration in a trainable fashion, as illustrated in Fig. 6.2.
4. Train the overall resulting network in an end-to-end fashion.

We next demonstrate how this rationale is translated into a concrete data-driven symbol detector architecture for flat channels (i.e., Eq. (6.1) with $L = 1$).

### Example: DetNet for Flat MIMO Channels

DetNet is a deep learning based symbol detector proposed in [32] for flat Gaussian MIMO channels. To formulate DetNet, we first detail the specific channel model for which it is designed and then show how it is obtained by unfolding the projected gradient descent method for recovering the maximum a-posteriori probability (MAP) estimate.

*Flat Gaussian MIMO Channel*

As we focus on stationary memoryless channels, we drop the subscript $i$ representing the time instance and write the input-output relationship of a flat Gaussian MIMO channel as

$$Y = HS + W, \tag{6.6}$$

where $H$ is a known deterministic $N \times K$ channel matrix, and $W$ consists of $N$ independent and identically distributed (i.i.d.) Gaussian random variables (RVs). Consider the case in which the symbols are generated from a binary phase shift keying (BPSK) constellation in a uniform i.i.d. manner (i.e., $\mathcal{S} = \{\pm 1\}$). In this case the MAP rule in Eq. (6.2) given an observation $Y = y$ becomes the minimum distance estimate, given by

$$\hat{s} = \arg\min_{s \in \{\pm 1\}^K} \| y - Hs \|^2. \tag{6.7}$$

*Project Gradient Descent Optimization*

While directly solving Eq. (6.7) involves an exhaustive search over the $2^K$ possible symbol combinations, it can be tackled with affordable computational complexity using the iterative projected gradient descent algorithm. Let $\mathcal{P}_{\mathcal{S}}(\cdot)$ denote the projection into the $\mathcal{S}$ operator, which for BPSK constellations is the sign function. The projected gradient descent iteratively refines its estimate, which at iteration index $q+1$ is obtained recursively as

$$\hat{s}_{q+1} = \mathcal{P}_{\mathcal{S}} \left( \hat{s}_q - \eta_q \left. \frac{\partial \| y - Hs \|^2}{\partial s} \right|_{s=\hat{s}_q} \right)$$
$$= \mathcal{P}_{\mathcal{S}} \left( \hat{s}_q - \eta_q H^T y + \eta_q H^T H \hat{s}_q \right), \tag{6.8}$$

where $\eta_q$ denotes the step size at iteration $q$, and $\hat{s}_0$ is set to some initial guess.

*Unfolded DetNet*

DetNet unfolds the projected gradient descent iterations in Eq. (6.8) into a DNN, which learns to carry out this optimization procedure from data. To formulate DetNet, we first fix a number of iterations $Q$. Next, we design a DNN with $Q$ layers, where each layer imitates a single iteration of Eq. (6.8) in a trainable manner.

   In particular, DetNet builds upon the observation that each projected gradient descent iteration consists of two stages: gradient descent computation (i.e., $\hat{s}_q - \eta_q H^T y + \eta_q H^T H \hat{s}_q$) and projection, namely, applying $\mathcal{P}_{\mathcal{S}}(\cdot)$. Therefore, each unfolded iteration is represented as two sublayers: The first sublayer learns to compute the gradient descent stage by treating the step-size as a learned parameter and applying a conventional fully connected layer with ReLU activation to the obtained value. For iteration index $q$, this results in

$$z_q = \text{ReLU} \left( W_{1,q} \left( \hat{s}_{q-1} - \delta_{1,q} H^T y + \delta_{2,q} H^T H \hat{s}_{q-1} \right) + b_{1,q} \right), \tag{6.9}$$

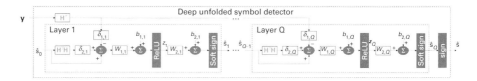

**Figure 6.5** Deep unfolded symbol detector. Parameters in black fonts are learned in training, while those in gray fonts are externally provided.

in which $\{W_{1,q}, \boldsymbol{b}_{1,q}, \delta_{1,q}, \delta_{2,q}\}$ are learnable parameters. The second sublayer learns the projection operator by approximating the sign operation with a soft sign activation proceeded by a fully connected layer, leading to

$$\hat{s}_q = \text{soft sign} \left( W_{2,q} z_q + \boldsymbol{b}_{2,q} \right). \tag{6.10}$$

Here, the learnable parameters are $\{W_{2,q}, \boldsymbol{b}_{2,q}\}$. The resulting deep network is depicted in Fig. 6.5, in which $\hat{s}_0$ is set to some initial guess, and the output after $Q$ iterations, denoted $\hat{s}_Q$, is used as the estimated symbol vector by taking the sign of each element.

Let $\boldsymbol{\theta} = \{W_{1,q}, W_{2,q}, \boldsymbol{b}_{1,q}, \boldsymbol{b}_{2,q}, \delta_{1,q}, \delta_{2,q}\}_{q=1}^{Q}$ be the trainable parameters of Det-Net.[1] To tune $\boldsymbol{\theta}$, the overall network is trained in an end-to-end manner to minimize the empirical weighted $\ell_2$ norm loss over its intermediate layers. In particular, by letting $\{\boldsymbol{s}_t, \boldsymbol{y}_t\}_{t=1}^{n_t}$ denote the training set consisting of channel outputs and their corresponding transmitted symbols, the loss function used for training DetNet is given by

$$\mathcal{L}(\boldsymbol{\theta}) = \frac{1}{n_t} \sum_{t=1}^{n_t} \sum_{q=1}^{Q} \log(q) \| \boldsymbol{s}_t - \hat{s}_q(\boldsymbol{y}_t; \boldsymbol{\theta}) \|^2, \tag{6.11}$$

where $\hat{s}_q(\boldsymbol{y}_t; \boldsymbol{\theta})$ is the output of the $q$th layer of DetNet with parameters $\boldsymbol{\theta}$ and input $\boldsymbol{y}_t$. This loss measure accounts for the interpretable nature of the unfolded network, in which the output of each layer is a further refined estimate of $S$.

## Summary
Deep unfolding incorporates model-based domain knowledge to obtain a dedicated DNN design, which follows an iterative optimization algorithm. Compared to the conventional DNNs discussed in the previous section, unfolded networks are typically interpretable, and tend to have a smaller number of parameters, and can thus be trained quicker [10]. Nonetheless, these deep networks are still highly parameterized and require a large volume of training data. For instance, DetNet is trained in [32] using approximately 250 million labeled samples.

One of the key properties of unfolded networks is their reliance on model knowledge. For example, the unfolded receiver must know that the channel input-output

---

[1] The formulation of DetNet in [32] includes an additional sublayer in each iteration intended to further lift its input into higher dimensions and introduce additional trainable parameters, as well as reweighing of the outputs of subsequent layers. As these operations do not follow directly from unfolding the projected gradient descent method, they are not included in the description here.

relationship takes the form in Eq. (6.6) in order to formulate the projected gradient iterations in Eq. (6.8), which in turn are unfolded into DetNet. The model-awareness of deep unfolding has its advantages and drawbacks. When the model is accurately known, deep unfolding essentially incorporates it into the DNN architecture, as opposed to conventional DNNs that must learn this from data. However, this approach does not exploit the model-agnostic nature of deep learning and thus may achieve degraded performance when the true channel conditions deviate from the model assumed in design (e.g., Eq. (6.6)). In particular, a key advantage of deep unfolding over the model-based optimization algorithm is in inference speed. For instance, DetNet requires fewer layers to reliably detect compared to the number of iterations required for projected gradient descent to converge.

Another important advantage of unfolded networks is their ability to improve the accuracy compared to the iterative optimization algorithm from which they originate. In particular, the set $\{\pm 1\}^K$ over which the optimization problem in Eq. (6.7) is formulated is not convex, and thus projected gradient descent is not guaranteed to recover its solution, regardless of the number of iterations. By unfolding the application of projected gradient descent for solving Eq. (6.7) into a DNN with trainable parameters, the resulting network is often able to overcome this difficulty and converge to the true solution of the optimization problem when properly trained, despite the nonconvexity. Finally, in the context of receiver design, most unfolded networks to date, including DetNet as well as OAMP-net [33] that unfolds the orthogonal approximate message passing optimization algorithm, require channel state information (CSI). For example, knowledge of the matrix H is utilized in the architecture depicted in Fig. 6.5. This implies that additional mechanisms for estimating the channel must be incorporated into the receiver architecture [34]. Nonetheless, while the aforementioned deep unfolding based receivers require CSI, unfolded networks can be designed without such knowledge [9]. For example, one can unfold the optimization algorithm assuming CSI is available and then treat H which appears in the unfolded network as part of its trainable parameters.

## 6.3.4    Symbol Detection via DNN-Aided Algorithms

DNN-aided hybrid algorithms combine domain knowledge in the form of a model-based inference algorithm for the problem at hand. This strategy allows the design of model-based ML systems with varying levels of domain knowledge, in which deep learning is used to robustify and remove model-dependence of specific components of the algorithm. In the following we first review the rationale when designing DNN-aided symbol detectors, after which we detail three concrete examples arising from different symbol detection algorithms.

**Overview of Design Process**

DNN-aided algorithms aim to carry out model-based inference methods in a data-driven fashion. These hybrid systems thus utilize deep learning not for the overall inference task, but for robustifying and relaxing the model-dependence of established model-based inference algorithms. Consequently, the design of DNN-aided hybrid systems consists of the following steps:

1. First, a proper inference algorithm is chosen. In particular, the domain knowledge is encapsulated in the selection of the algorithm that is learned from data. For example, the Viterbi algorithm is a natural candidate for symbol detection over finite-memory channels when seeking a symbol detector capable of operating in real time, or alternatively, the BCJR scheme is the suitable choice for carrying out MAP inference over such channels. When designing receivers for flat MIMO channels, interference cancellation methods may be the preferable algorithmic approach for symbol detection. We show how these methods are converted into DNN-aided algorithms in the sequel.
2. Once a model-based algorithm is selected, we identify its model-specific computations and replace them with dedicated compact DNNs.
3. The resulting DNNs can be either trained individually, or the overall system can be trained in an end-to-end manner.

Since the implementation of DNN-aided algorithms highly varies with the selection of the learned model-based method, we next present three concrete examples in the context of symbol detection: *ViterbiNet*, which learns to carry out Viterbi detection [14]; *BCJRNet*, which implements the BCJR algorithm of [15] in a data-driven fashion; and *DeepSIC*, which is based on the soft iterative interference cancellation methods for MIMO symbol detection [35].

## Example: ViterbiNet for Finite-Memory Channels

ViterbiNet proposed in [11] is a data-driven implementation of the Viterbi detection algorithm [14], which is one of the most common workhorses in digital communications. This DNN-aided symbol detection algorithm is suitable for finite-memory channels of the form in Eq. (6.1), without requiring prior knowledge of the channel conditional distributions $p_{Y_i|\{S_l\}_{l=i-L+1}^i}$. For simplicity, we assume in our description that the input cardinality $K$ is $K = 1$. As a preliminary step to presenting ViterbiNet, we now briefly review conventional model-based Viterbi detection.

### Viterbi Algorithm

The Viterbi algorithm recovers the maximum likelihood sequence detector:

$$\hat{s}^T\left(y^T\right) := \arg\max_{s^T \in \mathcal{S}^T} p_{Y^T|S^T}\left(y^T|s^T\right)$$

$$= \arg\min_{s^T \in \mathcal{S}^T} -\log p_{Y^T|S^T}\left(y^T|s^T\right). \tag{6.12}$$

Using Eq. (6.1) the optimization problem Eq. (6.12) becomes

$$\hat{s}^T\left(y^T\right) = \arg\min_{s^T \in \mathcal{S}^T} \sum_{i=1}^T -\log p_{Y_i|\{S_l\}_{l=i-L+1}^i}\left(y_i|\{s_l\}_{l=i-L+1}^i\right). \tag{6.13}$$

---

**Algorithm 6.1** Viterbi Algorithm [14].

---

**Init:** Fix an initial path $p_0(s) = \varnothing$ and path cost $c_0(\bar{s}) = 0, \bar{s} \in \mathcal{S}^L$.

1 **for** $i = 1, 2, \ldots, T$ **do**

2      For each state $\bar{s} \in \mathcal{S}^L$, compute previous state with shortest path, denoted $u_{\bar{s}}$, via

$$u_{\bar{s}} = \underset{\bar{s}' \in \mathcal{S}^L : p_{\bar{S}_i | \bar{S}_{i-1}}(\bar{s} | \bar{s}') > 0}{\arg\min} \left( c_{i-1}(\bar{s}') - \log p_{Y_i | \bar{S}_i}(y_i | \bar{s}) \right). \tag{6.15}$$

3      Update cost and path via

$$c_i(\bar{s}) = c_{i-1}(u_{\bar{s}}) - \log p_{Y_i | \bar{S}_i}(y_i | \bar{s}), \tag{6.16}$$

     and $p_i(\bar{s}) = [p_{i-1}(u_{\bar{s}}), u_{\bar{s}}]$;

4 **end**

**Output:** $\hat{s}^T = p_T(\bar{s}^*)$ where $\bar{s}^* = \arg\min_{\bar{s}} c_T(\bar{s})$.

---

To proceed, we define a state variable $\bar{S}_i := [S_{i-L+1}, \ldots, S_i] \in \mathcal{S}^L$. Since the symbols are i.i.d. and uniformly distributed, it follows that $p_{\bar{S}_i | \bar{S}_{i-1}}(\bar{s}_i | \bar{s}_{i-1}) = M^{-1}$ when $\bar{s}_i$ is a shifted version of $\bar{s}_{i-1}$; that is, the first $L - 1$ entries of $\bar{s}_i$ are the last $L - 1$ entries of $\bar{s}_{i-1}$, and zero otherwise. We can now write Eq. (6.13) as

$$\hat{s}^T(y^T) = \underset{s^T \in \mathcal{S}^T}{\arg\min} \sum_{i=1}^{T} - \log p_{Y_i | \bar{S}_i}\left( y_i | \{s_l\}_{l=i-L+1}^{i} \right). \tag{6.14}$$

The optimization problem in Eq. (6.14) can be solved recursively using dynamic programming, by iteratively updating a *path cost* $c_i(\bar{s})$ for each state $\bar{s} \in \mathcal{S}^L$. The resulting scheme, known as the Viterbi algorithm, is given as Algorithm 6.1, and illustrated in Fig. 6.6(a).

The Viterbi algorithm has two major advantages: First, it solves Eq. (6.12) at a computational complexity that is linear in the blocklength $T$. For comparison, the computational complexity of solving Eq. (6.12) directly grows exponentially with $T$. Second, the algorithm produces estimates sequentially during runtime. In particular, while in Eq. (6.12) the estimated output $\hat{s}^T$ is computed using the entire received block $y^T$, Algorithm 6.1 computes $\hat{s}_i$ once $y_{i+L-1}$ is received.

*ViterbiNet*

ViterbiNet proposed in [11] learns to implement the Viterbi algorithm from data in a model-agnostic manner. Following the rationale of DNN-aided algorithms, this is achieved by identifying the model-based components of the algorithm, which for Algorithm 6.1 boils down to the computation of the log-likelihood function $\log p_{Y_i | \bar{S}_i}(y_i | \bar{s})$. Once this quantity is computed for each $\bar{s} \in \mathcal{S}^L$, the Viterbi algorithm only requires knowledge of the memory length $L$. This requirement is much easier to satisfy compared to full CSI.

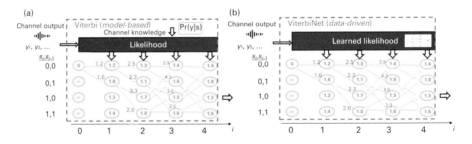

**Figure 6.6** Operation of the (a) Viterbi algorithm and (b) ViterbiNet. Here, the memory length is $L = 2$ and the constellation is $\mathcal{S} = \{0, 1\}$, implying that the number of states is $M^L = 4$. The values in inside the gray ovals are the negative log-likelihoods for each state, while the quantities adjacent to the gray ovals represent the updated path loss $c_i(\cdot)$. When multiple paths lead to the same state, as occurs for $i = 3, 4$, the path with minimal loss is maintained.

Since the channel is stationary, it holds that the log-likelihood function depends only on the realizations of $y_i$ and of $\bar{s}$, and not on the time index $i$. Therefore, to implement Algorithm 6.1 in a data-driven fashion, ViterbiNet replaces the explicit computation of the log-likelihoods with an ML-based system that learns to evaluate this function from training data. In this case, the input of the system is the channel output realization $y_i$ and the output is an estimate of $\log p_{Y_i|\bar{S}_i}(y_i|\bar{s})$ for each $\bar{s} \in \mathcal{S}^L$. The rest of the Viterbi algorithm remains intact, and the detector implements Algorithm 6.1 using the learned log-likelihoods. The proposed architecture is illustrated in Fig. 6.6(b).

Two candidate architectures are considered for learning to compute the log likelihood, one based on classification networks and one using density estimation networks.

**Learned likelihood using classification networks:** Since $y_i$ is given and may take continuous values while the desired variables take discrete values, a natural approach to evaluate $p_{Y_i|\bar{S}_i}(y_i|\bar{s})$ for each $\bar{s} \in \mathcal{S}^L$ is to estimate $p_{\bar{S}_i|Y_i}(\bar{s}|y_i)$ and then use Bayes rule to obtain

$$p_{Y_i|\bar{S}_i}(y_i|\bar{s}) = p_{\bar{S}_i|Y_i}(\bar{s}|y_i)\, p_{Y_i}(y_i)\, M^L. \tag{6.17}$$

A parametric estimate of $p_{\bar{S}_i|Y_i}(\bar{s}|y_i)$, denoted $\hat{P}_\theta(\bar{s}|y_i)$, is obtained for each $\bar{s} \in \mathcal{S}^L$ by training classification networks with softmax output layers to minimize the cross entropy loss. Here, for a labeled set $\{s_t, y_t\}_{t=1}^{n_t}$, the loss function is

$$\mathcal{L}(\theta) = \frac{1}{n_t} \sum_{t=1}^{n_t} -\log \hat{P}_\theta([s_{t-L+1}, \ldots, s_t]|y_t). \tag{6.18}$$

In general, the marginal PDF of $Y_i$ can be estimated from the training data using mixture density estimation via, e.g., expectation minimization (EM) [36, Ch. 2], or any other finite mixture model fitting method. However, obtaining an accurate density estimation becomes challenging when $Y_i$ is high-dimensional. Since $p_{Y_i}(y_i)$ does not depend on the variable $\bar{s}$, setting $p_{Y_i}(y_i) \equiv 1$ does not affect the decisions in

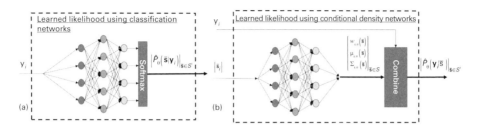

**Figure 6.7** Learned likelihood architectures based on (a) classification and (b) conditional density estimation networks.

Algorithm 6.1 due to the arg min arguments, which are invariant to scaling the conditional distribution by a term that does not depend on $\bar{s}$. The resulting structure is illustrated in Fig. 6.7(a).

**Learned likelihood using conditional density networks:** An additional strategy is to directly estimate the conditional $p_{Y_i|\bar{S}_i}\left(y_i|\bar{s}\right)$ from data. This can be achieved using conditional density estimation networks [37, 38] that are specifically designed to learn such PDFs, or alternatively, using normalizing flow networks to learn complex densities [39]. For example, mixture density networks [37] model the conditional PDF $p_{Y_i|\bar{S}_i}\left(y_i|\bar{s}\right)$ as a Gaussian mixture and train a DNN to estimate its mixing parameters, mean values, and covariances, denoted $w_{k,\theta}(\bar{s})$, $\mu_{k,\theta}(\bar{s})$ and $\Sigma_{k,\theta}(\bar{s})$, respectively, by maximizing the likelihood $\hat{P}_{\theta}\left(y_i|\bar{s}\right) = \sum_k w_{k,\theta}(\bar{s})\mathcal{N}\left(y_i|\mu_{k,\theta}(\bar{s}),\Sigma_{k,\theta}(\bar{s})\right)$, as illustrated in Fig. 6.7(b).

Both approaches can be utilized for learning to compute the likelihood in ViterbiNet. When the channel outputs are high-dimensional (i.e., $N$ is large) directly learning the conditional density is difficult and likely to be inaccurate. In such cases, the classification-based architecture, which avoids the need to explicitly learn the density, may be preferable. When the state cardinality $M^L$ is large, conditional density networks are expected to be more reliable. However, the Viterbi algorithm becomes computationally infeasible when $M^L$ grows, regardless of whether it is implemented in a model-based or data-driven fashion, making ViterbiNet unsuitable for such setups.

### Example: BCJRNet for Finite-Memory Channels

Factor graph methods, such as the sum-product algorithm, exploit the factorization of a joint distribution to efficiently compute a desired quantity [40]. In particular, the application of the sum-product algorithm for the joint input-output distribution of finite-memory channels allows for computing the MAP rule, an operation whose burden typically grows exponentially with the block size, with complexity that only grows linearly with $T$. This instance of the sum-product algorithm is exactly the BCJR detector proposed [15]. In the following we show how the BCJR method can be extended into the DNN-aided BCJRNet. As in our description of ViterbiNet, we again focus on the case of $K = 1$ and begin by presenting the model-based BCJR algorithm.

*BCJR Algorithm*

The BCJR algorithm computes the MAP rule in Eq. (6.2) for finite-memory channels with complexity that grows linearly with the block size. To formulate this method, we recall the definition of $\bar{S}_i$, and define the function

$$
\begin{aligned}
f\left(y_i, \bar{s}_i, \bar{s}_{i-1}\right) &:= p_{Y_i|\bar{S}_i}\left(y_i|\bar{s}_i\right) p_{\bar{S}_i|\bar{S}_{i-1}}\left(\bar{s}_i|\bar{s}_{i-1}\right) \\
&= \begin{cases} \frac{1}{M} p_{Y_i|\bar{S}_i}\left(y_i|\bar{s}_i\right) & (\bar{s}_i)_j = (\bar{s}_{i-1})_{j-1}, \quad \forall j \in \{2, \ldots, L\}, \\ 0 & \text{otherwise.} \end{cases}
\end{aligned}
\tag{6.19}
$$

Combining Eq. (6.19) and Eq. (6.1), we obtain a factorizable expression of the joint distribution $p_{Y^T, S^T}(\cdot)$, given by

$$
\begin{aligned}
p_{Y^T, S^T}\left(y^T, s^T\right) &= \prod_{i=1}^{T} \frac{1}{M} p_{Y_i|\bar{S}_i}\left(y_i|[s_{i-L+1}, \ldots, s_i]\right) \\
&= \prod_{i=1}^{T} f\left(y_i, [s_{i-L+1}, \ldots, s_i], [s_{i-L}, \ldots, s_{i-1}]\right).
\end{aligned}
\tag{6.20}
$$

The factorizable expression of the joint distribution in Eq. (6.20) implies that it can be represented as a factor graph with $T$ function nodes $\{f\left(y_i, \bar{s}_i, \bar{s}_{i-1}\right)\}$, in which $\{\bar{s}_i\}_{i=2}^{T-1}$ are edges while the remaining variables are half-edges.[2]

Using its factor graph representation, one can compute the joint distribution of $S^T$ and $Y^T$ by recursive message passing along this factor graph. In particular,

$$
p_{\bar{S}_k, \bar{S}_{k+1}, Y^T}(\bar{s}_k, \bar{s}_{k+1}, y^T) = \overrightarrow{\mu}_{\bar{S}_k}(\bar{s}_k) f(y_{k+1}, \bar{s}_{k+1}, \bar{s}_k) \overleftarrow{\mu}_{\bar{S}_{k+1}}(\bar{s}_{k+1}),
\tag{6.21}
$$

where the forward path messages satisfy

$$
\overrightarrow{\mu}_{\bar{S}_i}(\bar{s}_i) = \sum_{\bar{s}_{i-1}} f(y_i, \bar{s}_i, \bar{s}_{i-1}) \overrightarrow{\mu}_{\bar{S}_{i-1}}(\bar{s}_{i-1}),
\tag{6.22}
$$

for $i = 1, 2, \ldots, k$. Similarly, the backward messages are

$$
\overleftarrow{\mu}_{\bar{S}_i}(\bar{s}_i) = \sum_{\bar{s}_{i+1}} f(y_{i+1}, \bar{s}_{i+1}, \bar{s}_i) \overleftarrow{\mu}_{\bar{S}_{i+1}}(\bar{s}_{i+1}),
\tag{6.23}
$$

for $i = T - 1, T - 2, \ldots, k + 1$. This message passing is illustrated in Fig. 6.8.

The ability to compute the joint distribution in Eq. (6.21) via message passing results in the MAP detector in Eq. (6.2), an operation whose burden typically grows exponentially with the block size, with complexity that only grows linearly with $T$. This is achieved by noting that the MAP estimate satisfies

$$
\begin{aligned}
\hat{s}_i\left(y^T\right) = \arg\max_{s \in \mathcal{S}} \sum_{\bar{s}_{i-1} \in \mathcal{S}^L} &\overrightarrow{\mu}_{\bar{S}_{i-1}}(\bar{s}_{i-1}) f(y_i, [s_{i-L+1}, \ldots, s_i], \bar{s}_{i-1}) \\
&\times \overleftarrow{\mu}_{\bar{S}_i}([s_{i-L+1}, \ldots, s_i]),
\end{aligned}
\tag{6.24}
$$

---

[2] Here we use Forney style factor graphs [41], where variables are represented as edges or half-edges. However, it is also possible to represent variables as variables notes.

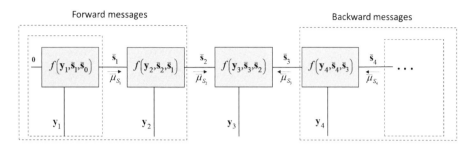

**Figure 6.8** Message passing over the factor graph of a finite-memory channel.

---

**Algorithm 6.2** BCJR algorithm.

---

**Init:** Fix an initial forward message $\overrightarrow{\mu}_{\bar{S}_0}(\bar{s}) = 1$ and a final backward message
$\overleftarrow{\mu}_{\bar{S}_T}(\bar{s}) \equiv 1$.

1 **for** $i = T - 1, T - 2, \ldots, 1$ **do**

2  | For each $\bar{s} \in \mathcal{S}^L$, compute backward message $\overleftarrow{\mu}_{\bar{S}_i}(\bar{s})$ via Eq. (6.23)

3 **end**

4 **for** $i = 1, 2, \ldots, T$ **do**

5  | For each $\bar{s} \in \mathcal{S}^L$, compute forward message $\overrightarrow{\mu}_{\bar{S}_i}(\bar{s})$ via Eq. (6.22)

6 **end**

**Output:** $\hat{s}^T = [\hat{s}_1, \ldots, \hat{s}_T]^T$, each obtained using Eq. (6.24)

---

for each $i \in \mathcal{T}$, where the summands can be computed recursively. When the block size $T$ is large, the messages may tend to zero, and are thus commonly scaled [42], e.g., $\overleftarrow{\mu}_{\bar{S}_i}(\bar{s})$ is replaced with $\gamma_i \overleftarrow{\mu}_{\bar{S}_i}(\bar{s})$ for some scale factor that does not depend on $\bar{s}$, and thus does not affect the MAP rule. The BCJR algorithm is summarized as Algorithm 6.2.

*BCJRNet*
BCJRNet is a receiver method that learns to implement MAP detection from labeled data. BCJRNet exploits the fact that in order to implement Algorithm 6.2, one must be able to specify the factor graph representing the underlying distribution. In particular, the stationarity assumption implies that the complete factor graph is encapsulated in the single function $f(\cdot)$ in Eq. (6.19) *regardless of the block size $T$*. Building upon this insight, BCJRNet utilizes DNNs to learn the mapping carried out at the function node separately from the inference task. The resulting learned stationary factor graph is then used to recover $\{S_i\}$ by message passing, as illustrated in Fig. 6.9. As learning a single function node is expected to be a simpler task compared to learning the overall inference method for recovering $S^T$ from $Y^T$, this approach allows using relatively compact DNNs, which in turn can be learned from a relatively small set of labeled data. Furthermore, the learned function node describes the factor graph for different values of $T$. When the learned function node is an accurate estimate of the

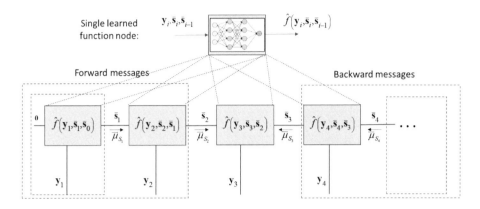

**Figure 6.9** BCJRNet with a learned stationary factor graph.

true one, BCJRNet effectively implements the MAP detection rule in Eq. (6.2), and thus approaches the minimal probability of error.

The function node that encapsulates the factor graph of stationary finite-memory channels is given in Eq. (6.19). The formulation in Eq. (6.19) implies that it can be estimated by training an ML-based system to evaluate $p_{Y_i|\bar{S}_i}(\cdot)$ from which the corresponding function node value is obtained via Eq. (6.19). Once the factor graph representing the channel is learned, symbol recovery is carried out using Algorithm 6.2. As the mapping of Algorithm 6.2 is invariant to scaling $f(y_i, \bar{s}_i, \bar{s}_{i-1})$ with some factor that does not depend on the states, it follows that a parametric estimate of the function $f(\cdot)$, denoted $\hat{f}_\theta(\cdot)$, can be obtained using the same networks utilized for learning the log-likelihood in ViterbiNet. Specifically, the learned log-likelihood is used in Eq. (6.19) to obtain $\hat{f}_\theta(\cdot)$. The resulting receiver, referred to as *BCJRNet*, thus implements BCJR detection in a data-driven manner.

### BCJRNet versus ViterbiNet

The same DNN architecture can be applied, once trained, to carry out multiple inference algorithms in a hybrid model-based/data-driven manner, including the BCJR scheme (as BCJRNet) as well as the Viterbi algorithm (via ViterbiNet). Since both BCJRNet and ViterbiNet utilize the same learned models, one can decide which inference system to apply (i.e., BCJRNet or ViterbiNet) by considering the differences in the algorithms from which they are derived (i.e., the BCJR algorithm and the Viterbi algorithm) respectively. The main advantages of Algorithm 6.1 over Algorithm 6.2, and thus of ViterbiNet over BCJRNet, are its reduced complexity and real-time operation. In particular, both algorithms implement recursive computations, involving $|\mathcal{S}|^L$ evaluations for each sample, and thus the complexity of both algorithms grows linearly with the block size $T$. Nonetheless, the Viterbi scheme computes only a forward recursion and can thus provide its estimations in real time within a given delay from each incoming observation, while the BCJR scheme implements both forward and backward recursions, and can thus infer only once the complete block is observed, while involving twice the computations carried out by the Viterbi detector.

The main advantage of Algorithm 6.2 over Algorithm 6.1 (i.e., of BCJRNet over ViterbiNet) stems from the fact that it implements the MAP rule in Eq. (6.2), which minimizes the error probability. The Viterbi algorithm is designed to compute the maximum likelihood *sequence* detector:

$$\hat{s}^T\left(y^T\right) = \arg\max_{s^T \in \mathcal{S}^T} p_{Y^T|S^T}\left(y^T|s^T\right), \tag{6.25}$$

which is not equivalent to the symbol-level MAP rule in Eq. (6.2). To see this, we focus on the case where the symbols are equiprobable, as in such scenarios the sequence-wise maximum likelihood rule coincides with the sequence-wise MAP detector. Here, the decision rule implemented by the Viterbi algorithm in Eq. (6.25) can be written as

$$\hat{s}^T\left(y^T\right) = \arg\max_{s^T \in \mathcal{S}^T} p_{S^T|Y^T}\left(s^T|y^T\right). \tag{6.26}$$

For a given realization $Y^T = y^T$, the function $p_{S^T|Y^T}\left(s^T|y^T\right)$ maximized by the sequence-wise detector in Eq. (6.26) is a joint distribution measure. The functions $\{p_{S_i|Y^T}\left(s_i|y^T\right)\}_{i=1}^T$, which are the individually maximized by the symbol-wise MAP rule computed by the BCJR scheme in Eq. (6.2), are the marginals of the aforementioned joint distribution. Furthermore, given $Y^T = y^T$ the elements of $S^T$ are statistically dependent in finite-memory channels. As a a result, the maxima of the joint distribution $p_{S^T|Y^T}$ is not necessarily the individual maximas of each of its marginals; that is, the elements of the vector $\hat{s}^T\left(y^T\right)$ in Eq. (6.25) are not necessarily the symbol-wise MAP estimates $\{\hat{s}_i\left(y^T\right)\}_{i=1}^T$ in Eq. (6.2). To conclude, the Viterbi algorithm does not implement the symbol-wise MAP even in the presence of equal priors. This explains the difference in their performance, since, unlike the BCJR scheme, the Viterbi algorithm does not minimize the error probability.

### Example: DeepSIC for Flat MIMO Channels

DeepSIC proposed in [43] is a DNN-aided hybrid algorithm that is based on the iterative soft interference cancellation (SIC) method [35] for symbol detection in flat MIMO channels. However, unlike its model-based counterpart, and alternative deep MIMO receivers such as DetNet, it is not tailored for linear Gaussian channels of the form in Eq. (6.6). The only assumption required is that the channel is memoryless (i.e., $L = 1$), and thus we drop the time index subscript in this example. As in our previous DNN-aided examples, we first review iterative SIC, after which we present its DNN-aided implementation.

*Iterative Soft Interference Cancellation*

The iterative SIC algorithm proposed in [35] is a MIMO detection method that combines multistage interference cancellation with soft decisions. The detector operates in an iterative fashion where, in each iteration, an estimate of the conditional PMF of $S_k$, which is the $k$th entry of $S$, given the observed $Y = y$, is generated for every symbol $k \in \{1, 2, \ldots, K\} := \mathcal{K}$ using the corresponding estimates of the interfering symbols $\{S_l\}_{l \neq k}$ obtained in the previous iteration. Iteratively repeating this procedure refines

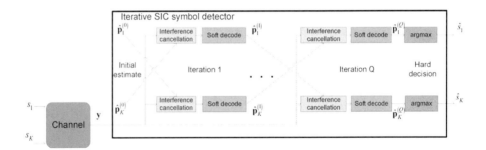

**Figure 6.10** Iterative soft interference cancellation.

the conditional distribution estimates, allowing the detector to accurately recover each symbol from the output of the last iteration. This iterative procedure is illustrated in Fig. 6.10.

To formulate the algorithm, we consider the flat Gaussian MIMO channel in Eq. (6.6). Iterative SIC consists of $Q$ iterations. Each iteration indexed $q \in \{1, 2, \ldots, Q\} \triangleq Q$ generates $K$ distribution vectors $\hat{\boldsymbol{p}}_k^{(q)}$ of size $M \times 1$, where $k \in \mathcal{K}$. These vectors are computed from the channel output $\boldsymbol{y}$ as well as the distribution vectors obtained at the previous iteration, $\{\hat{\boldsymbol{p}}_k^{(q-1)}\}_{k=1}^K$. The entries of $\hat{\boldsymbol{p}}_k^{(q)}$ are estimates of the distribution of $S_k$ for each possible symbol in $\mathcal{S}$, given the channel output $\boldsymbol{Y} = \boldsymbol{y}$ and assuming that the interfering symbols $\{S_l\}_{l \neq k}$ are distributed via $\{\hat{\boldsymbol{p}}_l^{(q-1)}\}_{l \neq k}$. Every iteration consists of two steps, carried out in parallel for each user: *interference cancellation* and *soft decoding*. Focusing on the $k$th user and the $q$th iteration, the interference cancellation stage first computes the expected values and variances of $\{S_l\}_{l \neq k}$ based on $\{\hat{\boldsymbol{p}}_l^{(q-1)}\}_{l \neq k}$. Letting $\{\alpha_j\}_{j=1}^M$ be the indexed elements of the constellation set $\mathcal{S}$, the expected values and variances are computed via $e_l^{(q-1)} = \sum_{\alpha_j \in \mathcal{S}} \alpha_j (\hat{\boldsymbol{p}}_l^{(q-1)})_j$, and $v_l^{(q-1)} = \sum_{\alpha_j \in \mathcal{S}} (\alpha_j - e_l^{(q-1)})^2 (\hat{\boldsymbol{p}}_l^{(q-1)})_j$, respectively. The contribution of the interfering symbols from $\boldsymbol{y}$ is then canceled by replacing them with $\{e_l^{(q-1)}\}$ and subtracting their resulting term. Letting $\boldsymbol{h}_l$ be the $l$th column of H, the interference canceled channel output is given by

$$\boldsymbol{Z}_k^{(q)} = \boldsymbol{Y} - \sum_{l \neq k} \boldsymbol{h}_l e_l^{(q-1)} = \boldsymbol{h}_k S_k + \sum_{l \neq k} \boldsymbol{h}_l (S_l - e_l^{(q-1)}) + \boldsymbol{W}. \tag{6.27}$$

Substituting the channel output $\boldsymbol{y}$ into Eq. (6.27), the realization of the interference canceled $\boldsymbol{Z}_k^{(q)}$, denoted $\boldsymbol{z}_k^{(q)}$, is obtained.

To implement soft decoding, it is assumed that $\tilde{\boldsymbol{W}}_k^{(q)} \triangleq \sum_{l \neq k} \boldsymbol{h}_l (S_l - e_l^{(q-1)}) + \boldsymbol{W}$ obeys a zero-mean Gaussian distribution, independent of $S_k$, and that its covariance is given by $\Sigma_{\tilde{W}_k^{(q)}} = \sigma_w^2 I_K + \sum_{l \neq k} v_l^{(q-1)} \boldsymbol{h}_l \boldsymbol{h}_l^T$, where $\sigma_w^2$ is the noise variance. Combining this assumption with Eq. (6.27), the conditional distribution of $\boldsymbol{Z}_k^{(q)}$ given $S_k = \alpha_j$ is multivariate Gaussian with mean value $\boldsymbol{h}_k \alpha_j$ and covariance $\Sigma_{\tilde{W}_k^{(q)}}$. Since $\boldsymbol{Z}_k^{(q)}$ is given by a bijective transformation of $\boldsymbol{Y}$, it holds that $p_{S_k|\boldsymbol{Y}}(\alpha_j|\boldsymbol{y}) = p_{S_k|\boldsymbol{Z}_k^{(q)}}(\alpha_j|\boldsymbol{z}_k^{(q)})$ for each $\alpha_j \in \mathcal{S}$ under the earlier assumptions. Consequently, the conditional

---

**Algorithm 6.3** Iterative Soft Interference Cancellation Algorithm [35].

---

**Init:** Set $q = 1$, and generate an initial guess of $\{\hat{\boldsymbol{p}}_k^{(0)}\}_{k=1}^K$.

1 **for** $q = 1, 2, \ldots, Q$ **do**

2    |  Compute the expected values $\{e_k^{(q-1)}\}$ and variances $\{v_k^{q-1}\}$;

3    |  *Interference cancellation:* For each $k \in \mathcal{K}$ compute $z_k^{(q)}$ via Eq. (6.27) ;

4    |  *Soft decoding:* For each $k \in \mathcal{K}$, estimate $\hat{\boldsymbol{p}}_k^{(q)}$ via Eq. (6.28)

5 **end**

**Output:** Hard decoded output $\hat{\boldsymbol{s}}$, obtained via Eq. (6.29)

---

distribution of $S_k$ given $Y$ is approximated from the conditional distribution of $Z_k^{(q)}$ given $S_k$ via Bayes theorem. Since the symbols are equiprobable, this estimated conditional distribution is computed as

$$
\left(\hat{p}_k^{(q)}\right)_j = \frac{\exp\left\{-\frac{1}{2}\left(z_k^{(q)} - \boldsymbol{h}_k \alpha_j\right)^T \Sigma_{\tilde{W}_k^{(q)}}^{-1}\left(z_k^{(q)} - \boldsymbol{h}_k \alpha_j\right)\right\}}{\sum\limits_{\alpha_{j'} \in \mathcal{S}} \exp\left\{-\frac{1}{2}\left(z_k^{(q)} - \boldsymbol{h}_k \alpha_{j'}\right)^T \Sigma_{\tilde{W}_k^{(q)}}^{-1}\left(z_k^{(q)} - \boldsymbol{h}_k \alpha_{j'}\right)\right\}}. \tag{6.28}
$$

After the final iteration, the symbols are decoded by taking the symbol that maximizes the estimated conditional distribution for each user:

$$
\hat{s}_k = \underset{j \in \{1, \ldots, M\}}{\arg\max} \left(\hat{p}_k^{(Q)}\right)_j. \tag{6.29}
$$

The overall joint detection scheme is summarized as Algorithm 6.3. The initial estimates $\{\hat{\boldsymbol{p}}_k^{(0)}\}_{k=1}^K$ can be arbitrarily set. For example, these may be chosen based on a linear separate estimation of each symbol for $\boldsymbol{y}$, as proposed in [35].

Iterative SIC has several advantages compared to both joint decoding as well as separate decoding: In terms of computational complexity, it replaces the joint exhaustive search over $\mathcal{S}^K$, required by the MAP decoder, with a set of computations carried out separately for each user. Hence, its computational complexity only grows linearly with the number of users [16], making it feasible with large values of $K$. Unlike conventional separate decoding, in which the symbol of each user is recovered individually while treating the interference as noise, the iterative procedure refines the separate estimates sequentially, and the usage of soft values mitigates the effect of error propagation. Algorithm 6.3 is thus capable of approaching the performance of the MAP detector, which is only feasible for small values of $K$.

### DeepSIC

Iterative SIC is specifically designed for linear channels of the form in Eq. (6.6). In particular, the interference cancellation in Step 2 of Algorithm 6.3 requires the contribution of the interfering symbols to be additive. Furthermore, it requires accurate CSI. To circumvent these limitations in the model-based approach, the DNN-aided DeepSIC learns to implement the iterative SIC from data.

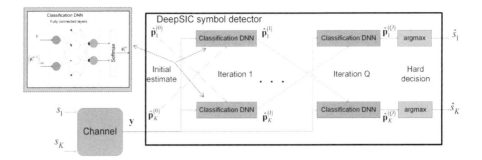

**Figure 6.11** DeepSIC.

**Architecture:** DeepSIC builds upon the observation that iterative SIC can be viewed as a set of interconnected basic building blocks, each implementing the two stages of interference cancellation and soft decoding (i.e., Steps 2 and 3 in Algorithm 6.3). While the high level architecture in Fig. 6.10 is ignorant of the underlying channel model, the basic building blocks are channel-model-dependent. In particular, interference cancellation requires the contribution of the interference to be additive, i.e., a linear model channel as in Eq. (6.6), as well as full CSI, in order to cancel the contribution of the interference. Soft decoding requires complete knowledge of the channel input-output relationship in order to estimate the conditional probabilities via Eq. (6.28).

Although each of these basic building blocks consists of two sequential procedures that are completely channel-model-based, we note that the purpose of these computations is to carry out a classification task. In particular, the $k$th building block of the $q$th iteration, $k \in \mathcal{K}$, $q \in \mathcal{Q}$, produces $\hat{\boldsymbol{p}}_k^{(q)}$, which is an estimate of the conditional PMF of $S_k$ given $\boldsymbol{Y} = \boldsymbol{y}$ based on $\{\hat{\boldsymbol{p}}_l^{(q-1)}\}_{l \neq k}$. Such computations are naturally implemented by classification DNNs (e.g., fully connected networks with a softmax output layer). Embedding these ML-based conditional PMF computations into the iterative SIC block diagram in Fig. 6.10 yields the overall receiver architecture depicted in Fig. 6.11. The initial estimates $\{\hat{\boldsymbol{p}}_k^{(0)}\}_{k=1}^K$ can be set to represent a uniform distribution; that is, $\left(\hat{\boldsymbol{p}}_k^{(0)}\right)_j \equiv \frac{1}{M}$.

A major advantage of using classification DNNs as the basic building blocks in Fig. 6.11 stems from the fact that such ML-based methods are capable of accurately computing conditional distributions in complex nonlinear setups without requiring a priori knowledge of the channel model and its parameters. Consequently, when these building blocks are trained to properly implement their classification task, the receiver essentially realizes iterative SIC for arbitrary channel models in a data-driven fashion.

**Training methods:** In order for the DNN-aided receiver structure of Fig. 6.11 to reliably implement joint decoding, its building block classification DNNs must be properly trained. Here, we consider two possible approaches to train the receiver based on a set of $n_t$ pairs of channel inputs and their corresponding outputs, denoted $\{\boldsymbol{s}_t, \boldsymbol{y}_t\}_{t=1}^{n_t}$: *end-to-end training* and *sequential training*.

*End-to-end training*: The first approach jointly trains the entire network (i.e., all the building block DNNs). Since the output of the deep network is the set of conditional distributions $\{\hat{p}_k^{(Q)}\}_{k=1}^K$, where each $\hat{p}_k^{(Q)}$ is used to estimate $S_k$, we use the sum cross entropy as the training objective. Let $\boldsymbol{\theta}$ be the network parameters, and $\hat{p}_k^{(Q)}(\boldsymbol{y}, \alpha; \boldsymbol{\theta})$ be the entry of $\hat{p}_k^{(Q)}$ corresponding to $S_k = \alpha$ when the input to the network parameterizd by $\boldsymbol{\theta}$ is $\boldsymbol{y}$. The sum cross entropy loss is

$$\mathcal{L}_{\text{SumCE}}(\boldsymbol{\theta}) = \frac{1}{n_t} \sum_{t=1}^{n_t} \sum_{k=1}^{K} -\log \hat{p}_k^{(Q)}\big(\boldsymbol{y}_t, (\boldsymbol{s}_t)_k; \boldsymbol{\theta}\big). \tag{6.30}$$

Training the receiver in Fig. 6.11 in an end-to-end manner based on the loss Eq. (6.30) jointly updates the coefficients of all the $K \cdot Q$ building block DNNs. For a large number of users, training so many parameters simultaneously is expected to require a large labeled set.

*Sequential training*: To allow the network to be trained with a reduced number of training samples, we note that the goal of each building block DNN does not depend on the iteration index: The $k$th building block of the $q$th iteration outputs a soft estimate of $S_k$ for each $q \in \mathcal{Q}$. Therefore, each building block DNN can be trained individually, by minimizing the conventional cross entropy loss. To formulate this objective, let $\boldsymbol{\theta}_k^{(q)}$ represent the parameters of the $k$th DNN at iteration $q$, and write $\hat{p}_k^{(q)}\big(\boldsymbol{y}, \{\hat{p}_l^{(q-1)}\}_{l \neq k}, \alpha; \boldsymbol{\theta}_k^{(q)}\big)$ as the entry of $\hat{p}_k^{(q)}$ corresponding to $S_k = \alpha$ when the DNN parameters are $\boldsymbol{\theta}_k^{(q)}$ and its inputs are $\boldsymbol{y}$ and $\{\hat{p}_l^{(q-1)}\}_{l \neq k}$. The cross entropy loss is given by

$$\mathcal{L}_{\text{CE}}(\boldsymbol{\theta}_k^{(q)}) = \frac{1}{n_t} \sum_{t=1}^{n_t} -\log \hat{p}_k^{(q)}\big(\tilde{\boldsymbol{y}}_t, \{\hat{p}_{t,l}^{(q-1)}\}_{l \neq k}, (\tilde{\boldsymbol{s}}_t)_k; \boldsymbol{\theta}_k^{(q)}\big), \tag{6.31}$$

where $\{\hat{p}_{t,l}^{(q-1)}\}$ represent the estimated probabilities associated with $\boldsymbol{y}_t$ computed at the previous iteration. The problem with training each DNN individually is that the soft estimates $\{\hat{p}_{t,l}^{(q-1)}\}$ are not provided as part of the training set. This challenge can be tackled by training the DNNs corresponding to each layer in a sequential manner, where for each layer the outputs of the trained previous iterations are used as the soft estimates fed as training samples.

Sequential training uses the $n_t$ input-output pairs to train each DNN individually. Compared to the end-to-end training that utilizes the training samples to learn the complete set of parameters, which can be quite large, sequential training uses the same dataset to learn a significantly smaller number of parameters, reduced by a factor of $K \cdot Q$, multiple times. Consequently, this approach is expected to require much fewer training samples, at the cost of a longer learning procedure for a given training set, due to its sequential operation, and possible performance degradation as the building blocks are not jointly trained.

**Summary**

DNN-aided algorithms implement hybrid model-based/data-driven inference by integrating ML into established model-based methods. As such, it is particularly suitable for digital communications setups, in which a multitude of reliable model-based algorithms exist, each tailored to a specific structure. The implementation of these techniques in a data-driven fashion thus has three main advantages as a model-based ML strategy: First, when properly trained, the resulting system effectively implements the model-based algorithm from which it originated, thus benefiting from its proven performance and controllable complexity, while being robust to CSI uncertainty and operable in complex environments, due to the usage of DNNs. This behavior is numerically illustrated in the simulation study detailed in Section 6.3.5.

Second, the fact that DNN-aided algorithms use ML tools as intermediate components in the overall end-to-end inference tasks allows the use of relatively compact networks which can be trained with small training sets. Even when the overall system consists of a large set of DNNs, as is the case in DeepSIC, their interpretable operation that follows from the model-based method facilitates their training with small datasets (e.g., via sequential training techniques).

Finally, DNN-aided algorithms can utilize different levels of domain knowledge, depending on what prior information one has on the problem at hand. For example, BCJRNet requires only prior knowledge that the channel has finite memory to learn to carry out MAP detection from data. When additional domain knowledge is available, such as an underlying stationarity or some partial CSI, it can be incorporated into the number and structure of the learned function nodes, further reducing the number of training data required to tune the receiver. The resulting ability of DNN-aided symbol detectors to adapt with small training sets can be exploited to facilitate channel tracking via periodic retraining using existing pilots and other forms of structures present in digital communications protocols, as demonstrated in [11, 43].

## 6.3.5    Numerical Study

In this section, we present a numerical study of the aforementioned symbol detection mechanisms. We begin with considering finite-memory channels, for which we evaluate the data-driven SBRNN receiver, ViterbiNet, and BCJRNet, comparing them to the model-based detection methods for such channels. Then we consider memoryless MIMO channels, where we compare the data-driven DetNet and DeepSIC to model-based detection.

**Finite-Memory Channel**

We first numerically evaluate the performance of the DNN-aided ViterbiNet and BCJRNet and compare this performance to that of the conventional model-based Viterbi algorithm and BCJR detector, as well as to that of the SBRNN receiver detailed in Section 6.3.2. Both ViterbiNet and BCJRNet are implemented using the classification architecture in Fig. 6.7(a) with three fully connected layers: a $1 \times 100$

layer followed by a $100 \times 50$ layer and a $50 \times 16(= |M|^L)$ layer, using intermediate sigmoid and ReLU activation functions, respectively. For the SBRNN receiver, we use BRNN length of $B = 10$ with three layers of LSTM cell blocks of size 100 and a dropout rate of 0.1. The networks are trained using 5,000 training samples, which is of the same order and even smaller compared to typical preamble sequences in wireless communication systems.

We consider two finite-memory channels: an AWGN channel and a Poisson channel, both with memory length of $L = 4$. For the additive white Gaussian noise (AWGN) channel, we let $W[i]$ be a zero-mean unit variance AWGN independent of $S[i]$, and let $\boldsymbol{h}(\gamma) \in \mathcal{R}^L$ be the channel vector obeying an exponentially decaying profile $(\boldsymbol{h})_\tau \triangleq e^{-\gamma(\tau-1)}$ for $\gamma > 0$. The input-output relationship is given by

$$Y[i] = \sqrt{\rho} \cdot \sum_{\tau=1}^{L} (\boldsymbol{h}(\gamma))_\tau \, S[i - \tau + 1] + W[i], \tag{6.32}$$

where $\rho > 0$ represents the SNR. The channel input is randomized from a BPSK constellation; that is, $\mathcal{S} = \{-1, 1\}$. For the Poisson channel, the symbols represent on-off keying, namely, $\mathcal{S} = \{0, 1\}$, and the channel output $Y[i]$ is generated via

$$Y[i]|\boldsymbol{S}^T \sim \mathbb{P}\left(\sqrt{\rho} \cdot \sum_{\tau=1}^{L} (\boldsymbol{h}(\gamma))_\tau \, S[i - \tau + 1] + 1\right), \tag{6.33}$$

where $\mathbb{P}(\lambda)$ is the Poisson distribution with parameter $\lambda > 0$, and $X \sim f(X)$ indicates that the random variable $X$ is distributed according to $f(X)$.

For each channel, we numerically compute the symbol error rate (SER) for different values of the SNR parameter $\rho$. For each SNR $\rho$, the SER values are averaged over 20 different channel vectors $\boldsymbol{h}(\gamma)$, obtained by letting $\gamma$ vary in the range $[0.1, 2]$. For comparison, we numerically compute the SER of the Viterbi and BCJR algorithms. In order to study the resiliency of the data-driven detectors to inaccurate training, we also compute the performance when the receiver only has access to a noisy estimate of $\boldsymbol{h}(\gamma)$, and specifically, to a copy of $\boldsymbol{h}(\gamma)$ whose entries are corrupted by i.i.d. zero-mean Gaussian noise with variance $\sigma_e^2$. In particular, we use $\sigma_e^2 = 0.1$ for the Gaussian channel in Eq. (6.32), and $\sigma_e^2 = 0.08$ for the Poisson channel in Eq. (6.33). We consider two cases: *perfect CSI*, in which the channel-model-based detectors have accurate knowledge of $\boldsymbol{h}(\gamma)$, while the data-driven receivers are trained using labeled samples generated with the same $\boldsymbol{h}(\gamma)$ used for generating the test data; and *CSI uncertainty*, where the model-based algorithms are implemented with the log-likelihoods (for Viterbi algorithm) and function nodes (for BCJR detection) computed using the noisy version of $\boldsymbol{h}(\gamma)$, while the training data is generated with the noisy version of $\boldsymbol{h}(\gamma)$ instead of the true one. In all cases, the information symbols are uniformly randomized in an i.i.d. fashion from $\mathcal{S}$, and the test samples are generated from their corresponding channel with the true channel vector $\boldsymbol{h}(\gamma)$.

The numerically computed SER values, averaged over 50,000 Monte Carlo simulations, versus $\rho \in [-6, 10]$ dB for the AWGN channel are depicted in Fig. 6.12, while the corresponding performance versus $\rho \in [10, 30]$ dB for the Poisson channel

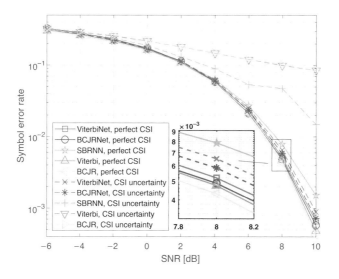

**Figure 6.12** Symbol error rate of different receiver structures for the AWGN channel with exponentially decaying taps.

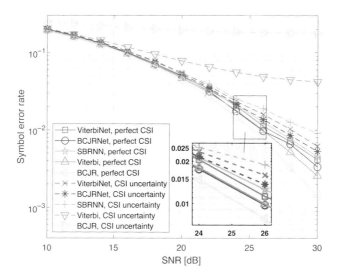

**Figure 6.13** Symbol error rate of different receiver structures for the Poisson channel.

are depicted in Fig. 6.13. Observing Figs. 6.12 and 6.13, we note that the performance of the data-driven receivers approaches that of their corresponding CSI-based counterparts. We also observe that the SBRNN receiver, which was shown in [28] to approach the performance of the CSI-based Viterbi algorithm when sufficient training is provided, is outperformed by ViterbiNet and BCJRNet here due to the small training set size. These results demonstrate that our DNN-aided detectors, which use compact

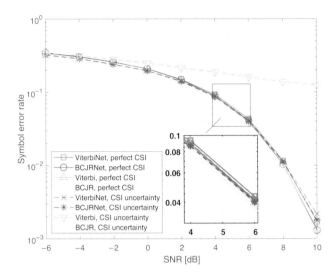

**Figure 6.14** Symbol error rate of different receiver structures for the AWGN channel with taps generated from the COST2100 model.

DNN structures embedded into model-based algorithms, require significantly less training compared to symbol detectors based on using established DNNs for end-to-end inference.

In the presence of CSI uncertainty, it is observed in Figs. 6.12 and 6.13 that both ViterbiNet and BCJRNet significantly outperform the model-based algorithms from which they originate. In particular, when ViterbiNet and BCJRNet are trained with a variety of different channel conditions, they are still capable of achieving relatively good SER performance under each of the channel conditions for which they are trained, while the performance of the conventional Viterbi and BCJR algorithms is significantly degraded in the presence of imperfect CSI. While the SBRNN receiver is shown to be more resilient to inaccurate CSI compared to the Viterbi and BCJR algorithms, it is outperformed by ViterbiNet and BCJRNet with the same level of uncertainty, and the performance gap is more notable in the AWGN channel.

Finally, we evaluate the application of ViterbiNet and BCJRNet for practical channel models. To that aim, we generate 10 realizations from the established COST2100 model [44], which is a widely used model for current cellular communication channels. In particular, we use the semi-urban 300 MHz line-of-sight configuration evaluated in [45] with a single antenna element. The channel output is corrupted by AWGN, and the symbol detectors operate assuming that the channel has $L = 4$ taps. The remaining simulation parameters are the same as those used in Fig. 6.12. The results, depicted in Fig. 6.14, demonstrate that the ability of ViterbiNet and BCJRNet to approach their model-based counterparts with perfect CSI, as well as achieve improved performance in the presence of CSI uncertainty, holds for practical channel models.

**Memoryless MIMO Channel**

Next, we numerically compare DeepSIC and DetNet for symbol detection in memoryless MIMO channels. In the implementation of the DNN-based building blocks of DeepSIC, we used a different fully connected network for each training method: For end-to-end training, where all the building blocks are jointly trained, we used a compact network consisting of a $(N + K - 1) \times 60$ layer followed by ReLU activation and a $60 \times M$ layer. For sequential training, which sequentially adapts subsets of the building blocks and can thus tune more parameters using the same training set (or alternatively, requires a smaller training set) compared to end-to-end training, we used three fully connected layers: An $(N + K - 1) \times 100$ first layer, a $100 \times 50$ second layer, and a $50 \times M$ third layer, with a sigmoid and a ReLU intermediate activation functions, respectively. In both iterative SIC as well as DeepSIC, we set the number of iterations to $Q = 5$. Following [32], DetNet is implemented with $Q = 90$ layers with a hidden sublayer size of $8K$. The data-driven receivers are trained with a relatively small dataset of $5,000$ training samples and tested over $20,000$ symbols.

We first consider a linear AWGN channel as in Eq. (6.6) with a relatively small $K$. Recall that iterative SIC as well as DetNet are all designed for such channels. Consequently, the following study compares the performance of DeepSIC and DetNet to that of the model-based iterative SIC as well as the MAP rule in Eq. (6.2) in a scenario for which all schemes are applicable. The model-based MAP and iterative SIC detectors, as well as DetNet, all require CSI, and specifically, accurate knowledge of the channel matrix H. DeepSIC operates without a priori knowledge of the channel model and its parameters, learning the decoding mapping from a training set sampled from the considered input-output relationship. In order to compare the robustness of the detectors to CSI uncertainty, we also evaluate them when the receiver has access to an estimate of H with entries corrupted by i.i.d. additive Gaussian noise whose variance is given by $\sigma_e^2$ times the magnitude of the corresponding entry, where $\sigma_e^2 > 0$ is referred to as the *error variance*. For DeepSIC, which is model-invariant, we compute the SER under CSI uncertainty by using a training set whose samples are randomized from a channel in which the true H is replaced with its noisy version.

We simulate the $6 \times 6$ linear Gaussian channel (i.e., $K = 6$ users and $N = 6$ receive antennas). The symbols are randomized from a BPSK constellation, namely, $\mathcal{S} = \{-1, 1\}$ and $M = |\mathcal{S}| = 2$. The channel matrix H models spatial exponential decay, and its entries are given by $(\mathsf{H})_{i,k} = e^{-|i-j|}$, for each $i \in \{1, \ldots, N\}, k \in \mathcal{K}$. For each channel, the SER of the receivers is evaluated for both perfect CSI (i.e., $\sigma_e^2 = 0$) as well as CSI uncertainty, for which we use $\sigma_e^2 = 0.1$. The evaluated SER versus the SNR, defined as $1/\sigma_w^2$, is depicted in Fig. 6.15.

Observing Fig. 6.15, we note that the performance of DeepSIC with end-to-end training approaches that of the model-based iterative SIC algorithm, which is within a small gap of the MAP performance. This demonstrates the ability of DeepSIC to implement iterative SIC in a data-driven fashion. The sequential training method, whose purpose is to allow DeepSIC to train with smaller data sets compared to end-to-end training, also achieves SER values comparable to iterative SIC. DetNet, which trains a large number of parameters in an end-to-end fashion, requires 100 times

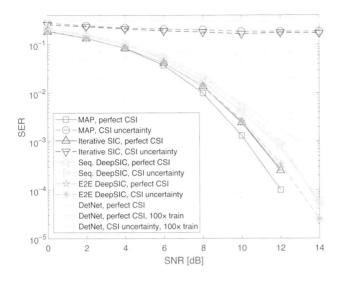

**Figure 6.15** Symbol error rate of different receiver structures for the $6 \times 6$ AWGN channel.

more training to approach such performance. In the presence of CSI uncertainty, DeepSIC is observed to substantially outperform the model-based iterative SIC and MAP receivers, as well as DetNet operating with a noisy version of H and trained with a hundred times more samples. In particular, it follows from Fig. 6.15 that a relatively minor error of variance $\sigma_e^2 = 0.1$ severely deteriorates the performance of the model-based methods, while DeepSIC is hardly affected by the same level of CSI uncertainty.

Next, we consider a Poisson channel. We use $K = 4$ and $N = 4$. Here, the symbols are randomized from an on-off keying for which $\mathcal{S} = \{0, 1\}$. The entries of the channel output are related to the input via the conditional distribution

$$(\boldsymbol{Y}[i])_j \, | \boldsymbol{S}[i] \sim \mathbb{P}\left(\frac{1}{\sqrt{\sigma_w^2}} \, (\mathsf{H}\boldsymbol{S}[i])_j + 1\right), \qquad j \in \{1, \dots, N\}. \tag{6.34}$$

As DetNet is designed for linear Gaussian channels, DeepSIC is the only data-driven receiver evaluated for this channel.

The achievable SER of DeepSIC versus SNR under both perfect CSI as well as CSI uncertainty with error variance $\sigma_e^2 = 0.1$ is compared to the MAP and iterative SIC detectors in Fig. 6.16. Observing Fig. 6.16, we again note that the performance of DeepSIC is only within a small gap of the MAP performance with perfect CSI and that the data-driven receiver is more robust to CSI uncertainty compared to the model-based MAP. In particular, DeepSIC with sequential training, which utilizes a deeper network architecture for each building block, outperforms here end-to-end training with basic two-layer structures for the conditional distribution estimation components. We conclude that under such non-Gaussian channels, more complex DNN models are required to learn to cancel interference and carry out soft detection accurately.

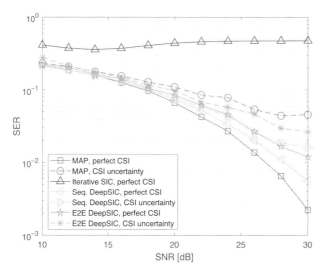

**Figure 6.16** Symbol error rate of different receiver structures for the $4 \times 4$ Poisson channel.

Furthermore, iterative SIC, which is designed for linear Gaussian channels ni Eq. (6.6) where interference is additive, achieves poor performance when the channel model is substantially different from Eq. (6.6). These results demonstrate the ability of DeepSIC to achieve excellent performance through learning from data for statistical models where model-based interference cancellation is effectively inapplicable.

## 6.4      Conclusion

Deep learning brings forth capabilities that can substantially contribute to future communications systems in tackling some of their expected challenges. In particular digital communications systems can significantly benefit from properly harnessing the power of deep learning and its model-agnostic nature. A successful integration of deep learning into communication devices can thus pave the way to reliable and robust communications in various setups, including environments where accurate statistical channel models are scarce or costly to obtain. However, digital communications setups are fundamentally different from applications in which deep learning has been extremely successful to date, such as computer vision. In particular, digital communication exhibit an extremely large number of possible outputs, as these outputs grow exponentially with the modulation order and the block length. They also have channel conditions that vary dynamically and require low computation complexity when used on small battery-powered devices. Consequently, in order to achieve the potential benefits of DNN-aided communications, researchers and system designers must go beyond the straight-forward application of DNNs designed for computer vision and natural language processing. A candidate strategy to utilize DNNs while accounting for the unique characteristics of digital communications setups, as well as the established knowledge of model-based communication methods accumulated over the last decades, is based on model-based ML, as detailed in this chapter.

We reviewed two main strategies for combining data-driven deep learning with model-based methods for digital communications, as well as discussing the need for such hybrid schemes due to the shortcomings of the extreme cases of purely data-driven and solely model-based methods. For each strategy, we presented the main steps in the design of the data-driven systems and provided concrete examples, all in the context of the basic communication task of symbol detection. We first discussed how established DNN architectures can be utilized as symbol detectors, presenting the SBRNN receiver of [28] as an example. Then we detailed how the framework of deep unfolding, which designs DNNs based on iterative optimization algorithms, can give rise to hybrid model-based/data-driven receivers, presenting DetNet of [32] as an example. We identified that the main drawback of these aforementioned techniques in the context of digital communications stems from their usage of highly parameterized DNNs applied in an end-to-end fashion, which directly results in the need for massive data sets for training. Then, we presented DNN-aided algorithms, where DNNs are integrated into existing model-based algorithms. We identified the latter as being extremely suitable for digital communications due to the wide variety of model-based algorithms designed for such setups, combined with its ability to incorporate different levels of domain knowledge as well as utilize compact DNNs as intermediate components in the inference process.

The DNN-aided symbol detectors presented as examples here (i.e., ViterbiNet [11], BCJRNet [12], and DeepSIC [43]) all numerically demonstrated improved performance over competing strategies when a limited amount of training data is available. In particular, it is demonstrated that these DNN-aided symbol detectors, which are designed to operate in a hybrid model-based/data-driven fashion, learn to approach the performance achieved by purely model-based approaches operating with perfect knowledge of the underlying channel model and its parameters. Furthermore, the DNN-aided symbol detectors were shown to be notably more resilient to CSI uncertainty compared to model-based schemes, carrying out accurate detection in the presence of inaccurate CSI. Finally, it was demonstrated that model-based deep learning enables DNN-aided receivers to learn their mapping from relatively small data sets, making it an attractive approach to combine with tracking of dynamic channel conditions.

## References

[1] Y. LeCun, Y. Bengio, and G. Hinton, "Deep learning," *Nature*, vol. 521, no. 7553, p. 436, 2015.

[2] G. Hinton et al., "Deep neural networks for acoustic modeling in speech recognition: The shared views of four research groups," *IEEE Signal Processing Magazine*, vol. 29, no. 6, pp. 82–97, 2012.

[3] Y. Bengio, "Learning deep architectures for AI," *Foundations and Trends in Machine Learning*, vol. 2, no. 1, pp. 1–127, 2009.

[4] K. Gregor and Y. LeCun, "Learning fast approximations of sparse coding," in *Proc. 27th Int. Conf. on Machine Learning*, pp. 399–406, 2010.

[5] L. Zheng et al., "Radar and communication coexistence: An overview: A review of recent methods," *IEEE Signal Processing Magazine*, vol. 36, no. 5, pp. 85–99, 2019.

[6] P. Singya, N. Kumar, and V. Bhatia, "Effect of non-linear power amplifiers on future wireless communication networks," *IEEE Microwave Magazine*, vol. 18, no. 5, 2017.

[7] N. Shlezinger et al., "Model-based deep learning," *arXiv preprint*, arXiv:2012.08405,

[8] J. R. Hershey, J. L. Roux, and F. Weninger, "Deep unfolding: Model-based inspiration of novel deep architectures," *arXiv preprint*, arXiv:1409.2574.

[9] V. Monga, Y. Li, and Y. C. Eldar, "Algorithm unrolling: Interpretable, efficient deep learning for signal and image processing," *IEEE Signal Processing Magazine*, vol. 38, no. 2, pp. 18–44, 2021.

[10] A. Balatsoukas-Stimming and C. Studer, "Deep unfolding for communications systems: A survey and some new directions," *arXiv preprint*, arXiv:1906.05774.

[11] N. Shlezinger et al., "ViterbiNet: A deep learning based Viterbi algorithm for symbol detection," *IEEE Trans. on Wireless Communications*, vol. 19, no. 5, pp. 3319–3331, 2020.

[12] N. Shlezinger et al., "Data-driven factor graphs for deep symbol detection," in *Proc. IEEE Int. Symp. on Information Theory (ISIT)*, 2020.

[13] I. Goodfellow, Y. Bengio, and A. Courville, *Deep Learning*. MIT Press, 2016.

[14] A. Viterbi, "Error bounds for convolutional codes and an asymptotically optimum decoding algorithm," *IEEE Trans. on Information Theory*, vol. 13, no. 2, pp. 260–269, 1967.

[15] L. Bahl et al., "Optimal decoding of linear codes for minimizing symbol error rate," *IEEE Trans. on Information Theory*, vol. 20, no. 2, pp. 284–287, 1974.

[16] J. G. Andrews, "Interference cancellation for cellular systems: a contemporary overview," *IEEE Wireless Communications*, vol. 12, no. 2, pp. 19–29, 2005.

[17] C. Tian et al., "Deep learning for image denoising: A survey," in *Int. Conf. on Genetic and Evolutionary Computing*, pp. 563–572, 2018.

[18] Z. Wang, J. Chen, and S. C. Hoi, "Deep learning for image super-resolution: A survey," *IEEE Trans. on Pattern Analysis and Machine Intelligence*, 2020.

[19] S. Minaee et al., "Image segmentation using deep learning: A survey," *arXiv preprint*, arXiv:2001.05566, 2020.

[20] G. Hinton et al., "Deep neural networks for acoustic modeling in speech recognition: The shared views of four research groups," *IEEE Signal Processing Magazine*, vol. 29, no. 6, pp. 82–97, 2012.

[21] A. Graves and N. Jaitly, "Towards end-to-end speech recognition with recurrent neural networks," in *Proc. 31st Int. Conf. on Machine Learning (ICML-14)*, pp. 1764–1772, 2014.

[22] D. Amodei et al., "Deep speech 2: End-to-end speech recognition in english and mandarin," in *Int. Conf. on Machine Learning*, pp. 173–182, 2016.

[23] D. Bahdanau, K. Cho, and Y. Bengio, "Neural Machine Translation by Jointly Learning to Align and Translate," *arXiv preprint*, arXiv:1409.0473, 2014.

[24] K. Cho et al., "Learning phrase representations using RNN encoder-decoder for statistical machine translation," *arXiv preprint*, arXiv:1406.1078, 2014

[25] S. Yang, Y. Wang, and X. Chu, "A survey of deep learning techniques for neural machine translation," *arXiv preprint*, arXiv:2002.07526, 2020.

[26] Z. Li and Y. Yu, "Protein Secondary Structure Prediction Using Cascaded Convolutional and Recurrent Neural Networks," *arXiv preprint*, arXiv:1604.07176, 2016.

[27] K. Lan et al., "A survey of data mining and deep learning in bioinformatics," *Journal of Medical Systems*, vol. 42, no. 8, p. 139, 2018.

[28] N. Farsad and A. Goldsmith, "Neural network detection of data sequences in communication systems," *IEEE Trans. on Signal Processing*, vol. 66, no. 21, pp. 5663–5678, 2018.

[29] M. Schuster and K. K. Paliwal, "Bidirectional recurrent neural networks," *IEEE Trans. on Signal Processing*, vol. 45, no. 11, pp. 2673–2681, 1997.

[30] B. Karanov et al., "End-to-end optimized transmission over dispersive intensity-modulated channels using bidirectional recurrent neural networks," *Optics Express*, vol. 27, no. 14, pp. 19650–19663, 2019.

[31] B. Karanov et al., "Experimental investigation of deep learning for digital signal processing in short reach optical fiber communications," *arXiv preprint*, arXiv:2005.08790, 2020.

[32] N. Samuel, T. Diskin, and A. Wiesel, "Learning to detect," *IEEE Trans. on Signal Processing*, vol. 67, no. 10, pp. 2554–2564, 2019.

[33] H. He et al., "A model-driven deep learning network for MIMO detection," in *Proc. IEEE GlobalSIP*, 2018.

[34] H. He et al., "Model-driven deep learning for joint MIMO channel estimation and signal detection," *arXiv preprint*, arXiv:1907.09439.

[35] W.-J. Choi, K.-W. Cheong, and J. M. Cioffi, "Iterative soft interference cancellation for multiple antenna systems." in *Proc. WCNC*, pp. 304–309, 2000.

[36] G. McLachlan and D. Peel, *Finite mixture models*. John Wiley & Sons, 2004.

[37] C. M. Bishop, "Mixture density networks," 1994; http://publications.aston.ac.uk/id/eprint/373/.

[38] J. Rothfuss et al. "Conditional density estimation with neural networks: Best practices and benchmarks," *arXiv preprint*, arXiv:1903.00954.

[39] I. Kobyzev, S. Prince, and M. A. Brubaker, "Normalizing flows: Introduction and ideas," *arXiv preprint*, arXiv:1908.09257, 2019.

[40] F. R. Kschischang, B. J. Frey, and H.-A. Loeliger, "Factor graphs and the sum-product algorithm," *IEEE Trans. on Information Theory*, vol. 47, no. 2, pp. 498–519, 2001.

[41] G. D. Forney, "Codes on graphs: Normal realizations," *IEEE Trans. on Information Theory*, vol. 47, no. 2, pp. 520–548, 2001.

[42] H.-A. Loeliger, "An introduction to factor graphs," *IEEE Signal Processing Magazine*, vol. 21, no. 1, pp. 28–41, 2004.

[43] N. Shlezinger, R. Fu, and Y. C. Eldar, "Deepsic: Deep soft interference cancellation for multiuser mimo detection," *IEEE Transactions on Wireless Communications*, vol. 20, no. 2, pp. 1349–1362, 2020.

[44] L. Liu et al., "The COST 2100 MIMO channel model," *IEEE Wireless Communications*, vol. 19, no. 6, pp. 92–99, 2012.

[45] M. Zhu, G. Eriksson, and F. Tufvesson, "The COST 2100 channel model: Parameterization and validation based on outdoor MIMO measurements at 300 MHz," *IEEE Trans. on Wireless Communications*, vol. 12, no. 2, pp. 888–897, 2013.

# 7 Constrained Unsupervised Learning for Wireless Network Optimization

Hoon Lee, Sang Hyun Lee, and Tony Q. S. Quek

## 7.1    Introduction

An optimization task in a wireless network is often formulated as a distributed decision-making problem. Figure 7.1 illustrates a generic distributed setup of wireless network optimization tasks. Multiple wireless nodes are equipped with computing devices and distributed over a coverage region. Each of the nodes independently makes its decision with a certain objective. The decision is often associated with an individual networking policy to serve its mobile users. Any network element responsible for such computations, including base stations and access points, can be designated as a wireless node. The objective function associated with an individual node is described as a network cost function characterizing the performance evaluated by the set of the decisions. Several strict regulations on wireless resources impose the corresponding design constraints on communication systems. It is necessary to control the behavior of a wireless node so that the collective network dynamics meets the restrictions. This is mathematically formalized by a set of multiple constraint functions. Therefore, the network optimization problem commonly boils down to the optimization minimizing the cost function subject to the set of design constraints. This approach includes a variety of management challenges of wireless networks, such as the resource allocation policy and the transceiver technique design.

The nonconvexity of the cost function and constraints incurred by a technical nuisance, such as internode interference and nonlinear transceiver architecture, poses issues in designing wireless networks with off-the-shelf convex optimization software packages. To resolve them, there have been intensive efforts in sophisticated nonconvex programming techniques for signal processing and wireless communication applications [1, 2]. The convergence dynamics and its optimality guarantee of such methods have been examined and verified extensively in diverse types of the networking setup. However, their dependence on special mathematical structures of the cost function and constraints leaves generic optimization tasks based on arbitrary networking setups unaddressed.

Another fundamental challenge is raised by the distributed nature of wireless networks. Independent decisions of wireless nodes depend on the collection of isolated measurements (e.g., the channel state information of a communication link) obtained

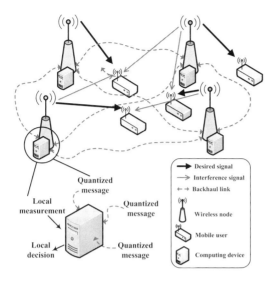

**Figure 7.1** A generic distributed wireless setup with interconnected wireless nodes.

from local sensing and detection technology. Typically, wireless nodes are deployed spatially over the target coverage region for the support of distributed mobile users. Local statistics of individual nodes are mostly available partially. Thus, such deficiency results in incorrect decisions in the joint optimization's perspectives. Existing distributed optimization techniques [3, 4] address this issue via distributed coordination among wireless nodes interconnected with backhaul infrastructures. Solutions are identified locally at individual nodes through message transfer strategies. These approaches can handle only formulations defined in separable forms of the cost function and constraints. Furthermore, the requirements on high precision of continuous-valued messages need impractically large bandwidths of the backhaul interface for unlimited cooperation capability. Therefore, it is necessary to develop a universal optimization framework handling various nonconvex formulations of cost functions and constraints that jointly designs distributed decision and quantized coordination policies over limited-bandwidth backhaul interfaces.

This chapter addresses constrained and unsupervised learning strategies that tackle the machine-learning-oriented design of distributed wireless network management. First, a centralized learning mechanism is developed so that lossless information sharing is allowed among wireless nodes over an ideal infinite-capacity backhaul configuration. A suite of neural network (NN) units is designed to produce a set of individual decisions based on a collection of local measurements of the nodes. Nonconvex constraints are considered in learning a dual formulation of the optimization problem with a guarantee of the strong duality [5]. Second, finite-capacity backhauls are reflected to taken a realistic networking setup into account. To do so, a distributed inference is obtained for a group of centrally trained NNs. In particular, a pair of component NN units, referred to as a quantizer NN and an optimizer NN, is introduced at individual nodes. In addition, an efficient neural quantization strategy is developed so that the collection of all NNs incorporated at wireless nodes are jointly trained in an end-to-end manner.

The rest of this chapter is organized to describe the details of distributed optimization in wireless networks as follows: A constrained training algorithm is developed in Section 7.3. An optimality guarantee of the NN-based formalism is verified with mathematical analysis. In Section 7.4, a distributed arrangement of the NN-based optimization inferences is introduced. The unsupervised learning approaches is tested with various networking applications as presented in Sections 7.5 and 7.6.

## 7.2     Distributed Management of Wireless Networks

### 7.2.1     System Model

An optimization task in distributed wireless networks is introduced first as depicted in Fig. 7.1. A group of $N$ wireless nodes participates in the joint minimization of a certain common network cost function. The nodes are interconnected via backhaul links, where they can exchange relevant statistics that a reliable distributed decision relies on. Let $\mathbf{a}_i$ be an $A_i$-dimensional vector of local measurements exclusively obtained at node $i$. It is connected with its neighbors via the set of links $\mathbf{E}_i \triangleq \{(i, j) : \forall j \neq i\}$. Local measurements collectively form a global observation set denoted by $\mathbf{a} \triangleq \{\mathbf{a}_i : 1 \leq i \leq N\}$, which describes the current overall network state. With only locally observable information at hand, node $i$ determines a vector of $X_i$-dimensional local decision variables denoted by $\mathbf{x}_i$ among a decision vector set $\mathcal{X}_i$.

The macroscopic performance of the network depends on the microscopic decisions of all wireless nodes. Since the network state affects the choices at the nodes, the performance metric and its associated cost function also depend on the global measurement $\mathbf{a}$. Consequently, a global decision set aggregating the local decisions, denoted by $\mathbf{x} \triangleq \{\mathbf{x}_i : 1 \leq i \leq N\}$, is regarded as the result of the optimization problem associated with specific state $\mathbf{a}$. A network cost function is expressed as $f(\mathbf{a}, \mathbf{x})$, or a function of the network state $\mathbf{a}$ and its corresponding decision results $\mathbf{x}$. The target is to determine $\mathbf{x}$ that minimizes the network cost function by combining local decisions $\mathbf{x}_i$ made independently at each node. Most wireless networking management tasks typically focus on the network cost $\mathbb{E}_{\mathbf{a}}[f(\mathbf{a}, \mathbf{x})]$ averaged over random instances of $\mathbf{a}$, rather than instantaneous cost value $f(\mathbf{a}, \mathbf{x})$ with respect to a specific realization $\mathbf{a}$. The average cost describes a long-term behavior of the network management technique by encompassing inherent stochastic properties, such as the randomness in fading channels and the network topology. The computation methods for the average cost rely on the availability of the distribution $p(\mathbf{a})$, or accurate statistics of channel propagation environment. The average cost function is evaluated as

$$\mathbb{E}_{\mathbf{a}}[f(\mathbf{a}, \mathbf{x})] = \int_{\mathbf{a}} f(\mathbf{a}, \mathbf{x}) p(\mathbf{a}) \mathrm{d}\mathbf{a}. \tag{7.1}$$

On the other hand, the distribution $p(\mathbf{a})$ is unavailable in most machine-learning applications. The integral in Eq. (7.1) would not be tractable in an analytical way. A data-driven approach can be adopted for numerical evaluation with a random realization set of $\mathbf{a}$. It is designated as a training dataset $\mathcal{A} \triangleq \{\mathbf{a}^{(d)} : 1 \leq d \leq D\}$ containing $D$ distinct samples $\mathbf{a}^{(d)}$ from the global measurement $\mathbf{a}$. For the training dataset, Eq. (7.1) is calculated using

$$\mathbb{E}_{\mathbf{a}}[f(\mathbf{a}, \mathbf{x})] = \frac{1}{D} \sum_{d=1}^{D} f(\mathbf{a}^{(d)}, \mathbf{x}^{(d)}), \tag{7.2}$$

where $\mathbf{x}^{(d)}$ is the decision corresponding to the $d$th realization of the network state $\mathbf{a}^{(d)}$.

For the minimization of the cost functions in Eqs. (7.1) and (7.2), the nodes need perfect access to the global network state information. However, they only observe their local state information that does not suffice for the optimal decision. This issue promotes the need for a cooperative optimization mechanism by means of the information sharing via their interconnecting backhaul links. The local information $\mathbf{a}_i$ dedicated to node $i$ is forwarded to its neighborhood. In practice, the data transmission capacity of the backhaul link is restricted for the limitation on wireless resource. Thus, perfect global sharing of the local information is normally prohibited. Such a restriction is characterized by the capacity of the backhaul link defined by the ratio of transmitted bits to channel uses. Also, the capacity of the link from node $i$ to node $j$ is denoted by $B_{ij}$. To accommodate this into the network optimization framework, an efficient quantization technique is necessary for the local measurements as well as a distributed decision strategy at an individual node.

## 7.2.2 Cooperative Mechanism

To develop a cooperative mechanism as indicated in the preceding discussion, the distributed decision rule at each node is made with two different computation units: a quantizer unit and a distributed optimizer unit (see Fig. 7.2). Upon measurement of the local information, the quantizer unit at node $i$ produces a discrete embedding vector of $\mathbf{a}_i$, called a *message*, to facilitate the information sharing over capacity-constrained backhaul links. The message transferred from node $i$ to node $j$ is denoted by $\mathbf{m}_{ij}$ and represented in a bipolar vector of length $B_{ij}$; that is, its element takes a discrete value of either $-1$ or $+1$. The optimizer unit makes an independent decision $\mathbf{x}_i$ with the available local information: the state measurement vector $\mathbf{a}_i$ and the collection of quantized messages received from neighboring node $j$ denoted by $\check{\mathbf{m}}_i \triangleq \{\mathbf{m}_{ji} : j \in \mathbf{E}_i\}$. In addition, the quantizer and optimizer units of node $i$ are represented in functional representations of $\mathcal{Q}_i(\cdot)$ and $\mathcal{D}_i(\cdot)$, respectively. In addition, let $\hat{\mathbf{m}}_i \triangleq \{\mathbf{m}_{ij} : j \in \mathbf{E}_i\}$ be a collection of the messages transmitted from node $i$. The quantizer unit at node $i$ is expressed as

$$\hat{\mathbf{m}}_i = \mathcal{Q}_i(\mathbf{a}_i) \in \{-1, +1\}^{B_i}, \tag{7.3}$$

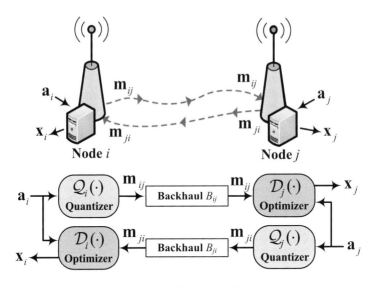

**Figure 7.2** Distributed optimization inference with quantized cooperation.

where $B_i \triangleq \sum_{j \neq i} B_{ij}$ is the number of bits that amounts to the total information sent by node $i$. The resulting message $\mathbf{m}_{ij}$ contains useful information required for the decision of node $j$ out of the state measurement observable at node $i$. Therefore, the role of the quantizer unit $\mathcal{Q}_i(\cdot)$ is not only simple discretization of the local measurement, but also message transmission for cooperative optimization using the backhaul links. The distributed optimizer unit $\mathcal{D}_i(\cdot)$ calculates the local decision variable as

$$\mathbf{x}_i = \mathcal{D}_i(\mathbf{c}_i), \tag{7.4}$$

where $\mathbf{c}_i$ is the concatenation of the local measurement $\mathbf{a}_i$ and the message set $\check{\mathbf{m}}$. The optimizer unit in Eq. (7.4) makes a distributed decision for $\mathbf{x}_i$ based on the received messages.

The cost minimization problem now boils down to the development of the learning technique for $\mathcal{Q}_i(\cdot)$ and $\mathcal{D}_i(\cdot)$, $(1 \leq i \leq N)$. The network optimization imposes design constraints so that the dynamic behaviors of wireless nodes are under control of the network management. Regularization constraints on local decision $\mathbf{x}_i$, such as the peak signal level and the transmit power budget, are reflected by a convex set $\mathcal{X}_i$ for $i$ $(1 \leq i \leq N)$. Additional sophisticated network controls associated with the state information and the global decision are enforced by constraint functions $g_k(\mathbf{a}, \mathbf{x})$ for $1 \leq k \leq K$. In particular, the constraints on long-term network behaviors (e.g., the average quality-of-service and the average transmit power budget), are ensured by the inequality on the average constraint function for some capacity level $G_k$ (i.e., $\mathbb{E}_{\mathbf{a}}[g_k(\mathbf{a}, \mathbf{x})] \leq G_k$). Since the average values of a function can be approximated using sample mean with respect to the training dataset, the resulting network cost minimization is formulated as

$$\min_{\{\mathcal{D}_i(\cdot),\mathcal{Q}_i(\cdot):1\leq i\leq N\}} \quad \frac{1}{D}\sum_{d=1}^{D} f(\mathbf{a}^{(d)},\mathbf{x}^{(d)}) \tag{7.5}$$

$$\text{subject to} \quad \frac{1}{D}\sum_{d=1}^{D} g_k(\mathbf{a}^{(d)},\mathbf{x}^{(d)}) \leq G_k,\ 1\leq k\leq K, \tag{7.6}$$

$$\mathbf{x}_i^{(d)} = \mathcal{D}_i(\mathbf{c}_i^{(d)}) \in \mathcal{X}_i,\ 1\leq i\leq N,\ 1\leq d\leq D, \tag{7.7}$$

$$\hat{\mathbf{m}}_i^{(d)} = \mathcal{Q}_i(\mathbf{a}_i^{(d)}) \in \{-1,+1\}^{B_i},\ 1\leq i\leq N,\ 1\leq d\leq D, \tag{7.8}$$

where $\hat{\mathbf{m}}_i^{(d)}$ and $\mathbf{c}_i^{(d)}$ are the outgoing messages in Eq. (7.3) and the input vector in Eq. (7.4) obtained from the $d$th measurement sample $\mathbf{a}^{(d)}$, respectively. This formulation allows to generalize design tasks encountered in various wireless communication applications, such as interference management, power control, and transceiver design. The binary constraint in Eq. (7.8) requires the quantization of messages for information sharing via finite-capacity backhaul links.

The global optimum of a generalized objective function in Eq. (7.5) with the optimality guarantees in wireless applications [6–9] has not been properly investigated. Major challenges lie in the design of operators $\mathcal{Q}_i(\cdot)$ and $\mathcal{D}_i(\cdot)$. However, the design problem may become intractable by traditional optimization techniques involved with a model-based formulation, since two operators can be chosen as any continuous mappings with unknown computation structure.

A machine-learning-based approach is employed to design two units. According to the universal approximation theorem [10, 11], two units handling unknown computational rules $\mathcal{Q}_i(\cdot)$ and $\mathcal{D}_i(\cdot)$ are modeled by appropriate fully connected NNs. The detailed descriptions on the machine-learning-based strategies for the constrained formulation in Eq. (7.5) are introduced focusing on an unsupervised learning technique that does not call for prior knowledge of the solution.

## 7.3    Centralized Approach

This section considers a centralized approach for the solution of Eq. (7.5) and investigates an efficient training technique addressing the solution of a constrained optimization. In an ideal case of infinite-capacity backhaul, the bipolar constraint on the signal value in Eq. (7.8) no longer exists, while the messages can take continuous-valued vectors. An optimum coordination strategy is to share $\mathbf{a}_i$ directly through lossless backhaul links. In this case, the best message generator unit $\mathcal{Q}_i(\cdot)$ in Eq. (7.3) becomes the identity map; that is, $\mathbf{m}_{ij} = \mathbf{a}_i,\ \forall j \in \mathbf{E}_i$. Thus, node $i$ obtain the perfect knowledge of the state information $\mathbf{a}$. It points out that a common form of a centralized optimization rule suffices for all nodes. That is, the global decision $\mathbf{x}$ is readily found by a central unit $\mathcal{C}(\cdot)$ as

$$\mathbf{x} = \mathcal{C}(\mathbf{a}). \tag{7.9}$$

Based on this operation, the distributed formulation Eq. (7.5) further reduces to

$$\min_{\mathcal{C}(\cdot)} \quad \frac{1}{D} \sum_{d=1}^{D} f(\mathbf{a}^{(d)}, \mathcal{C}(\mathbf{a}^{(d)})) \tag{7.10}$$

$$\text{subject to} \quad \frac{1}{D} \sum_{d=1}^{D} g_k(\mathbf{a}^{(d)}, \mathcal{C}(\mathbf{a}^{(d)})) \leq G_k, \ 1 \leq k \leq K, \tag{7.11}$$

$$\mathbf{x}_i^{(d)} \in \mathcal{X}_i, \ 1 \leq i \leq N, \ 1 \leq d \leq D. \tag{7.12}$$

Note that the training algorithm is much simplified to obtain a single function $\mathcal{C}(\cdot)$ instead of training all individual functions $\mathcal{D}_i(\cdot)$.

### 7.3.1 Learning to Optimize

The problem now becomes the design of a single operator $\mathcal{C}(\cdot)$ commonly used at all nodes. However, it is still challenging to solve the problem for its nonconvexity of the formulation and intractable computation structure. To tackle this, an NN-based strategy, called the *learning to optimize* approach [12–20], is considered. As illustrated in Fig. 7.3, this approach introduces an NN to model an unknown mapping for $\mathcal{C}(\cdot)$ in Eq. (7.9). A central optimizer NN $\phi_C(\cdot; \theta_C)$ with trainable parameter $\theta_C$ aims at approximating the solution of the problem in Eq. (7.10). The input to the central optimizer NN depends solely on the global measurement vector $\mathbf{a}$, and the resulting output is the decision $\mathbf{x}$. For simplicity, the description focuses on the fully connected NN architecture with $R$ layers. The output of layer $r$ ($1 \leq r \leq R$) is given as $U_r$-dimensional real vector $\mathbf{u}_r$. For convenience of the representation, the input and the output are denoted by $\mathbf{u}_0 = \mathbf{a}$ and $\mathbf{u}_{R+1} = \mathbf{x}$, respectively. Then, the output of each layer $r$ is expressed as

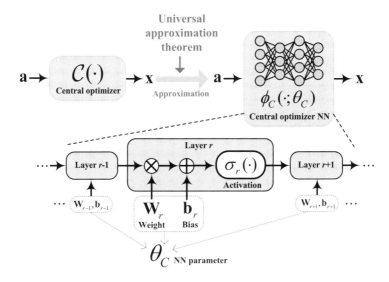

**Figure 7.3** An NN approximation of the central optimizer.

$$\mathbf{u}_r = \sigma_r(\mathbf{W}_r\mathbf{u}_{r-1} + \mathbf{b}_r), \ r = 1, \dots, R, \tag{7.13}$$

where $\mathbf{W}_r$ and $\mathbf{b}_r$ are an $U_r \times U_{r-1}$ weight matrix and an $U_r$-dimensional bias vector employed at layer $r$, respectively. Furthermore, $\sigma_r(\cdot)$ stands for a nonlinear activation function. The weights and biases are collected as a set of trainable parameters $\theta_C \triangleq \{\mathbf{W}_r, \mathbf{b}_r : 1 \le r \le R\}$. The cascaded neural calculation of Eq. (7.13) defines an end-to-end NN mapping $\mathbf{x} = \phi_C(\mathbf{a}; \theta_C)$ between $\mathbf{a}$ and $\mathbf{x}$, which is given by

$$\begin{aligned}
\mathbf{x} &= \phi_C(\mathbf{a}; \theta_C) \\
&= \sigma_R(\mathbf{W}_R\sigma_{R-1}(\mathbf{W}_{R-1} \times \cdots \times \sigma_1(\mathbf{W}_1\mathbf{a} + \mathbf{b}_1) + \cdots + \mathbf{b}_{R-1}) + \mathbf{b}_R).
\end{aligned} \tag{7.14}$$

The design target for $\phi_C(\cdot; \theta_C)$ is to approximate the operator $\mathcal{C}(\cdot)$ as

$$\mathbf{x} = \mathcal{C}(\mathbf{a}) \simeq \phi_C(\mathbf{a}; \theta_C). \tag{7.15}$$

Using the approximation in Eq. (7.15), a complicated optimization in Eq. (7.10) that identifies a valid function out of unaccountably many candidates can reduce to the training task of NN parameter set $\theta_C$ having a finite dimension. The universal approximation theorem guarantees an arbitrary precision of the approximation in Eq. (7.15) [10]. Precisely, it states that an NN having a finite number of fully connected layers with sigmoid activation can model a continuous-valued function within arbitrary small error. Furthermore, its extended results [11] validate the universal approximation theorem for a discrete function as well. Hence, the optimal solution for the nonconvex formulation in Eq. (7.10) is successfully characterized by a well-trained NN, whether the output of the target mapping $\mathcal{C}(\cdot)$ is continuous or discrete. However, the theorems ensure the existence of such NNs only, while their specific design techniques are left unaddressed.

It is essential to determine the structure and parameter set $\theta_C$ of an efficient NN so that the resulting outputs $\mathbf{x}^{(d)} = \phi_C(\mathbf{a}^{(d)}; \theta_C)$ for each $\mathbf{a}^{(d)}$ becomes optimal for $1 \le d \le D$. To make a feasible decision $\mathbf{x}_i \in \mathcal{X}_i$, the output activation can be designed as a projection operation onto convex feasible set $\mathcal{X}_i$ [21]. Thus, it is defined as $\Pi_{\mathcal{X}_i}(\mathbf{u}) \triangleq \arg\min_{\mathbf{x}_i \in \mathcal{X}_i} \|\mathbf{u} - \mathbf{x}\|^2$, which becomes a convex quadratic program [22]. As a result, an implicit constraint Eq. (7.12) can be removed for simplification. The substitution of Eq. (7.15) into Eq. (7.10) results in the overall formulation expressed as

$$\min_{\theta_C} \mathcal{F}(\theta_C) \tag{7.16}$$

$$\text{subject to } \mathcal{G}_k(\theta_C) \le G_k, \ 1 \le k \le K, \tag{7.17}$$

where new cost and constraint functions $\mathcal{F}(\theta_C)$ and $\mathcal{G}_k(\theta_C)$, are defined, respectively, as

$$\mathcal{F}(\theta_C) \triangleq \frac{1}{D} \sum_{d=1}^{D} f(\mathbf{a}^{(d)}, \phi_C(\mathbf{a}^{(d)}; \theta_C)) \tag{7.18}$$

$$\mathcal{G}_k(\theta_C) \triangleq \frac{1}{D} \sum_{d=1}^{D} g_k(\mathbf{a}^{(d)}, \phi_C(\mathbf{a}^{(d)}; \theta_C)). \tag{7.19}$$

Therefore, the nonconvex network management problem in Eq. (7.10) boils down to a tractable task of identifying the NN parameter $\theta_C$.

## 7.3.2    Constrained Unsupervised Learning

The prior knowledge of the optimal solution is, in general, unavailable since the problem in Eq. (7.10) has a nonconvex objective and constraints. In this configuration, a simple supervised learning approach [12], which trains an NN to memorize optimal solutions, is not a suitable strategy. To handle Eq. (7.16) without numerous samples of optimal solutions, developing an unsupervised NN training algorithm is essential by incorporating the network constraint in Eq. (7.17). However, it is challenging since most training techniques are originally intended for unconstrained learning tasks. To this end, the Lagrange duality method [5] can be considered as a valid alternative. The Lagrange duality formulation transforms the original constrained formulation to an unconstrained one associated with a dual function called a Lagrangian. Regardless of convexity properties of the original problem, the Lagrange duality approach for Eq. (7.16) establishes its optimality in terms of the strong duality [19, 20, 23]. The Lagrangian of Eq. (7.16) is defined as

$$\mathcal{L}(\theta_C, \lambda) = \mathcal{F}(\theta_C) + \sum_{k=1}^{K} \lambda_k (\mathcal{G}_k(\theta_C) - G_k), \qquad (7.20)$$

where a nonnegative $\lambda_k$, referred to as a Lagrange multiplier, is a dual variable corresponding to an individual constraint in Eq. (7.17), and a vector representing their collection is denoted by $\lambda \triangleq \{\lambda_k : 1 \leq k \leq K\} \in \mathbb{R}^K$. The resulting dual problem is given as

$$\max_{\lambda} \ \min_{\theta_C} \mathcal{L}(\theta_C, \lambda) \qquad (7.21)$$

$$\text{subject to } \lambda_k \geq 0, \ 1 \leq k \leq K. \qquad (7.22)$$

The dual formulation sets forth simultaneous maximization and minimization of the Lagrangian with respect to the dual variable $\lambda$ and the NN parameter $\theta_C$, respectively. Such an optimization problem is normally dealt with the primal-dual method [24]. It adjusts the feasibility of the NN output with respect to the cost function in cooperation with a careful control for the dual variable during the NN parameter training process. Alternating updates are conducted between the primal variable (i.e., the NN parameter $\theta_C$, and the dual variable $\lambda$. The dual variable $\lambda$ helps optimizing the NN parameter $\theta_C$ by providing a tight lower bound of the optimal network cost function. To be precise, the real-time inference that uses the NN computation rule in Eq. (7.14) does not take the dual variable as its input. Thus, the dual variable is discarded upon the completion of the optimization of the NN parameter $\theta_C$, and the NN computation rule in Eq. (7.14) is used only, which saves the computational cost in the real-time inference of the solution.

Let $z^{[t]}$ be the state of variable $z$ evaluated at the $t$th iteration. The primal update for $\theta_C^{[t]}$ is derived using a gradient descent (GD) technique given by

$$\theta_C^{[t]} = \theta_C^{[t-1]} - \eta \nabla_{\theta_C} \mathcal{L}(\theta_C^{[t-1]}, \boldsymbol{\lambda}^{[t-1]})$$

$$= \theta_C^{[t-1]} - \eta \left( \nabla_{\theta_C} \mathcal{F}(\theta_C^{[t-1]}) + \sum_{k=1}^{K} \lambda_k^{[t-1]} (\nabla_{\theta_C} \mathcal{G}_k(\theta_C^{[t-1]}) - G_k) \right), \quad (7.23)$$

where a positive real number $\eta$ is a learning rate, and $\nabla_z$ represents the gradient operator with respect to variable $z$. Using the chain rule, the gradient $\nabla_{\theta_C} \mathcal{F}(\theta_C)$ in Eq. (7.23) is explicitly expressed as

$$\nabla_{\theta_C} \mathcal{F}(\theta_C) = \frac{1}{D} \sum_{d=1}^{D} \nabla_{\theta_C} f(\mathbf{a}^{(d)}, \phi_C(\mathbf{a}^{(d)}; \theta_C))$$

$$= \frac{1}{D} \sum_{d=1}^{D} \nabla_{\mathbf{x}} f(\mathbf{a}^{(d)}, \mathbf{x}) \Big|_{\mathbf{x} = \phi_C(\mathbf{a}^{(d)}; \theta_C)} \cdot \nabla_{\theta_C} \phi_C(\mathbf{a}^{(d)}; \theta_C), \quad (7.24)$$

and, likewise, $\nabla_{\theta_C} \mathcal{G}_k(\theta_C)$ is obtained as

$$\nabla_{\theta_C} \mathcal{G}_k(\theta_C) = \frac{1}{D} \sum_{d=1}^{D} \nabla_{\mathbf{x}} g_k(\mathbf{a}^{(d)}, \mathbf{x}) \Big|_{\mathbf{x} = \phi_C(\mathbf{a}^{(d)}; \theta_C)} \cdot \nabla_{\theta_C} \phi_C(\mathbf{a}^{(d)}; \theta_C). \quad (7.25)$$

To keep a dual variable nonnegative for the complementary slackness [5], the projected subgradient method [24] is applied to the dual update for $\lambda_k^{[t]}$. Since the subgradient of the Lagrangian with $\lambda_k$ is simply linear, the explicit expression for new $\lambda_k^{[t]}$ is obtained as

$$\lambda_k^{[t]} = \left[ \lambda_k^{[t-1]} + \eta \partial_{\lambda_k} \mathcal{L}(\theta_C^{[t-1]}, \boldsymbol{\lambda}^{[t-1]}) \right]_+$$

$$= \left[ \lambda_k^{[t-1]} + \eta (\mathcal{G}_k(\theta_C^{[t-1]}) - G_k) \right]_+, \quad (7.26)$$

where $\partial_z$ represents the subgradient with respect to variable $z$ and $[x]_+ \triangleq \max\{0, x\}$ is a nonnegative projection function equivalent to rectified linear unit (ReLU) activation. The primal-dual update pair with Eqs. (7.23) and (7.26) forms a joint GD-based training algorithm for the NN parameter set $\theta_C$ and the corresponding dual solution $\boldsymbol{\lambda}$. Parallel computations of sample gradients can proceed with the aids of general-purpose graphical processing units (GPGPUs). The training rule of Eqs. (7.23) and (7.26) requests no prior information of the optimal decision. As a result, it can be realized in a fully unsupervised manner without labeled training dataset.

Major challenges in the implementation of the primal-dual method are involved with the computation of the gradient for Eqs. (7.18) and (7.19). These functions, in fact, require a prohibitively high complexity for simultaneous evaluation of $2D$ gradients Eqs. (7.24) and (7.25) even in a single primal-dual update. Thus, a stochastic optimization algorithm in machine-learning area, in particular, mini-batch stochastic gradient descent (SGD) method, is used to reduce a computational load

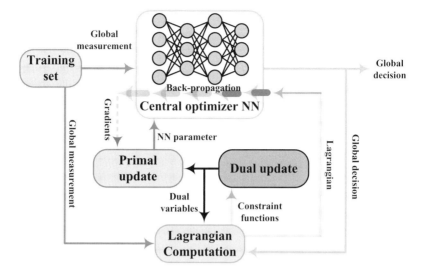

**Figure 7.4** Primal-dual training strategy for constrained unsupervised learning.

for the updates in Eqs. (7.23) and (7.26). It exploits a subset set of the training data $\mathcal{A} = \{\mathbf{a}^{(1)}, \ldots, \mathbf{a}^{(D)}\}$, referred to as a mini-batch set. The gradients in Eqs. (7.24) and (7.25) are evaluated over mini-batch set $\mathcal{S} \subset \mathcal{A}$ of size $S$ that contains much smaller number of realizations than the training data (i.e., $S \ll D$).

Figure 7.4 illustrates the overall the primal-dual training algorithm for the constrained NN optimization. Initially, a mini-batch set is sampled at random from the training dataset. In forward pass calculation, the samples are passed to the central optimizer NN $\phi_C(\cdot; \theta_C)$ to obtain the corresponding global decision vector $\mathbf{x} = \phi_C(\mathbf{a}; \theta_C)$. The Lagrangian in Eq. (7.20) is subsequently computed using a set of pairs $(\mathbf{a}, \mathbf{x})$ and dual variables determined at the previous iteration. The Lagrangian calculation result, which contains the information about the corresponding constraint function, is transferred to the dual update in Eq. (7.26). The updated dual variables reflect the extent of the constraint violation. Thus, they control the direction of the primal update so that the forward pass decision output maintains the feasibility for the network constraint Eq. (7.17). The backward pass calculation or back propagation through the central optimizer NN proceeds with the Lagrangian to calculate its gradient. The primal update Eq. (7.23) in turn updates the NN parameter and feeds it back to the central optimizer NN for the next iteration. The detailed procedure of the overall algorithm is summarized in Algorithm 7.1.

The central optimizer NN is trained along with the dual variable updates. Once the optimization of the NN parameter is completed, the dual variable becomes unnecessary since the trained central optimizer NN $\phi_C(\cdot; \theta_C)$ is able to compute the global decision $\mathbf{x}$ alone. As depicted in Fig. 7.5, the real-time implementation is accomplished by means of the parameter $\theta_C$ stored in individual nodes. With the global observation $\mathbf{a}$ at hands, each node can obtain the global decision $\mathbf{x} = \phi_C(\mathbf{a}; \theta_C)$ using

---

**Algorithm 7.1** Primal-dual training algorithm.

Initialize $t = 0$, $\theta_C^{[0]}$ and $\lambda^{[0]}$.
**repeat**
    Set $t \leftarrow t + 1$.
    Sample a mini-batch set $\mathcal{S} \subset \mathcal{A}$ from the training dataset $\mathcal{A}$.
    Update $\theta_C^{[t]}$ and $\lambda^{[t]}$ from (7.23) and (7.26), respectively.
**until** convergence

---

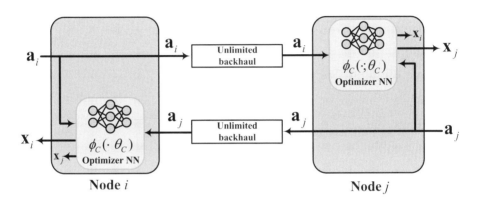

**Figure 7.5** Centralized learning structure with central optimizer NNs.

the trained parameter $\theta_C$. The desired solution $\mathbf{x}_i$ is obtained at node $i$ by masking other part $\mathbf{x}_j$ in the central solution, $j \neq i$.

## 7.4   Distributed Approach

The constrained unsupervised learning strategy can be extended to a distributed network setup in Eq. (7.5). Unlike the centralized learning structure that shares uncoded (or raw) measurement directly among nodes, the distributed approach needs additional quantization operation $\mathcal{Q}_i(\cdot)$. It is in charge of encoding of the local observation $\mathbf{a}_i$ into a discrete message for the capacity-limited backhaul link transmission. Two component functions $\mathcal{Q}_i(\cdot)$ and $\mathcal{D}_i(\cdot)$ of node $i$ are realized via quantizer NN $\phi_{Q_i}(\cdot; \theta_{Q_i})$ and distributed optimizer NN $\phi_{D_i}(\cdot; \theta_{D_i})$, respectively. Figure 7.6 shows the structure of the node computation comprised of two different NNs. Thus, there exist $2N$ NNs to be trained in total over the the network. The message quantizer in Eq. (7.3) and the distributed optimizer in Eq. (7.4) are approximated, respectively, as

$$\hat{\mathbf{m}}_i = \mathcal{Q}_i(\mathbf{a}_i) \simeq \phi_{Q_i}(\mathbf{a}_i; \theta_{Q_i}) \in \{-1, +1\}^{B_i}, \tag{7.27}$$

$$\mathbf{x}_i = \mathcal{D}_i(\mathbf{c}_i) \simeq \phi_{D_i}(\mathbf{c}_i; \theta_{D_i}). \tag{7.28}$$

The universal approximation theorem also holds for Eqs. (7.27) and (7.28). To design these NNs, the collection of the corresponding NN parameters is denoted by

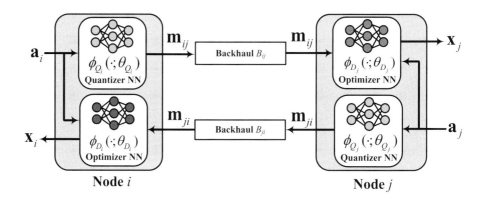

**Figure 7.6** Distributed learning structure with quantizer NNs and optimizer NNs.

$\boldsymbol{\theta} = \{\theta_{D_i}, \theta_{Q_i} : 1 \leq i \leq N\}$. Accordingly, the resulting distributed optimization in Eq. (7.5) is recast as

$$\min_{\boldsymbol{\theta}=\{\theta_{D_i}, \theta_{Q_i}:1 \leq i \leq N\}} \mathcal{F}(\boldsymbol{\theta}) \tag{7.29}$$

$$\text{subject to} \quad \mathcal{G}_k(\boldsymbol{\theta}) \leq G_k, \ 1 \leq k \leq K, \tag{7.30}$$

$$\phi_{Q_i}(\mathbf{a}_i^{(d)}; \theta_{Q_i}) \in \{-1, +1\}^{B_i}, \ 1 \leq i \leq N, 1 \leq d \leq D, \tag{7.31}$$

where cost and constraint functions $\mathcal{F}(\boldsymbol{\theta})$ and $\mathcal{G}_k(\boldsymbol{\theta})$ are expressed with a slight abuse of the notation, respectively, as

$$\mathcal{F}(\boldsymbol{\theta}) \triangleq \frac{1}{D} \sum_{d=1}^{D} f(\mathbf{a}^{(d)}, \{\phi_{D_i}(\mathbf{c}_i^{(d)}; \theta_{D_i}) : 1 \leq i \leq N\}) \tag{7.32}$$

$$\mathcal{G}_k(\boldsymbol{\theta}) \triangleq \frac{1}{D} \sum_{d=1}^{D} g_k(\mathbf{a}^{(d)}, \{\phi_{D_i}(\mathbf{c}_i^{(d)}; \theta_{D_i}) : 1 \leq i \leq N\}). \tag{7.33}$$

Note that the distributed formulation in Eq. (7.29) is viewed as a generalization of the centralized counterpart in Eq. (7.16). The binary constraint in Eq. (7.31) can be relaxed to allow continuous-valued message transmission. As discussed earlier, the quantizer NN $\phi_{Q_i}(\cdot; \theta_{Q_i})$ uses the local measurement $\mathbf{a}_i$ as the message output $\mathbf{m}_{ij}$. In this case, the quantizer NNs can be removed by setting $\mathbf{m}_{ij} = \mathbf{a}_i, \forall j \in \mathbf{E}_i$. This results in a simplified distributed optimization rule $\mathbf{x}_i = \phi_{D_i}(\mathbf{c}_i; \theta_{D_i}) = \phi_{D_i}(\mathbf{a}; \theta_{D_i})$. As a result, the distributed learning problem in Eq. (7.29) reduces to the centralized task in Eq. (7.16), where the central optimizer NN $\phi_C(\cdot; \theta_C)$ is represented by a distributed optimizer NN group defined as $\phi_C(\mathbf{a}; \theta_C) = \{\phi_{D_i}(\mathbf{a}; \theta_{D_i}) : 1 \leq i \leq N\}$ with the parameter set $\theta_C = \{\theta_{D_i} : 1 \leq i \leq N\}$.

The output layer of the distributed optimizer NN has an activation function of a projection operator for the convex condition $\mathbf{x}_i = \phi_{D_i}(\mathbf{a}_i, \check{\mathbf{m}}_i; \theta_{D_i}) \in \mathcal{X}_i$, thereby ensuring the decision alphabet constraint. On the other hand, it is not simple to address a nonconvex binary constraint in Eq. (7.31) with the projection activation (e.g., a

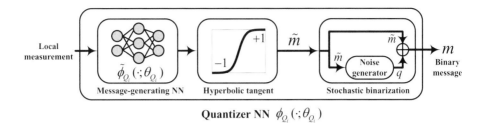

**Quantizer NN** $\phi_{Q_i}(\cdot;\theta_{Q_i})$

**Figure 7.7** Structure of quantizer NN with stochastic binarization.

signum function) since its gradient becomes zero for all input ranges of the output layer. Such a deterministic binarization incurs a critical vanishing gradient problem. In such a case, the NN parameters are no longer updated after few iterations of the gradient-based algorithms, possibly ending up failing to learn a valid binary solution. Therefore, a sophisticated binarization technique that improves the SGD update of the quantizer NN is necessary.

## 7.4.1 Stochastic Message Binarization

To develop a new NN-based quantization technique compatible to existing SGD training algorithms, this subsection focuses on generating a scalar bipolar message $m \in \{-1, +1\}$. For simple representation, all subscripts are ignored. The results are easily extended to a general type of discrete vector-valued messages with the element-wise quantization. A key idea is to model the randomness of the actual quantized value with a carefully controlled output activation. As depicted in Fig. 7.7, the quantizer NN $\phi_{Q_i}(\cdot;\theta_{Q_i})$ consists of three component units: a message-generating NN $\tilde{\phi}_{Q_i}(\cdot;\theta_{Q_i})$, a hyperbolic tangent activation $\tanh(x) \triangleq \frac{e^{2x}-1}{e^{2x}+1}$, and a stochastic binarization activation. Let $\tilde{m} \triangleq \tanh(\tilde{\phi}_{Q_i}(\mathbf{a}_i;\theta_{Q_i}))$ denote a relaxed binary message output that takes a value within the bounded signal range $[-1, +1]$. A bipolar message $m \in \{-1, +1\}$ is regarded as a quantized value of this continuous-valued message $\tilde{m}$. The resulting quantization NN output is expressed as

$$m = \tilde{m} + q = \tanh(\tilde{\phi}_{Q_i}(\mathbf{a}_i;\theta_{Q_i})) + q, \tag{7.34}$$

where $q$ reflects a quantization noise added deliberately to the relaxed message $\tilde{m}$ to guarantee the discreteness of message $m$. Since $m$ takes on either $-1$ or $+1$, the corresponding values of $q$ are given by $-1 - \tilde{m}$ and $1 - \tilde{m}$, respectively. To obtain a nontrivial quantization output, the stochastic nature of the quantization is applied by characterizing it with binary noise probabilities of $\Pr\{q = -1-\tilde{m}\} = p$ and $\Pr\{q = 1-\tilde{m}\} = 1 - p$. The probability value $p \in [0,1]$ is a hyperparameter set beforehand. Therefore, the quantization noise is modeled as a Bernoulli random variable given by

$$q = \begin{cases} -1 - \tilde{m}, & \text{with probability } p, \\ 1 - \tilde{m}, & \text{with probability } 1 - p. \end{cases} \tag{7.35}$$

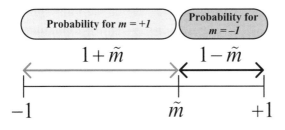

**Figure 7.8** Interpretation of stochastic binarization

A carefully controlled noise in Eq. (7.35) guarantees the output message $m$ to have a value of either $-1$ or $+1$. Such a binarization strategy is referred to as a *stochastic binarization*. The frequency of the final value relies on the noise statistics, and the identification of hyperparameter $p$ is crucial for good quantization results. An efficient design principle is to hold the zero mean property $\mathbb{E}_q[q] = 0$, leading to a simple implementation of the back propagation since it lowers a corrupting impact of intentional noise on the gradient value for training. The mean of the noise is expressed as

$$\mathbb{E}_q[q] = \tilde{m} + p(-1 - \tilde{m}) + (1 - p)(1 - \tilde{m}) = 1 - 2p. \qquad (7.36)$$

Equating it to zero obtains the value of $p$ as

$$p = \frac{1 - \tilde{m}}{2}. \qquad (7.37)$$

Figure 7.8 sketches the intuition behind the stochastic binarization. Provided that $\tilde{m}$ lies within the interval $[-1, +1]$, the stochastic binarization activation yields a discrete output of $-1$ or $+1$ with probability $\frac{1-\tilde{m}}{2}$ and $\frac{1+\tilde{m}}{2}$, respectively. These probabilities are proportional to the distances from the unquantized value $\tilde{m}$ to binary points of $+1$ and $-1$.

An end-to-end training strategy is now addressed subject to the stochastic binarization technique. For the SGD updates, the gradient of the binary message $m$ with respect to the quantizer NN parameter $\theta_{Q_i}$, denoted by $\nabla_{\theta_{Q_i}} m$, is necessary. However, the randomizing noise in Eq. (7.35) inhibits the approach from obtaining an analytical expression for the gradient $\nabla_{\theta_{Q_i}} m$. To resolve this, gradient estimation techniques [25, 26] are employed so that an intractable stochastic gradient of the NN is approximated with its average value with respect to noise probability distribution (i.e., the binary noise distribution in Eq. (7.35)). For a given NN output $\tilde{m}$, the gradient of the binary message $m$ is approximated as

$$\nabla_{\theta_{Q_i}} m \simeq \nabla_{\theta_{Q_i}} \mathbb{E}_q[m] \qquad (7.38)$$

$$= \nabla_{\theta_{Q_i}} \tilde{m} + \nabla_{\theta_{Q_i}} \mathbb{E}_q[q] \qquad (7.39)$$

$$= \nabla_{\theta_{Q_i}} \tilde{m} \qquad (7.40)$$

$$= \nabla_{\theta_{Q_i}} \tanh(\tilde{\phi}_{Q_i}(\mathbf{a}_i; \theta_{Q_i})) \tag{7.41}$$

$$= \left( \frac{2e^{2x}}{e^{2x}+1} \right)^2 \Bigg|_{x=\tilde{\phi}_{Q_i}(\mathbf{a}_i; \theta_{Q_i})} \cdot \nabla_{\theta_{Q_i}} \tilde{\phi}_{Q_i}(\mathbf{a}_i; \theta_{Q_i}) \tag{7.42}$$

Note here that the zero mean noise property removes the gradient of $\mathbb{E}_q[q]$ in Eq. (7.40). It also indicates in Eq. (7.40) that the gradient of the continuous-valued message $\tilde{m}$ can replace that of the binary message $m$. Since $\tilde{m}$ is the output of the hyperbolic tangent activation, its gradient never disappears and becomes well-defined in the entire range of the NN output value. As a result, the quantizer NNs can be efficiently trained with existing mini-batch SGD algorithms together with the optimizer NNs.

Figure 7.9 depicts the overall training strategy of the quantizer NN with the stochastic binarization activation. By combining Eqs. (7.34) and (7.40), the stochastic binarization layer has a hybrid structure for forward-pass and backward-pass operation. In the forward-pass computation, the quantizer NN produces the binary message $m = \tilde{m} + q$. The message-generating NN unit $\tilde{\phi}_{Q_i}(\mathbf{a}_i; \theta_{Q_i})$ first preprocesses the local measurement input $\mathbf{a}_i$. The following hyperbolic tangent activation bounds the NN output within the interval $[-1, +1]$. The resulting output of continuous-valued message $\tilde{m} = \tanh(\tilde{\phi}_{Q_i}(\mathbf{a}_i; \theta_{Q_i}))$ is fed into the stochastic binarization Eq. (7.23). The forward-pass stochastic binarization adds a Bernoulli random value $q$ in Eq. (7.35) to $\tilde{m}$. The final output of the quantizer NN $\phi_{Q_i}(\cdot; \theta_{Q_i})$ becomes a bipolar message $m$ obtained as

$$m = \phi_{Q_i}(\mathbf{a}_i; \theta_{Q_i}) = \tanh(\tilde{\phi}_{Q_i}(\mathbf{a}_i; \theta_{Q_i})) + q. \tag{7.43}$$

On the other hand, the backward-pass operation calculates the gradient with respect to the NN parameter $\theta_{Q_i}$. It is noted in Eq. (7.40) that the randomness of the stochastic operation is ignored in the gradient computation, thereby using $\tilde{m} = \tanh(\tilde{\phi}_{Q_i}(\mathbf{a}_i; \theta_{Q_i}))$ as the final output in the backward-pass operation. Combining this with Eq. (7.43), the operation of the quantizer NN is summarized as

$$\phi_{Q_i}(\mathbf{a}_i; \theta_{Q_i}) = \begin{cases} \tanh(\tilde{\phi}_{Q_i}(\mathbf{a}_i; \theta_{Q_i})) + q & \text{in forward pass,} \\ \tanh(\tilde{\phi}_{Q_i}(\mathbf{a}_i; \theta_{Q_i})) & \text{in backward pass.} \end{cases} \tag{7.44}$$

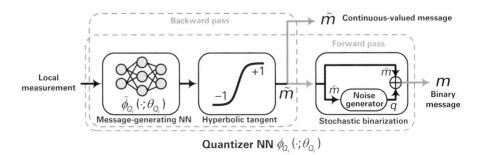

**Figure 7.9** Training flow of quantizer NN.

In the forward-pass operation, the quantizer NN estimates the network cost function to examine the current performance of the optimizer for training NNs. In addition, it produces a binary message in Eq. (7.43) which each of neighboring nodes uses to compute its own local decision when deployed in the network. On the contrary, the backward-pass value calculated with the random perturbation removed is used in the evaluation of an uncorrupted estimate on the gradient to update the NN parameter during the training process.

The stochastic binarization can be implemented in Tensorflow using a stop gradient operator `tf.stop_gradient()`. This command directly outputs an input tensor without explicit evaluation of the derivative. To be precise, the binarization output is realized as `tf.stop_gradient`$(m - \tilde{m}) + \tilde{m}$. It blocks the gradient calculation of $m - \tilde{m}$ in the backward pass, whereas it produces a binary value of the message $m$ in the forward pass since `tf.stop_gradient()` releases the input value intact as the output.

## 7.4.2    Centralized Training and Distributed Deployment

Figure 7.10 illustrates an end-to-end training algorithm for a group of wireless nodes equipped with a pair of optimizer NN and quantizer NN. The cost function of this group training algorithm is the Lagrangian function obtained from Eqs. (7.29) and (7.30). Note that the binary constraint Eq. (7.31) is not included in the Lagrangian since it is implicitly reflected at the output activation of the quantizer NN.

As discussed previously, the strong duality holds for Eq. (7.29). Thus, the primal-dual training method in Eqs. (7.23) and (7.26) can be used for joint optimization of the NN parameter set $\theta$ as well as dual variable vector $\lambda$. A centralized training algorithm

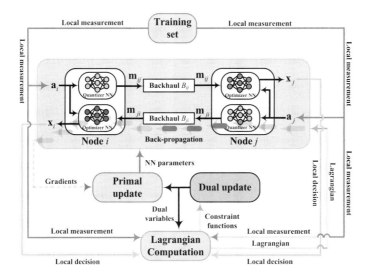

**Figure 7.10**  Centralized training method of distributed NNs.

is run at a cloud server that collects all NNs and data samples. In contrast to the training process the central optimizer NN shown in Fig. 7.4, in a distributed setup, the global measurement vector $\mathbf{a}$ is split into $N$ local observations $\mathbf{a}_i$ ($1 \leq i \leq N$), each provided into component NN units incorporated at the corresponding node. All nodes exchange binary messages while getting trained. As a result, the quantizer NNs are able to learn efficient coordination policies over capacity-constrained backhauls. At the same time, the optimizer NNs can find their own local decisions that minimize the network cost function calculated by collecting binary information transferred from the neighborhood. Node $i$ configures its own NNs $\phi_{D_i}(\cdot; \theta_{D_i})$ and $\phi_{Q_i}(\cdot; \theta_{Q_i})$ with the trained parameters to make an inference on $\mathbf{x}_i$. The NNs do not need dual variable for obtaining the solution. Thus, only the node network in a gray box of Fig. 7.10 is deployed and other components in the training algorithm are removed for the real-time distributed inference. Thus, the resulting deployed structure can save the computational cost of the network management.

## 7.5      Case Study 1: Cognitive Multiple Access Channels

This section presents an application of the constrained unsupervised learning technique to a distributed transmit power control task. To this end, a cognitive multiple access channel (C-MAC) network [7] is considered first. The C-MAC describes a spectrum sharing problem of the long-term evolution unlicensed (LTE-U) system. In this application, wireless access links of unlicensed secondary nodes (SNs) are managed in a license-assisted manner under the supervision of licensed primary networks. An important design challenge for transmission technology is a protocol that guarantees certain levels of quality of service (QoS) both for primary and secondary networks. This special feature imposes a nontrivial design constraint that an individual SN does not interfere with the licensed system communication. One possible solution is a power allocation policy for the secondary network carefully optimized so that its system capacity is enhanced via spectrum sharing, while ensuring a certain level of the QoS for the licensed nodes. Thus, a distributed power control method among SNs separated geometrically is necessary.

## 7.5.1      System Description

Consider an uplink setup of a wireless cognitive radio network with $N$ SNs. Each SN desires to send its data to a secondary base station (SBS). However, no legitimate resource block is assigned to a wireless node in this secondary network. A license-assisted access allows for the uplink communication of the secondary network by sharing resource blocks with an SN assigned to a licensed primary node (PN). Thus, the communication in the secondary network naturally interferes with the PN. Each SN is designated as a computing node in a distributed wireless network shown in Fig. 7.1. The channel qualities of wireless links from SN $i$ to the SBS and the PN are denoted by $a_{S,i}$ and $a_{P,i}$, respectively. For simple representation, a local measurement vector of SN $i$ is defined as $\mathbf{a}_i \triangleq \{a_{S,i}, a_{P,i}\} \in \mathbb{R}^2$. Furthermore, the global measurement vector $\mathbf{a}$ is defined as $\mathbf{a} \triangleq \{a_{S,i}, a_{P,i} : 1 \leq i \leq N\} \in \mathbb{R}^{2N}$.

In the C-MAC system, a local decision variable $x_i$ corresponds to the transmit power level of SN $i$. Their collection, denoted by $\mathbf{x}$, forms the global solution of the cognitive radio network; that is, $\mathbf{x} \triangleq \{x_i : 1 \leq i \leq N\}$. These variables are collectively identified to maximize the ergodic sum capacity of the secondary network (i.e., the sum capacity averaged over channel states $\mathbf{a}$), while maintaining the level of the interference to the PN below a predefined threshold. Thus, SN $i$ aims to make a local decision on its state $x_i$ based on its observation $\mathbf{a}_i$ obtained via distributed coordination among the neighborhood. The resulting problem is formulated as [7]

$$\max \quad \frac{1}{D} \sum_{d=1}^{D} \log \left( 1 + \sum_{i=1}^{N} a_{S,i}^{(d)} x_i^{(d)} \right) \tag{7.45}$$

$$\text{subject to} \quad \frac{1}{D} \sum_{d=1}^{D} x_i^{(d)} \leq P, \ 1 \leq i \leq N, \tag{7.46}$$

$$\frac{1}{D} \sum_{d=1}^{D} \left( \sum_{i=1}^{N} a_{P,i}^{(d)} x_i^{(d)} \right) \leq \Gamma, \tag{7.47}$$

$$x_i^{(d)} \geq 0, \ 1 \leq i \leq N, \ 1 \leq d \leq D, \tag{7.48}$$

where $P$ and $\Gamma$ are the transmit power budget at each SN and the interference limit for the PN, respectively. Furthermore, $a_{S,i}^{(d)}$ and $a_{P,i}^{(d)}$ are the $d$th realization of the channel gains $a_{S,i}$ and $a_{P,i}$, respectively. SN $i$ identifies the corresponding decision $x_i^{(d)}$ subject to various realizations of the collected information. Meanwhile, the nonnegativity of Eq. (7.48) is guaranteed by means of the ReLU activation at the output layer of an optimizer NN.

## 7.5.2    Implementation

The implementation issues of NNs and hyperparameters are described briefly. For the case of the infinite-capacity backhaul link, the centralized approach uses four hidden layers of fully connected NN architecture. On the other hand, the optimizer NN and the quantizer NN for the distributed configuration are constructed with three and single hidden layers, respectively. Such an NN arrangement allows for manageable computational complexity at each node regardless of the backhaul capacity. All hidden layers have $10N$ neurons with the ReLU activation. The training algorithm can be accelerated by leveraging the batch normalization technique applied at each layer. The Adam optimizer [27] is applied with learning rate $\eta = 5 \times 10^{-5}$ and mini-batch size $S = 5 \times 10^3$ for the backward-pass operation. The Xavier initialization [28] is consideration for the initial random choice of NN parameters, whereas the dual variables are all initialized to zero. The inference performance of the trained NNs is evaluated using $10^4$ independent test samples.

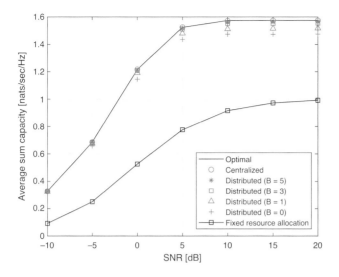

**Figure 7.11** Average sum capacity performance with respect to SNR. (© 2019 IEEE. Reprinted, with permission, from [20])

### 7.5.3 Numerical Results

An example of the C-MAC system with $N = 2$, $\Gamma = 1$, and a constant limit of $B_{ij} = B$ is considered. Figure 7.11 presents the average sum capacity performance examined with the test dataset of randomly generated samples. For simplicity, the noise level is set to unity, and the resulting average signal-to-noise ratio (SNR) is SNR $= P$. Once the target SNR is set, the corresponding average power budget is determined to apply for the inequality in Eq. (7.46). The Rayleigh fading environment is considered to generate the dataset of channel gains for training, validation, and inference processes.

The optimal scheme is implemented using the algorithm in [7] and a simple power control solution $x_i = \max\left\{P, \frac{\Gamma}{aP,i}\right\}$ is obtained for the fixed resource allocation as benchmarks. The centralized approach achieves almost identical performance to the optimal scheme. This verifies the optimality of the constrained unsupervised learning algorithm. Although there is a slight performance gap between the distributed approach and the optimal method, it becomes small as the backhaul capacity $B$ grows. The 3-bit resolution of messages is sufficient for wireless node coordinations to reach the optimal performance. It is noted in the distributed learning method with $B = 0$ that no information exchange occurs among SNs. Nevertheless, such a technique exhibits the sum-capacity performance close to the optimum, outperforming the simple power control method. It is justified that, in the centralized training process at the cloud, each built-in NN of an individual SNs indirectly observe the average behavior of other NNs (i.e., the Lagrangian function gradients which evidence the decisions of other NNs), via the primal-dual training update strategy. Trained with numerous samples, distributed NNs are able to learn a statistical behavior of the C-MAC system. This trained ability allows the real-time inference even in the case of no node coordination.

The convergence dynamics of the primal-dual training algorithm for the centralized approach are illustrated in Fig. 7.12. The convergence of the primal update is analyzed in Fig. 7.12(a). The average sum capacity is plotted with respect to the training iteration for SNR $= 0$ and 5 dB. A zero duality gap is achieved by the NNs are trained with the primal-dual update rules to achieve the zero duality gap. Thus, the Lagrangian evaluated from the NN solution is plotted together for comparison. The strong duality ensures that the Lagrangian value provides the tight upper bound of the solution. Regardless of the value of SNR, the average sum capacity obtained by the NN reaches to the optimal performance specified by the Lagrangian value.

Next, the convergence of the stochastic dual update Eq. (7.26) are depicted in Fig. 7.12(b) for SNR $= 5$ dB. The optimal dual variables computed by the algorithm in [7] is presented together as a reference. Two dual variables $\lambda_1$ and $\lambda_2$ are related to the average power constraint for two SNs in Eq. (7.46), and the third one $\lambda_3$ is associated with the interference constraint in Eq. (7.47). The primal-dual training algorithm identifies the optimal dual variables even if no prior knowledge of the optimal points is available.

The feasibility of all constraints in Eqs. (7.46) and (7.47) are demonstrated in Fig. 7.12(c). The extent of the constraint violation is measured using the difference between both sides in inequalities calculated for the test dataset; that is, $\frac{1}{D}\sum_{d=1}^{D} x_i^{(d)} -$ $P$ for $i = 1, 2$ and $\frac{1}{D}\sum_{d=1}^{D}\left(\sum_{i=1}^{N} a_{P,i}^{(d)} x_i^{(d)}\right) - \Gamma$. Thus, a positive value indicates the violations of the corresponding constraint. At the initial stage of the training process, the decisions of individual SNs are normally infeasible since they cannot meet the interference constraint; that is, $\frac{1}{D}\sum_{d=1}^{D}(\sum_{i=1}^{N} a_{P,i}^{(d)} x_i^{(d)}) - \Gamma > 0$. As observed in Fig. 7.12(b), dual variable $\lambda_3$ grows rapidly in the first $1.1 \times 10^5$ training iterations. In turn, $\lambda_3$ decreases and converges eventually to its optimum point, and the associated interference constraint now becomes satisfied as $\frac{1}{D}\sum_{d=1}^{D}(\sum_{i=1}^{N} a_{P,i}^{(d)} x_i^{(d)}) - \Gamma = 0$. The power constraints are initially inactive for the first a few iterations, as $\frac{1}{D}\sum_{d=1}^{D} x_i^{(d)} - P < 0$. This leads to null dual variables $\lambda_1 = \lambda_2 = 0$ to guarantee the complementary slackness condition [5]; that is, $\lambda_i(\frac{1}{D}\sum_{d=1}^{D} x_i^{(d)} - P) = 0$. As the transmit power level approaches close to the power budget $P$, the corresponding dual variables achieve the optimal values at the $1.1 \times 10^5$th training iteration for meeting this condition. Therefore, the NN solution obtained with the primal-dual learning strategy holds the strong duality after the $1.1 \times 10^5$th iteration for SNR $= 5$ dB.

The convergence for the distributed learning solution is tested in Fig. 7.13 for SNR $= 5$ dB with the backhaul capacity $B = 0$ and 5 bits. The distributed approach shows the convergence both in the primal and dual domains with comparable convergence speeds to the centralized solution. As discussed in Fig. 7.11, the backhaul capacity of $B = 5$ bits is sufficient for the distributed NN to approach to the optimum performance. This is again verified in Fig. 7.13(a) and 7.13(b) by exhibiting the convergence behaviors in the primal and dual domains, respectively. When no coordination is allowed as $B = 0$, the performance gaps show up in the primal performance and the value of dual variables. The dual variable $\lambda_3$ for the IT constraint fails to converge to the optimal value since the SNs cannot exchange their local channel gains

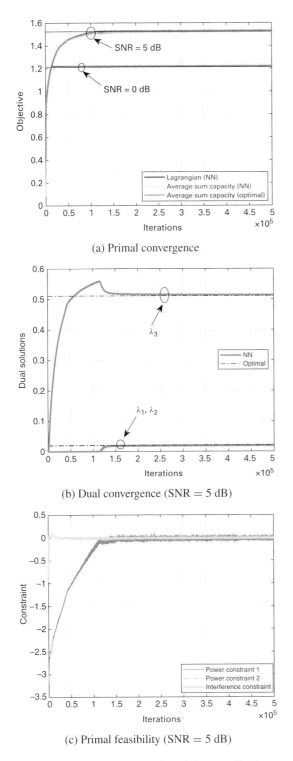

(a) Primal convergence

(b) Dual convergence (SNR = 5 dB)

(c) Primal feasibility (SNR = 5 dB)

**Figure 7.12** Convergence behavior of the centralized approach. (© 2019 IEEE. Reprinted, with permission, from [20])

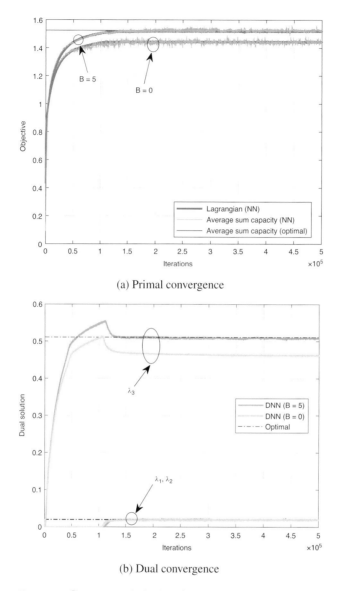

(a) Primal convergence

(b) Dual convergence

**Figure 7.13** Convergence behavior of the distributed approach for SNR = 5 dB.

in the online inference step. This confirms the importance of the cooperation among the SNs in the distributed transmit power control task.

The overall results show that the centralized approach indeed achieves the known optimal point. The convergence behavior of the primal-dual training algorithm can be analyzed by the optimality condition of the dual formulation such as the Karush-Turcker-Kuhn condition [24]. The effectiveness of the distributed learning solution can also be examined by comparing it with the optimal algorithm.

## 7.6 Case Study 2: Interference Channels

This section considers an application of the network management in a wireless interference channel (IFC) setup. In this configuration, the wireless network contains multiple nodes that interfere with each other. Mutual interference among nodes incurs the nonconvexity in the network cost function. Unfortunately, the global optimal performance of most IFC networks remains unidentified. Popular applications of the IFC system include traditional cellular networks, heterogeneous networks, and ad hoc networks. Coordinate multipoint transmission (CoMP) systems in 4G multicell technology can also be described by the IFC network configuration since transmission points that are interconnected for cooperative transmission are distributed over the network. Thus, the distributed management of the interference among multiple transmission points with the limited coordination is essential in optimizing the overall system performance of the network.

### 7.6.1 System Description

An IFC network configuration is investigated with $N$ transmitter-receiver pairs. The transmitter of the $i$th pair ($1 \leq i \leq N$) sends its data to the dedicated destination. All transmitters share all wireless resource blocks, thereby inherently incurring interference to neighboring destinations. Thus, a distributed power control mechanism may be applied to mitigate the multiuser interference. Transmitter $i$ is responsible for the optimal selection of its transmit power $x_i$. The global decision vector, representing the set of all transmit power levels, is denoted by $\mathbf{x} \triangleq \{x_i : 1 \leq i \leq N\}$. Let $a_{ij}$ be a channel gain of the wireless propagation channel from the transmitter of the $i$th pair to the receiver of the $j$th pair. The transmitter of the $i$th pair is informed of a local measurement vector denoted by $\mathbf{a}_i \triangleq \{a_{ji} : j \in \mathbf{E}_i\}$ by its dedicated receiver that collects channel gains associated with its transmitter and other transmitters of all neighboring pairs. The global measurement is defined as $\mathbf{a} \triangleq \{a_{ji} : j \in \mathbf{E}_i, 1 \leq i \leq N\}$. In this configuration of wireless networks, two different objective functions are considered for the distributed network optimization: The first objective is the sum rate, and the optimization maximizing it is formulated as

$$\max \quad \frac{1}{D} \sum_{d=1}^{D} \left( \sum_{i=1}^{N} \log \left( 1 + \frac{a_{ii}^{(d)} x_i^{(d)}}{1 + \sum_{j \in \mathbf{E}_i} a_{ji}^{(d)} x_j^{(d)}} \right) \right) \tag{7.49}$$

$$\text{subject to} \quad x_i^{(d)} \in [0, P], \ 1 \leq i \leq N, \ 1 \leq d \leq D, \tag{7.50}$$

where $a_{ji}^{(d)}$ is the $d$th realization of the channel propagation gain $a_{ji}$, and $x_i^{(d)}$ stands for the corresponding power level decision. The maximum limit $P$ is imposed as a peak power constraint in Eq. (7.50). The second task is the consideration of the fairness in the power control. The corresponding optimization problem that maximizes the minimum rate performance is expressed as

$$\max \quad \frac{1}{D} \sum_{d=1}^{D} \left( \min_i \log \left( 1 + \frac{a_{ii}^{(d)} x_i^{(d)}}{1 + \sum_{j \in \mathbf{E}_i} a_{ji}^{(d)} x_j^{(d)}} \right) \right) \tag{7.51}$$

$$\text{subject to} \quad x_i^{(d)} \in [0, P], \ 1 \le i \le N, \ 1 \le d \le D. \tag{7.52}$$

The main difference between two formulations is that the objective function of the first one can be separately considered with all individual terms, whereas the objective function of the second one is addressed as the sum of several functions defined with respect to the same set of variables, which usually makes the distributed optimization of the latter much more challenging than the former. However, in centralized training and distributed deployment strategy, two problems can be handled in a common machine-learning framework.

## 7.6.2    Numerical Results

An IFC system is considered with $N = 3$ and $B_{ij} = B$. For simplicity, the unit noise variance is assumed, leading to a simple definition of SNR as the peak power budget $P$ (i.e., SNR $= P$). The simulation results for the sum rate maximization task in Eq. (7.49) is illustrated in Fig. 7.14. A local optimum solutions of Eq. (7.49) obtained by the weighted minimum mean square (WMMSE) algorithm [29] are presented together. Furthermore, the sum rate performances of naive power allocation strategies, such as peak power setup $x_i = P$ and random transmission in which generates $x_i$ uniformly in $[0, P]$, are also provided as benchmarks. Figure 7.14(a) shows the average sum rate with respect to the SNR. The centralized learning approach shows the performance over the previous local optimum solution, validating the viability of the NN-assisted optimization approach for handling a nonconvex objective function in Eq. (7.49). The distributed NN with $B = 1$ achieves the performance of the WMMSE algorithm that requests the infinite-capacity backhaul link for the information sharing among transmitter-receiver pairs. Similar to the C-MAC network, the distributed approach with no mutual interaction ($B = 0$) is superior to existing naive power allocation methods. The impact of the backhaul link capacity on the average sum rate performance is demonstrated in Fig. 7.14(b). Regardless of the value of the SNR, the distributed learning architecture exhibits the real-time inference performance approaching the centralized one. This verifies that the cooperation via binary message exchanges is a crucial feature for the distributed power control in the IFC network.

The results for the minimum rate maximization in Eq. (7.51) are also investigated with the same configuration. To deal with this nonsmooth objective function, an NN architecture with additional hidden layers and neurons is applied. The centralized optimizer NN has $20N$-neuron five hidden layers. For the distributed structure, the optimizer NN and the quantizer NN exploit four and single hidden layers, respectively, each with $20N$ neurons. Figure 7.15 plots the average maximized minimum (max-min) rate performance. The optimum performance is reproduced using the algorithm in [30]. The average max-min rate performance is provided in Fig. 7.15(a) for various values of the SNR. The centralized NN approach provides almost identical

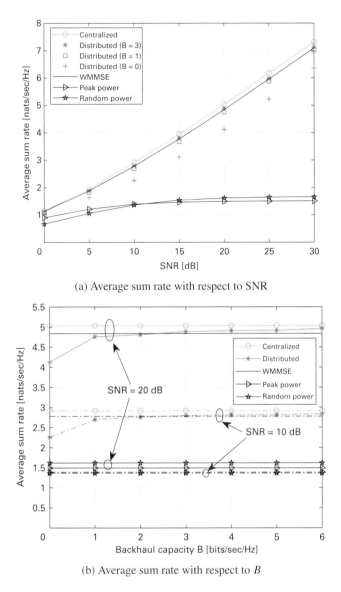

(a) Average sum rate with respect to SNR

(b) Average sum rate with respect to $B$

**Figure 7.14** Average sum rate performance. (© 2019 IEEE. Reprinted, with permission, from [20])

performance to the optimal scheme. Furthermore, the max-min rate in the case of no coordination ($B = 0$) also outperforms simple power control methods. By comparison with the sum rate maximization applications in Fig. 7.14(a), it is noticed that the max-min formulation in Eq. (7.51) requires sharing of higher dimensional messages among transmitter-receiver pairs to achieve the performance close to the centralized solution. Similar trends can be observed in Fig. 7.15(b) illustrating the average max-min rate with respect to the value of $B$. Nevertheless, the distributed learning approach

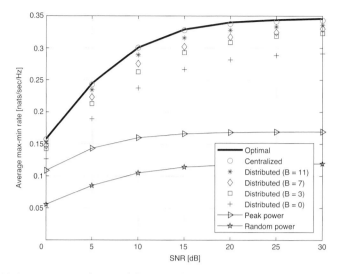

(a) Average max-min rate of the centralized approach with respect to SNR

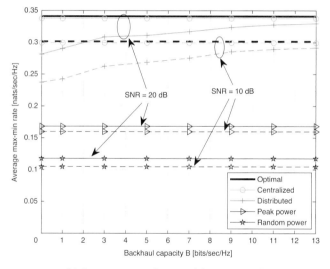

(b) Average max-min rate with respect to $B$

**Figure 7.15** Average max-min rate performance. (© 2019 IEEE. Reprinted, with permission, from [20])

gradually approaches to the performance of the centralized method as the backhaul link capacity grows.

The overall results demonstrate the NN's ability to address nonconvex formulations. For the sum rate maximization, it is revealed that the constrained unsupervised learning technique is superior to a local optimal performance. A nonsmooth objective in the minimum rate maximization, albeit nontrivial to address via existing distributed optimization techniques, is efficiently addressed by the NN-based approach.

The impact of the backhaul link capacity on the distributed power control tasks is also investigated. For all configurations of the simulation, the performance monotonically increases as the backhaul capacity grows.

## 7.7     Conclusion

This chapter presents an overview of machine-learning approaches for identifying a distributed management policy of arbitrary wireless systems with nonconvex design constraints. A key idea of recent approaches on learning-based network management is to replace exiting model-based algorithms with a group of NNs. A combination of expert knowledge in optimization theory and state-of-the-art learning techniques promotes the development of constrained unsupervised learning strategies. With some mathematical analysis, such a NN-assisted network optimization strategy establishes the optimality. A binarization technique of the NN outputs is developed to accommodate finite-capacity backhaul links. A cooperative learning mechanism among wireless nodes equipped with incorporated NNs presents a remarkable success in the optimization of distributed decision-making policies.

## References

[1] M. Hong et al., "A unified algorithmic framework for block-structured optimization involving big data: With applications in machine learning and signal processing," *IEEE Signal Processing Magazine*, vol. 33, no. 1, pp. 57–77, Jan. 2016.

[2] Y. Sun, P. Babu, and D. P. Palomar, "Majorization-minimization algorithms in signal processing, communications, and machine learning," *IEEE Trans. on Signal Processing*, vol. 65, no. 3, pp. 794–816, Feb. 2017.

[3] S. Boyd et al., "Distributed optimization and statistical learning via the alternating direction method and multipliers," *Foundat. Trends Machine Learning*, vol. 3, no. 1, pp. 1–122, 2010.

[4] F. R. Kschischang, B. J. Frey, and H.-A. Loeliger, "Factor graphs and the sum-product algorithm," *IEEE Trans. on Information Theory*, vol. 47, no. 2, pp. 498–519, Feb. 2001.

[5] S. Boyd and L. Vandenberghe, *Convex Optimization*. Cambridge University Press, 2004.

[6] M. Mohseni, R. Zhang, and J. M. Cioffi, "Optimized transmission of fading multiple-access and broadcast channels with multiple antennas," *IEEE Journal on Selected Areas in Communications*, vol. 24, no. 8, pp. 1627–1639, Aug. 2006.

[7] R. Zhang, S. Cui, and Y.-C. Liang, "On ergodic sum capacity of fading cognitive multiple-access and broadcast channels," *IEEE Trans. on Information Theory*, vol. 55, no. 11, pp. 5161–5178, Nov. 2009.

[8] L. Liu, R. Zhang, and K.-C. Chua, "Wireless information and power transfer: A dynamic power splitting approach," *IEEE Trans. on Communications*, vol. 61, no. 9, pp. 3990–4001, Sep. 2013.

[9] J. Xu, L. Duan, and R. Zhang, "Proactive eavesdropping via cognitive jamming in fading channels," *IEEE Trans. Wireless Communications*, vol. 16, no. 5, pp. 2790–2806, May 2017.

[10] K. Hornik, M. Stinchcombe, and H. White, "Multilayer feedforward networks are universal approximators," *Neural Networking*, vol. 2, no. 5, pp. 359–366, 1989.

[11] Z. Lu et al., "The expressive power of neural networks: A view from the width," in *Proc. Neural Information Processing Systems (NIPS)*, pp. 6231–6239, 2017.

[12] H. Sun et al., "Learning to optimize: Training deep neural networks for interference management," *IEEE Trans. on Signal Processing*, vol. 66, no. 20, pp. 5438–5453, Oct. 2018.

[13] W. Lee, D.-H. Cho, and M. Kim, "Deep power control: transmit power control scheme based on convolutional neural network," *IEEE Communications Letters*, vol. 22, no. 6, pp. 1276–1279, June 2018.

[14] P. de Kerret, D. Gesbert, and M. Filippone, "Team deep neural networks for interference channels," in *Proc. IEEE Int. Conf. Communications (ICC)*, pp. 1–6, May. 2018.

[15] M. Kim, P. de Kerret, and D. Gesbert, "Learning to Cooperate in Decentralized Wireless Networks," in *Proc. IEEE Asilomar Conf. Signals, Systems, and Computers (AC-SSC)*, pp. 281–285, 2018.

[16] W. Lee, "Resource allocation for multi-channel underlay cognitive radio network based on deep neural network," *IEEE Communications Letters*, vol. 22, no. 9, pp. 1942–1945, Sep. 2018.

[17] W. Lee, M. Kim, and D.-H. Cho, "Transmit power control using deep neural network for underlay device-to-device communication," *IEEE Wireless Communications Letters*, to be published.

[18] W. Lee, M. Kim, and D.-H. Cho, "Deep learning based transmit power control in underlaid device-to-device communication," *IEEE Systems Journal*, vol. 13, no. 3, pp. 2551–2554, 2019.

[19] M. Eisen et al., "Learning optimal resource allocations in wireless systems," *IEEE Trans. on Signal Processing*, vol. 67, no. 10, pp. 2775–2790, May 2019.

[20] H. Lee et al., "Deep learning framework for wireless systems: Applications to optical wireless communications," *IEEE Communications Magazine*, vol. 57, no. 3, pp. 35–41, Mar. 2019.

[21] H. Lee, I. Lee, and S. H. Lee, "Deep learning based transceiver design for multi-colored VLC systems," *Opt. Express*, vol. 26, no. 5, pp. 6222–6238, Feb. 2018.

[22] B. Amos and J. Z. Kolter, "Optnet: Differentiable optimization as a layer in neural networks," in *Proc. Int. Conf. Learning Representations (ICLR)*, 2017.

[23] W. Yu and R. Lui, "Dual methods for nonconvex spectrum optimization of multicarrier systems," *IEEE on Communications*, vol. 54, no. 7, pp. 1310–1321, July 2006.

[24] S. Boyd, L. Xiao, and A. Mutapcic, "Subgradient methods," tech. report EE364b, Stanford University, Oct. 2014.

[25] Y. Bengio, N. Leonard, and A. Courville, "Estimating or propagating gradients through stochastic neurons for conditional computation," *arXiv preprint*, arXiv:1305.2982.

[26] T. Raiko et al., "Techniques for learning binary stochastic feedforward neural networks," in *Proc. Int. Conf. Learning Representations (ICLR)*, 2015.

[27] D. Kingma and J. Ba, "Adam: A method for stochastic optimization," in *Proc. Int. Conf. Learning Representations (ICLR)*, 2015.

[28] X. Glorot and Y. Bengio, "Understanding the difficulty of training deep feedforward neural networks," in *Proc. 13th Int. Conf. Artificial Intelligence and Statistics*, pp. 249–256, 2010.

[29] Q. Shi et al., "An iteratively weighted MMSE approach to distributed sum-utility maximization for a MIMO interfering broadcast channel," *IEEE Trans. on Signal Processing*, vol. 59, no. 9, pp. 4331–4340, July 2011.

[30] D. W. H. Cai et al., "Max-min SINR coordinated multipoint downlink transmission-duality and algorithms," *IEEE Trans. on Signal Processing*, vol. 60, no. 10, pp. 5384–5395, Oct. 2012.

# 8  Radio Resource Allocation in Smart Radio Environments

Alessio Zappone and Merouane Debbah

## 8.1  Introduction

In the timeframe between 2020 and 2030, the demand for wireless connectivity is expected to grow by 55 percent each year, reaching 5,016 exabytes, with data-rate requirements up to 1 Tb/s [1]. Moreover, future wireless networks will have to provide many innovative heterogeneous services, each with diverse and specific requirements, ranging from extreme reliability, very low latency, very high rate, and energy efficiency. This calls for a new approach to network management and operation, which is able to cope with the extreme complexity that future networks will have to face. New network infrastructures will have to be fully integrated with the environment through transmitters embedded in walls, data caching, and wireless sensors everywhere, a trend that leads to the new paradigm of smart radio environments [2–4]. Intelligence will be placed not only in the network core, but also distributed across all network segments. Rather than waiting for instructions from the core network, network devices should be self-organizing and capable of independent decision-making. A fully self-configuring network can reduce capital expenditure (CAPEX) and operational expenditures (OPEX) by at least five times [5]. However, the huge number of devices to serve, the infrastructure heterogeneity, and the inherent randomness of wireless traffic evolutions make the design of such a system impossible to handle using only traditional approaches based on mathematical models, strongly motivating the use of data-driven techniques. In this direction, a promising approach lies in the use of deep learning by Artificial Neural Networks (ANNs) [6], which are able to learn from previous experience and infer in real-time the resource allocation policy to employ, with a much lower computational complexity than traditional model-based designs. Nevertheless, despite the strong need to consider data-driven approaches, artificial intelligence should not replace, but rather enable and improve, traditional design approaches. This chapter will show that the theoretical expertise and insight about wireless networks built through years of scientific research can be embedded into artificial intelligence algorithms. It will be shown how theoretical models provide useful a priori knowledge to guide deep learning methods, thus reducing the amount of data that is required to successfully train neural networks. This is a major advantage as far as wireless networks are concerned, since acquiring and storing massive amounts of live traffic data usually involves expensive measurement campaigns and privacy issues.

The rest of this chapter is organized as follows. Section 8.2 describes the considered ANN model and how to embed neural networks into wireless networks. A brief literature overview about the use of deep learning for radio resource allocation in wireless networks is provided, too. Then, Section 8.3 develops a radio resource allocation framework based on the joint use of theoretical models and deep learning methodologies, while Section 8.4 addresses the complexity of performing resource allocation by ANNs. Finally, concluding remarks are provided in Section 8.5.

## 8.2     Artificial Neural Networks

The elementary building block of ANNs is the so-called neuron, which performs specific elaborations on the input features. The specific computations performed by each neuron depend on the type of neural network that is considered. In the following, we focus our attention on fully connected ANNs. In this model, the neurons are organized in successive layers, with $N_\ell$ denoting the number of neurons in layer $\ell$. The input of the ANN is a vector of real numbers, the dimension of which is equal to the number of neurons in the *input layer*. Then, the input data is forwarded to the following layers, with each neuron in any given layer receiving as input the outputs of all the neurons of the preceding layer. Each neuron $n$ in layer $\ell$ takes an affine combination of the input vector using neuron-dependent weights $\mathbf{w}_{n,\ell}$ and bias term $b_{n,\ell}$, and then it applies a nonlinear function $f_{n,\ell}$, called the activation function, to produce its output. Thus, the output $x_{n,\ell}$ of neuron $n$ in layer $\ell$ corresponding to the input $\mathbf{x}_{\ell-1}$ is written as

$$z_{n,\ell} = \mathbf{w}_{n,\ell}^T \mathbf{x}_{\ell-1} + b_{n,\ell}, \tag{8.1}$$

$$x_{n,\ell} = f_{n,\ell}(z_{n,\ell}). \tag{8.2}$$

This processing is performed sequentially by each layer, until the *output layer*, the output of which provides the output vector of the ANN. All layers between the input and the output layer are called *hidden layers*. If only one hidden layer is present, the ANN is called *shallow*, whereas if more than one hidden layer is present, the ANN is called *deep* [6]. Deep neural networks are usually preferred, since they typically perform better than shallow neural networks. For this reason, typically the term deep neural networks is used in place of the more general ANNs, which refers to both deep and shallow neural networks. However, in this chapter, the more general term ANN will be used. Moreover, it should be mentioned that empirical evidence shows that deep learning outperforms other machine learning frameworks when a large amount of training data is available [7], whereas no significant gains are observed when smaller training sets are used.

The most important feature of ANNs, which makes them a popular tool in many fields of science, is their universal function approximation property. By tuning the weights and bias applied in the affine combinations of each neuron, it is possible to implement any continuous mapping between the input and output vector of the ANN

[8, 9]. This makes ANNs an extremely versatile tool, which can perform a variety of different tasks based on how the weights and bias are tuned during the training process. The coming section provides more details about training ANNs.

### 8.2.1    Training ANNs

The training process of a neural network aims at tuning the weights $w_{n,\ell}$ and biases $b_{n,\ell}$ of each neuron $n$ in layer $\ell$, in order for the input-output map of the ANN to be as close as possible to the desired map. To elaborate, let us consider a training set composed of $N_{TR}$ input samples with the corresponding desired output, namely

$$\mathscr{S}_{TR} = \left\{ \left( \mathbf{x}_0^{(1)}, \mathbf{x}_{L+1}^{(1)} \right), \ldots, \left( \mathbf{x}_0^{(N_{TR})}, \mathbf{x}_{L+1}^{(N_{TR})} \right) \right\} . \tag{8.3}$$

For each layer $\ell = 1, \ldots, L+1$, let us define the matrix $\mathbf{W}_\ell = \left[ \mathbf{w}_{1,\ell}, \ldots, \mathbf{w}_{N_\ell,\ell} \right]$ and the vector $\mathbf{b}_\ell = \left[ b_{1,\ell}, \ldots, b_{N_\ell,\ell} \right]^T$.

The task of the training algorithm is to tune the weights and biases of the ANN in order to minimize the error between actual output of the ANN, denoted by $\widehat{\mathbf{x}}_{L+1}$, and the desired output $\mathbf{x}_{L+1}$, over the training set. This amounts to solving the following minimization problem:

$$\min \frac{1}{N_{TR}} \sum_{nt=1}^{N_{TR}} \mathscr{L} \left( \mathbf{x}_{L+1}^{(nt)}, \widehat{\mathbf{x}}_{L+1}^{(nt)} \, (\mathbf{W}, \mathbf{b}) \right) \tag{8.4a}$$

$$\text{s.t. } \mathbf{W}_\ell \in \mathbb{R}^{N_{\ell-1} \times N_\ell} , \ \forall \ell = 1, \ldots, L+1 \tag{8.4b}$$

$$\mathbf{b}_\ell \in \mathbb{R}^{N_\ell \times 1} , \ \forall \ell = 1, \ldots, L+1 , \tag{8.4c}$$

where $\mathbf{W} = \{\mathbf{W}_\ell\}_{\ell=1}^L$, $\mathbf{b} = \{\mathbf{b}_\ell\}_{\ell=1}^L$. In principle, $\mathscr{L}$ can be any measure of distance between the actual and the desired output samples, with the mean squared error and the cross entropy being two canonical choices.

Equation (8.4) is usually tackled by means of stochastic gradient descent type methods, coupled with the use of the back-propagation algorithm for the computation of the derivatives of the cost function. Both these tools have the advantage of considerably limiting the computational complexity of the training algorithm, which is a desirable feature because training sets are typically on the order of hundreds of thousands of samples.

**Back-Propagation Algorithm**

In large ANNs with many neurons and large training sets, the direct computation of the derivatives of the training error in Eq. (8.4a) with respect to all network weights and bias terms would require an unmanageable complexity. To ease this computational burden, the back-propagation algorithm provides a fast way of computing the gradient of the training error in a recursive way [10].

To elaborate, let us observe that the derivative of Eq. (8.4a) is the average of the derivatives of the loss function $\mathscr{L}(\mathbf{x}_{L+1}, \widehat{\mathbf{x}}_{L+1}(\mathbf{W}, \mathbf{b}))$ over the training set. The back propagation algorithm computes the derivatives of $\mathscr{L}(\mathbf{x}_{L+1}, \widehat{\mathbf{x}}_{L+1}(\mathbf{W}, \mathbf{b}))$. Specifically, given a training sample $\mathbf{x}_0$, the algorithm performs first a forward propagation to compute the corresponding actual output $\hat{\mathbf{x}}_{L+1}(\mathbf{W}, \mathbf{b})$. Next, the derivative of the error function with respect to $z_{n,L+1}$ is given by

$$\frac{\partial \mathscr{L}}{\partial z_{n,L+1}} = \frac{\partial \mathscr{L}}{\partial \mathbf{x}_{L+1}(n)} f'_{n,L+1}(z_{n,L+1}), \ \forall \, n = 1, \ldots, N_{L+1}. \tag{8.5}$$

At this point, the derivatives of the error function with respect to $z_{n,\ell}$, for all $\ell = L$, $L-1, \ldots, 1$, can be recursively computed, proceeding from the last to the first layer. Specifically, it holds[1]

$$\frac{\partial \mathscr{L}}{\partial z_{n,\ell}} = \sum_{k=1}^{N_\ell+1} \frac{\partial \mathscr{L}}{\partial z_{k,\ell+1}} \frac{\partial z_{k,\ell+1}}{\partial z_{n,\ell}}$$

$$= \sum_{k=1}^{N_\ell+1} \frac{\partial \mathscr{L}}{\partial z_{k,\ell+1}} w_{k,\ell+1}(n) f'(z_{n,\ell}), \tag{8.6}$$

which is obtained from the derivatives with respect to $z_{k,\ell+1}$ for all $k$. Finally, based on Eq. (8.6) and recalling Eq. (8.1), the derivatives with respect to the weights and bias terms are readily obtained as

$$\frac{\partial \mathscr{L}}{\partial w_{n,\ell}(k)} = \frac{\partial \mathscr{L}}{\partial z_{n,\ell}} \mathbf{x}_{\ell-1}(k), \tag{8.7}$$

$$\frac{\partial \mathscr{L}}{\partial b_{n,\ell}} = \frac{\partial \mathscr{L}}{\partial z_{n,\ell}}. \tag{8.8}$$

**Stochastic Gradient Descent**

While back propagation has a lower complexity than direct derivative computation, it still requires averaging the derivatives over the complete training set, which poses computational complexity issues. For this reason, training algorithms for ANNs do not employ classical gradient searches, but rather they resort to the method of stochastic gradient descent [11], which is based on the computation of an estimate of the true gradient, based on a randomly selected subset of the entire training set, called a mini-batch. Mathematically speaking, denoting by $\mathscr{S}_{SGD}$ the set of indexes associated with the training samples in the selected mini-batch, and by $N_S$ the cardinality of $\mathscr{S}_{SGD}$, the stochastic gradient descent method follows the same steps as the traditional gradient algorithm, but it replaces the true gradient with the following estimate:

$$\widehat{\nabla L}(\mathbf{W}, \mathbf{b}) = \frac{1}{N_S} \sum_{nt \in \mathscr{S}_{SGD}} \nabla \mathscr{L}\left(\mathbf{x}_{L+1}^{(nt)}, \hat{\mathbf{x}}_{L+1}^{(nt)}(\mathbf{W}, \mathbf{b})\right). \tag{8.9}$$

Each time a gradient descent step is taken, a different, randomly selected mini-batch $\mathscr{S}_{SGD}$ is used to estimate the true gradient.

---

[1] Recall that the derivative with respect to $x$ of the function $g(\mathbf{y}(x))$, with $\mathbf{y}(x) = [y_1(x), \ldots, y_I(x)]$, is given by $\sum_{i=1}^{I} (\nabla_y g)^T J_x \mathbf{y}$, where $J_x$ denotes the Jacobian operator with respect to $x$.

The computational complexity of stochastic gradient descent depends on the size $N_S$ of the mini-batches. If $N_S = N_{TR}$, then the algorithm reduces to the usual gradient descent method, which usually converges in fewer iterations since the true gradient is used, but it requires more complex iterations, since the size of the batch is equal to the complete training set size. If instead $N_S = 1$, the algorithm normally requires more iterations to converge, but each iteration has a limited complexity. In general, stochastic gradient descent might converge before the complete training set has been employed, even if several passes through the training set are usually required to obtain low enough training errors. Each pass through the complete training set is referred to as an *epoch*.

All modern algorithms for ANN training are refinements of stochastic gradient descent with back propagation. The specific training algorithm that has been used to produce the numerical results to follow, is the *Adam algorithm* [12], which applies a momentum technique to speed up the convergence of the standard stochastic gradient descent method [13], exploiting both first and second moments of the gradient.

## 8.2.2    Embedding ANNs into Wireless Networks

In order to successfully use deep learning for wireless communications, a key question is how to integrate ANNs into existing and future wireless network topologies. The main open question is where to train and run neural networks. Normally, in other fields of science, neural networks follow a centralized approach in which a single *centralized brain* manages the complete system. However, in the context of wireless networks, this approach poses the following issues:

1. **Latency.** One major goal of future wireless networks is a low end-to-end communication latency, which for some applications, is required to be on the order of a millisecond. Thus, it is impractical for distributed devices to wait for the cloud to run the ANN and feed back the results.
2. **Privacy.** Privacy and security will be critical issues of future wireless networks. Especially in some vertical applications, it is not desirable for end-users to share information with the cloud, thus making cloud-based deep learning impractical.
3. **Connectivity.** One major goal of future wireless networks is to provide ubiquitous connectivity at all times even in areas or at times when no reliable connection to the cloud exists.

Therefore, wireless networks will need a combination of both centralized and distributed approaches. This points toward a scenario in which, like in human society, a system should possess both a collective and shared intelligence, but also an individual intelligence. Wireless networks should possess a *cloud intelligence* that should be accessible to all nodes, together with a *device intelligence*, that belongs to each individual device. As discussed in the coming section, centralized approaches have received more attention in the open literature. For this reason, the rest of this chapter will focus on centralized methodologies for resource management in wireless networks. Nevertheless, we stress the need to develop distributed methodologies able to operate in future wireless networks.

## 8.2.3    State-of-the-Art Review

The topic of ANN-based radio resource allocation in wireless networks has started to receive attention in the last couple of years. Overviews on this topic are provided in [2, 14, 15], which review supervised, unsupervised, and reinforcement learning based approaches.

In [16], a fully connected ANN is used for sum-rate maximization, by training it to learn the input-output map of the iterative weighted minimum mean square (WMMSE) power control algorithm [17]. It is shown that the performance of the WMMSE method can be mimicked with a good degree of accuracy, while at the same time reducing the computational complexity significantly. In [18, 19], a fully connected ANN is used for energy efficiency maximization in interference networks. The neural network is trained on the optimal energy-efficient power allocation, which is computed offline by means of a novel optimization procedure also proposed in [18]. The results indicate that ANNs are able to learn optimal resource allocation policies with a negligible error. Similar results are obtained in [20, 21] with reference to power control and user-cell association in massive multiple-in multiple-out (MIMO) multicell systems. Instead, a different approach is taken in [22], where sum-rate maximization is performed by means of a fully connected ANN, which is trained using the system sum-rate as training cost function. This reduces the complexity of building the training set, but at the same time it produces a suboptimal training set.

In [23] a cloud-RAN system with caching capabilities is considered. Echo-state neural networks are used to enable base stations to predict the content request distribution and mobility pattern of each user, thus determining the best content to cache. It is shown that the use of deep learning increases the network sum effective capacity of around 30 percent compared with baseline approaches based on random caching. In [24], deep reinforcement learning is used to develop a power control algorithm for a cognitive radio system in which a primary and secondary user share the spectrum. It is shown that both users can meet their quality of service (QoS) requirements despite the fact that the secondary user has no information about the primary user's transmit power. The use of deep reinforcement learning is also considered in [25], where it is used to develop a power control algorithm for weighted sum-rate maximization in interference channels subject to maximum power constraints. The proposed algorithm exhibits fast convergence and satisfactory performance. A decentralized robust precoding scheme in a network MIMO system is developed in [26] by ANNs. In [27], online power allocation policies for a large and distributed system with energy-harvesting nodes are developed by merging deep reinforcement learning and mean field games. It is shown that the proposed method outperforms all other available online policies and suffers a limited gap compared to the use of noncausal off-line policies.

## 8.3      ANN-Based Resource Management

Any resource allocation problem in a wireless network can be formulated as the problem of finding the optimal map from the set of system parameters (e.g., propagation channels, number of active nodes) to the resource allocation policy that maximizes the performance metric of interest. Formally speaking, denoting $\mathbf{d} \in \mathbb{R}^N$ the ensemble of system parameters and $\mathbf{x}^* \in \mathscr{S}$ the optimal resource allocation, with $\mathscr{S}$ denoting the set of feasible allocations, the resource allocation problem can be seen as the map:

$$\mathscr{F} : \mathbf{d} \in \mathbb{R}^N \to \mathbf{x}^* \in \mathscr{S} . \tag{8.10}$$

Based on this formulation, the approach is to exploit the universal function approximation property of fully connected ANNs to train an ANN to learn the map in Eq. (8.10). To elaborate, the approach follows the following steps:

1. **Offline training.** $N_{TR}$ training samples can be generated by resorting to optimization-theoretic techniques to find the optimal resource allocation $\mathbf{x}_{nt}^*$ corresponding to the realization $\mathbf{d}_{nt}$ of the system parameters, for $nt = 1, \ldots, N_{TR}$. Otherwise stated, this provides $N_{TR}$ points of the map in Eq. (8.10). Provided that a sufficient number of training samples are generated, the training set can be used to train an ANN to learn Eq. (8.10). It is important to remark that this step can be performed off-line (i.e., before the ANN-based wireless network starts operating) and sporadically (i.e., the training phase needs to be updated only when the conditions in which the ANN will operate are significantly different from the training conditions, or when the statistical distribution of $\mathbf{d}$ has significantly changed with respect to that assumed for the generation of the training set.

2. **Online inference.** Once the ANN has been trained, and until a new training phase is required, the ANN can be used to compute the resource allocation to employ during the online operation of the wireless network. This only requires to feed the ANN with the current configuration of the system parameters (i.e., the current realization of $\mathbf{d}$) and perform a forward propagation to obtain an estimate of the corresponding optimal allocation $\mathbf{x}^*$. The universal approximation property ensures that the estimation error can be made small at will by properly configuring the ANN. Thus, the trained ANN enables to perform near-optimal resource allocation, without actually having to solve any optimization problem during the online operation of the wireless network. When a system parameter changes, it is only required to feed the new realization of $\mathbf{d}$ to update the resource allocation to employ.

**Remark 8.1** *It is important to stress that the method described here operates using only synthetic data (i.e., computer-generated data), without the need of acquiring empirical data (i.e., live data obtained through field measurements). This is possible because the resource allocation problem can be theoretically modeled and formulated, thus showing how theoretical modeling provides a valuable a priori expertise that can simplify the acquisition of training data. It should also be mentioned that in many* *cases existing theoretical models provide only a coarse approximation of reality. In this case, it is possible to merge a large dataset of synthetic data and a small dataset of empirical data, employing the framework of deep transfer learning [2]. In the rest of this chapter, it will be always assumed that a theoretical model of the resource allocation problem to solve is available.*

The rest of this section describes a case study that employs this ANN-based technique for the energy-efficient allocation of the transmit powers in a wireless interference network.

### 8.3.1    Energy Efficiency Maximization by ANNs

Let us consider the uplink of a multicell wireless network with $M$ base stations (BSs) and $K$ mobile users. Each BS is equipped with $N$ antennas, whereas the mobile users have a single antenna. Let $\mathbf{h}_{k,m}$ be the $N \times 1$ channel from user $k$ to BS $m$, $p_k$ be the $k$th user's transmit power, $\mathbf{c}_k$ the $N \times 1$ receive vector for user $k$, and $\sigma_m^2$ the received noise power at BS $m$. Then, the signal-to-interference and noise ratio (SINR) enjoyed by user $k$ at its intended receiver $m_k$ is

$$\gamma_k = \frac{p_k |\mathbf{c}_k^H \mathbf{h}_{k,m_k}|^2}{\sigma^2 + \sum_{j \neq k} p_j |\mathbf{c}_k^H \mathbf{h}_{j,m_k}|^2} = \frac{p_k d_{k,k}}{\sigma^2 + \sum_{j \neq k} p_j d_{k,j}}, \tag{8.11}$$

where $d_{k,j} = |\mathbf{c}_k^H \mathbf{h}_{j,m_k}|^2$, for all $k, j$. The global energy efficiency of the network is defined as the ratio between the network achievable rate and the total power consumption [28], namely

$$\text{GEE} = \frac{B \sum_{k=1}^{K} \log_2(1 + \gamma_k)}{P_c + \sum_{k=1}^{K} \mu_k p_k}, \tag{8.12}$$

where $B$ the communication bandwidth, $P_c$ the hardware static power consumed in the whole system, and $\mu_k$ the inverse of the efficiency of the power amplifier used by transmitter $k$. Then, the power control global energy efficiency (GEE) maximization problem is cast as

$$\max_{\{p_k\}_{k=1}^K} \frac{B \sum_{k=1}^{K} \log_2(1 + \gamma_k)}{P_c + \sum_{k=1}^{K} \mu_k p_k} \tag{8.13a}$$

$$\text{s.t. } P_{min,k} \leq p_k \leq P_{max,k}, \forall\, k = 1, \ldots, K \tag{8.13b}$$

wherein $P_{max,k}$ and $P_{min,k}$ are the maximum feasible and minimum acceptable transmit powers for user $k$, respectively. In [28, 29] it is shown how Eq. (8.13) requires in general an exponential complexity to be solved optimally. Thus, relying only on optimization theory would require sustaining an exponential complexity to solve Eq. (8.13) whenever any propagation channel realization changes, which is clearly impractical in modern wireless communication systems.

Resorting to the ANN-based method described earlier, let us observe that Eq. (8.13) can be seen as the map

$$\mathscr{F} : \mathbf{d} = \{d_{k,\ell}, P_{min,k}, P_{max,k}\}_{k,\ell} \in \mathbb{R}^{K(M+2)} \to \mathbf{p}^* \in \mathbb{R}^K. \qquad (8.14)$$

Then, an ANN can be trained to learn the unknown map $\mathscr{F}$, thus providing the desired power allocation for any given realization of $\mathbf{d}$. Once trained, computing the output of the ANN $\mathbf{p}^*$, given an input $\mathbf{d}$, only requires computing a forward propagation of the trained ANN. Thus, whenever a propagation channel realization changes, it is only required to perform a forward propagation, without having to solve Eq. (8.13) anew. This approach was successfully applied for power control for the optimization of different performance measures of wireless networks, such as achievable rate and energy efficiency [2, 18].

In order to generate a training set of size $N_{TR}$, Eq. (8.13) must be solved $N_{TR}$ times. As already discussed, this can be accomplished offline, and thus a much higher computational burden can be sustained. To this end, tailor-made global optimization algorithms for energy efficiency maximization are available [18], which enable to generate large training sets with affordable offline complexity. On the other hand, first-order optimal methods are also available, which have a practical computational complexity, and have been numerically shown to enjoy global optimality in some special cases [29].

## Numerical Analysis

The performance of the proposed method is numerically addressed considering the uplink of a MIMO system in which $K = 10$ users are randomly placed in a circular area with radius 500 m. The path loss has been modeled following [30], with power decay factor equal to 4.5, while fast fading terms have been modeled as realizations of zero-mean, unit-variance complex Gaussian random variables. The circuit power consumption term is equal to $P_c = 1$ W, while $\mu_k = 10$ for all $k$, and maximum ratio combining is adopted at all base stations. The noise power at the base station is $\sigma^2 = F \mathcal{N}_0 B$, where $F = 3$ dB is the receive noise figure, $B = 180$ kHz, and $\mathcal{N}_0 = -174$ dBm/Hz is the noise spectral density.

A fully connected ANN has been considered, having $L = 10$ hidden layers. Layers 1 and 2 have 18 neurons, and the number of neurons of the other layers decreases by two every two layers. Thus, the output layer has 10 neurons, providing the users' transmit powers. A training set and a test set of $10^4$ samples each have been independently generated employing the sequential fractional programming framework from [29]. The ADAM training algorithm has been employed, with the mean square error as the training cost function.

Figure 8.1 shows a numerical comparison between the energy-efficient power allocation by the sequential fractional programming method from[2] [29] and the GEE value obtained by ANN-based power control versus the maximum transmit power $P_{max}$. The results are averaged over the $10^4$ samples of the test set. As a benchmark, Fig. 8.1

---

[2] The method from [29] is provably first-order optimal but not provably globally optimal.

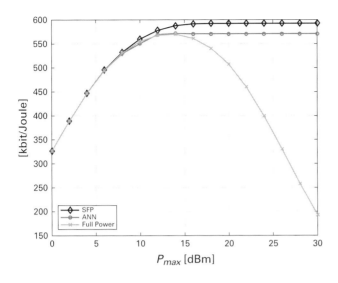

**Figure 8.1** Achieved GEE versus $P_{max}$ for ANN-based power allocation, sequential fractional programming from [29], and Uniform power allocation.

also shows the energy efficiency obtained when all users transmit with power $P_{max}$. Despite the much lower complexity, the ANN-based scheme is optimal for low $P_{max}$, and near optimal for larger $P_{max}$, achieving 95 percent of the GEE value obtained by the sequential method from [29]. It should be mentioned that the ANN would even outperform the first-order optimal method from [29] by increasing the size of the training set by a factor 10 [18]. However, this also increases the complexity of the training phase.

The results indicate that ANN-based resource allocation can provide near-optimal performance while requiring only minimal computations during the online operation of the wireless network. On the other hand, it requires a heavy computational burden during the off-line training phase, which is not needed by traditional optimization-theoretic methods. Although the training phase needs to be executed only sporadically, it is interesting to perform a complexity-aware comparison between the energy-efficient performance of ANN-based resource allocation and of traditional optimization-theoretic methods, considering an observation window embracing multiple channel coherence blocks. This is the objective of the coming section.

## 8.4     Complexity-Aware ANN-Based Resource Management

Let us consider again the problem of maximizing the GEE of a wireless network. As a specific case study, let us consider the downlink channel of a single-cell system in which a base station with $N$ antennae serves $K$ single-antenna mobile users. The received signal at user $k$ can be expressed as

$$r_k = \mathbf{h}_k^H \sum_{\ell=1}^{K} \sqrt{p_\ell} \mathbf{q}_\ell s_\ell + w_k \,, \tag{8.15}$$

where $\mathbf{h}_k$ is the $N \times 1$ channel between the base station and user $k$, $\mathbf{q}_k$ is the beamforming vector serving user $k$, $p_k$ is the transmit power carrying the information symbol $s_k$ intended for user $k$, and $w_k$ is the thermal noise power, modeled as a Gaussian random variable with zero mean and variance $\sigma^2$. As a result, the $k$th user's achievable rate is expressed as

$$R_k = B \log_2 \left( 1 + \frac{p_k |\mathbf{h}_k^H \mathbf{q}_k|^2}{\sigma^2 + \sum_{\ell \neq k} p_\ell |\mathbf{h}_k^H \mathbf{q}_\ell|^2} \right) , \tag{8.16}$$

with $B$ the communication bandwidth, while the network GEE is expressed as

$$\text{GEE} = B \frac{\sum_{k=1}^{K} R_k(p_1, \ldots, p_K)}{P_c + \mu \sum_{k=1}^{K} p_k} , \tag{8.17}$$

where $\mu$ denotes the inverse of the base station amplifier efficiency and $P_c$ models the hardware static power consumption of the whole system.

### 8.4.1    Problem Formulation

The GEE definition in Eq. (8.17) refers to a specific channel coherence interval. However, in order to perform a long-term GEE resource allocation, the expression in Eq. (8.17) must be modified to embrace a time interval of $L$ consecutive channel coherence blocks. In addition, the GEE in does not explicitly model the energy consumption that is required to solve the GEE maximization problem and obtain the power allocation policy to be used. In order to include these points in the analysis, let us denote by $\{\mathbf{h}_{k,n}\}_{k=1}^{K}$ and $\{\mathbf{q}_{k,n}\}_{k=1}^{K}$ the channels and beamforming vectors in the $n$th coherence interval. Similarly, $\{\bar{\mathbf{p}}_k^{(n)}\}$ represents the optimal power allocation in the coherence interval $n$; that is, the solution of the problem

$$\max_{\{p_k\}_{k=1}^{K}} \frac{(T_c - T_{op})B \log_2 \left( 1 + \frac{p_k a_k^{(n)}}{1 + \sum_{\ell \neq k} p_\ell b_{k,\ell}^{(n)}} \right)}{T_c P_c + T_{op} P_{op} + (T_c - T_{op})\mu \sum_{k=1}^{K} p_k} \tag{8.18a}$$

$$\text{s.t.} \sum_{k=1}^{K} p_k \leq P_{max}, \tag{8.18b}$$

wherein $a_k^{(n)} = |\mathbf{h}_{k,n}^H \mathbf{q}_k|^2 / \sigma^2$ and $b_{k,\ell}^{(n)} = |\mathbf{h}_{k,n}^H \mathbf{q}_{\ell,n}|^2 / \sigma^2$. The GEE in Eq. (8.18a) differs from the definition in Eq. (8.17) as it also accounts for the energy that is consumed for solving Eq. (8.18). Specifically, the channel coherence block is divided into a time interval of duration $T_{op}$, during which the solution of Eq. (8.18) is determined, consuming a power $P_{op}$, and a time $T_c - T_{op}$, during which data communication occurs, and transmit power is consumed. Finally, the static power $P_c$ is consumed for the whole duration $T_c$ of the coherence block. Based on Eq. (8.18a), the long-term GEE after $L$ coherence intervals can be defined as

$$
\text{GEE}_L = \frac{(T_c - T_{op})B \sum_{n=1}^{L} \sum_{k=1}^{K} R_k(\bar{p}_1^{(n)}, \ldots, \bar{p}_K^{(n)})}{LT_c P_c + LT_{op} P_{op} + \mu(T_c - T_{op}) \sum_{n=1}^{L} \sum_{k=1}^{K} \bar{p}_k^{(n)}}. \tag{8.19}
$$

In the rest of this chapter, the function in Eq. (8.19) will be evaluated using: (a) the power allocation obtained by solving Eq. (8.18) in each coherence interval by means of traditional optimization theory, and (b) the power allocation obtained by first training an ANN at the beginning of the whole time frame of $L$ coherence blocks, and then computing the power allocation to use in each coherence block by performing a forward propagation of the trained ANN. These two cases will be treated separately in Sections 8.4.2 and 8.4.3, respectively.

### 8.4.2 Optimization-Oriented Power Allocation

Traditional power allocation approaches require solving Eq. (8.18) in each of the $L$ coherence intervals. In this section, we do not consider global optimization methods, which require an exponential complexity, but rather we focus on more practical algorithms that tackle Eq. (8.18) by solving a sequence of convex or pseudo-convex relaxations of Eq. (8.18). This is the case, for example, of sequential fractional fractional programming algorithms [29] as well as of other approaches based on the iterative use of alternating optimization and of fractional programming [31].

To elaborate, let us denote by $N_{it}$ the number of (pseudo-)convex relaxations of Eq. (8.18) that are to be solved in each coherence interval. Then, since Eq. (8.18) has $K$ optimization variables, and recalling that the asymptotic complexity of a generic convex optimization problem scales at most with the fourth power of the number of variables [32], the complexity required to solve Eq. (8.18) scales as $N_{it}\mathscr{O}(K^4)$. Therefore, the time to compute the power allocation in each channel coherence interval and the related power consumption are given by

$$
T_{op} \approx K^4 N_{it} T_{DSP}, \tag{8.20}
$$

$$
P_{op} \approx K^4 N_{it} P_{DSP}, \tag{8.21}
$$

where $T_{DSP}$ denotes the time that the digital signal processor takes to compute one real multiplication, and $P_{DSP}$ is the associated power consumption. Plugging Eqs. (8.20) and (8.21) into Eq. (8.19) yields the long-term GEE over the $L$ coherence intervals.

### 8.4.3 ANN-Based Power Allocation

ANN-based power control requires a preliminary phase at the start of the $L$ channel coherence blocks, during which the ANN is trained employing a training set generated from $N_{tr}$ solutions of Eq. (8.18), corresponding to $N_{tr}$ different realizations of $\{a_k, b_{k,\ell}\}_{k,\ell}$. Once the ANN is trained, it can be used to infer the power allocation to employ in each of the $L$ channel coherence blocks. In other words, the training phase must be executed only once per $L$ coherence blocks, while a new inference phase is required in each of the $L$ coherence blocks.

Thus, the expression in Eq. (8.19) must be adapted to account for the fact that, when ANN-based power control is used, the total time during which the system is online is not $LT_c$, but $LT_c - T_{tr}$, with $T_{tr}$ the time required for the ANN training phase. In turn, this implies that the time used for communication in each coherence block is $(T_c - T_{op} - \frac{T_{tr}}{L})$, and that the energy consumption at the denominator of the energy efficiency must also include the energy consumption during the training phase. As a result, in the case of ANN-based power control, Eq. (8.19) takes the expression in Eq. (8.22),

$$
\text{GEE}_L^{ANN} = \frac{\left(T_c - T_{op} - \dfrac{T_{tr}}{L}\right) B \displaystyle\sum_{n=1}^{L}\sum_{k=1}^{K} R_k(\bar{p}_1^{(n)}, \ldots, \bar{p}_K^{(n)})}{LT_c P_c + LT_{op} P_{op} + T_{tr} P_{tr} + \mu\left(T_c - T_{op} - \dfrac{T_{tr}}{L}\right)\displaystyle\sum_{n=1}^{L}\sum_{k=1}^{K}\bar{p}_k^{(n)}}.
$$

$$(8.22)$$

with $P_{tr}$ the power required to train the ANN.

Dividing numerator and denominator of Eq. (8.22) by $\frac{(T_c - T_{op} - \frac{T_{tr}}{L})}{L}$, Eq. (8.22) becomes

$$
\frac{\dfrac{B}{L}\displaystyle\sum_{n=1}^{L}\sum_{k=1}^{K} R_k(\bar{p}_1^{(n)}, \ldots, \bar{p}_K^{(n)})}{\dfrac{T_c P_c + T_{op} P_{op} + T_{tr} P_{tr}/L}{T_c - T_{op} - T_{tr}/L} + \dfrac{\mu}{L}\displaystyle\sum_{n=1}^{L}\sum_{k=1}^{K}\bar{p}_k^{(n)}},
$$

$$(8.23)$$

which shows that the time and energy overhead of the initial training phase become negligible in the long term. Indeed, in the limit of $L \to \infty$, both $T_{tr} P_{tr}/L$ and $T_{tr}/L$ tend to zero, while all other terms tend to positive quantities, provided the ANN does not output zero transmit powers.[3] Unfortunately, setting $L \to \infty$ might not be the best choice. Indeed, the wireless scenario is time-varying, thus implying that the training scenario grows increasingly different from the current operation scenario. Thus, an inherent trade-off exists in the choice of the number $L$ of coherence intervals to be considered before the ANN is trained again.

Another fundamental trade-off is the size of the ANN architecture. A larger ANN might provide better inference capabilities, but on the other hand, more layers and neurons lead to a larger energy consumption and longer training and prediction times.

---

[3] This appears as an event having zero probability. It is reasonable to expect that in the worst case the ANN will output random transmit powers, which will still lead to a positive energy efficiency.

The rest of this section will derive expressions for $T_{op}$, $P_{op}$ and for $T_{tr}$, $P_{tr}$ as a function of the size of the ANN. In turn, this enables computing Eq. (8.19) and analyzing the behavior of the system energy efficiency over time.

**Inference Phase**

During this phase, the trained ANN is used to compute the power allocation to use in each channel coherence interval. For all $n = 1, \ldots, L$, this requires feeding the ANN with the channel parameters in the $n$th coherence interval and performing a forward propagation. Neglecting the complexity related to additions, from Eq. (8.1), we see that this requires performing $\sum_{\ell=1}^{L+1} N_{\ell-1} N_\ell$ real multiplications and evaluating $\sum_{\ell=1}^{L+1} N_\ell$ activation functions. Recalling that activations are usually elementary functions, such as the ReLU function that only requires determining whether the input is positive or negative, it is reasonable to assume that the complexity of evaluating one activation functions is not larger than the complexity of one real multiplication. As a result, the duration and power consumption of each inference phase are on the order of

$$T_{op} \approx \left( \sum_{\ell=1}^{L+1} N_{\ell-1} N_\ell + N_\ell \right) T_{DSP}, \tag{8.24}$$

$$P_{op} \approx \left( \sum_{\ell=1}^{L+1} N_{\ell-1} N_\ell + N_\ell \right) P_{DSP}. \tag{8.25}$$

Referring to the single-cell system model in Section 8.4.1, it is seen that, in Eqs. (8.24) and (8.25), the number of users $K$ appears in the size of the output layer $N_{L+1} = K$ and in the size of the input layer $N_0 = K^2$. Instead, the number of neurons in the other layers does not scale with $K$. Thus, that the exact complexity of the inference phase is quadratic in $K$, as opposed to the asymptotic complexity of traditional optimization-oriented approach that scales with the fourth power of $K$.

**Training Phase**

The training phase requires running the stochastic gradient descent algorithm, using the back-propagation method to compute the gradient in each iteration. Let us denote by $d_{mb}$ the size of the mini-batches, by $N_{mb}$ the number of mini-batches, and by $N_{ep}$ the number of training epochs. Thus, for each of the $d_{mb}$ samples of a mini-batch, a total of $N_{mb} N_{ep}$ stochastic gradient steps are performed each time by the back-propagation algorithm. On the other hand, the back-propagation algorithm requires one forward propagation and one back propagation for each sample of the mini-batch and gradient step. Thus, defining by $N_\ell$ the number of neurons in layer $\ell$, the number of operations required by the training algorithm is $C_{tr} = C(N_1, \ldots, N_{L+1}) d_{mb} N_{mb} N_{ep}$, where $C(N_1, \ldots, N_{L+1})$ is the number of operations required by a single forward and backward propagation. As discussed in the last section, a forward propagation requires $\sum_{\ell=1}^{L+1} N_{\ell-1} N_\ell + N_\ell$. As for the backward propagation, it requires computing, for each layer $\ell$, the derivatives with respect to $z_{n,\ell}$, with $n = 1, \ldots, N_\ell$. From Eq. (8.6), it is seen that this requires $2N_\ell N_{\ell+1}$

real multiplications plus the evaluation of $N_\ell$ derivatives of the activation functions, for $n = 1, \ldots, N_\ell$. Moreover, from Eq. (8.7) it follows that other $N_\ell$ multiplications are required to compute the derivatives with respect to $w_{n,\ell}$. Therefore, the total computational complexity of the training phase is

$$
\begin{aligned}
C_{tr} &= \left( \sum_{\ell=0}^{L} 2N_\ell N_{\ell+1} + \sum_{\ell=1}^{L+1} 3N_\ell \right) d_{mb} N_{mb} N_{ep} \\
&= \left( \sum_{\ell=1}^{L+1} 2N_{\ell-1} N_\ell + 3N_\ell \right) d_{mb} N_{mb} N_{ep} .
\end{aligned}
\tag{8.26}
$$

Thus, the duration and power consumption of the training phase can be expressed as

$$
T_{tr} = \left( \sum_{\ell=1}^{L+1} 2N_{\ell-1} N_\ell + 3N_\ell \right) d_{mb} N_{mb} N_{ep} T_{DSP}
\tag{8.27}
$$

$$
P_{tr} = \left( \sum_{\ell=1}^{L+1} 2N_{\ell-1} N_\ell + 3N_\ell \right) d_{mb} N_{mb} N_{ep} P_{DSP}.
\tag{8.28}
$$

### 8.4.4    Numerical Results

In order to perform a complexity-aware numerical analysis of ANN-based power control, let us consider the donwlink of a multiuser system in which $K = 8$ mobile users communicate with a single base station equipped with $M = 10$ antennas. A communication bandwidth of $B = 1$ MHz is used, and maximum ratio transmission is employed at the base station. The maximum feasible transmit power is $P_{max} = 20$ dBW, the hardware-dissipated power is $P_c = 10$ dBW, and the receive noise power is $\sigma^2 = FBN_0$, with $F = 3$ dB and $N_0 = -174$ dBm/Hz.

The channel coherence time is $T_c = 1$ ms and a total observation window of $L = 3.6 * 10^6$ coherence intervals have been considered for simulation, which amounts to a total time interval of one hour. In each coherence interval, fading channels have been independently generated following the Rayleigh model. The mobile users' positions at the beginning of the observation window have been randomly generated in a circle of radius 500 m around the base station. Then, the users' positions have been updated assuming that each user moves at a speed that is randomly and uniformly generated in $[0, 1]$ m/s, which is realistic for pedestrian networks, and with a direction that is randomly and uniformly updated every 100 s in $[0, 2\pi]$. Path-loss effects have been modeled after [30].

In order to solve Eq. (8.18), a feed-forward, fully connected ANN with four hidden layers has been used, equipped with 256, 128, 64, and 32 neurons with ReLU activation functions, while the output layer yields the 8 users' transmit powers and employs a linear activation function. A mini-batch size $d_{mb} = 128$ has been used, and the training process has been run for $N_{ep} = 10$ epochs, using the mean squared error

**Table 8.1.** Training and validation errors versus training epoch.

|          | Training MSE | Validation MSE |
|----------|--------------|----------------|
| Epoch 1  | 0.214        | 0.216          |
| Epoch 3  | 0.0103       | 0.0124         |
| Epoch 5  | 0.0074       | 0.0094         |
| Epoch 7  | 0.0065       | 0.0077         |
| Epoch 10 | 0.0065       | 0.0077         |

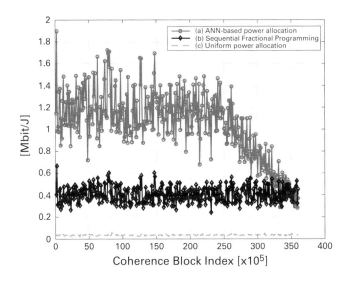

**Figure 8.2** GEE in Eq. (8.18a) versus the coherence block index $n$ for ANN-based power allocation, sequential fractional programming from [29], and Uniform power allocation.

as a training cost function. The training set has been generated considering, for each user $k$, 12,500 random realizations of Rayleigh fading and of positions within a circle of 200 m around the initial position of user $k$. Thus, the total training set for the $K = 8$ users comprises $10^5$ samples. The resulting training and validation errors are shown in Table 8.1, which shows that neither underfitting nor overfitting occurs.

Figure 8.2 shows the global energy efficiency (GEE) in Eq. (8.18a) for the $L$ simulated coherence intervals, considering $T_{DSP} = 1$ ns and $P_{DSP} = 1$ mW, for the following algorithms:

1. ANN-based power allocation wherein the ANN described and trained as described here is used to predict the power allocation in each coherence interval.
2. Sequential fractional programming method from [29] applied in each coherence interval.
3. Uniform power allocation in which every user is served with power $p_k = P_{max}/K$ in each coherence interval.

The results show that ANN-based power allocation is significantly more energy-efficient than both sequential fractional programming and uniform power allocation for a long part of the observation window. This is due to the fact that, like all optimization-oriented methods, sequential programming requires solving the GEE maximization problem in each coherence interval in the observation window, which leads to a significant power consumption, whereas the uniform power allocation is a too heuristic power allocation method, even though it consumes no energy for computation purposes. Instead, ANN-based power allocation strikes an excellent trade-off between these two approaches, thanks to the lower complexity required by a forward propagation compared to the use of convex optimization algorithms. On the other hand, as expected, as time passes, the performance of the ANN-based approach tends to degrade, due to the users' mobility, which eventually brings the users out of the 200 m circle radius that was used to generate the training set. This shows that the training of the ANN needs to be updated only after a long window, during which only the inference phase is required to update the power allocation policy. Of course, the main factor affecting the frequency of update of the training phase is the users' mobility. Due to users' mobility, the testing conditions will eventually be too different from the training conditions, requiring a new training phase. In the pedestrian environment considered here, this happens after a significant amount of time. On the other hand, a more frequent training phase might be required for scenarios with higher mobility, such as vehicular networks.

Finally, with reference to the $L$ coherence blocks simulated in Fig. 8.2, the long-term energy efficiency in Eq. (8.19) achieved by the ANN-based power control is 1.012 Mb/J, while sequential fractional programming achieved 0.411 Mb/J and uniform power allocation 0.012 Mb/J, respectively, which shows again that ANN-based power control is the most energy-efficient scheme over the complete observation window.

## 8.5     Conclusions

This chapter has dealt with the use of deep learning for resource allocation in smart radio environments. The issue of integrating ANNs into the architecture of future wireless networks has been addressed, and a framework for radio resource allocation based on the use of ANNs has been presented. Moreover, the long-term performance of ANN-based energy-efficient resource allocation has been analyzed, also accounting at the modeling stage for the power consumption due to the computation of the optimized power allocation in each channel coherence block.

The results have confirmed that ANN-based power control significantly reduces the power consumption compared to state-of-the-art methods based on optimization-theoretic techniques, while at the same time achieving near-optimal performance. Therefore, ANN-based resource allocation provides a suitable framework for online radio resource allocation in wireless networks.

# References

[1] ITU, "IMT traffic estimates for the years 2020 to 2030," tech. report, ITU-RM.2370-0, 2015.

[2] A. Zappone, M. Di Renzo, and M. Debbah, "Wireless networks design in the era of deep learning: Model-based, AI-based, or both?" *IEEE Trans. on Communications*, vol. 67, no. 10, pp. 7331–7376, Oct. 2019.

[3] M. Di Renzo et al., "Smart radio environments empowered by reconfigurable AI meta-surfaces: An idea whose time has come," *EURASIP Journal on Wireless Communincations and Networking*, vol. 129, 2019.

[4] M. Di Renzo et al., "Smart radio environments empowered by reconfigurable intelligent surfaces: How it works, state of research, and road ahead," *IEEE Journal on Selected Areas in Communications*, 2020.

[5] Telus and Huawei, "Next generation SON for 5G," white paper, 2016.

[6] I. Goodfellow, Y. Bengio, and A. Courville, *Deep Learning.* MIT Press, 2016.

[7] J. Schmidhuber, "Deep learning in neural networks: An overview," *arXiv preprint*, arXiv:1404.7828, 2014.

[8] K. Hornik, M. Stinchcombe, and H. White, "Multilayer feedforward networks are universal approximators," *Neural Networks*, vol. 2, pp. 359–366, 1989.

[9] M. Leshno et al., "Multilayer feedforward networks with a nonpolynomial activation function can approximate any function," *Neural Networks*, vol. 6, pp. 861–867, 1993.

[10] D. E. Rumelhart, G. E. Hinton, and R. J. Williams, "Learning representations by back-propagating errors," *Nature*, vol. 323, 1986.

[11] L. Bottou, *Algorithms and Stochastic Approximations*, D. Saad, ed. Cambridge University Press, 1998.

[12] D. P. Kingma and J. L. Ba, "Adam: A method for stochastic optimization," in *Int. Conf. on Learning Representation*, 2015.

[13] B. T. Polyak, "Some methods of speeding up the convergence of iteration methods," *USSR Computational Mathematics and Mathematical Physics*, vol. 4, no. 5, pp. 1–17, 1964.

[14] R. Li et al., "Intelligent 5G: When cellular networks meet artificial intelligence," *IEEE Wireless Communications*, vol. 24, no. 5, pp. 175–183, Oct. 2017.

[15] F. D. Calabrese et al., "Learning radio resource management in 5G networks: Framework, opportunities and challenges," *arXiv preprint*, arXiv:1611.10253, 2017.

[16] H. Sun et al., "Learning to optimize: Training deep neural networks for wireless resource management," *IEEE Trans. on Signal Processing*, vol. 66, no. 20, pp. 5438–5453, 2018.

[17] Q. Shi et al., "An iteratively weighted MMSE approach to distributed sum-utility maximization for a MIMO interfering broadcast channel," *IEEE Trans. on Signal Processing*, vol. 59, no. 9, pp. 4331–4340, Sep. 2011.

[18] B. Matthiesen et al., "A globally optimal energy-efficient power control framework and its efficient implementation in wireless interference networks," *IEEE Trans. on Signal Processing*, vol. 68, pp. 3887–3902, 2020.

[19] A. Zappone et al., "Model-aided wireless artificial intelligence: Embedding expert knowledge in deep neural networks towards wireless systems optimization," *arXiv preprint*, arXiv:1808.01672, 2019.

[20] A. Zappone, L. Sanguinetti, and M. Debbah, "User association and load balacing for massive MIMO through deep learning," in *Asilomar Conf. on Signals, Systems, and Computers*, 2018.

[21] L. Sanguinetti, A. Zappone, and M. Debbah, "A deep-learning framework for energy-efficient resource allocation in massive MIMO systems," in *Asilomar Conf. on Signals, Systems, and Computers*, 2018.

[22] F. Liang et al., "Towards optimal power control via ensembling deep neural networks," *arXiv preprint, arXiv:1807.10025*, 2018.

[23] M. Chen et al., "Echo state networks for proactive caching in cloud-based radio access networks with mobile users," *IEEE Trans. on Wireless Communications*, vol. 16, no. 6, pp. 3520–3535, June 2017.

[24] J. Fang et al., "Intelligent power control for spectrum sharing: A deep reinforcement learning approach," *arXiv preprint*, arXiv:1712.07365, 2018.

[25] Y. S. Nasir and D. Guo, "Deep reinforcement learning for distributed dynamic power allocation in wireless networks," *arXiv preprint*, arXiv:1808.00490, 2018.

[26] P. De Kerret and D. Gesbert, "Robust decentralized joint precoding using team deep neural network," in *IEEE Int. Symp. on Wireless Communication Systems (ISWCS)*, 2018.

[27] M. K. Sharma et al., "Distributed power control for large energy harvesting networks: A multi-agent deep reinforcement learning approach," *arXiv preprint*, arXiv:1904.00601, 2019.

[28] A. Zappone and E. Jorswieck, "Energy efficiency in wireless networks via fractional programming theory," *Foundations and Trends in Communications and Information Theory*, vol. 11, no. 3-4, pp. 185–396, 2015.

[29] A. Zappone et al., "Globally optimal energy-efficient power control and receiver design in wireless networks," *IEEE Trans. on Signal Processing*, vol. 65, no. 11, pp. 2844–2859, June 2017.

[30] G. Calcev et al., "A wideband spatial channel model for system-wide simulations," *IEEE Trans. on Vehicular Technology*, vol. 56, no. 2, March 2007.

[31] S. He et al., "Coordinated multicell multiuser precoding for maximizing weighted sum energy efficiency," *IEEE Trans. on Signal Processing*, vol. 62, no. 3, pp. 741–751, Feb. 2014.

[32] Y. Nesterov, *Introductory Lectures on Convex Optimization : A Basic Course*. Kluwer Academic Publisher, 2004.

# 9 Reinforcement Learning for Physical Layer Communications

Philippe Mary, Christophe Moy, and Visa Koivunen

## Acronyms

| | |
|---|---|
| ACK | Acknowledgment |
| ANN | Artificial Neural Network |
| AWGN | Additive White Gaussian Noise |
| BPSK | Binary Phase Shift Keying |
| BS | Base Station |
| DL | Deep Learning |
| DRL | Deep Reinforcement Learning |
| FNN | Feedforward Neural Network |
| HF | High Frequency |
| HMM | Hidden Markov Model |
| KL-UCB | Kullback-Leibler Upper Confidence Bound |
| MAB | Multi-Armed Bandit |
| MDP | Markov Decision Process |
| NN | Neural Network |
| OFDM | Orthogonal Frequency Division Multiple |
| OSA | Opportunistic Spectrum Access |
| PHY | Physical layer of the OSI stack |
| POMDP | Partially Observable Markov Decision Process |
| ML | Machine Learning |
| QoS | Quality of Service |
| RL | Reinforcement Learning |
| SARSA | current State, current Action, next Reward, next State, next Action |
| SINR | Signal to Interference plus Noise Ratio |
| SU | Secondary User(s) |
| UCB | Upper Confidence Bound |

## Notation and symbols

| | |
|---|---|
| $A(t)$ | Action random variable at $t$ |
| $a$ | A realization of the random variable action $A$ |
| $\mathbb{E}_P[X(t)]$ | Expectation of $X(t)$ under the distribution $P$ |
| $\mathbb{P}_X[X(t) = x]$ | Probability measure of the event $X(t) = x$, under the distribution of X |
| $P_{X(t+1)\mid X(t)}(x' \mid x)$ | Conditional probability measure of $X$ at $t+1$ knowing $X$ at $t$ |
| $\pi(a \mid s)$ | Policy $\pi$, i.e. probability to choose action $a$, while observing state s |
| $\pi^*$ | Optimal policy |
| $q_\pi(a,s)$ | Action-state value of the pair $(a,s)$ and following hereafter the policy $\pi$ |
| $q_*(a,s)$ | Optimal action-state value of the pair $(a,s)$ and following hereafter $\pi^*$ |
| $R(t)$ | Reward random variable at $t$ |
| $r$ | A realization of the random variable reward $R$ |
| $S(t)$ | State random variable at $t$ |
| $s$ | A realization of the random variable state $S$ |
| $v_\pi(s)$ | Value function of the state $s$ under the policy $\pi$ |
| $v_*(s)$ | Optimal value function of the state $s$ under $\pi^*$ |
| | |
| $[\cdot]^+$ | $\max(\cdot, 0)$ |
| $\mathbb{1}\{a\}$ | Indicator function equals to 1 if $a$ is true and 0 otherwise |
| $\mathbf{1}_n$ | Column vector full of ones of length $n$ |
| | |
| $\mathcal{A}$ | Set of possible actions |
| $\mathbb{N}$ | Set of positive integers |
| $\mathcal{R}$ | Set of possible rewards |
| $\mathbb{R}$ | Set of real numbers |
| $\mathbb{R}_+$ | Set of real and positive numbers |
| $\mathcal{S}$ | Set of possible states |

## 9.1     Introduction

Wireless communication systems have to be designed in order to cope with time-frequency-space varying channel conditions and a variety of interference sources. In cellular wireless systems for instance, a channel is estimated regularly by mobile terminals and base stations (BS) using dedicated pilot signals. This allows for adapting the transmitters and receivers to the current channel conditions and interference scenario. Powerful adaptive signal processing algorithms have been developed in the last few decades in order to cope with the dynamic nature of the wireless channel, including the least mean square and recursive least square algorithms for channel equalization or estimation, the Kalman filtering in multiple-input multiple-output channel matrix and frequency offset tracking. These techniques rely on well-established mathematical models of physical phenomena that allow systems to derive the optimal processing for a given criterion (e.g., mean square error and assumed noise and interference distribution models).

Any mathematical model has trade-offs between complexity and tractability. A complete, and hence complex, model may be useless if any insight on the state of the system cannot be drawn easily. For instance, the wireless propagation channel is absolutely deterministic and the signal received at any point in the space at any time can be precisely predicted by the Maxwell equations. However, this would require a prohibitive amount of computation and memory storage for a receiver to calculate at any point the value of the electric and magnetic fields using detailed and explicit knowledge of the physical characteristics of scatterers in the propagation environment, such as the dielectric and permittivity constants of the walls and other obstacles. It is much more efficient to design receivers that perform well in environments that have been stochastically characterized instead of using explicit deterministic model of each particular propagation environment.

Modern and emerging wireless systems are characterized by massive amounts of connected mobile devices, BSs, sensors, and actuators. Modeling such large-scale wireless systems has become a formidable task because of, for example, very small cell sizes, channel-aware link adaptation and waveform deployment, diversity techniques, and optimization of the use of different degrees of freedom in tranceivers. Consequently, it may not be feasible to build explicit and detailed mathematical models of wireless systems and their operational environments. In fact, there is a serious modeling deficit that calls for creating awareness of the operational wireless environment through sensing and learning.

Machine learning (ML) refers to a large class of algorithms that aim to give a machine the capability to acquire knowledge or behavior. If the machine is a wireless system, which is man-made, then the goal of ML in that case is to let the system choose its signal processing techniques and protocols to perform communication without being programmed a priori. The learning can be *supervised*, requiring labeled data, *unsupervised*, or based on *reinforcement learning* requiring trial-and-error approach. Supervised learning refers to methods that learn from a training set for which the desired output is provided by an external supervisor (i.e., with labeled data) and then perform a task on data that are not present in the training set. Unsupervised learning refers to methods that attempt to find hidden structures or patterns in a large dataset. Reinforcement learning (RL) is a class of machine learning techniques that aim at maximizing a cumulative *reward* signal over a finite or infinite time horizon for the selected actions. RL refers to the interaction through trial and error between an *agent*, the entity that learns, and its operational *environment*. This RL technique, which is the focus of this chapter, is particularly useful when the agent wants to acquire knowledge on the characteristics of its environment while making minimal assumptions on it. In this chapter, we focus on RL for physical layer (PHY) communications. The rewards in learning are then typically associated with a high data rate or high signal-to-interference and noise ratio (SINR). Even if ML in general and RL in particular are what we may call *data-driven approaches*, mathematical and physics-based modeling should not be completely abandoned. Indeed, wireless systems are man-made with plenty of structures that can be exploited to make the learning faster, transparent, and explainable.

Three main reasons for RL to be applied to a wireless communication problem are (a) the mathematical modelling of the environment is far too complex to be implemented in an agent, or such a mathematical model does not exist; (b) the signalling for acquiring the useful data needed to properly run the system is too complex; and (c) the desired outcome or goal of the learning can be described as a scalar reward. For instance, the massive machine-type communication or small-cell deployment where detailed network planning is not feasible are interesting scenarios where RL would be attractive technology. Moreover, machine-type communications typically involve some cheap battery operated devices where complex base band processing is not possible, which is in line with condition (a). Moreover, excessive signalling is excluded because of the huge number of devices to be connected (i.e., condition (b)). Hence, transmission strategies, such as selection of channel, transmit power, and beam patterns, can be *learned* from scratch or at least with very few a priori information in order to maximize the number of received packets (i.e., condition (c)).

*Adaptability* is the key benefit of RL techniques, as well as classical adaptive signal processing techniques such as the least mean square filter. However, unlike adaptive filtering, RL does not rely on well-established mathematical models of the physical phenomena that occur during a wireless transmission (see Section 9.2). RL techniques select suitable actions based on the feedback received from the surrounding environment in order to maximize a *reward* function over time. An important example of RL applied to physical layer communication is the flexible and opportunistic use of the underutilized frequency bands.

RL techniques involve a set of possible actions, a set of states, and a reward signal. These notions will be rigorously defined in the next section. Some examples of typical actions are whether to transmit on a given band or not or select among the predesigned beamformers or precoders. The states of the environment the agent observes are, for instance, whether the selected band is idle or not or feedback from the receiver that the observed SINR value is below a threshold on certain channels or the transmission has been successful ($+1$) or not ($0$ or $-1$), respectively.

In the particular example of the opportunistic spectrum access, the cardinality of action and state sets is small. Hence, the learning or convergence phase is faster than in a scenario where the action space would be much larger. If we imagine a smart radio device that is able to do link adaptation by choosing the modulation technique, transmit power, and select a channel to transmit its signal, the search space would become much larger. Hence, the convergence toward the optimal result that maximizes a given reward function would take much longer time and may become complex for RL approaches. However, recent advances in deep RL methods have alleviated this problem. An example on this scenario, link adaptation, will be given in this chapter. Another practical example considered later in this chapter is smart radio access network with the goal of optimizing the energy consumption of a cluster of BSs. This is a demanding problem with a high-dimensional state space that makes finding the optimal solution using classical optimization techniques difficult. RL may help to find a good solution through trial and error, and consequently reducing the complexity.

Deep learning (DL) strategies involve multiple layers of artificial neural networks (ANNs) that are able to extract hidden structures of labeled (*supervised learning*) or unlabeled data (*unsupervised learning*) [1]. These techniques prove to be efficient in many applications such as image classification, speech recognition and synthesis, and all applications involving a large amount of data to be processed. Recently, researchers in wireless communications have shown interest for using ML and DL in wireless networks design (see [2, 3] and references therein), mainly for networking issues, but more recently also for PHY layer, such as user-channel matching or performance optimization in large scale randomly deployed networks. This is particularly useful when a complete mathematical model of the behavior of a network is intractable due for example to the large dimensionality of the state or action spaces. RL and DL can be combined into *deep reinforcement learning* (DRL) approach when the number of observable states or possible actions are too large for conventional RL. This technique relies on the RL principle (i.e., an agent interacting with its environment), but the action to be taken is obtained through a nonlinear processing involving neural networks (NN) [4].

In this chapter, we give comprehensive examples of applying RL in optimizing the physical layer of wireless communications by defining a different class of problems and the possible solutions to handle them. In Section 9.2, we present the basic theories needed to address a RL problem, including Markov decision process (MDP) and partially observable Markov decision process (POMDP), but also two very important and widely used algorithms for RL: the Q-learning and SARSA algorithms. We also introduce the DRL paradigm and the section ends with an introduction to the multiarmed bandits (MAB) framework. Section 9.3 focuses on some toy examples to illustrate how the basic concepts of RL are employed in communication systems. We present applications extracted from literature with simplified system models using similar notation as in Section 9.2. In Section 9.3, we also focus on modeling RL problems, that is, how action and state spaces and rewards are chosen. The chapter concludes in Section 9.4 with thoughts on RL trends and ends with a review of a broader state of the art in Section 9.5.

*Notations*

The table at beginning of the chapter summarizes the notation used in the following. We review the main ones here. Random variables, and stochastic processes that depend on time, are denoted in capital font, such as $X(t)$, while their realizations in normal font, such as $x(t)$. Random vectors are denoted in capital bold font, such as $\mathbf{X}(t)$, and their realizations in small bold font, such as $\mathbf{x}(t)$. The conditional probability distribution of a random variable $X$ at time $t+1$ given another random variable $Y$ at $t$, is denoted as $P_{X(t+1)|Y(t)}(x \mid y)$ and when no ambiguity is possible on the underlying distribution function, simply by $p(x \mid y)$, being understood that the first and the second arguments in $p(\cdot \mid \cdot)$ rely to the values of the random variables at time $t+1$ and $t$, respectively, except otherwise mentioned. Vectors and matrices are denoted with bold font. Moreover, $\mathbb{1}\{a\}$ is the indicator function, which is equal to $1$ if $a$ is true and $0$ otherwise. $D\,(P \parallel Q)$ is the Kullback-Leibler divergence between the distributions $P$ and $Q$ and is defined as

$$D\,(P \parallel Q) = \int_{\mathcal{I}} \log \frac{dP}{dQ} dP \triangleq \mathbb{E}_P \left[ \log \frac{dP}{dQ} \right],$$

where $dP/dQ$ is called the Radon-Nikodym derivative and is defined if $P \ll Q$; that is, the measure $P$ is absolutely continuous with respect to (w.r.t.) the measure $Q$. This means that for all $x \in \mathcal{I}$, if $Q(x) = 0$ then $P(x) = 0$. This definition holds if the probability measures are continuous but also if they are discrete. In the former case, the Radon-Nikodym derivative is simply the ratio between the density probability functions, and in the latter case, it is the ratio of the probability mass functions that can be denoted as $p$ and $q$.

## 9.2      Reinforcement Learning: Background

### 9.2.1      Overview

RL may be considered as a sequential decision-making process. Unlike supervised learning, it does not need annotation or labeling of input-output pairs. The decision-maker in RL is called an agent. An agent has to sequentially make decisions that affect future decisions. The agent interacts with its environment by taking different actions. There are a number of different alternative actions, the agent has to choose from. Selecting the best action requires considering not only immediate but also long-term effects of its actions. After taking an action, the agent observes the environment state and receives a reward. Given the new state of the environment, the agent takes the next action. A trajectory or history of subsequent states, actions, and rewards is created over time. The reward quantifies the quality of the action and consequently captures what is important for the learning system. The objective of the learning is typically to maximize the sum of future rewards but also some other objective that is a function of the rewards may be employed. The learning takes place by trial and error so that the rewards reinforce decisions that move the objective toward its optimum. As a consequence, the system learns to make good decisions and can independently improve its performance by probing different actions for different states. Hence, consequences to future actions are learned. Commonly, the immediate rewards are emphasized and the future rewards are discounted over time. Sometimes actions with a small immediate reward can lead to a higher payoff in a long term. One would like to choose the action that trades off between the immediate rewards and the future payoffs in the best possible way.

   RL are typically modeled using an MDP framework. The MDP provides a formal model to design and analyze RL problems as well as a rigorous way to design algorithms that can perform optimal decision making in sequential scenarios. If one models the problem as an MDP, then there exists a number of algorithms that will be able to automatically solve the decision problem. However, real practical problems cannot be strictly modeled with an MDP in general, but this framework can be a good approximation of the physical phenomena that occur in the problem such that this model may perform well in practice.

## 9.2.2 Markov Decision Process

An MDP model is comprised of four components: a set of states, a set of actions, the transitions (i.e., how actions change the state), and the immediate reward signal due to the actions. We consider a sequence of discrete time steps $t = \{0, 1, 2, \ldots\}$ and briefly describe each component. The state describes the set of all possible values of dynamic information relevant to the decision process. In a broad sense, one could think that a state describes the way the world currently exists and how an action may change the state of the world. Obviously, we are considering only an abstract and narrow view of the world that is relevant to the learning task at hand and our operational environment. The state at time instance $t$ is represented by a random variable $S(t) \in \mathcal{S}$, where $\mathcal{S}$ is the state space, or the set of all possible values of dynamic information relevant to the learning process. In the context of a physical layer of wireless communications, the state could describe, for example, the occupancy of the radio spectrum or the battery charging level of the user equipment.

The environment in which the agent acts is modeled as a Markov chain. A Markov chain is a stochastic model for describing the sequential state transitions in memoryless systems. A memoryless transition means that

$$P_{S(t+1)|\mathbf{S}(t)}\left(s' \mid \mathbf{s}\right) = P_{S(t+1)|S(t)}\left(s' \mid s\right), \tag{9.1}$$

with

$$P_{S(t+1)|S(t)}\left(s' \mid s\right) \triangleq \mathbb{P}\left[S(t+1) = s' \mid S(t) = s\right], \tag{9.2}$$

if $\mathcal{S}$ is a discrete set. Moreover $\mathbf{S}(t) = [S(0), \ldots, S(t)]$ and $\mathbf{s} = [s(0), \ldots, s(t)]$, $S(t)$ is the state at time instance $t$ and $P_{S(t+1)|S(t)}$ is the conditional distribution of $S(t+1)$ given $S(t)$. In other words, the probability of the state transition to state $S(t+1)$ is only dependent on the current state $S(t)$. Hence, there is no need to remember the history of the past states to determine the current state transition probability.

The actions are the set of possible alternatives we can take. The problem is to know which of these actions to choose in a particular state of the environment. The possible actions the agent can choose at state $S(t)$ is represented with the random variable $A(t) \in \mathcal{A}$, where $\mathcal{A}$ is the action space. The possible actions may depend on the current state. Typical actions in a wireless communication context can be giving access to the channel, sensing the channel, selecting a beam pattern, or adjusting the power, for instance.

The action the agent takes activates a state transition of the Markov chain. The state transition function specifies in a probabilistic sense the next state of the environment as a function of its current state and the action the agent selected. It defines how the available actions may change the state. Since an action could have different effects, depending upon the state, we need to specify the action's effect for each state in the MDP. The state transition probabilities depend on the action such as

$$P_{S(t+1)|S(t), A(t)}(s' \mid s, a) \triangleq p(s' \mid s, a). \tag{9.3}$$

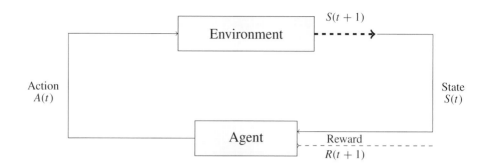

**Figure 9.1** Reinforcement learning principle in MDP.

If we want to make the decision-making process automatic and run it in a computer, then we must be able to quantify the value associated to the actions taken. The value is a scalar and called a reward. The reward from action $A(t)$ is represented by a random variable $R(t) \in \mathcal{R}$, where $\mathcal{R}$ is the set of rewards, and it specifies the immediate value for performing each action in each state.

The reward provides the means for comparing different actions and choosing the right one for achieving the objective of the learning process. The reward distribution depends on which action was selected and on the state transition. The conditional probability distribution defining the dynamics of an MDP process is

$$P_{R(t+1)S(t+1)| S(t)A(t)} \left( r, s' \mid s, a \right) \triangleq p(r, s' \mid s, a). \qquad (9.4)$$

An MDP can last for a finite or infinite number of time steps, distinguishing between finite and infinite horizon models and methods. The interconnection between the different elements of an MDP can be illustrated with the well-known diagram in Fig. 9.1.

A reward function $R()$ gives a payoff $R(t+1)$ for choosing action $A(t)$ in state $S(t)$ resulting in new state $S(t+1)$. After selecting an action the agent obtains the reward and the next state, but no information on which action would have been the best choice toward its objective in the long term. Hence, the agent will perform active probing and obtain experience about the possible system states, available actions, transitions, and rewards to learn how to act optimally. A widely used objective is to maximize the expected sum of discounted rewards over an infinite horizon [5, 6]:

$$J_\pi = \mathbb{E}_\pi \left[ \sum_{t=0}^{\infty} \gamma^t R(t+1) \right], \qquad (9.5)$$

where $\gamma < 1$ is the discount factor emphasizing more immediate rewards and the expectation is taken over the distribution of the rewards, following a certain *policy* $\pi$ whose the meaning will be detailed hereafter. Discounting is the most analytically tractable and hence the most commonly used approach. Other objective functions may be employed, including expected reward over a finite time horizon and regret, which measures the expected loss in the learning compared to an optimal policy.

The solution to an MDP is called a policy, and it simply specifies the best sequence of actions to take during the learning process. A policy maps states to actions (i.e., $\pi : S \rightarrow A$) and can be *deterministic* (i.e., a single or a set of deterministic actions is performed when the agent encounters the state $S(t)$), or it can be *stochastic* and is defined as the conditional probability measure of $A(t)$ given $S(t)$, or $P_{A(t)| S(t)}\,(a \mid s)$, often simply denoted as $\pi(a \mid s)$.[1] It is basically a sequence of the decision rules to be used at each decision time instance. For the infinite-horizon discounted model, there exists an optimal deterministic stationary policy. The finite-horizon model is appropriate when the system has a hard deadline or the agent's lifespan is known.

An optimal policy would need to consider all the future actions. Such policies are called nonmyopic policies. In comparison, a policy where the agent maximizes the immediate reward is called a myopic policy. Typically, the myopic policies are suboptimal for $\gamma > 0$. The goal is to derive a policy that gives the best actions to take for each state, for the considered horizon.

## Value Functions

In order to be able to derive the optimal policy that maximizes the function $J$ in Eq. (9.5), one needs to evaluate the value of the states under a certain policy $\pi$ (i.e., a function $v_\pi : S \rightarrow \mathbb{R}$). This is defined as the expectation, over the policy $\pi$, of the discounted reward obtained when starting from the state $s \in S$, that is to say [5]

$$v_\pi(s) = \mathbb{E}_\pi \left[ \sum_{t=0}^{\infty} \gamma^t R(t + 1) \mid S(0) = s \right], \forall s \in S. \tag{9.6}$$

Since the state value function depends on the state $s$ taken for the computation, this quantity may be averaged over the states in order to obtain the average reward in Eq. (9.5). The state value function represents the average of the discounted reward that would be obtained starting from an initial state $s$ and following the policy $\pi$.

Similarly, one can define the *action-state* function, $q_\pi : S \times A \rightarrow \mathbb{R}$, by averaging the discounted reward when starting at $t = 0$ from state $s \in S$, taking the action $a \in A$, and then following the policy $\pi$ thereafter. We hence have now an expectation conditioned on the state and the action, that is [5, 6]

$$q_\pi(s, a) = \mathbb{E}_\pi \left[ \sum_{t=0}^{\infty} \gamma^t R(t + 1) \mid S(0) = s, A(0) = a \right], \forall(s, a) \in S \times A. \tag{9.7}$$

## Bellman Equations

A remarkable property of the state value function in Eq. (9.6) is that it follows a recursive relation that is widely known as the Bellman equation [7] of the state value function. For all $s \in S$, one can prove [5]

---

[1] Note in that case, $a$ and $s$ refer to the value of the action and the state, respectively, at time $t$, because action is immediately chosen while observing a given state and not delayed to the next time slot.

$$v_\pi(s) = \sum_{a \in \mathcal{A}} \pi(a \mid s) \sum_{(s',r) \in \mathcal{S} \times \mathcal{R}} p(s',r \mid s,a) \left[ r + \gamma v_\pi(s') \right]. \tag{9.8}$$

The expression in Eq. (9.8) is known as the Bellman's equation for the state value function and has to be seen as the expectation of the random variable $R(t + 1) + \gamma v_\pi(S(t + 1))$ over the joint distribution $P_{A(t)|S(t)} P_{S(t+1)R(t+1)|S(t)A(t)}$. The Bellman's equation links the state value function at the current state with the next state value function averaged over all possible states and rewards knowing the current state and the policy $\pi$. The value of a state $S(t)$ is the expected instantaneous reward added to the expected discounted value of the next states, when the policy $\pi$ is followed. The proof of this relation relies on separating Eq. (9.6) into two terms, $t = 0$ and $t > 0$, and using essentially the Markovian property and the Bayes' rule in the second term to make appear $v_\pi(s')$.

The optimal state value and action-state value functions are obtained by maximizing $v_\pi(s)$ and $q_\pi(s,a)$ over the policies, that is [6]

$$v_*(s) = \sup_{\pi \in \Phi} v_\pi(s), \tag{9.9}$$

and

$$q_*(s,a) = \sup_{\pi \in \Phi} q_\pi(s,a), \tag{9.10}$$

where $\Phi$ is the set of all stationary policies (i.e., policies that do not evolve over time).[2] Moreover, Eq. (9.9) can be written w.r.t. Eq. (9.10) as $v_*(s) = \sup_{a \in \mathcal{A}} q_*(s,a)$. The optimal state value function also obeys to the Bellman recursion and one can show that [5, 6]

$$v_*(s) = \sup_{a \in \mathcal{A}} \left\{ \sum_{(r,s') \in \mathcal{S} \times \mathcal{R}} p(r,s' \mid s,a) \left[ r + \gamma v_*(s') \right] \right\} \quad \forall s \in \mathcal{S}. \tag{9.11}$$

This last equation means that the expected return from a given state $s$ and following the optimal policy $\pi^*$ is equal to the expected return following the best action from that state. Moreover, substituting $v_*(s')$ in Eq. (9.11) with the supremum over the actions of $q_*(s',a)$, we obtain the iterative property on the action-state value function:

$$q_*(s,a) = \sum_{(s',r) \in \mathcal{S} \times \mathcal{R}} p(s',r \mid s,a) \left[ r + \gamma \sup_{a' \in \mathcal{A}} q_*(s',a') \right]. \tag{9.12}$$

Equations (9.11) and (9.12) are the Bellman optimality equations for $v_*$, and $q_*$, respectively.

The policy should mostly select the actions that maximize $q_*(s,\cdot)$. A policy that chooses only actions that maximize a given action-state function $q(s,\cdot)$ for all $s$ is called *greedy* w.r.t. $q$. A greedy policy is a decision procedure based on the immediate reward without considering the long-term payoff. In general, only considering local

---

[2] This applies to fixed over-time *deterministic* policy or a stationary *stochastic* policy, in the strict sense. That is, the conditional law of $A(t)$ given $S(t)$ does not depend on $t$.

or immediate reward may prevent from finding the highest payoff on the long term. However, since $v_*$ or $q_*$ already contain the reward consequences of all possible states, then a greedy policy w.r.t. $q_*$ is hence optimal on the long run [6]. Hence, if one is able to compute $q_*$, the optimal policy follows.

## Policy Evaluation

RL algorithms aim to find the best sequences of actions in order to get close to Eq. (9.11). As we will see later with the Q-learning, the idea is to keep an estimate of the optimal state or action state value functions at each time step and find a way to make them converge to the optimal value functions. The expressions in Eq. (9.11) and Eq. (9.12) are fixed-point equations and can be solved using dynamic programming. Given a policy $\pi$, the policy evaluation algorithm first finds the value function for immediate rewards and then extends the time horizon one by one. Basically, one just adds the immediate reward of each of the available actions to the value function computed in previous step. This way one builds the value function in each iteration based on the previous one. Let consider a deterministic policy $a = \pi(s)$. A sequence of value functions can be obtained for all states $s$ such as

$$v_{t+1}(s) = \underbrace{r(s,a)}_{\text{immediate average reward}} + \gamma \underbrace{\sum_{s' \in \mathcal{S}} p(s' \mid s,a) v_t(s')}_{\text{value function of previous steps}}, \qquad (9.13)$$

where the immediate average reward, $r(s,a)$, is the expectation over the distribution of the rewards knowing the action $\pi(s)$ taken in the state $s$ at time $t$, or $r(s,a) = \mathbb{E}_{P_{R\mid SA}}[R(t+1) \mid S(t) = s, A(t) = a]$. This iterative equation can be computed for each action taken in each state. This equation converges because the linear operator linking the value function of state $S(t)$ to the value function of state $S(t+1)$ is a maximum norm contraction [6].

Similarly, the action-state value iteration algorithm aims at iteratively solving the following equation for a deterministic policy:

$$q_{t+1}(s,a) = \underbrace{r(s,a)}_{\text{immediate average reward}} + \gamma \underbrace{\sum_{s' \in \mathcal{S}} p(s' \mid s,a) q_t(s',a)}_{\text{action-state value function of previous steps}}. \qquad (9.14)$$

The problem of this procedure is that the agent needs to know the complete dynamic of the MDP, $p(r, s' \mid s, a)$, which is rarely the case in practice. Fortunately, there are algorithms like Q-learning that converge to the optimal policy without knowing the dynamic of the physical process.

## Policy Improvement

The iterative procedures described so far allow us to evaluate the performance of a policy. Indeed, by choosing a given policy $\pi_0$ one can compute Eqs. (9.13) or (9.14) until convergence, for example, $|v_{t+1}(s) - v_t(s)| \leq \epsilon$, where $\epsilon$ is an arbitrary small number. When the procedure has converged, the state value is obtained; that is, $v_{\pi_0}(s)$ is the state value function achieved by following the policy $\pi_0$ for state $s$. Hence, one has the *evaluation* of the performance of the policy $\pi_0$. But we still do not know if it is the best policy that leads to the maximal value of the state function $v_*(s)$.

In order to progress to the optimal policy, one needs a *policy improvement* step. The idea is to create a new policy $\pi'$ that differs from $\pi$ by a different action taken when being in state $s$, for example, $\pi'(s) \neq \pi(s)$. If we are able to choose an action when being in state $s$ such that $q_\pi(s, \pi'(s)) \geq v_\pi(s)$, then $v_{\pi'}(s) \geq v_\pi(s)$, $\forall s \in \mathcal{S}$ [5]. Basically, this means that by taking an action $a'$ such that $a' \neq \pi(s)$ and following the policy $\pi$ hereafter such that $q_\pi(s, a') \geq v_\pi(s)$, the policy $\pi'$ should not be worse than $\pi$ on the performance of the learning task. Let $\pi_t$ denote the policy obtained at iteration $t$. Once Eq. (9.14) has converged under $\pi_t$, we create $\pi_{t+1}$ by choosing the greedy policy w.r.t. $q_{\pi_t}$ which is obtained by choosing the action that maximizes the right-hand side of Eq. (9.14) for all states $s$, and we repeat the process until having no policy improvements.

**Combination of Policy Evaluation and Improvement**

In practice, the two procedures we have introduced, the policy evaluation and improvement, can be combined into one algorithm called *value iteration*. This algorithm consists of taking the action that maximizes the (action-)state value function at each time step, or taking the supremum over $\mathcal{A}$ of the right-hand side of Eq. (9.13) for the state value function and replacing $q_t(s', a)$ by $\sup_{a' \in \mathcal{A}} q_t(s', a')$ in Eq. (9.14) for the action-state value function. The new equations obtained also converge to $v_*$ and $q_*$ thanks to the fixed-point theorem and the property of the Bellman equation. This means that a greedy policy w.r.t. $v_t$ ($q_t$) is allowed to converge to the optimum (action-)state value.

## 9.2.3   Partially Observable Markov Decision Process

Complete observability is required for MDP-based learning. When the states of the underlying Markovian process are not directly observable, the situation is referred as a POMDP. This model adds sensor measurements that contain information about the state of the MDP model as well as an observation function describing the conditional observation probabilities. Instead of observing the current state of the process directly, we may just have access to a noisy version or estimate of the state. Typically, we use physical sensors to acquire a set of noisy observations containing relevant information about the state. The principle of reinforcement learning under POMDP is illustrated on Fig. 9.2. For example, in physical layer wireless communication we need to estimate key system parameters such as the channel impulse or frequency response, channel matrix and its covariance, SINR, perform time and frequency synchronization as well as design transmit or receive beamformers, precoders, or decoders from data acquired by the receivers. The received signals contain unobservable random noise and harmful interferences. Moreover, in order to adapt our transmitters and optimize the resources usage, we may need to use feedback from receivers or exploit channel reciprocity that are both subject to error. Most of the parameters characterizing the state of a wireless

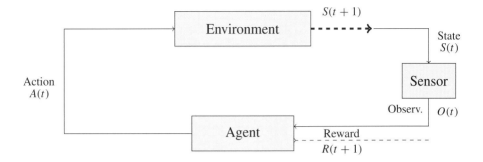

**Figure 9.2** Reinforcement learning principle in POMDP. Unlike the MDP in Fig. 9.1, the state of the environment is observed through a sensor that gives a partial state of the environment. For instance, the sensor may be the receiver in PHY layer communication and actions would be adapting the operational parameters of the transmitter/receiver pair.

system at the physical layer are defined in a continuous space. Hence, the conventional MDP model is not necessarily applicable. The errors caused by noisy sensors are unobservable and random and need to be described using probability distributions. Consequently, the states are also described in terms of probability distributions.

In MDP-based learning our goal is to find a mapping from states to actions. In the case of POMDPs, we are looking for a mapping from probability distributions associated with the states to actions. The probability distribution over all possible model states is called a *belief state* in POMDP jargon and the entire probability space (i.e., the set of all possible probability distributions) is called the *belief space*. The observations acquired by physical sensors contain unobservable and random noise. Hence, we need to specify a probabilistic observation model called an observation function in POMDP jargon. This observation model simply tells us the probability of each observation for each state in the model. It is formally given as a conditional probability of an observation given a state-action couple. As a result, we will have uncertainty about the state due to incomplete or uncertain information. That means we have a set of states, but we can never be certain which state we are in. The uncertainties are handled by associating a probability distribution over the set of possible states. The distribution is defined by the set of observations, observation probabilities, and underlying MDP. The POMPD model considers *beliefs* of the states instead of the actual states in MDP. The belief is a measure in interval [0, 1] that describes the probability of being in a specific state. The agent interacts with the environment and receives observations and then updates its belief state by updating the probability distribution of the current state. In practice, the agent updates its belief state by using a state estimator based on the last action, current acquired observation, and previous belief state. By solving a POMDP problem, we find a policy that tells which action is optimal for each belief state.

A POMDP is a tuple $< \mathcal{S}, \mathcal{A}, \mathcal{R}, \Omega, O >$. The environment is represented by a set of states $\mathcal{S} = \{s_1, s_2, \ldots, s_N\}$ with $|\mathcal{S}| = N$, a set of possible actions denoted by

$\mathcal{A} = \{a_1, a_2, \dots, a_K\}$ with $|\mathcal{A}| = K$. The transition from state $s$ to new state $s'$ given the action $a$ is governed by a probabilistic transition function as in Eq. (9.3) that is the conditional distribution $P_{S(t+1)|S(t), A(t)}$. As an immediate result, the agent receives a reward $R(t + 1)$ that depends on the state $s$ the agent was in and the action $a$ it took. These components of the POMDP-tuple are as in a conventional MDP. Since state $s'$ is not directly observable, it can be estimated by using observations that contain information about the state. A finite set of observations emitted by the state is denoted by $z \in \Omega$ with $|\Omega| = M$. An observation function defines a probability distribution for each action $A(t)$ and the resulting state $S(t + 1)$. The observation $Z(t + 1)$ is then a random variable and the probability distribution of the observations conditioned on the new state $S(t + 1)$ and action $A(t)$ is $P_{Z(t+1)|S(t+1), A(t)}(z \mid s', a) \triangleq O(z, s', a)$.

In a POMDP, the following cycle of events takes place. An environment is in state $S$ and the agent computes the belief state $B$. The agent takes an action $A$ using the policy $\pi(\cdot \mid B)$ and the environment transitions to a new state $S'$ according to the state transition distribution. The new state is not directly observable, but the environment emits an observation $Z$ according to the conditional distribution $O(z, s', a)$. Then the agent receives a reward $R$ for action taken from belief state $B$. Finally, the agent updates the belief state and runs the cycle again. As a result, a trajectory or history of subsequent belief states, actions, observation, and rewards is created over time. The goal of an agent is to learn a policy $\pi$ that finds actions that maximize the policy's value. There are many ways to measure the quality. The most common choice is the discounted sum of rewards as in the case of MDP.

The belief state is a sufficient statistic that contains all the relevant information about the past observations and enjoys the Markov property. The next belief state depends only on the current belief state and the current action and observation. Hence, there is no need to remember the whole history of actions and observations. Updating the distribution requires using the transition and observation probabilities and a formula stemming from the Bayes rule. Let us denote $P_B(s) \triangleq \mathbb{P}_B[S = s]$ the distribution of the belief state, which is the probability that the environment is in state $s$ under the distribution of the belief. The belief state update that can be considered as a state estimation step in POMDP framework is given by

$$P_B(s') = \frac{O(z, s', a)}{p(z|b, a)} \sum_{s \in \mathcal{S}} p(s' \mid s, a) P_B(s). \tag{9.15}$$

The denominator is a normalizing constant that makes the sum equal to one; that is, $p(z|b, a) = \sum_{s' \in \mathcal{S}} O(z, s', a) \sum_{s \in \mathcal{S}} p(s'|s, a) P_B(s)$. The distributions are updated simply by applying the Bayes' Rule and using the model parameters. Similarly to MDP, the agent wishes to choose its actions such that it learns an optimal policy $\pi^*(b)$.

The main practical difficulty in POMDP models is finding a solution that is a policy that chooses optimal action for each belief state. The state space is just defined in terms of probability distributions. Algorithms used for solving POMDP typically stem from dynamic programming. It can be solved using value iteration, policy iteration, or a variety of other techniques developed for solving MDP. The optimal value function $v_*$

is then the value function associated with an optimal policy $\pi^*$. The early approach for solving POMDPs was using belief-MDP formulation. The value iteration follows the Bellman's equation already introduced for MDP:

$$v_0(b) = 0. \tag{9.16}$$

$$v_{t+1}(b) = \sup_{a \in \mathcal{A}} [r(b,a) + \gamma \sum_{z \in \Omega} p(z \mid b,a) v_t(b')], \tag{9.17}$$

where $r(b,a) = \sum_{s \in \mathcal{S}} P_B(s) r(s,a)$ is the average reward for action $a$ in the belief state $b$ and where $r(s,a)$ is the expected reward obtained in state $s$ and taking action $a$ that has been defined in Eqs. (9.13) and (9.14). Moreover, $p(z \mid b,a)$ has been defined previously and denotes the conditional distribution of the observation $z$ at time $t+1$, given the belief state $b$ and taking action $a$ at $t$. A value function may be modeled with a structure that can be exploited in making the problem solving easier or even feasible. For example, a piecewise linear and convex approximation may be used in finite-horizon scenarios. Value iteration provides an exact way of determining the value function for POMDP. Hence, the optimal action can be found from the value function for a belief state. Unfortunately, the complexity of solving a POMDP problem via value iteration is exponential in the number of observations and actions. Moreover, the dimensionality of the belief space grows proportionally to the number of states. The dimensionality of belief space can be reduced using a parametric model such as a Gaussian mixture model. Also point-based value iteration has been proposed for solving POMDP. In such methods, a small set of reachable belief points are selected, and the Bellman updates are done at these points while storing values and gradients. Heuristic methods employing search trees have been developed, too. The methods build an AND/OR tree of reachable belief states from the current belief and perform a search over the tree using well-known methods such as branch-and-bound.

Value iteration is commonly used since the classical policy iteration algorithm for POMDP proposed in [8] has a high complexity and is thus less popular. There are, however, algorithms of lower complexity. One can also use the policy iteration methods we described earlier for solving POMDP problems.

## 9.2.4    Q-learning and SARSA Algorithm

Q-learning and the current State, current Action, next Reward, next State, and next Action (SARSA) algorithms are iterative procedures to learn the optimal policy that maximizes the expected reward from any starting state, without the knowledge of the MPD dynamics. Both algorithms exhibit some similarities but differ in some key points that we detail hereafter. Q-learning and SARSA are called tabulated algorithms; that is, they are based on the construction of a look-up table, a.k.a Q-table, that allows the algorithm to trace and update the expected action-value function for all action-state pairs $(s,a) \in \mathcal{S} \times \mathcal{A}$. Once the Q-table has been built, the optimal policy is to choose the action with the highest score. The basic idea of both algorithms is to build a new estimate from an old estimate, which is updated by an incremental difference between a target and the old estimate. This can be formalized as follows:

$$
\underbrace{q_t\left(S(t), A(t)\right)}_{\text{new estimate}} \leftarrow \underbrace{q_t\left(S(t), A(t)\right)}_{\text{old estimate}} + \alpha_t \left[ \underbrace{T_{t+1}}_{\text{target}} - \underbrace{q_t\left(S(t), A(t)\right)}_{\text{old estimate}} \right], \tag{9.18}
$$

where $\alpha_t$ is the *learning rate* at time $t$, which is a scalar between 0 and 1. The learning rate tells us how much we want to explore something new and how much we want to exploit the current choice. Indeed, in each of the RL algorithms, exploitation and exploration have to be balanced in order to trade-off between the desire for the agent to increase its immediate reward, by exploiting actions that gave good results so far, and the need to explore new combinations to discover strategies that may lead to larger rewards in the future. At the beginning of the procedure, one may expect to spend more time exploring while the Q-table fills up. Later the agent exploits more than it explores in order to increase its reward. The learning rate should satisfy the following conditions $\sum_{t=0}^{\infty} \alpha_t = \infty$ and $\sum_{t=0}^{\infty} \alpha_t^2 < \infty$ (e.g., $\alpha_t = 1/(t+1)$). Finally, note that $\alpha$ can also be taken as a constant less than one, and hence not satisfy the conditions. However in practice, learning tasks occur over a finite time horizon, and hence the conditions are satisfied.

### Q-learning

Q-learning has first been introduced by Watkins in 1989 [9]. In this algorithm, the target is equal to

$$
T_{t+1} = R(t+1) + \gamma \max_{a' \in \mathcal{A}} q_t\left(S(t+1), a'\right), \tag{9.19}
$$

which is nothing but the algorithmic form of the Bellman equation in Eq. (9.12). When the algorithm has converged, $T_{t+1}$ should be equal to $q_t(S(t), A(t))$, nullifying the difference term in Eq. (9.18). Let us more closely examine how the Q-learning algorithm works.

The Q-table is a table with the states along the rows and columns representing the different actions we can take. At time step $t = 0$, the Q-table is initialized to 0; that is, $q_0(s(0), a(0)) = 0, \forall (s, a) \in \mathcal{S} \times \mathcal{A}$. A starting state is chosen randomly with a certain probability: $\mathbb{P}\left[S(0) = s\right]$. At the beginning, the agent does not know the reward that each action will provide, so it chooses one action randomly. This action leads to a new state $s'$ and a reward $r$ in a stochastic manner according to $p(r, s' \mid s, a)$ if the problem is stochastic or according to deterministic rules otherwise. The immediate reward the agent receives by taking the action $a(t)$ at time $t$ is the variable $R(t+1)$ in Eq. (9.19). Then the algorithm looks at the line represented by the state $S(t+1) = s'$ in the Q-table and chooses the value that is maximum in the line, which corresponds to the term $\max_{a' \in \mathcal{A}} q_t(S(t+1), a')$ in Eq. (9.19); among all possible actions from the next state we end up at $t+1$, or $S(t+1)$, one chooses the one that leads to the maximum expected return. The difference in Eq. (9.18) acts as a gradient that allows some actions in a given state to be reinforced or, on the contrary to dissuade the agent to take some other actions being in a given state. By choosing the learning rate, $\alpha_t = 1/(t+1)$

for instance, the agent will pay less and less attention over time to future expected returns in the update of the current estimation of the Q-table. If $\alpha = 0$, the agent does not learn anymore, while $\alpha = 1$ means that the agent remains in an active learning behavior state. Moreover, $\gamma$ in Eq. (9.18) is the discounting factor that defines how much the agent cares about future rewards (i.e., the ones that will be obtained starting from $S(t + 1)$), compared to the immediate reward, $R(t + 1)$.

Q-learning is an algorithm that explores and exploits at the same time. But which policy should be followed when updating the Q-table? That is, how are the actions chosen at each time step? Actually, this is the strength of Q-learning, as it does not (so much) matter for the algorithm convergence. The $\epsilon$−greedy policy is, however, a widely used technique. It consists of randomly choosing an action with probability $\epsilon$ and the action that maximizes the current action-state value at time $t$ with probability $1 - \epsilon$. Of course, $\epsilon$ can be kept constant or may vary during the learning in order to explore more at the beginning, $\epsilon \approx 1$, and exploit more after a while, $\epsilon \ll 1$. However, it has been shown that Q-learning makes the Q-table converge to $q_*$, and hence to the optimal policy as soon as every action-state pair has been visited an infinite number of times, irrespective to the policy followed during the training [10, 11]. This property makes Q-learning an *off-policy* procedure.

In some practical problems, MDP may present some terminal states, or absorbing states, and the learning is done over several episodes. Once the agent reaches the terminal state, the episode ends and the algorithm restarts at a random state, and keep going to estimate the Q-table.

### SARSA algorithm

SARSA is an algorithm that has been proposed in [12] and differs from Q-learning simply by the target definition in Eq. (9.18). In SARSA, we use

$$T_{t+1} = R(t + 1) + \gamma q_t \left( S(t + 1), A(t + 1) \right). \tag{9.20}$$

The difference with the Q-learning approach is that the next action $A(t + 1)$ to be taken when the agent observes the next state $S(t + 1)$ starting from the state $S(t)$ and having taken the action $A(t)$ is no longer the one that maximizes the next expected return from the next state $S(t + 1)$. The action $A(t + 1)$ has to be taken according to a policy, which is why SARSA is called an *on-policy* method. The policy the agent follows when in state $S(t)$, and hence the choice of the action at time $t$, $A(t)$, is the *behavior policy*, while the choice of the action to be taken when in state $S(t + 1)$, or $A(t + 1)$, characterizes the *target policy*. In Q-learning, the target policy was greedy w.r.t. $q_t$; that is, the action that maximizes the next expected action-state value is chosen. In SARSA on the other hand, the agent updates the Q-table from the quintuple $(S(t), A(t), R(t + 1), S(t + 1), A(t + 1))$ where both behavior and target policies are $\epsilon$−greedy; that is, the next action to be taken while observing state $S(t + 1)$ is chosen randomly with the probability $\epsilon$ and the action maximizing $q_t(S(t+1), a')$ is taken with probability $1 - \epsilon$. Under the assumption that the actions taken under the target policy are, at least occasionally, also taken under the behavior policy, SARSA converges to the optimal policy, or the greedy policy w.r.t. $q_*$.

## 9.2.5 Deep RL

The previous $Q$-learning or SARSA approaches are suitable when the dimensions of the state and action spaces of the problem are small or moderate. In that case, a look-up table can be used to update the $Q$ values. However, when the number of states or possible actions becomes large, the complexity and the storage needs of $Q$-learning become prohibitive. Instead of using a table for updating the action-state value function, one may search for approximating the action-state values with a suitable function $q_\theta : S \times A \rightarrow \mathbb{R}$ with the vector parameter $\theta$. The simplest approximation function is the linear one. This consists of finding a suitable mapping, $\psi : S \times A \rightarrow \mathbb{R}^d$, used to represent the *features* of the action-state couple, where $d$ is the dimension of this feature representation [6]. The entries of the vector $\psi(s, a)$ are the *basis functions*, and they span the space in which the Q function is approached. The linear approximation of the action-value function is hence $q_\theta(s, a) = \theta^T \psi(s, a)$ and will be used to approximate the optimal action-state value.

If an oracle would give us the action-state function under the policy $\pi$, one could compute an error function between a target and a prediction as follows:

$$L(\theta) = \mathbb{E}_\pi \left[ \left( q_\pi(S, A) - \theta^T \psi(S, A) \right)^2 \right], \tag{9.21}$$

where the expectation is taken over the joint distribution of the state, reward, and action. The weights $\theta$ can be updated using the gradient of the loss function such as

$$\theta_{t+1} = \theta_t + \alpha_t \mathbb{E}_\pi \left[ \left( q_\pi(S, A) - \theta^T \psi(S, A) \right) \psi(S, A) \right]. \tag{9.22}$$

However, the agent never knows in general the true value of the objective (i.e., the true action-state value function under the policy $\pi$). It can only estimate this value. For the Q-learning algorithm, $q_\pi(S, A)$ in Eq. (9.22) is substituted by expression in Eq. (9.19), where the estimation of the action-state function is replaced by its linear approximation such that

$$\theta_{t+1} = \theta_t + \alpha_t \left( R(t+1) + \gamma \max_{a' \in A} \theta^T \psi(S(t+1), a') - \theta^T \psi(S, A) \right) \psi(S, A). \tag{9.23}$$

However, the features extraction phase, or constructing the function $\psi$, can be complex if the problem dimension is very large [6].

One may observe that when passing from Eq. (9.22) to Eq. (9.23), we did not limit ourselves to substitute $q_\pi(S, A)$ by $R(t+1) + \gamma \max_{a' \in A} \theta^T \psi(S(t+1), a')$ but the expectation operator also vanished. There are only random samples from a given database (a batch) to compute the sequence $\{\theta\}_t$. The latter is a random sequence, and hence we are not sure that this will decrease the value of the loss at each step. However, one can show that this will decrease the loss in *average*. This is the principle of the stochastic gradient descent [13].

*Deep Q-learning*

The approximation function can be nonlinear, but it has to be differentiable w.r.t. the parameters of the approximation function. The principle of the deep Q-learning is simple and consists of designing a neural network that outputs all the action-state values for all possible actions in a given state $s$, $q_\theta(s, \cdot)$, $\forall s \in \mathcal{S}$. In other words, if the state space is of dimension $n$, meaning each state is represented by a vector of $n$ components, and if there are $m$ possible actions for each state, the neural network is a mapping from $\mathbb{R}^n$ to $\mathbb{R}^m$. The idea to use a deep neural network to approximate the Q-function in a RL setting has been first introduced by Mnih et al. in [14], where the authors proposed using a deep convolution neural network to learn to play Atari games.

We will not discuss here in detail the different neural networks and deep learning in general, and the interested reader may refer to reference books dealing with neural networks and the basics of deep learning, such as the book of Courville, Goodfellow and Bengio [1]. An ANN is made of (artificial) neurons that are linked through several layers. There exists several kind of neural networks, such as the feedforward neural network (FNN), the recurrent neural network, convolutional neural network as mentioned earlier, and many others (see for instance [3] for a classification of ANN). In the following, we will just refer to FNN. The main ingredients of an FNN in our case are an input layer that accepts the states of the decision process, or more specifically, a *representation* of the states, an output layer that returns the estimated action-state values for all actions for a given state at the input and several hidden layers between both as illustrated on Fig. 9.3 where two hidden layers has been considered. The input layer is made of $n_0$ neurons, and takes the samples $s_0 \in \mathbb{R}^n$ as an entry. The layer $\ell = 1, \ldots, L$ has $n_\ell$ neurons. The $L+1$ layer is the output layer and has $n_{L+1}$ neurons. It outputs the approximation of the Q-function, for the entry $s_0$; that is, $q_\theta(s_0, a_i)$ of each action $a_i$, $i = 1, \ldots, n_{L+1}$. $\Theta$ represents the parameters of the neural network that will be explained hereafter.

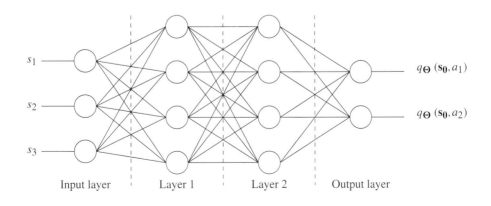

Figure 9.3  Feedforward neural network.

For all $\ell = 1, \ldots, L + 1$, the vector output at the layer $\ell$, $s_\ell \in \mathbb{R}^{n_\ell}$, is

$$s_\ell = f_\ell \left( \boldsymbol{\theta}_\ell s_{\ell-1} + \boldsymbol{b}_\ell \right) \tag{9.24}$$

where $\boldsymbol{\theta}_\ell \in \mathbb{R}^{n_{\ell-1} \times n_\ell}$ is the matrix of weights between the neurons of layer $\ell - 1$ and the layer $\ell$, $\theta_{i,j}^\ell$ is the weight between the $i$th neuron of layer $\ell - 1$ and the $j$th neuron of layer $\ell$, $s_{\ell-1}$ is the signal output of the layer $\ell - 1$. Moreover, $f_\ell$ is the *activation function* at the layer $\ell$ whose objective is to perform a nonlinear operation on the linear combination of the signal output of the previous layer. The function applies at each neuron of layer $\ell$; the function applies to each entry of the vector in the argument of $f_\ell$ in Eq. (9.24). The activation function aims at enabling the signal output. Several kinds of activation functions exist, such as sigmoidal, hyperbolic tangent, rectified linear unit (ReLU), and exponential linear unit (ELU); see [2] for a more detailed description. Finally, $\boldsymbol{b}_\ell$ is the bias term of the neurons of the layer $\ell$ that can change the threshold of the input signal value at which the neurons enable the output signal. The parameters of the FNN is made of the succession of matrix weights and bias terms between each layers and is denoted: $\boldsymbol{\Theta} = \{\boldsymbol{\theta}_\ell, \boldsymbol{b}_\ell\}_{\ell=1}^{L}$.

In order to make a neural network learn, a loss function should be defined between a target (a desired output) and the actual output obtained at iteration $t$. It can be defined as in Eq. (9.21) by using the target in Q-learning, Eq. (9.19);

$$L_t(\boldsymbol{\Theta}_t) = \mathbb{E}_{S,A,R,S'} \left[ \left( R(t+1) + \gamma \max_{a'} q_{\boldsymbol{\Theta}_t^-} \left( S(t+1), a' \right) - q_{\boldsymbol{\Theta}_t} \left( S(t), A(t) \right) \right)^2 \right], \tag{9.25}$$

where $\boldsymbol{\Theta}_t$ is the set of parameters at iteration $t$ and $\boldsymbol{\Theta}_t^-$ is the network parameters used to compute the target at iteration $t$. Deriving the loss function w.r.t. the network parameters we obtained a generalized stochastic gradient descent for nonlinear Q-function approximation as

$$\boldsymbol{\Theta}_{t+1} = \boldsymbol{\Theta}_t + \alpha_t \left( R(t+1) + \gamma \max_{a' \in \mathcal{A}} q_{\boldsymbol{\Theta}_t^-}(S(t+1), a') - q_{\boldsymbol{\Theta}_t}(S(t), A(t)) \right)$$
$$\times \nabla_{\boldsymbol{\Theta}_t} q_{\boldsymbol{\Theta}_t}(S(t), A(t)), \tag{9.26}$$

which decreases the expected loss under certain conditions. The gradient in Eq. (9.26) should be understood as the gradient w.r.t. the weights keeping the bias constant and also the gradient w.r.t. the bias keeping the weights constant.

However, this method may diverge if naively implemented. Indeed, if the samples $(s, a, r, s')$ are obtained by sampling the Markov chain at each successive transition, the data given to the neural network will be correlated and non-i.i.d. and will be not ideal for training the neural network. A solution is to store the experience tuples $(s(t), a(t), r(t+1), s(t+1))$ into a memory pooled over many episodes (an episode ends when a terminal state is reached) and to uniformly choose at each time instant a tuple from the memory in order to update Eq. (9.26). This is called *experience replay*, and it breaks correlation among the samples [14]. Moreover, if the target network $\boldsymbol{\Theta}^-$, used to retrieve the Q-values, is updated after each iteration with the new network computed at time $t$, then the policy may oscillate and data values may switch from an extreme to another and the parameters could diverge. The idea is to freeze the target network $\Theta^-$ during a certain number of time steps $T$ and to update the target network with $\Theta$ in Eq. (9.26) after $T$ steps. This allows us to reduce the oscillations or the risk of divergence.

There is another unwanted effect of the Q-learning that is the overestimation of the action-state values. This bias leads the algorithm to perform poorly in some stochastic environments and comes from the fact that the max operator is used to estimate the maximum expected value. Indeed, the max operator in Eq. (9.19) aims at estimating the value in Eq. (9.7) for the next state $S(t+1)$, which is an expectation. This method, often called the *single estimator*, has a positive bias that can be shown to follow from Jensen's inequality. Van Hasselt proposed in [15] to use a *double estimator* technique to unbiase the estimation of the maximum expected value that occurs in the Bellman's equation of the action-state value function. The idea is to create two sets of unbiased estimators w.r.t. the expectation, $q^A$ and $q^B$, that will be applied on two sets of samples, $\mathcal{A}$ and $\mathcal{B}$, such that the estimators are unbiased w.r.t. the mean on these samples. $\mathcal{A}$ and $\mathcal{B}$ contain the samples associated to the random variables $q^A(S(t), \cdot)$ and $q^B(S(t), \cdot)$, respectively. The maximum expected value of the first estimator $q^A$ is estimated with the max operator on the set of the experiences $\mathcal{A}$, $\max_a q^A(s', a) = q^A(s', a^*)$, as in the regular Q-learning algorithm. Then, we use the action $a^*$ on the estimator of the Q-function on the set $\mathcal{B}$ as an estimation of the maximum expected value of $\max_a \mathbb{E}\left[q^A(S', a)\right]$. A similar update is performed with $b^*$ on $q^B$ and using $q^A$. The iterative system of equations in the double Q-learning are such as [15, Algorithm 1]

$$q_t^A(S(t), A(t)) \leftarrow q_t^A(S(t), A(t)) + \alpha_t\left[R(t+1) + \gamma q_t^B(S(t+1), a^*) - q_t^A(S(t), A(t))\right],$$

$$q_t^B(S(t), A(t)) \leftarrow q_t^B(S(t), A(t)) + \alpha_t\left[R(t+1) + \gamma q_t^A(S(t+1), b^*) - q_t^B(S(t), A(t))\right].$$

The principle of the *double Q-learning* can be applied to any approximation techniques of the action-state function, in particular when a deep neural network is employed [16].

## 9.2.6    Multiarmed Bandits

### Fundamentals

A MAB model holds its name from a slot machine with several levers[3] that the player/user activates in the hope of hitting the prize. Each machine has a certain probability of delivering the money to the player. The goal of the agent, to keep the terminology used in the RL framework, is to play the machine that gives the maximum expected gain in the long run. Mathematically, a MAB model is a collection of $K$ random variables $R_i$, $i = 1, \ldots, K$, where $i$ denotes the "arm" of the bandit, each distributed as $P_{R_i}$, unknown to the agent. The player sequentially chooses an arm, the action of the agent $\{A(t)\}_{t \geq 0}$, and collects rewards over time $\{R(t+1)\}_{t \geq 0}$. When the

---

[3] Or equivalently several one-armed gambling machines.

agent pulls arm $a$ as its action at time $t$, $A(t) = a$, it gets a reward randomly drawn from the distribution $P_{R_a}$; that is, $R(t) \sim P_{R_a}$. The goal of the agent is to maximize the expected rewards obtained up the time horizon $T$; that is, $\mathbb{E}_{P_R}\left[\sum_{t=1}^{T} R(t)\right]$. By denoting $\mu_i$ the expectation of arm $i$, $\mu_i = \mathbb{E}[R_i]$, the agent should play as much as possible the arm with the maximum mean reward, $\mu^* = \arg\max_i \mu_i$, and the suboptimal arms the least amount possible. However, the arm with the maximum mean reward is of course unknown to the agent, in general, and the agent has to make decisions, defining its policy, based only on the past observations. To do so, the agent has to explore sufficiently in order to accumulate information on the rewards given by the arms, and also has to exploit the best arms (i.e., those that have given the highest cumulated rewards so far). This is the famous exploration-exploitation trade-off we mentioned earlier that each RL algorithm has to deal with.

One could consider the MAB as a special case of MDP described in Section 9.2.2 where only a single state is considered. The conditional probability distribution defining the dynamic of this particular MDP in Eq. (9.4) reduces to

$$P_{R(t+1),S(t+1)|S(t),A(t)}(r,s' \mid s,a) = P_{R(t)|A(t)}(r \mid a) \triangleq P_{R_a}. \tag{9.27}$$

In this definition, the state transition vanishes because there is only one state in the MDP and the reward is immediately obtained after pulling the lever $a$.

There are two schools of thought in MAB, or two approaches: the Bayesian, proposed by Thompson [17], and the frequentist approach [18, 19]. When the distribution of arms depends on a parameter $\theta$, the MAB framework is said to be *parametric*; that is, the distribution of the MAB is $P_{R_\theta} = \left(P_{R_{\theta_1}}, \ldots, P_{R_{\theta_K}}\right)$ and $\theta = (\theta_1, \ldots, \theta_K)$ is the vector parameter of the MAB. In the Bayesian setting, $\theta$ is a random variable drawn from a prior distribution $P_\theta$. In an i.i.d. scenario where the rewards are drawn from a Bernouilli distribution[4] and conditional to $\theta_a$, $R_a(t) \sim P_{R_a}$ with mean $\mu_a$ and the elements of the series $\{R_a(t)\}_{t,a}$ are independent. In the frequentist approach, $R_a(t)$ has the same properties as in the Bayesian setting, but $\theta$ is no longer random. Instead it is an unknown deterministic parameter. We will not discuss the difference between Bayesian and frequentist approaches further, and the interested reader may consult the excellent treatise in [20] for more details about the both approaches. In the following, we will focus on the frequentist approach for which the notion of *regret* is defined [21].

### Regret

The regret under the time horizon $T$ can be understood rather intuitively: it the difference between the average expected reward one would obtain if one always plays the optimal arm and the average expected reward obtained following a policy $\pi$, which is the sequential series of actions $A(0), A(1), \ldots A(T-1)$, different from the optimal one. By denoting the index of the optimal arm as $o = \arg\max_i \mu_i$ and its expected reward $\mu^* = \max_i \mu_i \triangleq \mathbb{E}[R_o(t)]$, the regret may be written as

---

[4] We will see later the Markovian setting.

$$\mathcal{D}_{\mu}(T) = T\mu^* - \mathbb{E}_{P_R}\left[\sum_{t=1}^{T} R(t)\right].  \tag{9.28}$$

Note that the regret depends on the parameter $\mu = (\mu_1, \dots, \mu_K)$. The regret can also be seen in a different, but totally equivalent way. Let us consider several experiments of duration $T$, each arm $i = 1, \dots, K$ will be played $N_i(T)$, which is a random variable for all $i$, with the expected value $\mathbb{E}[N_i(T)]$. For instance, let us assume that $K = 3$ and $T = 9$. During the first experiment, arm 1 has been played four times, arm 2 has been played three times, and arm 3 has been played two times.[5] Assuming a stationary setting, where the distribution of the regret does not vary in time, the regret in Eq. (9.28) becomes $T\mu^* - 4\mathbb{E}[R_1] - 3\mathbb{E}[R_2] - 2\mathbb{E}[R_3]$ or $T\mu^* - 4\mu_1 - 3\mu_2 - 2\mu_3$. The second experiment gives $(5, 2, 2)$ for arms 1, 2, and 3 respectively and so on. If $n$ experiments are run, the empirical average of the regret over the experiments, $\overline{D_{\mu}(T)}$, in our example gives

$$\overline{D_{\mu}(T)} = T\mu^* - \left(\overline{N_1(T)}\mu_1 + \overline{N_2(T)}\mu_2 + \overline{N_3(T)}\mu_3\right),  \tag{9.29}$$

where $\overline{N_i(T)} = \frac{1}{n}\sum_{j=1}^{n} N_i(j; T)$ is the empirical average of the number of times arm $i$ has been pulled at time $T$. When $n \to \infty$, $\overline{N_i(T)} \to \mathbb{E}[N_i(T)]$, the regret in Eq. (9.28) can be expressed as a function of the average number of times each arm has been played:

$$\mathcal{D}_{\mu}(T) = \sum_{i=1}^{K} \left(\mu^* - \mu_i\right)\mathbb{E}[N_i(T)],  \tag{9.30}$$

where the expectation depends on $\mu$. From Eq. (9.30), it seems clear that a policy should play as much as possible the best arm and the suboptimal arms the least amount possible. Lai and Robbins have shown [19, Th. 2] that any policy cannot have a better regret than a logarithmic one asymptotically, when $T \to \infty$. It is equivalent to say that the average number of plays of suboptimal arms is logarithmically bounded when $T \to \infty$. That is, for all $i$ such as $\mu_i < \mu^*$ and if $\mu \in [0, 1]^K$ we have

$$\lim_{T\to\infty} \inf \frac{\mathbb{E}[N_i(T)]}{\log T} \geq \frac{1}{D\left(P_{R_i} \| P_{R_o}\right)}  \tag{9.31}$$

where $D\left(P_{R_i} \| P_{R_o}\right)$ is the Kullback-Leibler divergence between the reward distributions of arm $i$ and the optimal arm, respectively. This result gives the bound of fundamental performance of any sequential policy in MAB settings, but it does not give any insight on how to explicitly design a policy that achieves this bound. Several algorithms have been proposed that are known to be order optimal, where the regret behaves logarithmically with time. We will not detail all of them in this chapter, but we will focus on a particular class that is applicable to physical layer communications: the upper confidence bound (UCB) algorithm.

---

[5] The sequence of the arm selection depends on the policy the agent follows.

*UCB algorithm*

The UCB algorithms are based on the computation of an index for each arm and selecting the arm with the highest index. The index is composed of two terms. The first is an estimate of the average reward of arm $i$ at time $t$, $\hat{\mu}_i(t) = \frac{1}{N_i(t)} \sum_{n=1}^{N_i(t)} R_i(n)$, and the second term is a measure of the uncertainty of this estimation. The UCB algorithm consists of choosing the maximum of this uncertainty to compute the index. Auer et al. proposed several algorithms in [21] for bounded expected rewards in $[0, 1]$. The most famous algorithm is UCB1 that allows one to choose arm $a(t + 1)$ at the next time such that

$$a(t + 1) = \arg\max_i \left[ \hat{\mu}_i(t) + \sqrt{\frac{2 \log t}{N_i(t)}} \right], \tag{9.32}$$

where the second term in the square root can be obtained using Chernoff-Hoeffding inequality, which represents the upper bound of the confidence interval in the estimation of $\hat{\mu}_i$. The more arm $i$ is played, as $N_i(t)$ increases, the smaller the confidence interval, which means the index value relies on the average value of the cumulated reward obtained so far. When arm $i$ is played less the second term is larger, which encourages the agent to play another arm. The second term allows for exploration while the first term encourages the exploitation of the arm that has given the largest rewards so far. Moreover, thanks to the log term that is unbounded with time, all arms will be played asymptotically. Note that other UCB algorithms can perform well in practice such as the Kullback-Leibler UCB (KL-UCB) for which a nonasymptotic regret bound can be proved [22]. Other algorithms such as the recency-based exploration algorithm, have been proposed in order to deal with nonstationary environments in which the parameters of the distribution may change with time [23].

## 9.2.7    Markovian MAB

Historically, the first bandits studied were binary; that is, the rewards are drawn from a Bernouilli distribution. Actually the result of Lai and Robbins in [19] is quite general and holds for $\mu \in [0, 1]^K$ for any distribution of the arms $P_{R_i}$. However, the independence of the rewards is an important assumption that does not necessarily hold in many practical problems.

In particular, the case of Markovian rewards is of practical interest in wireless communication. Each arm in Markovian MAB is characterized by a nonperiodic Markov chain with finite state space $\mathcal{S}_i$. For all arm $i$, the agent receives a positive reward $R_i(s)$ that depends on the state observed for arm $i$. The change from a state to another follows a Markovian process, under the conditional probability $P_{S_i(t+1)|S_i(t)}(s_i' \mid s_i) \triangleq p(s_i' \mid s_i)$. The stationary distribution of the Markov chain of arm $i$ is denoted as $\boldsymbol{P}_i = \left\{ P_{S_i}(s) \triangleq p_i(s), S_i \in \mathcal{S}_i \right\}$. Each arm $i$ has an expected reward that is

$$\mu_i = \sum_{s \in \mathcal{S}_i} r_i(s) p_i(s) \triangleq \mathbb{E}_{P_{S_i}}[R_i]. \tag{9.33}$$

The goal of the agent is still to find a policy $\pi$, in which the sequential observation of arms minimizes the regret over the time, which is defined in Eqs. (9.28) or (9.30). When the agent observes an arm $i$ at $t$, it samples a Markovian process that evolves with time. A particular attention has to be paid on the status of Markov chains that are not observed that leads to the distinction between the *rested* and *restless* cases.

*Rested MAB*

A Markovian MAB is qualified as rested when only the Markov chain of the arm that is played evolves and the others remaining frozen. This assumption is strong since the Markov chains of the arms that are not observed do not evolve with time, and it does not matter how much time elapsed between two consecutive visits to a given arm. The authors in [24] were the first to be interested in the fundamental performance in terms of regret of Markovian MAB with multiple plays,[6] and they proposed a policy that is *asymptotically* efficient that achieves the regret lower bound asymptotically. The authors in [25] showed that a slightly modified UCB1 achieves a logarithmic regret uniformly over time in this setting.

*Restless MAB*

A Markovian MAB is considered restless if the Markov chains of all arms evolve with time, irrespective of which arm is played. This assumption implies radical changes in the regret analysis because the state we will observe when pulling an arm $i$ at $t + 1$ directly depends on the time elapsed since the last visit to this arm. This is because the reward distribution we get by playing an arm depends on the time elapsed between two consecutive plays of this arm, and since arms are not played continuously, the sample path experienced by the agent does not correspond to a sample path followed when observing a discrete time homogeneous Markov chain. The solution came from [26] where the authors proposed a *regenerative cycle algorithm* to deal with the discontinuous observation of evolving Markov chains. In practice, the agent still keeps going to apply the UCB1 algorithm, introduced earlier, but computes the index only on the samples the agent has collected when observing a given arm $i$. This structure requires one to observe an arm during a certain amount of time before computing the UCB1 index. The interested reader may refer to [26] for further details and [25] for the extension to multiple plays. This setting is particularly interesting because it finds a natural application in wireless communications with the opportunistic spectrum access scenario when the state of the bands the user has to access evolves independently of the action the user takes. For example, the band may be sporadically occupied by another system or the propagation condition may evolve with time.

**Contextual MAB**

Contextual MAB generalizes the classical concept introduced earlier toward more general RL. Conventional MAB does not take advantage of any knowledge about the

---

[6] This is the case when the player may pull more than one arm or if multiple players are considered (without collisions).

environment. The basic idea of contextual MAB is to condition the decision making on the state of the environment. This allows for making the decisions based both on the particular scenario we are in and the previous observations we have acquired. A contextual MAB algorithm observes a context in the form of useful side information, followed by a decision by choosing one action from the set of alternative ones. It then observes an outcome of that decision which defines the obtained reward. In order to benefit from the context information, there needs to exist dependency between the expected reward of an action and its context. The goal of learning is to maximize a cumulative reward function over the time span of interest. The side information in physical layer communications may be a feature vector containing for example location and device information, experienced interference, received signal strength fingerprint, channel state information (CSI) of a particular user, or potential priority information. Such side information would allow for the selection of a base station, service set, or antennas such that higher rewards are achieved. Extensions of well-known UCB algorithms have been developed for contextual MAB problems. Since the context plays an important role in the dynamic of contextual MAB, it may be crucial to detect the change of a context in order to adapt the policy to this change. Methods based on statistical multiple change-point detection and sequential multiple hypothesis testing may be used for that purpose [27, 28].

### Adversarial MAB

Adversarial bandit problems are defined using sequential game formulation. The adversarial model means that the decisions may lead to the worst possible payoff instead of an optimistic view of always making choices that lead to optimal payoff. The problems are modeled assuming a deterministic and uninformed adversary and that the payoffs or costs for all arms and all time steps of playing the arms are chosen in advance. The adversary is often assumed to be uninformed so that it makes its choices independent of the previous outcomes of the strategy. At each iteration, a MAB agent chooses an arm it plays while an adversary chooses the payoff structure for each arm. In other words, the reward of each arm is no longer chosen to be stochastic but they are *deterministically assigned* to an *unknown* sequence, $\mathbf{r}(1), \mathbf{r}(2), \ldots$ where $\mathbf{r}(t) = (r_1(t), \ldots, r_K(t))^T$ and $r_i(t) \in \mathcal{R} \subset \mathbb{R}$ is the reward of the $i$th arm at time $t$. By denoting the policy $\pi$ that maps a time slot to an arm index to play at the next time slot,[7] $\pi(1), \pi(2), \ldots$ is the sequence of plays of the agent. Considering a time-horizon $h$, the goal of the agent is to minimize the regret

$$\mathcal{D}_h = \max_{j \in \{1, \ldots, K\}} \sum_{t=1}^{h} r_j(t) - \mathbb{E}_\pi \left[ \sum_{t=1}^{h} R_{\pi(t)}(t) \right], \tag{9.34}$$

where the adversary chooses $r_1(1), r_2(1) \ldots, r_k(t)$ and the player strategy chooses actions of policy $\pi(t)$. Since the policy is commonly assumed to be random, the

---

[7] The policy is a mapping from the set of indices and rewards obtained so far to the next index, $\pi : (\{1, \ldots, K\} \times \mathcal{R})^{t-1} \to \{1, \ldots, K\}$.

regret is also random (so the notation $R_{\pi(t)}(t)$), and the expectation is taken over the distribution of the policy. Adversarial bandits are an important generalization of the bandit problem since no assumptions on the underlying distributions are made, hence the name adversarial. The player has access to the trace of rewards for the actions that the algorithm chose in previous rounds, but it does not know the rewards of actions that were not selected. Widely used methods such as UCB are not suitable for this problem formulation. The exponential-weight algorithm called *Exp3* for exploration and exploitation is widely used for solving adversarial bandits problems [29, 30].

## 9.3 RL at PHY Layer

In this section, the concepts of Q-learning, deep learning, and MAB are illustrated through some examples deriving from communication problems addressed in literature. The first example in Section 9.3.1 deals with power management under queuing delay constraint in a point-to-point wireless fading communication channel and derives from [31, 32]. This problem has been widely studied in literature under various formalization including with MDP (see the discussion in Section 9.5) and a rather simple toy example can be extracted from it. In Section 9.3.2, we present two examples to illustrate the reinforcement learning in large dimension (i.e., optimal caching in a single cell network) and the extension of the single user power-delay management problem dealt with in Section 9.3.1 to the multiuser case with large state and action spaces. Finally Section 9.3.3 illustrates the use of MAB with the opportunistic spectrum access (OSA) problem and green networking issue. This section ends with experimental results and a proof of concept to validate the MAB principle applied to the OSA issue. These examples do not aim to provide the full solution of the problem raised, which can be found in the related literature, but rather a simple problem statement with explicit action and state spaces and the reward function that can be chosen to solve the problem with a reinforcement learning strategy.

### 9.3.1 Example with Q-learning

Let us consider a point-to-point communication system over a block fading channel. The channel gain is assumed to be constant on a slot of duration $\Delta t$ and changes from one slot to another according to the distribution $P_{H(t+1)|H(t)}(h' \mid h)$ with $H \in \mathcal{H}$ where $\mathcal{H}$ is assumed to be a finite countable set. At each time slot, the transmitter can send a packet or remain silent, and it can also choose its transmit power as well as its modulation and coding scheme, for example, a 16-QAM with a convolutional code of rate 1/2. At each time slot, a certain number of bits is generated by the application layer and stored in a buffer waiting for their transmission. The transmitter aims at sending the highest number of bits as possible with minimal power consumption while limiting the waiting time in the buffer.

At each time slot $t$, $N(t)$ new bits are generated and stored in the buffer before being transmitted. $\{N(t)\}_{t \in \mathbb{N}}$ are i.i.d. random variables with the distribution $P_N(n)$.

The buffer state $B(t) \in \mathcal{B} = \{0, 1, \ldots, B_{\max}\}$ represents the number of bits stored in the queue at time $t$ and $B_{\max}$ is the maximal buffer size. At each time slot, transmitter chooses $\beta(t) \in \{1, \ldots, B(t)\}$ bits[8] to be transmitted and encodes them into a codeword of length $n_c$ channel uses, assumed to be fixed. The rough spectral efficiency of the transmission is hence $\rho(t) = \beta(t)/n_c$. Moreover, transmitter chooses its power level $P_{\text{tx}}(t) \in \{0, p_1, \ldots, p_{\max}\}$ and hence the total power consumed at $t$ is

$$P_{\text{tot}}(t) = p_{\text{on}} + \alpha^{-1} P_{\text{tx}}(t), \qquad (9.35)$$

where $p_{\text{on}}$ is the static power consumed by the electronic circuits and $\alpha \in ]0, 1]$ the efficiency of the power amplifier. One can define the state space of the power consumption as $\mathcal{P} = \{p_{\text{on}}, \ldots, p_{\text{on}} + \alpha^{-1} p_{\max}\}$.

The codeword error probability, a.k.a. *pairwise error probability*, $\epsilon$ is defined as

$$\epsilon = \mathbb{P}\left[\hat{m}(t) \neq m(t)\right], \qquad (9.36)$$

where $m(t)$ and $\hat{m}(t)$ are the message sent and estimated at the receiver at time $t$, respectively. This probability has a complex expression in general (not always available in closed form) that depends on the channel and modulation coding scheme, the transmit power, and the channel gain: $\epsilon \triangleq f(\beta, n_c, p_{\text{tx}}, h)$. Bounds and approximations in finite block length exist, when $n_c$ is finite, but remain complex to evaluate [33, 34].

The buffer state evolution, or the number of bits stored in the queue, can be described by a Markov chain with the dynamics

$$B(t+1) = \left[B(t) - \beta(t) \cdot \mathbb{1}\left\{\hat{m}(t) = m(t)\right\}\right]^+ + N(t). \qquad (9.37)$$

This equation states that the number of bits in the buffer in the next time slot, $B(t+1)$, is the number of bits that is stored at the current time slot, $B(t)$, plus the new bits arriving in the buffer, $N(t)$, minus the number of bits that has been sent through the channel if the transmission is successful; that is, the indicator function is equal to 1 if so and 0 otherwise. Otherwise, the packet remains in the queue and another attempt will occur in the next time slot.

*State space*

The state space can be defined as the space containing the channel state, the buffer state, and the power consumption state: $\mathcal{S} = \mathcal{H} \times \mathcal{B} \times \mathcal{P}$.

*Action space*

At each time slot, transmitter chooses a power, including the choice not to transmit, $P_{\text{tx}}(t) = 0$, and a number of bits $\beta(t)$ that is mapped into a codeword[9] with a fixed block length $n_c$ but with a variable rate $\rho(t) = \beta(t)/n_c$. The action space is then described as $\mathcal{A} = \{0, p_1, \ldots, p_{\max}\} \times \{1, \ldots, B(t)\}$.

---

[8] Only a maximum of $B(t)$ bits can be encoded and sent at time $t$.

[9] This transformation is the usual way to consider the encoding phase in information theory. In practice, a transmitter selects a channel coding rate, and the resulting bit-train is grouped into symbols in a chosen constellation.

*Reward/cost functions*

In this type of problem, one may be interested in transmitting bits with the minimal power while limiting the awaiting time in the buffer. In that case, the global reward can be expressed w.r.t. two cost functions: the power and the waiting time cost functions [31, 32], $c : \mathcal{A} \times \mathcal{S} \to \mathbb{R}_+$ and $w : \mathcal{A} \times \mathcal{S} \to \mathbb{R}_+$[10]. The power consumption depends on the transmit power at time $t$ that depends on the target error rate $\epsilon$, the channel state $h$, and the code rate $\rho$, or $\beta$ since $n_c$ is fixed. The power cost can be defined as

$$c : \begin{cases} \mathcal{A} \times \mathcal{S} & \longrightarrow \mathbb{R}_+ \\ (a, s) & \longmapsto p_{\text{tot}}(\epsilon, h, \beta) \end{cases}. \tag{9.38}$$

The buffer waiting time cost is defined as

$$w : \begin{cases} \mathcal{A} \times \mathcal{S} & \longrightarrow \mathbb{R}_+ \\ (a, s) & \longmapsto \eta \mathbb{1}\{b(t+1) > B_{\max}\} + (b(t) - \beta(t)\mathbb{1}\{\hat{m}(t) = m(t)\}) \end{cases}, \tag{9.39}$$

The first term represents the cost to be in overflow with $\eta$ a constant, for the sake of simplicity. It means that the cost to pay when the buffer is in overflow is independent of the amount of the overflow.[11] The second term is the holding cost, or the cost for keeping $b - \beta$ bits in the buffer if the transmission is successful.

*Policy*

The transmission scheduling policy consists in mapping the system state to an action at each time slot $t$; that is, according to the buffer state, the channel state observed at the receiver[12] and the target error rate desired, the policy tells us how many information bits stored in the queue we should encode at the next transmission slot and at which power.

Hence a desirable policy should solve an optimization problem. From the cost functions defined previously, the expected discounted power and waiting time costs, given an initial state $s_0 \stackrel{\triangle}{=} S(0)$, are defined as

$$C_\pi(s_0) = \mathbb{E}_\pi \left[ \sum_{t=0}^{\infty} \gamma^t c\left(A(t), S(t)\right) \mid S(0) = s_0 \right] \tag{9.40}$$

and

$$W_\pi(s_0) = \mathbb{E}_\pi \left[ \sum_{t=0}^{\infty} \gamma^t w\left(A(t), S(t)\right) \mid S(0) = s_0 \right]. \tag{9.41}$$

The expectation is taken over the distribution of the policy and the dynamic of the underlying MDP. The problem of finding the minimal power consumption while limiting the waiting time cost can be formally described as [31, 32][13]

---

[10] Beside the action space, the waiting time cost only depends on the buffer state $\mathcal{B}$, so $w$ is incentive to the other states in $\mathcal{S}$.

[11] One can make this cost dependent on the amount of overflow; see [32].

[12] The channel state information is assumed to be fed back to the transmitter.

[13] Note that one could have searched for minimizing the waiting time cost under a total power budget as studied in [35], which leads to an equivalent strategy.

$$\min_{\pi \in \Phi} C_\pi(s_0) \text{ s.t. } W_\pi(s_0) \leq \delta \ \forall s_0 \in \mathcal{S}. \tag{9.42}$$

The problem relies on a constrained optimisation problem with unknown dynamics. One can combine the power and waiting time cost functions $c$ and $w$ in Eqs. (9.38) and (9.39), respectively, in a dual Lagrangian expression such that $\ell(a, s; \lambda) = c(a, s) + \lambda w(a, s), \lambda \in \mathbb{R}_+$ as proposed in [31, 32]. One can hence write an expected discounted Lagrangian cost on the same model than in Eq. (9.40), for instance, but substituting $c$ by $\ell$.

One can apply the Q-learning algorithm detailed in Section 9.2.4 by replacing the reward $R$ in Eq. (9.19) by the average discounted Lagrangian cost to obtain the optimal policy that minimizes the average power consumed under a buffer delay constraint.

REMARK 9.1    *The naive implementation of the Q-learning algorithm may be inefficient in terms of the algorithm's convergence time as reported in [32]. Indeed, Q-learning does not assume any knowledge about the dynamic of the underlying MDP. Hence, the exploration part, which is fundamental in Q-learning, slows down the convergence time due to the large number of combination of states and actions. However, in wireless communication some dynamics may not be completely unknown. The authors in [32] proposed to use the concept of the post-decision states, presented in wireless communication literature in [36]. The concept consists of reducing the amount of state to explore to make good decisions in the long run by basing the actions to take on states that would be observed considering only the known dynamics.*

## 9.3.2    Example with Deep-RL

When the state and action spaces become large, the tabulated methods for SARSA or Q-learning are no longer practical. In that case, methods relying on the approximation of the Q-function are meaningful like the linear approximation and those based on deep neural networks.

### Cache-Enabled Communications

The mobile Internet allows anyone to access heterogeneous data from anywhere. However, not all content is requested by the same users; that is, the data do not have the same popularity and some videos, for instance, may be more requested than other files by a user. In order to reduce the data traffic on the backhaul link, and hence the network operating costs, the most requested files can be kept into the storage unit at the base station; this is what is called caching. Hence, the most "popular" files are stored at the base station and can be delivered quickly to the user when requested, reducing the cost to download the file from a distant server.

The problem of learning the optimal caching strategy to satisfy user demand or data offloading in various environments has been addressed in lot of works, such as those reported in [4, Table V]. In this section, we focus on a simple example to illustrate how the caching problem may be addressed with deep Q-learning. The example is an adaptation of [37] where this problem is studied in a more complex setting. We briefly summarize how to properly choose the action and state spaces and rewards in order to find the optimal caching strategy.

Let us consider a network with a single cell serving many users. Each user may ask for a file $f$ in the set $\mathcal{F} = \{1, \ldots, F\}$, and we assume that only $M \ll F$ files can be stored at the base station. The files are requested randomly according to a certain distribution characterizing their *popularity* at time $t$. The popularity of the files is modeled as a random vector $\mathbf{P}(t) = [\mathrm{P}_1(t), \ldots, \mathrm{P}_F(t)]^T$ where the distribution of each popularity can be modeled with Zipf's law.[14] This law gives the average number of occurrences of each file and can be estimated online.

The goal of the network is to decide the files to cache and those to remove from the cache storing unit at each time slot. Because of the large number of possible requested files the number of possible choices is huge: $2^M$ with $M \gg 1$. A classical tabulated Q-learning approach, like the example presented in Section 9.3.1, is not suitable.

*Action space*
Let $\mathcal{A}$ be the set of *caching action vectors* that contains the binary action vector $\mathbf{A}(t)$ at time $t$ such that $\mathcal{A} = \{\mathbf{A} \mid \mathbf{A} \in \{0, 1\}^F, \mathbf{A}^T \mathbf{1} = M\}$. $A_f(t) \in \{0, 1\}$, $f \in \{1, \ldots, F\}$ is a random variable that is equal to 1 if the file $f$ is cached in the BS and 0 otherwise at time $t$.

*State space*
The state space is made of the popularity profile and the caching action vector, the latter being also an indicator of the cache status at time $t$. The popularity profile vector $\mathbf{P}$ is assumed to evolve according to a Markov chain with $|\mathcal{P}|$ states taken in the set $\mathcal{P} = \{\mathbf{P}^1, \ldots, \mathbf{P}^{|\mathcal{P}|}\}$. The state space is hence $\mathcal{S} = \mathcal{P} \times \mathcal{A}$.

*Reward/cost function*
Similar to the previous example, the reward takes the form of a cost function as

$$c : \begin{cases} \mathcal{A} \times \mathcal{S} & \longrightarrow \mathbb{R}_+ \\ (a, p) & \longmapsto \lambda_1 a(t)^T \left(1 - a(t-1)\right) + \lambda_2 \left(1 - a(t)\right)^T p(t) \end{cases}. \tag{9.43}$$

The cost function is made of two parts: (a) a term related to the cost of not having a file cached in the previous time slot, term $1 - a(t-1)$, which needs to be cached in the current time slot and (b) an immediate cost for non caching the file with high popularity profile at time $t$. The constants $\lambda_1$ and $\lambda_2$ allow us to balance the importance of these two costs.

*Policy*
The goal of RL is to learn the optimal policy $\pi^* : \mathcal{S} \to \mathcal{A}$ that minimizes the long-term weighted average cost function

---

[14] Zipf's law has been first used to characterize the frequency of occurrence of a word according to its rank.

$$\pi^* = \arg\min_{\pi \in \Phi} \mathbb{E}_\pi \left[ \sum_{t=0}^{\infty} \gamma^t c\left( (A(t), P(t)) \right) \mid S(0) = s(0) \right], \qquad (9.44)$$

where the expectation is carried out through the distribution of the policy (if stochastic policy is used) and the distribution of the random variables $A(t)$ and $P(t)$. Given a state $s(t)$ at time $t$, the policy looks for the set of files to be stored at time $t + 1$, or $a(t + 1)$ according to the popularity profile observed so far.

By denoting the state transition probabilities as $P_{S(t)\mid S(t-1), A(t-1)} \overset{\triangle}{=} p(s' \mid s, a)$, the Q-function can be obtained using the Bellman's equation as

$$q_\pi(s, a) = \overline{c(s, a)} + \gamma \sum_{s' \in \mathcal{S}} p\left( s' \mid s, a \right) q_\pi(s', a) \qquad (9.45)$$

where $\overline{c(s, a)} = \mathbb{E}_{P_{S'\mid SA}} \left[ c\left( A(t), P(t) \right) \right]$. Finding the optimal policy in Eq. (9.44) that is the solution of Eq. (9.45), which is obtained after policy evaluation and improvement steps, requires one to know the dynamics of the underlying Markov chain. The authors in [37] proposed a Q-learning algorithm with linear function approximation as introduced in Section 9.2.5 in order to cope with the high dimensionality of the problem: $q(s, a) \approx \psi(s)^T (1 - a)$ where $\psi(s)$ is a state dependent feature vector that can be expressed as $\theta^P + \theta^R a$ in which $\theta^P$ represents the average cost of non caching files when the popularity is in state $p$, and $\theta^R$ is the average cache refreshing cost per file. By adapting the recursion in Eq. (9.23) to the problem, where the reward is replaced by the cost function in Eq. (9.43) and the maximization is replaced by a minimization operation, one can show that this technique is able to converge to the optimal caching strategy for large number of files and popularity profiles.

## Multiuser Scheduling and Power Allocation

Let us extend the problem presented in Section 9.3.1 by adding multiple users in the system. An orthogonal frequency division multiple access downlink network, with a single cell and $K$ users is considered. The whole time-frequency resource is divided in $N_{rb}$ resource blocks (RBs) and one RB is made of $N_s$ subcarriers. The base station handles $K$ queues, one for each user, and has to serve the users by allocating the suitable transmission powers and the number of RBs in order to transmit the maximum number of bits with the minimal total power consumption and under buffer waiting time constraints.

The channel gain is considered to be constant on one RB, over $N_s$ subcarriers and during $\Delta T_{RB}$, and varies from one RB to another according to the distribution $P_{H(t+1)\mid H(t)}(h' \mid h) = \prod_{k=1}^{K} \prod_{r=1}^{N_{rb}} P_{H_{kr}(t+1)\mid H_{kr}(t)}(h'_{kr} \mid h_{kr})$ with $H_{kr}(t) \in \mathcal{H}$ the random variable representing the channel gain of user $k \in \mathcal{K} = \{1, \ldots, K\}$ on RB number $r \in \mathcal{N}_{RB} = \{1, \ldots, N_{RB}\}$ at time $t$. This relation means that the channel gains are independent from one RB to another in frequency and from a user to another. Similarly to Section 9.3.1, the application layer of each user generates $N_k(t)$ bits/packets at time slot $t$ according to the distribution $P_N(n)$. The generated bits are stored in a buffer for each user characterized by its size $B_k(t) \in \mathcal{B}_k$, where $\mathcal{B}_k$ is defined for all users $k \in \mathcal{K}$, as in Section 9.3.1. We assume that only a packet of $L$ information bits can be sent per user and per time slot. BS can choose the modulation and coding scheme (MCS) for each user, $\text{mcs}_k \in \mathcal{MC} = \{\text{mcs}_1, \ldots, \text{mcs}_C\}$, i.e. a couple $(\chi_k, \rho_k)$, where $\chi_k$ and $\rho_k$ are the modulation order in a QAM constellation and the rate of channel encoder for user $k$, respectively. The MCSs are ordered from the lowest to the highest spectral efficiency. A set of MCS used for the LTE system can be found in [38, Table I].

The power and RBs allocation can be done at once by the BS by choosing the transmission power vector to allocate to user $k$ over all the RBs at time $t$, $\boldsymbol{P}_k(t)$, in the power state space $\mathcal{P}$ such that $\mathcal{P} = \{\boldsymbol{P} \mid \boldsymbol{P} \in \{0, p_1, \ldots, p_{\max}\}^{N_{\mathrm{RB}}}, \frac{1}{N_{\mathrm{RB}}} \boldsymbol{P}^T \mathbf{1}_{N_{\mathrm{RB}}} \leq \overline{p_{\mathrm{tot}}}\}$, where $\overline{p_{\mathrm{tot}}}$ is the maximum average power budget that can be allocated to a user over all the subcarriers. The power of user $k$ on RB $r$ at time $t$, $p_{kr}(t)$, is null if the RB is not used by the user. In papers where this kind of problem is handled with classical convex optimization tools, RB allocation is dealt with an auxiliary variable that is equal to one when user $k$ uses RB $r$ and 0 otherwise [38, 39].

The error probability of user $k$ is defined as in Eq. (9.36), $\epsilon_k \triangleq \mathbb{P}\left[\hat{\omega}_k(t) \neq \omega_k(t)\right]$, where $\omega_k(t)$ is the message sent by user $k$ at time $t$. It depends on the chosen MCS, the transmission power and the channel state experienced over each RB allocated to the user $k$. The queue dynamic of user $k$ is then

$$B_k(t+1) = \left[B_k(t) - L \cdot \mathbb{1}\left\{\hat{\omega}_k(t)\right\}\right]^+ = \omega_k(t)\right]^+ + N_k(t). \tag{9.46}$$

*State space*
The state space is made of the buffer state of each user, the channel gain state of each user on each RB allocated to it, and the power consumed by each user:
$\mathcal{S} = \mathcal{H}^{K N_{\mathrm{RB}}} \times \mathcal{B}^K \times \mathcal{P}^K$.

*Action space*
BS chooses the power and the MCS couple $\text{mcs}_k = (\chi_k, \rho_k)$ to allocate to all users. The action space is hence $\mathcal{A} = \mathcal{P}^K \times \mathcal{MC}^K$.

*Reward/cost functions*
The objective of the network operator may be to minimize the power consumed to serve all the $K$ users in the cell while guaranteeing a limited buffer waiting time. We assume that the power consumption is only made of the transmit power (i.e., static power consumption is neglected). The power consumed by user $k$ depends on the target error rate required, the observed channel state, and the used MCS. The total power consumption of the cell can be written as

$$c : \begin{cases} \mathcal{A} \times \mathcal{S} & \longrightarrow \mathbb{R}_+ \\ (\boldsymbol{a}, \boldsymbol{s}) & \longmapsto \sum_{k=1}^K \boldsymbol{p}_k(\epsilon_k, \boldsymbol{h}_k, \text{mcs}_k)^T \mathbf{1}_{N_{\mathrm{RB}}} \end{cases}. \tag{9.47}$$

The buffer waiting time cost is defined similarly than in Eq. (9.39), so the total cost of the waiting time over the cell is

$$w : \begin{cases} \mathcal{A} \times \mathcal{S} & \longrightarrow \mathbb{R}_+ \\ (\boldsymbol{a}, \boldsymbol{s}) & \longmapsto \sum_{k=1}^{K} \eta \mathbb{1} \{b_k(t+1) > B_{\max}\} + (b_k(t) - L \mathbb{1} \{\hat{\omega}_k(t) = \omega_k(t)\}) \end{cases}$$
(9.48)

The average Lagrangian discounted cost, which is identified to our Q-function, can be obtained similarly than in Section 9.3.1.

$$q_\pi (\boldsymbol{s}_0, \boldsymbol{a}_0; \lambda) = \mathbb{E}_\pi \left[ \sum_{t=0}^{\infty} \gamma^t \ell (\boldsymbol{A}(t), \boldsymbol{S}(t); \lambda) \mid \boldsymbol{S}(0) = \boldsymbol{s}_0, \boldsymbol{A}(t) = \boldsymbol{a}_0 \right],$$
(9.49)

where $\boldsymbol{s}_0, \boldsymbol{a}_0$ are the initial state and action vectors and $\ell (\boldsymbol{a}, \boldsymbol{s}; \lambda)$ is defined as in Section 9.3.1 but with the functions in Eqs. (9.47) and (9.48) and $\lambda$ is the Lagrange multiplier.

*Policy*
The problem is to find the policy $\pi^*$ that minimizes Eq. (9.49) for all $(\boldsymbol{s}_0, \boldsymbol{a}_0)$ for a given $\lambda$. This optimization problem deals with a huge number of variables, and the classical tabulated Q-learning requires to much time to converge and too much capacity storage. In this situation, DRL can be a suitable solution to approach the function in Eq. (9.49) by implementing, for instance, an FNN as illustrated in Fig. 9.3. For a given set of parameters of the deep network at time $t$, $\boldsymbol{\Theta}_t$, the loss function is defined as in Eq. (9.25) where $R(t) = \ell (\boldsymbol{A}(t), \boldsymbol{S}(t); \lambda)$ and the optimal Q-function is approached at time $t$ by $q_{\boldsymbol{\Theta}_t} (\boldsymbol{S}(t), \boldsymbol{A}(t); \lambda)$. The optimal set of weights of the neural network can be updated using Eq. (9.26) with proper variables.

## 9.3.3    Examples with MAB

### Multichannel Access Problem

Let us consider a set of $K$ independent channels, $\mathcal{K} = \{1, \ldots, K\}$, that can be used opportunistically by $U$ users, with $K \geq U$ to communicate with a BS. The channels may be also used by other users that belong to a primary network. The users can sense one or more channels to estimate if they are used or not. To perform this task, the users can rely on multiple signal processing techniques ranging from the simple energy detector to sophisticated signal classifiers [40]. If the channel is detected to be free, the user transmits in that band; otherwise the transmitter remains silent. The action space is hence $\mathcal{A} = \{\text{transmit}, \text{silent}\}$. When the channel is free however, it may be rated with a low or a high quality depending on the level of the received SINR for instance or the actual data rate a user experienced on it. The state space may be limited to $\mathcal{S} = \{\text{busy}, \text{free}\}$, but the quality of the band can be included in the reward function as it has been proposed in [23, 41]. In the case where a single user is considered, $U = 1$:

$$R_k(t) = (1 - S_k(t)) f(S_k(t)),$$
(9.50)

where $S_k(t) \in \{0, 1\}$ is the state of band $k$ at time $t$ where 1 means that the band is detected as occupied and 0 that is free. Moreover, $f(S_k(t))$ is the observed data rate on band $k$ when it is in the state $S_k(t)$, and it can be considered that $f(S_k(t)) \in [0, 1], \forall S_k(t)$.[15]

The goal for an agent is to select the band $k$ that maximizes the data rate on the long run. If the expected rewards of each band were known, the optimal strategy would be to sense and to transmit (when possible) always on the band presenting the maximal expected reward. In the absence of this knowledge, the agent has to *learn* which are the best bands and concentrate on them. An index-based policy can be proposed to solve this problem, like the UCB algorithm, which is order-optimal as explained in Section 9.2.6. For instance, in [23], the authors proposed computing the index $I_k(t)$ for each band $k$ and choosing the one with the highest value at the next round. The index is computed as

$$I_k(t) = \overline{r_k(t)} + g\left(\frac{t}{n_k(t)}\right), \tag{9.51}$$

where $n_k(t)$ is the number of times band $k$ has been sensed up to time $t$, $\overline{r_k(t)} = \frac{1}{n_k(t)} \sum_{t'=1}^{n_k(t)} r_k\left(p_k(t')\right)$ where $p_k(t')$ is a sensing time instant corresponding to the $t'$-th visit of the band $k$. Finally, $g$ is a concave function with $g(1) = 0$, e.g. $g(x) = \sqrt{\log(x)}$. One may notice that this function is different from the classical bias of UCB introduced in Section 9.2.6, which is in the form $\sqrt{\frac{\log(t)}{n_k(t)}}$. The authors proved that the proposed index policy has an aggressive exploration characteristic compared to the UCB where the bias increases slowly. An aggressive exploration statistic means that the optimal data rate will be reached faster than in a nonaggressive exploration characteristic (classical UCB), but it will also be lower than classic UCB.

According to the statistical model of the rewards, the problem described may fall into the classes of rested MAB or restless MAB. If the rewards, the experienced data rate of the users, are i.i.d. on each band, the problem is a rested-MAB. If however, the state $S_k(t)$ is described by a two-state Markov chain (i.e., the Gilbert-Elliot model) the problem can be classified as a restless MAB. In [23] the authors showed that the policy is order-optimal for i.i.d. and Markovian rewards as well.

In the previous example, the reward signal was a function of the state of the channel sensed and the data rate experienced in this channel by the user. In some applications, one may be interested in acquiring knowledge both on the channel availability and on channel quality. The channel quality has to be taken in a broad sense. It may be the noise level, including the average power density of the interference, the average data rate experienced in this band, etc. The dynamic spectrum access with different channel quality can be represented as in Fig. 9.4. In this example, the state space is extended to $\mathcal{S} = \{\text{busy}, \text{low quality}, \text{high quality}\}$. The action space is still the same, $\mathcal{A} = \{\text{transmit}, \text{silent}\}$, but the transmission now occurs on a state (i.e., channel is free) that can be explicitly rated with a high or low quality. One may solve this problem by considering a full RL approach, using value function as described in Section 9.2. However, a simple index-based policy can perform well in this context. The authors

---

[15] By convention, $f(1) = 0$. Moreover, the experienced data rate can be normalized w.r.t. the channel capacity achievable in that channel.

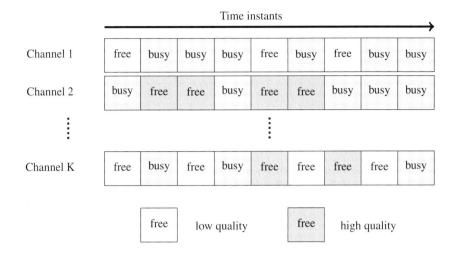

**Figure 9.4** Dynamic spectrum access with different channel qualities.

in [42] proposed that each user $u \in \{1, \ldots, U\}$ can compute the following index for each band and choose the maximum one as the next band to sense

$$I_k^u(t) = \frac{1}{n_k^u(t)} \sum_{t'=1}^{n_k^u(t)} s_k^u(p_k^u(t')) - q_k^u(t) + \sqrt{\frac{\alpha \log(t)}{n_k^u(t)}} \qquad (9.52)$$

where the first term is the empirical average of the reward obtained by user $u$ when sensing band $k$; that is, 0 if the channel is occupied and 1 if the channel is free. The second term is a function of the empirical average of the quality of band $k$ for user $u$ and is expressed as

$$q_k^u(t) = \frac{\beta m_k^u(t) \log(t)}{n_k^u(t)}, \qquad (9.53)$$

where $m_k^u(t) = g_*^u(t) - g_1^{u,k}(t)$ and $g_1^{u,k}(t)$ denotes the empirical average of the quality of band $k$, when available, sensed by user $u$ at time $t$ and is expressed as

$$g_1^{u,k}(t) = \frac{1}{n_k^u(t)} \sum_{t'=1}^{n_k^u(t)} r_1^{u,k}(p_k^u(t')), \qquad (9.54)$$

where $r_1^{u,k}(p_u^k(t'))$ is the reward obtained by user $u$ rating the quality of band $k$, between 0 and 1, at time $p_k^u(t')$. Finally, $g_*^u(t) = \max_{k \in \mathcal{K}} g_1^{u,k}(t)$. The last term in Eq. (9.52) is the classical exploration term of the UCB algorithm where the parameter $\alpha$ forces the exploration of other bands to find channels that are the most often available.

The parameter $\beta$ in Eq. (9.53) gives weight to the channel quality. At each iteration, the agent computes the empirical mean, up to the current time instant, on the quality observed if the band is free. At the same time, the best channel among those already tried is updated and the score is computed by weighting the difference between the estimated best channel and the current channel. If $\alpha$ and $\beta$ increase, the agent explores more than it exploits. When $\alpha$ and $\beta$ decrease, the empirical mean of the states dominates the index calculation and the exploitation of the best band computed in the previous iteration is favored.

Due to the restless nature of the problem, the index computation cannot be done when one just starts observing a Markov chain after being selected. Indeed, arms evolve independently irrespective of which band is selected or not (i.e., *restless* MAB), and the distribution of rewards that user $u$ gets from a band $k$ is a function of the time elapsed since the last time user $u$ sensed this band. The sequence of observations of a band that is not continuously sensed does not correspond to a Markov chain. To overcome this issue, when user $u$ observes a given band, the algorithm waits for encounter a predefined state, named a *regenerative state* $\xi_k$ [26]. Once $\xi_k$ is encountered, rewards is recorded until $\xi_k$ is observed a second time and the policy index $I_k^u$ is computed and another band selected according to the result. This structure is necessary to deal with the restless nature of the problem in order to recreate the condition of continuous observation of Markov chain. It is worth mentioning, however, that exploitation in this context occurs whenever a free band is detected.

Moreover, the multiplayer setting makes the problem rather involved since collisions between agents need to be handled. The random rank idea from [43] has been adapted to this problem. Each user maintains an ordered set of channel indexes (arms indexes), $\mathcal{K}_u = \sigma_u(\mathcal{K})$, where $\sigma_u$ is a permutation of $\{1, \ldots, K\}$ for user $u$, with $\sigma_u(1) > \cdots > \sigma_u(K)$ from the best to the worst rated. The rank $r$ for user $u$ corresponds to the $r$th entry in the set $\mathcal{K}_u$. If users $u$ and $u'$ choose the same channel to sense the next time slot, they collide. In that case, they draw a random number from their respective sets $\mathcal{K}_u$ and $\mathcal{K}_{u'}$ as their new rank and go for these new channels in the next time slot.

## Green Networking

The radio access networks are not used at their full capacity all the time. Experimental results in voice call information recorded by operators over one week exhibit periods with high traffic load and others with moderate to low traffic [44]. Hence, it may be beneficial for a network operator to dynamically switch off a BS that does not handle high traffic in its cell at a given time in order to maximize the energy efficiency of the network. However, the set of BS to be switched OFF should be chosen with care while maintaining a sufficient quality of service for users.

Let us consider an heterogeneous wireless cellular network made of macro and small cells where the set of BS $\mathcal{Y} = \{1, 2, \ldots, Y\}$ lies in a 2D area in $\mathbb{R}^2$, each serving a cell $k$. The decision to switch ON or OFF a BS is taken by a central controller and depends on the traffic load of each cell and its power consumption. The traffic load $\rho_k(t)$ of a cell $k$ at time $t$ depends on the statistic of the arrival and departure processes and on the data rate $\Theta_k(x, t)$ that can be provided by cell $k$ to the user positioned at $x$ at time $t$.

The maximization of the energy efficiency of the network by selecting the set of transmitting BS is an NP-hard problem and can be expressed as follows [45]:

$$\mathcal{Y}_{on^*}(t) = \arg\max_{\mathcal{Y}_{on}(t)} \left[ \sum_{k \in \mathcal{Y}_{on}(t)} \frac{\sum_{x \in \mathcal{C}_k(t)} \Theta_k(x,t)}{P_k(t)} \right] \text{ s.t.}$$

$$(c_1) \qquad 0 \le \rho_k(t) \le \rho_{th}, \forall k \in \mathcal{Y}_{on}(t) \qquad\qquad (9.55)$$

$$(c_2) \qquad \Theta_k(x,t) \ge \Theta_{min}, \forall x \in \mathcal{C}_k(t), \forall k \in \mathcal{Y}_{on}(t)$$

$$(c_3) \qquad \mathcal{Y}_{on}(t) \ne \emptyset$$

where $\mathcal{Y}_{on}(t)$ is the set of active BS at time $t$, $P_k(t)$, $\mathcal{C}_k(t)$ are the power consumed and the coverage of cell $k$ at time $t$, respectively. Moreover, $\rho_{th}$ and $\Theta_{min}$ are the traffic load upper limit and the minimum required data rate per user, respectively. Constraint $(c_1)$ is stated for stability reason,[16] $(c_2)$ states that each user has to be served with a minimum data rate, and $(c_3)$ ensures that at least one BS is active at each time slot. Finding the optimal configuration by an exhaustive search would be prohibitive in large networks since the optimal BS active set belongs to a set of $2^Y - 1$ combinations.

The authors in [45] have shown that this problem can be solved with MAB formulation. The problem can be illustrated with Fig. 9.5 where at each iteration the central controller chooses an action $\boldsymbol{a}$ among $|\mathcal{A}| = 2^Y - 1$ possible actions: $\boldsymbol{a}(t) = [a_1(t), \ldots, a_Y(t)]^T$ with $a_k(t) = 1$ if BS $k$ is switched ON at $t$ and 0 otherwise, and where $\mathcal{A}$ is the action space. The state is represented by a random variable $S(t) \in \{0, 1\}$ where $s(t) = 1$ if all constraints in Eq. (9.55) are satisfied and 0 otherwise. In other words, the value of the state of the Markov chain relies on the fact that the selected action leads to a feasible solution of Eq. (9.55). Observing the state $s(t)$ and taking the

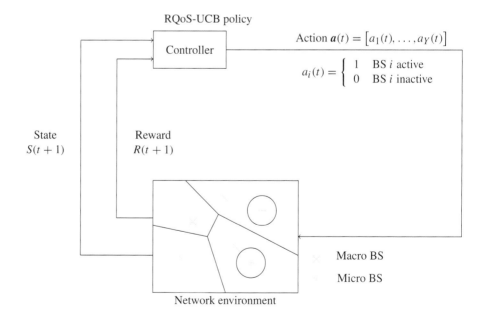

**Figure 9.5** RL framework for BS switching operation.

---

[16] A traffic load that is too high results in a diverging queue size in the network.

action $\mathbf{a}(t)$ at time $t$ lead to the network in the state $S(t+1)$ and give a reward $R(t+1)$ in the next time slot according to the conditional transition probability distribution introduced in Eq. (9.4). The reward is the energy efficiency computed as in the cost function in Eq. (9.55).

In [45] the authors proposed applying the same policy as in Eq. (9.52), where the index of the user can be dropped out, $k$ represents the number of the actions, the middle term $q_k(t)$ is expressed as in Eq. (9.53), and $r_1(t)$ in Eq. (9.54) is the energy efficiency of the network at time $t$ when the set of active BS is such that constraints in Eq. (9.55) are satisfied, $s(t) = 1$.

REMARK 9.2   *modeling the problem in Eq. (9.55) with MAB framework does not change that the problem dimensionality is exponential in the number of BS. However, MAB explores only once all network configurations to assign an index to each of them and then chooses the next configuration according to the highest index computed at the previous iteration with Eq. (9.52), instead of doing an exhaustive search at each time slot.*

REMARK 9.3   *This problem could also be addressed with DRL technique. Indeed, if the number of BS is too large, the convergence time of MAB algorithms is too large, which makes the problem unsolvable for large scale networks. DRL can be used instead and the set of BS to switch on may be obtained by a DNN for each state of the environment.*

## 9.3.4    Real-World Examples

We give details on a few concrete implementations of RL for communication, which rely on theoretical applications of RL using the UCB algorithms for single user cognitive radio MAB-modeled problems in [46] and for the opportunistic spectrum access (OSA) scenario in [47]. In these examples, UCB1 [21] algorithm is mostly used as a proof of the pertinence of bandit algorithms for free spectrum-access problems, but other bandit algorithms have also been considered and could be implemented as well.

In the spectrum access context, the goal is to be able to manage a large spectrum without adding complexity to the radio system by enlarging its bandwidth (and consequently without adding complexity to the transceivers). Figure 9.6 shows that learning is a mean to decrease the receivers architecture complexity while maintaining a legacy bandwidth to the OSA radio system with extended capabilities. RL enables to reconstruct the global bandwidth knowledge, thanks to successive small-scale investigations. RL offers a light solution, in terms of implementation complexity cost compared to wide (full)band OSA systems.

### Implementation at the Postprocessing Level
Due to the unique world wide covering characteristic of high-frequency (HF) transmissions, there is a high necessity to reduce collisions between users that act in a decentralized manner. RL algorithms have been applied on real measurements of the HF spectrum that has been recorded during a radio-ham contest [48], when HF radio

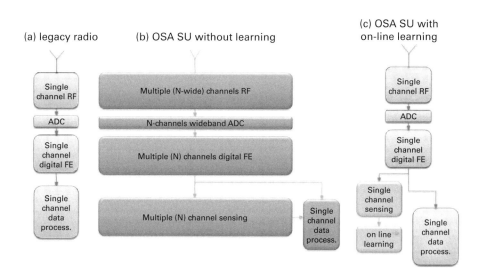

**Figure 9.6** Comparison of the receiver architectures for OSA with or without RL.

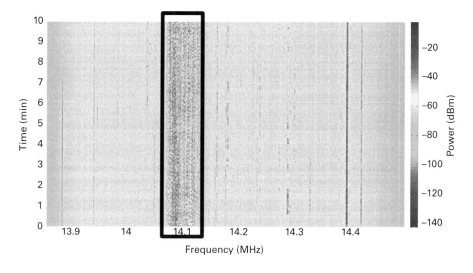

**Figure 9.7** HFSA_IDeTIC_F1_V01 database extraction for HF channel traffic during a radio-ham contest at 14.1 MHz (area highlighted in the rectangle) [48].

traffic is at its highest. Figure 9.7 shows the subset of the spectrum data the MAB algorithms considered in [48]. The goal of learning is to enable users to find unoccupied slots in time and frequency before transmitting and to maximize the probability that no collision occurs during the transmission duration. For instance, the cognitive device should avoid the frequencies delimited by the rectangle in Fig. 9.7, as it is a highly used bandwidth by primary users.

The solution in [48] proposes a new hybrid system that combines two types of machine learning techniques based on MAB and hidden Markov models (HMM).

**Table 9.1.** Summary for HF signals postprocessing.

| Characteristics | Comment |
|---|---|
| RL algorithm | HMM combined with UCB1 |
| Reward | channel availability |
| Implementation side | HF transceiver |
| method for RL feedback loop | Sensing |

This system can be seen as a meta-cognitive engine that automatically adapts its data transmission strategy according to the HF environment's behavior to efficiently use the spectrum holes. The proposed hybrid algorithm, which combines a UCB algorithm and HMM, increases the time the opportunistic user transmits when conditions are favorable, and is also able to reduce the required signalling transmissions between the transmitter and the receiver to inform which channels have been selected for data transmission.

Table 9.1 sums up the characteristics of the RL algorithm implementation for this use case. The reward is the channel availability detected by the cognitive users. More details about these measurements can be found in [48].

### Implementation in a Proof of Concept

The first real-time RL implementation on a real radio signal took place in 2013 for OSA scenario and was first published in 2014 [49] at the Karlsruhe Workshop and extended hereafter in [50]. This consisted of a proof of concept (PoC) in the laboratory conditions with one USRP[17] platform emulating the traffic generated by a set of primary users, users that own the frequency bands, and another USRP platform running the sensing and learning algorithm of one secondary user, a user that opportunistically exploits the licensed band. Both i.i.d and Markovian MAB traffic models have been tested. The UCB1 algorithm was used first in order to validate the RL approach, but then other bandit algorithms have been implemented later, such as Thompson Sampling, KL-UCB. The multiuser version has been implemented in [51]; moreover, several videos implementing the main UCB algorithms in a USRP-based platform demonstrating the real-time learning evolution under various traffic models (i.i.d. and Markovian) can be found on the Internet.[18] In order to help the experimental community to verify and develop new learning algorithms, an exhaustive Python code library and framework for simulations have been provided on GithHub[19, 20, 21] that encompasses a lot of MAB algorithms published until mid-2019.

---

[17] https://www.ettusresearch.com/
[18] https://www.youtube.com/channel/UC5UFCuH4jQ_s_4UQb4spt7Q/videos
[19] "SMPyBandits: an Open-Source Research Framework for Single and Multi-Players Multi-Arms Bandits (MAB) Algorithms in Python"
[20] The code is available at https://GitHub.com/SMPyBandits/SMPyBandits
[21] The documentation is available at https://SMPyBandits.GitHub.io/

**Table 9.2.** Summary for OSA proof-of-concept.

| Characteristics | Comment |
| --- | --- |
| RL algorithm | UCB1 (or any other bandit algorithm) |
| Reward | channel 'availability' |
| Implementation side | secondary user |
| method for RL feedback loop | Sensing |

**Table 9.3.** Summary for IoT proof-of-concept.

| Characteristics | Comment |
| --- | --- |
| RL algorithm | UCB1 and Thomson Sampling |
| Reward | channel 'availability' |
| Implementation side | embedded on device |
| method for RL feedback loop | Emulated ACK |

Table 9.2 summarizes the main characteristics of the RL algorithm implementation for this use case. More details about these measurements can be found in [49].

A PoC implementing MAB algorithms for Internet of Things (IoT) access has been done in [52]. It consists in one gateway, one or several learning IoT devices, embedding UCB1 and Thompson Sampling algorithms, and a traffic generator that emulates radio interferences from many other IoT devices. The IoT network access is modeled as a discrete sequential decision-making problem. No specific IoT standard is implemented in order to stay agnostic to any specific IoT implementation. The PoC shows that intelligent IoT devices can improve their network access by using low complexity and decentralized algorithms that can be added in a straightforward and cost-less manner in any IoT network (such as Sigfox, LoRaWAN, etc.), without any modification at the network side. Table 9.3 summarizes the characteristics of the RL algorithm implementation for this use case. More details about these measurements can be found in [52], but we present the main outcomes here.

Figure 9.8 shows the UCB parameters during an execution on four channels numbered as channels 2, 4, 6 and 8.[22, 23] We can see on the top left of Fig. 9.8 that the left channel (channel 2) has been only tried twice over 63 trials by the IoT device and it did not receive the ACK from the network at both times, as it can be seen at the top right panel that represents the number of successes on each channel. So the success rate for channel 2 is null, as seen at the bottom right panel. The more the channel index increases, from the left to the right, the better the success rate. That is why channel 8,

---

[22] The entire source code for this demo is available online, open-sourced under GPLv3 license, at https:// bitbucket.org/scee_ietr/malin-multi-armed-bandit-learning-for-iot-networks-with-grc/src/master/. It contains both the GNU Radio Companion flowcharts and blocks, with ready-to-use Makefiles to easily compile, install, and launch the demonstration.

[23] A six-minute video showing the demonstration is at https://www.youtube.com/watch?v= HospLNQhcMk&feature=youtu.be.

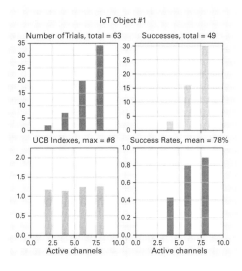

**Figure 9.8**  UCB parameters monitored during PoC execution [52].

with a 90 percent success rate has been preferably used by the algorithm with 35 trials over 63 (top left panel) and 30 successes (top right panel) over the 49 total successes obtained until the caption of this figure during the experiment.

### Real-World Experimentations

The ultimate step of experimentation is the real world, validating the MAB approaches for intelligent spectrum access. It has been done on a LoRaWAN network deployed in the licence free 868 MHz ISM band for Europe. The experiment has been conducted in two steps. The first step consists of emulating an artificial IoT traffic in controlled radio conditions, inside an anechoic chamber, in order to validate the devices and the gateway implementation itself [53]. The second step is to make it run in real-world conditions without being able to neither control the spectrum use, nor the propagation conditions in the area [54]. In step 1, seven channels have been considered, whereas in step 2, only three channels were used due to the configuration of the gateway, which was controlled by the LoRaWAN network provider. For the two measurement campaigns, Pycom equipped with Lopy4[24] shields have been used as devices and a standard gateway from Mutlitech.

**Step 1.** The characteristics of the seven channels are given in Table 9.4, which gives the index of the channel, the percentage of time occupancy (or jamming) the channels experience due to other IoT devices in the area, the center frequency of each channel (channel bandwidth is set to 125 kHz). USRP platforms have been used to generate the surrounding IoT traffic.

Figure 9.9 shows that, due to the surrounding IoT traffic, the channel number 6, the curve with star markers, has been much more played by the cognitive device, thanks

---

[24] https://pycom.io/

**Table 9.4.** Channels characteristics for step 1 experiments.

| Channel | % of jamming | Frequency (in MHz) |
|---------|-------------|--------------------|
| 0 | 30% | 866.9 |
| 1 | 25% | 867.1 |
| 2 | 20% | 867.3 |
| 3 | 15% | 867.5 |
| 4 | 10% | 867.7 |
| 5 | 5% | 867.9 |
| 6 | 0% | 868.1 |

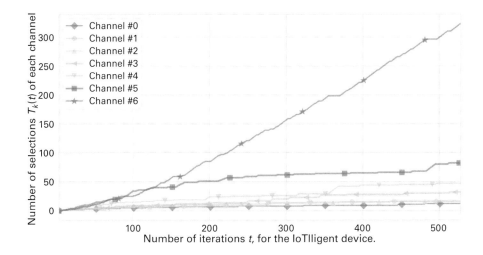

**Figure 9.9** Number of selections for each of the seven channels used in step 1 for IoTligent device [53].

to the learning algorithm it embeds. This device is therefore named IoTligent and is hence able to maximize the success rate of its transmission. A transmission is called a "success" when a message has been sent by the IoT device in uplink, received by the LoRaWAN gateway, and transmitted to the application server that sends an ACK back toward the IoT device, which is transmitted by the gateway in downlink at the same frequency used in uplink and finally received by the IoT device. A normal device following standard LoRaWAN features used by default, served as reference and the results in term of number of channel uses and success rate have been summarized in Table 9.5.

As we can see on Table 9.5, whereas the reference IoT device uniformly transmits on all channels (around 75 times during this experiment), we can see that the IoTligent device concentrates its transmissions on the most vacant channels, with a clear choice for channel 6. Over a total of 528 iterations, 323 transmissions have been done in this channel for IoTligent, which is more than four times compared to the reference IoT, with 75 transmissions. Moreover, IoTligent selects the channel 6 much more than

**Table 9.5.** Success rate and number of attempts for each channel of regular IoT device and IoTligent.

| Channel | Reference IoT % of success | nb of Tx | IoTligent % of success | nb of Tx |
|---|---|---|---|---|
| 0 | 21% | 76 | 8% | 12 |
| 1 | 20% | 76 | 25% | 16 |
| 2 | 24% | 75 | 25% | 16 |
| 3 | 49% | 76 | 50% | 32 |
| 4 | 62% | 74 | 62% | 47 |
| 5 | 76% | 76 | 74% | 82 |
| 6 | 96% | 75 | 94% | 323 |

**Table 9.6.** Summary for step 1 experiments

| Characteristics | Comment |
|---|---|
| RL algorithm | UCB1 (or any other bandit algorithm) |
| Reward | received ACK |
| Implementation side | embedded on device |
| method for RL feedback loop | Standard LoRaWAN ACK |

channel 0, almost 27 times more. Hence, the IoT device with learning capability is able to increase its global success rate drastically that reaches 80 percent (420 successful ACK received over 528) compared to 50 percent for the reference IoT device (266 successful ACK received only).

In this example, due to its ability to favor the use of less occupied channels, IoTligent demonstrates 2.5 times improvement in performance compared to standard IoT LoRaWAN device in terms of number of successes. Note that in this experimental setup, radio collisions are only considered as obstacles to the reception of ACK by the IoT devices.

It is worth mentioning that adding this learning capability does not impose changes to LoRaWAN protocols: no additional retransmissions to be sent, no additional power consumed, no data to be added in frames. The only condition is that the proposed solution should work with the acknowledged (ACK) mode for IoT. The underlying hypothesis, however, is that the channels occupancy by surrounding radio signals (IoT or not) is not equally balanced. In other words, some ISM subbands are less occupied or jammed than others, but it is not possible to predict it in time and space, so the need to learn on the field. Table 9.6 sums up the characteristics of the RL algorithm implementation for this use case. More details about these measurements can be found in [53].

**Step 2.** Real-world experiments have been done on a LoRa network deployed in the town of Rennes, France, with three channels 868.1 MHz, 868.3 MHz, and 868.5 MHz. IoTligent is completely agnostic to the number of channels and can be used in any

**Table 9.7.** Summary for step 2 experiments.

| Characteristics | Comment |
| --- | --- |
| RL algorithm | UCB1 (but could be any bandit algorithm) |
| Reward | channel 'availability' (collisions, jamming and propagation) |
| Implementation side | embedded on device |
| method for RL feedback loop | Standard LoRaWAN ACK |

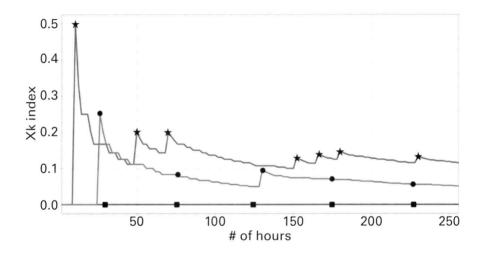

**Figure 9.10** Empirical mean evolution through time over 11 days [54].

country or ITU Region (i.e., 866 MHz and 915 MHz ISM bands as well). Since this experiment is run in real conditions, we have no means to determine exactly which of the four following possible phenomenon influences the IoTligent devices behavior: (a) collisions with other LoRaWAN IoT devices, (b) collisions with IoT devices running other IoT standards, (c) collisions with other radio jammers in the ISM band and (d) propagation issues.

We now look at the results obtained by IoTligent, for 129 transmissions done every two hours, over an 11 day period. Figure 9.10 shows the empirical mean experienced by the device on each of the three channels. This represents the average success rate achieved in each channel since the beginning of the experiment. The average success rates for the three channels – 0 (868.1 MHz), 1 (868.3 MHz), and 2 (868.5 MHz) – are represented by the curves with squares, stars, and bullets, respectively. Each peak corresponds to a LoRa successful bidirectional exchange between the device and the application server: from device transmission, to ACK reception by the device. Each peak in Fig. 9.10 reveals a successful transmission where ACK has been received by IoTligent device. We can see that channel 1, star markers, has been the most successful, before channel 2, while channel 0 always failed to send back an ACK to the device. During the experiment indeed, channel 0 has been tried 29 times with no success. IoTligent device uses channel 1 61 times with 7 successful bidirectional exchanges (7 peaks on the curve with stars) and channel 2 39 times with 2 successes (2 peaks on the curve with bullets). At the end of the experiment, one can observe 11.5 percent successful bidirectional connections for channel 1 and 5 percent for channel 2, whereas channel 0 never worked from the device point of view. For comparison, a regular IoT device, performing a random access, achieves a global average successful rate of 5.5 percent. Table 9.7 summarizes the characteristics of the RL algorithm implementation for this use case. More details about these measurements can be found in [54].

## 9.4 Conclusions and Future Trends

During recent years, we have witnessed a shift in the way wireless communication networks are designed, at least in the academic community, by introducing ML techniques to optimize the networks. With the complexity of wireless systems increasing, it becomes increasingly difficult to build an explainable mathematical model to predict the performance of large-scale systems. Exhaustive and complex simulations are not always an option because they are resource-intensive. In this sense, ML in general and RL in particular have the potential to overcome the network modelling deficit by learning the optimal network functioning with minimal assumptions on the underlying phenomena. This is of particular importance when nonlinear phenomena needs to be taken into account and for which it is difficult to get analytical insights.

In this chapter, we focused on RL for PHY layer communications. Even if this domain enjoys a long successful history of well-defined mathematical models, the heterogeneity of the use cases envisaged in 5G for instance, advocates for adding smartness in the PHY layer of upcoming wireless systems. All along this chapter, we have provided some PHY examples in which RL or deep RL can help to achieve good performance results. The key point to apply RL algorithms to a PHY layer problem is when one cannot solve the problem analytically or by writing an *explicit program* to do so. There are, without any doubt, many practical situations where one cannot *explicitly* solve the associated optimization problem related to the communication system design, such as nonlinear peak-to-average power ratio reduction, acoustic transmission, and transmission with a nonlinear power-amplifier.

The key benefit of RL is its adaptability to unknown system dynamics. However, the convergence time is crucial in wireless communications and real networks cannot afford to spend too much time learning a good strategy. For instance, the deep Q-learning in [37] converges around $10^4$ iterations, for two popularity profile states. This cannot be directly converted into time delay because it depends on the number of requests the users do, but if the network records one request per second, it takes almost 3 hours to converge toward the optimal strategy. The converging time is also around $10^3$–$10^4$ iterations in the problem of optimal buffer state and transmission power dealt with in [32]. In case of nonstationary environment, it can be prohibitive.

The feasibility of a PHY layer data-driven designed has been proved in [55, 56]. However, even if possible, learning an entire PHY layer without any *expert knowledge* is not necessarily desirable. For instance, the synchronization task requires an huge amount of computation and a dedicated NN in [56] to synchronize the transmission while it can be simply performed using well-known OFDM signal structures [57] may not be worthy. Hence it is apparent that DRL can be an efficient tool to design the PHY layer of future wireless systems, however, it would may gain in efficiency if cross-fertilization research between ML and model-based approach would be undertaken. In order to achieve this goal, explainable ANN with information-theoretic tools is a promising research direction, as attempted in [58]. The *transfer learning* concept, which consists in transferring the knowledge acquired in a previous task to a current one in order to speed up the converge, is also a promising research direction if a mixed transfer (model-based to data-driven) is considered.

## 9.5    Bibliographical Remarks

This chapter aims to be as self-contained as possible. It presents, in a tutorial way, the theoretical background of RL and some application examples, drawn from the PHY layer communication domain, where RL techniques can be applied. The theoretical part of Section 9.2 revisits the fundamentals of RL and is largely based on the great books of Sutton and Barto [5] and Csaba Szepesvari [6] that gathered in a comprehensive way the original results of Bellman [7], who introduced the notion of dynamic programming, and Watkins [9], who proposed the Q-learning algorithm.

The example in Section 9.3.1 follows from [31, 32]. The problem of joint power and delay management can be dated back to at least the early 2000s. In [59] the authors studied the power adaptation and transmission rate to optimize the power consumption and the buffer occupancy as well. They characterized the optimal power-delay trade-off and provided the former steps of a dynamic scheduling strategy achieving the Pareto front derived earlier. While the previous work investigated the power-delay Pareto front in block fading channels, the authors in [60] came back to AWGN channel but proposed more explicit schedulers to operate close to the fundamental bound. The authors in [35] revisited the same problem as in [59] and proved the existence of a stationary policy to achieve the power-delay Pareto front. The formulation of the joint delay-power management problem as a constrained MDP can be reported in [31, 32, 35, 36].

The example in Section 9.3.2 has been considered in [37] with a more complex setting, where the authors introduced two kind of popularity profiles: local and global. The local popularity profile allows to cache the requested files according to the local demand while the global one allows anticipate the local demand by monitoring the most wanted files over the network. We simplified the system model in order to make a toy example easily with this application. Due to the potentially large number of files to be cached, this problem can be addressed with deep neural network as well, see for instance [61, 62].

The examples of Section 9.3.3 follow from [23, 42] for the OSA problem with quality of the transmission and from [45] for the switch ON/OFF base station problem.

The opportunistic spectrum access problem has been considered as the typical use case for the application of MAB and RL algorithms in wireless communications since their inception in the field. The RL framework in general, and MAB in particular, are well-suited for describing the success and the failure of a user that tries to opportunistically access to a spectral resource that is sporadically occupied. The second reason is that the performance of the learning algorithm strongly depends on the ability of the device to detect good channels that motivates the reborn of the research on signal detection algorithms. A broad overview on signal detection and algorithms to opportunistically exploit the frequency resource has been provided in [63].

The examples we proposed in this chapter are not exhaustive since it was not the objective and at least two complete surveys have been provided recently on that topic [4, 64]. We invite the interested reader to consult these articles and the references therein to go further in the application of (deep)-RL techniques to some specific wireless communication problems. In this section, we give some articles that may be interesting to consult when addressing PHY layer communication challenges.

An interesting use case in which DRL can be applied is the IoT. The huge number of cheap devices to be connected with small signalling overheads make this application appealing for some learning approaches implemented at BS for instance. The authors in [65] revisited the problem of dynamic spectrum access for one user. Even if this set up has been widely studied in literature, the authors modeled the dynamic access issue as a POMDP with correlated states and they introduce a deep Q-learning algorithm to choose the best channels to access at each time slot. An extension to multiple devices has been provided in [66]. A relay with buffering capability is considered to forward the packets of the other nodes to the sink. At the beginning of each frame, the relay chooses packets from buffers to transmit on some channels with the suitable power and rate. They propose a deep Q-learning approach with a FNN to optimize the packet transmission rate. IoT networks with energy harvesting capability can also be addressed with deep Q-learning in order to accurately predict the battery state of each sensor [67].

These works consider centralized resource allocation; that is, the agent is run in the BS, relay, or another server but is located at one place. In large-scale heterogeneous networks, with multiple kind of BS, such as macro, small, or pico BS, this approach is no longer valid and distributed learning has to be implemented. In [68], the authors considered the problem of LTE access through WiFi small cells by allocating the communication channels and managing the interference. The small BS are in competition to access the resources and hence the problem is formulated as a non cooperative game which is solved using deep RL techniques. The same kind of problem has been tackled in [69] while in a different context and different utility function.

DRL has also been successfully applied to complex and changing radio environments with multiple conflicting metrics such as in satellite communications in [70]. Indeed, the orbital dynamics, the variable propagation environment, such as a cloudy or clear sky, the multiple optimization objectives to handle, such as a low bit error rate, throughput improvement, power and spectral efficiencies, make the analytical optimization of the global system untractable. The authors used a deep Q-learning with a deep NN to choose the actions to perform at each cognitive cycle, such as modulation and encoding rate, power transmission, and demonstrate that the proposed solution achieved good performance compared to the ideal case obtained with a brute force search.

There are still numerous papers that deal with RL and deep-RL for PHY layer communications, ranging from resource allocations to PHY layer security for instance. Since this is a hot topic rapidly evolving at the time we write this book, one can expect interesting and important contributions in this field in the upcoming years. Since the feasibility and the potential of RL techniques has been demonstrated for PHY layer communications, we encourage the research community to also address important issues such as the convergence time reduction of learning algorithms or the energy consumption reduction for training a deep NN instead of an expert-based design of PHY layer.

## References

[1] I. Goodfellow, Y. Bengio, and A. Courville, *Deep Learning*. The MIT Press, 2016.

[2] A. Zappone, M. Di Renzo, and M. Debbah, "Wireless networks design in the era of deep learning: Model-based, AI-based, or both?" *IEEE Trans. on Communications*, vol. 67, no. 10, pp. 7331–7376, 2019.

[3] M. Chen et al., "Artificial neural networks-based machine learning for wireless networks: A tutorial," *IEEE Communications Surveys Tutorials*, vol. 21, no. 4, pp. 3039–3071, 2019.

[4] N. C. Luong et al., "Applications of deep reinforcement learning in communications and networking: A survey," *IEEE Communications Surveys Tutorials*, vol. 21, no. 4, pp. 3133–3174, 2019.

[5] R. S. Sutton and A. G. Barto, *Reinforcement Learning: An Introduction*, 2nd ed. MIT Press, 2018.

[6] C. Szepsvari, *Algorithms for Reinforcement Learning*. Morgan and Claypool, 2010.

[7] R. E. Bellman, *Dynamic Programming*. Princeton University Press, 1957.

[8] R. D. Smallwood and E. J. Sondik, "The Optimal Control of Partially Observable Markov Processes over a Finite Horizon," *Operations Research*, vol. 21, no. 5, pp. 1071–1088, 1973.

[9] C. Watkins, "Learning from delayed rewards," Ph.D. dissertation, King's College, 1989.

[10] C. Szepsvari, "Learning and exploitation do not conflict under minimax optimality," in *Proc. 9th European Conf. on Machine Learning*, pp. 242–249, April 1997.

[11] E. Even-Dar and Y. Mansour, "Learning rates for q-learning," *Journal of Machine Learning Research*, vol. 5, Dec. 2004.

[12] G. A. Rummery and M. Niranjan, "On-line Q-learning using connectionist systems," tech. report CUED/FINFENG/ TR 166, Cambridge University Engineering Department, Sep. 1994.

[13] L. Bottou, *On-line Learning and Stochastic Approximations*. Cambridge University Press, p. 9–42, 1999.

[14] V. Mnih et al., "Human-level control through deep reinforcement learning," *Nature*, vol. 518, no. 7540, pp. 529–533, 2015.

[15] H. V. Hasselt, "Double q-learning," in *Advances in Neural Information Processing Systems 23*, J. D. Lafferty et al., eds., Curran Associates, Inc., pp. 2613–2621, 2010.

[16] H. V. Hasselt, A. Guez, and D. Silver, "Deep reinforcement learning with double q-learning," in *Proc. 13th Conf. on Artificial Intelligence (AAAI)*, pp. 2094–2100, 2016.

[17] W. R. Thompson, "On the likelihood that one unknown probability exceeds another in view of the evidence of two samples," *Biometrika*, vol. 25, no. 3/4, pp. 285–294, Dec. 1933.

[18] H. Robbins, "Some aspects of the sequential design of experiments," *Bulletin of the American Mathematics Society*, vol. 58, no. 5, pp. 527–535, 1952.

[19] T. L. Lai and H. Robbins, "Asymptotically efficient adaptive allocation rules," *Advances in Applied Mathematics*, vol. 6, no. 1, pp. 4–22, 1985.

[20] E. Kaufmann, "Analyse de stratégies bayésiennes et fréquentistes pour l'allocation séquentielle de ressources," Ph.D. dissertation, Telecom ParisTech, 2014.

[21] P. Auer, N. Cesa-Bianchi, and P. Fischer, "Finite-time analysis of the multiarmed bandit problem," *Machine Learning*, vol. 47, no. 2-3, pp. 235–256, May 2002.

[22] O. Cappé et al., "Kullback-Leibler Upper Confidence Bounds for Optimal Sequential Allocation," *Annals of Statistics*, vol. 41, no. 3, pp. 1516–1541, 2013.

[23] J. Oksanen and V. Koivunen, "An order optimal policy for exploiting idle spectrum in cognitive radio networks," *IEEE Trans. on Signal Processing*, vol. 63, no. 5, pp. 1214–1227, March 2015.

[24] V. Anantharam, P. Varaiya, and J. Walrand, "Asymptotically efficient allocation rules for the multiarmed bandit problem with multiple plays-part ii: Markovian rewards," *IEEE Trans. on Automatic Control*, vol. 32, no. 11, pp. 977–982, Nov. 1987.

[25] C. Tekin and M. Liu, "Online learning of rested and restless bandits," *IEEE Trans. on Information Theory*, vol. 58, no. 8, pp. 5588–5611, Aug. 2012.

[26] C. Tekin and M. Liu, "Online learning in opportunistic spectrum access: A restless bandit approach," in *Proc. IEEE INFOCOM*, pp. 2462–2470, 2011.

[27] H. V. Poor and O. Hadjiliadis, *Quickest Detection*. Cambridge University Press, 2008.

[28] E. Nitzan, T. Halme, and V. Koivunen, "Bayesian methods for multiple change-point detection with reduced communication," *IEEE Trans. on Signal Processing*, vol. 68, pp. 4871–4886, 2020.

[29] P. Auer et al., "The nonstochastic multiarmed bandit problem," *SIAM Journal on Computing*, vol. 32, no. 1, pp. 48–77, Jan. 2003.

[30] G. Stoltz, "Incomplete information and internal regret in prediction of individual sequences," Ph.D. dissertation, Université Paris XI, 2005.

[31] M. H. Ngo and V. Krishnamurthy, "Monotonicity of constrained optimal transmission policies in correlated fading channels with ARQ," *IEEE Trans. on Signal Processing*, vol. 58, no. 1, pp. 438–451, 2010.

[32] N. Mastronarde and M. van der Schaar, "Joint physical-layer and system-level power management for delay-sensitive wireless communications," *IEEE Trans. on Mobile Computing*, vol. 12, no. 4, pp. 694–709, 2013.

[33] Y. Polyanskiy, H. Poor, and S. Verdú, "Channel coding rate in the finite blocklength regime," *IEEE Trans. on Information Theory*, vol. 56, no. 5, pp. 2307–2359, Dec. 2010.

[34] D. Anade et al., "An upper bound on the error induced by saddlepoint approximations – applications to information theory." *Entropy*, vol. 22, no. 6, p. 690, June 2020.

[35] M. Goyal, A. Kumar, and V. Sharma, "Power constrained and delay optimal policies for scheduling transmission over a fading channel," in *IEEE 22nd Annual Joint Conf. of the IEEE Computer and Communications Societies (INFOCOM)*, vol. 1, pp. 311–320, 2003.

[36] N. Salodkar et al., "An on-line learning algorithm for energy efficient delay constrained scheduling over a fading channel," *IEEE Journal on Selected Areas in Communications*, vol. 26, no. 4, pp. 732–742, 2008.

[37] A. Sadeghi, F. Sheikholeslami, and G. B. Giannakis, "Optimal and scalable caching for 5g using reinforcement learning of space-time popularities," *IEEE Journal of Selected Topics in Signal Processing*, vol. 12, no. 1, pp. 180–190, 2018.

[38] M. Maaz, P. Mary, and M. Hélard, "Energy minimization in HARQ-I relay-assisted networks with delay-limited users," *IEEE Trans. on Vehicular Technology*, vol. 66, no. 8, pp. 6887–6898, 2017.

[39] D. S. Wing et al., "Cross-layer design for OFDMA wireless systems with heterogeneous delay requirements," *IEEE Trans. on Wireless Communications*, vol. 6, no. 8, pp. 2872–2880, 2007.

[40] E. Axell et al., "Spectrum sensing for cognitive radio : State-of-the-art and recent advances," *IEEE Signal Processing Magazine*, vol. 29, no. 3, pp. 101–116, May 2012.

[41] J. Oksanen, V. Koivunen, and H. V. Poor, "A sensing policy based on confidence bounds and a restless multi-armed bandit model," in *Conf. Record of the 46th Asilomar Conf. on Signals, Systems and Computers (ASILO-MAR)*, pp. 318–323, 2012.

[42] N. Modi, P. Mary, and C. Moy, "QoS driven channel selection algorithm for cognitive radio network: Multi-user multi-armed bandit approach," *IEEE Trans. on Cognitive Communications and Networking*, vol. 3, no. 1, pp. 49–66, March 2017.

[43] A. Anandkumar et al., "Distributed algorithms for learning and cognitive medium access with logarithmic regret," *IEEE Journal on Selected Areas in Communications*, vol. 29, no. 4, pp. 731–745, April 2011.

[44] E. Oh, K. Son, and B. Krishnamachari, "Dynamic base station switching-on/off strategies for green cellular networks," *IEEE Trans. on Wireless Communications*, vol. 12, no. 5, pp. 2126–2136, May 2013.

[45] N. Modi, P. Mary, and C. Moy, "Transfer restless multi-armed bandit policy for energy-efficient heterogeneous cellular network," *EURASIP Journal on Advances in Signal Processing*, vol. 2019, no. 46, 2019.

[46] W. Jouini et al., "Multi-Armed Bandit Based Policies for Cognitive Radio's Decision Making Issues," in *3rd conf. on Signal Circuits and Systems*, Nov. 2009.

[47] W. Jouini et al., "Upper confidence bound based decision making strategies and dynamic spectrum access," in *Int. Conf. on Communications (ICC)*, May 2010.

[48] L. Melian-Guttierrez et al., "Hybrid UCB-HMM: A machine learning strategy for cognitive radio in HF band," *IEEE Trans. on Cognitive Communications and Networking*, vol. 2016, no. 99, 2016.

[49] C. Moy, "Reinforcement learning real experiments for opportunistic spectrum access," in *Karlsruhe Workshop on Software Radios (WSR)*, Mar. 2014.

[50] S. J. Darak, C. Moy, and J. Palicot, "Proof-of-concept system for opportunistic spectrum access in multi-user decentralized networks," *EAI Endorsed Trans. on Cognitive Communications*, vol. 16, no. 7, Sep. 2016.

[51] S. Darak et al. "Spectrum utilization and reconfiguration cost comparison of various decision making policies for opportunistic spectrum access using real radio signals,"

in *Cognitive Radio Oriented Wireless Networks and Communications (CrownCom)*, May 2016.

[52] L. Besson, R. Bonnefoi, and C. Moy, "GNU radio implementation of malin: Multi-armed bandits learning for IoT networks," in *IEEE Wireless Communications and Networking Conf. (WCNC)*, April 2019.

[53] C. Moy et al., "Decentralized spectrum learning for radio collision mitigation in ultra-dense IoT networks: Lorawan case study and experiments," *Annals of Telecommunications*, Aug. 2020.

[54] C. Moy, "IoTligent: First world-wide implementation of decentralized spectrum learning for IoT wireless networks," in *URSI AP-RASC*, March 2019.

[55] T. O'Shea and J. Hoydis, "An introduction to deep learning for the physical layer," *IEEE Trans. on Cognitive Communications and Networking*, vol. 3, no. 4, pp. 563–575, 2017.

[56] S. Dörner et al., "Deep learning based communication over the air," *IEEE Journal of Selected Topics in Signal Processing*, vol. 12, no. 1, pp. 132–143, 2018.

[57] A. Felix et al., "OFDM-autoencoder for end-to-end learning of communications systems," in *IEEE 19th Int. Workshop on Signal Processing Advances in Wireless Communications (SPAWC)*, pp. 1–5, 2018.

[58] C. Louart, Z. Liao, and R. Couillet, "A random matrix approach to neural networks," *Annals of Applied Probability*, vol. 28, no. 2, pp. 1190–1248, April 2018.

[59] R. A. Berry and R. G. Gallager, "Communication over fading channels with delay constraints," *IEEE Trans. on Information Theory*, vol. 48, no. 5, pp. 1135–1149, 2002.

[60] D. Rajan, A. Sabharwal, and B. Aazhang, "Delay-bounded packet scheduling of bursty traffic over wireless channels," *IEEE Trans. on Information Theory*, vol. 50, no. 1, pp. 125–144, 2004.

[61] C. Zhong, M. C. Gursoy, and S. Velipasalar, "Deep reinforcement learning-based edge caching in wireless networks," *IEEE Trans. on Cognitive Communications and Networking*, vol. 6, no. 1, pp. 48–61, 2020.

[62] Y. He et al., "Deep-reinforcement-learning-based optimization for cache-enabled opportunistic interference alignment wireless networks," *IEEE Trans. on Vehicular Technology*, vol. 66, no. 11, pp. 10 433–10 445, 2017.

[63] J. Lunden, V. Koivunen, and H. V. Poor, "Spectrum exploration and exploitation for cognitive radio: Recent advances," *IEEE Signal Processing Magazine*, vol. 32, no. 3, pp. 123–140, May 2015.

[64] Q. Mao, F. Hu, and Q. Hao, "Deep learning for intelligent wireless networks: A comprehensive survey," *IEEE Communications Surveys Tutorials*, vol. 20, no. 4, pp. 2595–2621, 2018.

[65] S. Wang et al., "Deep reinforcement learning for dynamic multichannel access in wireless networks," *IEEE Trans. on Cognitive Communications and Networking*, vol. 4, no. 2, pp. 257–265, 2018.

[66] J. Zhu et al., "A new deep-q-learning-based transmission scheduling mechanism for the cognitive internet of things," *IEEE Internet of Things Journal*, vol. 5, no. 4, pp. 2375–2385, 2018.

[67] M. Chu et al., "Reinforcement learning-based multiaccess control and battery prediction with energy harvesting in iot systems," *IEEE Internet of Things Journal*, vol. 6, no. 2, pp. 2009–2020, 2019.

[68] U. Challita, L. Dong, and W. Saad, "Proactive resource management for LTE in unlicensed spectrum: A deep learning perspective," *IEEE Trans. on Wireless Communications*, vol. 17, no. 7, pp. 4674–4689, 2018.

[69] O. Naparstek and K. Cohen, "Deep multi-user reinforcement learning for distributed dynamic spectrum access," *IEEE Trans. on Wireless Communications*, vol. 18, no. 1, pp. 310–323, 2019.

[70] P. V. R. Ferreira et al., "Multiobjective reinforcement learning for cognitive satellite communications using deep neural network ensembles," *IEEE Journal on Selected Areas in Communications*, vol. 36, no. 5, pp. 1030–1041, 2018.

# 10   Data-Driven Wireless Networks: Scalability and Uncertainty

Feng Yin, Yue Xu, and Shuguang Cui

## 10.1   Introduction

This chapter presents a systematic discussion on futuristic scalable data-driven wireless networks with special considerations over learning uncertainty. Scalability has become a critical issue due to the ever-increasing data volume and model complexity since it is the major factor in determining the agility and response time of a data-driven wireless network. We first introduce the forward-looking architecture and computing framework of scalable data-driven systems from a global perspective. Then, we survey relevant learning models and their training algorithms to be performed at distributed local devices from a local perspective. We focus on Bayesian nonparametric learning and reinforcement learning as they have a higher potential to address the uncertainty issue than the traditional learning approaches based on deep neural networks. We also touch upon the model interpretability and adaptivity. Finally, we use three practical use cases to demonstrate the effectiveness of the proposed scalable data-driven wireless networks from various aspects and pinpoint several promising research directions.

### 10.1.1   Motivation

The next-generation wireless networks are migrating from traditional paradigms based on mathematical modeling to data-driven paradigms based on big data and machine learning techniques. On the one hand, the ever-expanding and context-rich wireless big data contain valuable information that can be exploited to customize futuristic wireless networks in almost all aspects [1], including architecture design, resource management, and task scheduling. On the other hand, machine learning, as one of the most powerful artificial intelligence tools, constitutes strong learning-from-data capabilities to discover valuable patterns from historical data for accurate future planning [2]. Therefore, data-driven wireless networks are anticipated to draw better understanding on both the networks and the mobile users in order to deliver personalized and adaptive services to embrace an intelligent future.

While the popularity of the term "data-driven" has been recently fueled by the growth of available data and computing power, scalability becomes increasingly important due to a manifold of driving demands, such as adaptivity, low latency, low complexity, and privacy preservation. Specifically, many existing data-driven solutions were developed in a centralized design by assuming that a single device has full access to the entire dataset and almost unlimited storage and computing power for decision making. However, increasingly many new breeds of intelligent devices and delay-sensitive applications require real-time reactions and high reliability, including brake control for self-driving vehicles, collision avoidance for drones, and motion perceptions for augmented/virtual reality (AR/VR). These newly emerging applications have sparked a major interest in developing scalable solutions to deliver lower latency and better robustness than the traditional centralized counterparts.

The ever-growing network size (e.g., due to network densification and exploding number of connected Internet-of-Things (IoT) devices), model size (e.g., large number of model parameters), and data volume together lead to optimization tasks with unprecedented complexity, which surpass the computing capability of a single device. These challenges can be addressed by scalable solution. In this chapter, "scalable solution" refers to any efficient learning algorithms that are ideally capable of processing and storing any amount of data generated in a large network. Here, we particularly focus on parallel learning algorithms that decompose a large global learning task into smaller pieces that can be handled locally with modest amount of data and affordable processing cost. In addition, privacy-sensitive data are preferably not logged into a centralized server merely for the purpose of model training. Such security and privacy concerns make the scalable solution a natural choice for keeping data safely stored at the local devices (e.g., user devices or third-party edge devices) while only exchanging model-parameter related updates (e.g., the gradients), for information sharing.

The idea of developing scalable solutions has been widely recognized in [1, 3–8], which mainly focus on developing dedicated scalable algorithms for specific wireless applications. However, developing an entirely scalable data-driven system requires us to address a myriad of fundamental challenges, including architecture design and computing framework adaptation from a *global* perspective, as well as on-device learning models and training algorithms from a *local* perspective. The overarching goal of this chapter is to provide a systematic discussion on the data-driven scalability.

Another unprecedented feature of the next-generation wireless network is full automation, which requires that all subsystems can not only conduct prediction based on AI and data-driven models but also make reasonable decisions. To this end, we need to carefully address the uncertainty issue in machine learning. In this chapter, we focus on the uncertainty in the data, in the selected learning model, and in the model prediction. Uncertainty is critical to decision making for wireless networks since the correctness of a decision may largely influence the service quality of the mobile users' experience in the end. For instance, when we want to control the on-off status of a base station based on our wireless traffic prediction, the uncertainty about our prediction is equally or even more important for making a decision. The system should be able to tell whether it is confident enough to make a decision or whether more data should be collected to obtain a better confidence. Thus, we consider Bayesian learning that is able to provide both an accurate prediction and a natural uncertainty region than a single-point estimate obtained from traditional deterministic learning approaches based on deep neural networks (DNNs). In addition, a valid uncertainty region can be used for actively learning the changing environment, which is critical to reinforcement learning (RL). Both Bayesian learning and RL are known to mimic the reasoning mechanisms of our brain, and they are promising to be deployed to build an artificial intelligence (AI) brain for next-generation wireless networks.

## 10.1.2    Scope and Organization

In this chapter, we focus on the following aspects:

- **Scalability:** Although previous works [1, 3, 9, 10] have mentioned the importance and urgency of developing a scalable wireless network, they lack in-depth discussions on the potential building components in both hardware and software. In contrast, this work provides a dedicated introduction and systematic justification on the wireless system scalability by addressing the scalable wireless architecture, theoretical scalable learning framework, and scalable learning algorithms in a row and explaining their interconnections.
- **Uncertainty:** This work draws public attention from the traditional learning approaches based on DNNs type models [4] to Bayesian nonparametric learning and reinforcement learning, which can better handle the growing system and environmental uncertainty in the context of data-driven management and control. In general, we discuss both the model uncertainty and the data uncertainty occurring in data-driven wireless networks.
- **Interpretability and Adaptivity:** We also touch upon these two aspects, which call for the use of Bayesian learning and RL.
- **Practical Use Cases:** This work provides three concrete use cases to quantitatively demonstrate how to use the proposed scalable learning paradigm for data-driven wireless network in practice. The effectiveness of the proposed scalable learning paradigm will be demonstrated from different aspects.

It is noteworthy that the scalability and prediction uncertainty of our learning paradigm introduced in this chapter are critical to decision making and vital to the wireless system reliability, according to 3GPP [11], in terms of system malfunction, delay, and transmission error rate.

The rest of this chapter is organized as follows. In Section 10.2, we first draw a futuristic architecture of the scalable data-driven wireless network, which cross-fertilizes the *cloud-based intelligence* and *on-device intelligence*. In Section 10.3, we present both parallel and fully distributed scalable learning frameworks from a theoretical standpoint, which analytically specifies how distributed local devices are supposed to collaborate with each other to solve a joint learning task. In Section 10.4, we discuss both the learning models and algorithms implemented at each local device. In Section 10.5, we introduce three representative wireless use cases to showcase the effectiveness of applying the proposed scalable learning paradigm to specific wireless applications. Finally, we pinpoint some future directions along this line of research in Section 10.6.

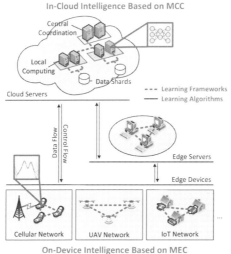

**Figure 10.1** An integrated scalable data-driven wireless architecture. The learning frameworks specify how the distributed devices collaborate with each other, and the learning algorithms specify how a local device learns from its collected data.

## 10.2     Scalable Network Architecture

In this section, we investigate how to integrate scalable intelligence into the existing wireless architectures. In particular, the scalable intelligence orchestrates the *in-cloud intelligence* and *on-device intelligence* built upon the mobile cloud computing (MCC) architecture and mobile edge computing (MEC) architecture, respectively. Therefore, scalable intelligence is able to facilitate accurate, timely, and end-to-end response to cutting-edge communication technologies such as infrastructure densification, mmWave communication, massive MIMO, and energy-efficient network management. This scalable wireless architecture also lays the infrastructural foundation for scalable learning frameworks and scalable learning algorithms to be introduced in later sections. Figure 10.1 illustrates a futuristic scalable wireless architecture to approach the blueprint of intelligence-everywhere in data-driven systems.

### 10.2.1     Cloud-Based Intelligence

Cloud-based intelligence aims to oversee all learning and management tasks across the whole network and making decisions from a global view. One tempting choice is to construct a global AI brain at the cloud centre of the MCC architecture. Specifically, MCC proposes gathering all available computing resources at one place for joint management and coordination [7]. Hence, the cloud-based intelligence can make full use of the abundant cloud computing resources to give a deeper analysis of the network by training a sophisticated learning model with big data. Moreover, the cloud-based environment supports on-demand computing resource provisioning and fast internal information transmission, which lays a good foundation for scalable computation.

In addition, the cloud-based intelligence should integrate the network slicing techniques such as software defined networking (SDN) to attain required configuration flexibility. In particular, the SDN technique promises to logically decouple the control plane and data plane [12]. Thus, the control functions can be aggregated in a software-based cloud server to construct a programmable and software-oriented global AI brain, such that the cloud-based intelligence is able to reconfigure the network in a smart and timely manner according to the latest learning outcomes. Meanwhile, the network slicing technique also enables the global AI brain to slice the physical network into multiple virtual networks, such that each of them can be customized to satisfy the specifications of different learning tasks.

## 10.2.2    On-Device Intelligence

The on-device intelligence in the scalable architecture aims to bring intelligence closer to terminal devices with less or even no dependency on the remote cloud and therefore exhibits the following merits. First, performing learning on distributed local devices largely reduces the latency caused by interacting with a remote cloud, which is crucial for delay-sensitive applications such as autonomous driving and collision avoidance for drones. Second, keeping datasets securely stored on each device can largely relieve privacy and security concerns. This in turn motivates cooperation among devices. Third, on-device learning alleviates the dependency on the connectivity to a remote cloud server, which makes on-device intelligence more robust against harsh scenarios. In this context, the emerging MEC techniques [8], which promise to bring the computing and storage capability closer to the devices at the edge can be exploited to construct the on-device intelligence in the scalable architecture. In particular, each edge device in the MEC architecture is able to acquire its own dataset through local sensing and train its own learning model to individually tackle small-scale problems. As such, the scalable architecture can trade accuracy for latency (or response time) to local events on different devices. Moreover, neighboring devices can collaborate with each other to contribute a superior collective intelligence through, for example, multi-agent learning. Hence, the on-device intelligence is able to complement the in-cloud intelligence to offer end users with prompt reactions, secured privacy, and seamless connectivity.

## 10.3    Scalable Learning Frameworks

Scalable wireless architecture needs to make good use of multiple computing units in order to deliver better performance than the traditional all-in-one solution. Although the aforementioned scalable architecture has demonstrated in Section 10.2 how to integrate intelligence into the wireless infrastructure, it is still unclear how a batch

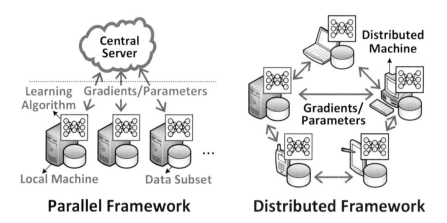

**Figure 10.2**  A parallel learning framework with two layers and a fully distributed learning framework.

of distributed local devices should collaborate to solve a learning task. Hence, in this section, we present the scalable learning frameworks and specify the learning protocol for distributed devices. Here, "scalable learning" refers to two different paradigms of distributed optimization, one called "parallel learning" with centralized consensus to be introduced in Section 10.3.1 and the other one called "fully distributed learning" without centralized coordination to be introduced in Section 10.3.2.

## 10.3.1    Parallel Learning Framework

Parallel learning speeds up the learning process by decoupling a large-scale learning problem into many small subproblems and handling them in a distributed manner. Hence, the computational burden can be distributed to multiple local devices, which improves the overall processing speed. As shown in Fig. 10.2, a representative scalable computation topology comprises two layers: (a) a bottom layer with multiple local devices, each of which learns from a subset of the complete data, and (b) a top layer with one central device, which coordinates the learning at the bottom layer and fuses the local results into a global one. The distributed local devices should have sufficient computing power, and they can be smartphones, drones, connected vehicles, and sensors in different wireless applications. Moreover, the central server should have enough communication, computing, and storage capacity, which can be provided by base stations, cloud centers, etc.

It is noteworthy that the malfunction of one (or several) local device(s) will only cause certain information loss instead of a complete breakdown. Therefore, the scalable architectures are more reliable than the traditional centralized architectures. Next we present three promising scalable computing schemes.

## ADMM

The alternating direction method of multipliers (ADMM) was first introduced in the mid-1970s and has been largely extended to currently handle a wide range of optimization problems in machine learning [13]. Specifically, ADMM takes the form of a decomposition-coordination procedure, which blends the benefits of dual decomposition and augmented Lagrangian methods for constrained optimization [13]. The basic idea behind ADMM is to solve a large global problem by alternating between local optimization and global consensus. By using ADMM on each local device, the local parameters are individually updated by solving a small-scale subproblem decomposed from the original large-scale problem and constrained to be close to the global parameters. The local devices need to synchronize their local estimates of the global parameters with each other by periodically communicating with the server in the top layer as shown in Fig. 10.2.

The overall complexity of ADMM depends on the number of iterations toward convergence as well as the complexity for solving the local optimization problem. Although ADMM may slowly converge to the desired solution even for convex problems [13], a few iterations is often sufficient to attain a satisfactory solution in practice. It is well acknowledged that ADMM is effective in terms of solving large-scale optimization problems [13, 14]. Hence, it is particularly promising to be used for scalable learning in data-driven wireless systems.

After having introduced the background, we next introduce two ADMM-based parameter optimization schemes, which can effectively balance the computation and communication efficiency. The first scheme, namely, the classical ADMM (cADMM) based scheme, approximates the original centralized optimization problem as a nonconvex consensus problem [13] by introducing a set of local parameters $\{\theta_1, \theta_2, \dots, \theta_K\}$ and the global model parameters $z$, namely,

$$\min \quad \sum_{i=1}^{K} l^{(i)}(\theta_i),$$

$$\text{s.t.} \quad \theta_i - z = 0, \quad \forall i = 1, 2, \dots, K,$$

(10.1)

where $K$ is the total number of local devices and $l^{(i)}(\theta_i)$ is nonconvex in terms of the local model parameter $\theta_i$ in general.

The augmented Lagrangian function for Eq. (10.1) is given by

$$\mathcal{L}(\{\theta_i\}, z, \{\beta_i\}) = \sum_{i=1}^{K} l^{(i)}(\theta_i) + \beta_i^T(\theta_i - z) + \frac{\rho_i}{2} \|\theta_i - z\|_2^2,$$

(10.2)

where $\beta_i$ is a dual variable and $\rho_i$ stands for a predetermined regularization parameter. The $(r + 1)$th iteration of the cADMM for solving Eq. (10.1) can be decomposed into

$$z^{r+1} = \frac{1}{K} \sum_{i=1}^{K} \left( \theta_i^r + \frac{1}{\rho_i} \beta_i^r \right), \tag{10.3a}$$

$$\theta_i^{r+1} = \arg \min_{\theta_i} l^{(i)}(\theta_i) + (\beta_i^r)^T (\theta_i - z^{r+1}) + \frac{\rho_i}{2} \| \theta_i - z^{r+1} \|_2^2, \tag{10.3b}$$

$$\beta_i^{r+1} = \beta_i^r + \rho_i(\theta_i^{r+1} - z^{r+1}). \tag{10.3c}$$

This workflow is shown in Fig. 10.3(b) for clarity.

The optimality conditions for the ADMM solution are determined by the primal residuals $\Delta_p$ and dual residuals $\Delta_d$ [13]. They can be given for each local model as

$$\Delta_{i,p}^{r+1} := \theta_i^{r+1} - z^{r+1}, \quad i = 1, 2, \ldots, K, \tag{10.4a}$$

$$\Delta_d^{r+1} := \rho(z^{r+1} - z^r). \tag{10.4b}$$

These residuals will converge to zero as ADMM iterates. Hence, we can adopt the stopping criteria $\| \Delta_p^r \|_2 \le \epsilon^{\text{pri}}$ and $\| \Delta_d^r \|_2 \le \epsilon^{\text{dual}}$, where $\epsilon^{\text{pri}}$ and $\epsilon^{\text{dual}}$ are the feasibility tolerance constants for the primal and dual residuals, respectively.

Next, we introduce a more recent proximal ADMM (pxADMM) based scheme, which was proposed in [14] and is capable of reducing the communication overhead and the computational time at the same time. Unlike Eq. (10.3b), where the local model parameters $\theta_i$ are updated by exactly minimizing the augmented Lagrangian function, the pxADMM takes a proximal step with respect to (w.r.t.) $\theta_i$ by applying the first-order Taylor expansion to $l^{(i)}(\theta_i)$, leading to

$$\theta_i^{r+1} = \arg \min_{\theta_i} \left( \nabla l^{(i)}(z^{r+1}) + \beta_i^r \right)^T (\theta_i - z^{r+1}) + \left( \frac{\rho_i + L_i}{2} \right) \| \theta_i - z^{r+1} \|_2^2, \tag{10.5}$$

where $L_i$ is a newly introduced positive constant to make the following inequality

$$\| \nabla l^{(i)}(\theta_i) - \nabla l^{(i)}(\theta_i') \| \le L_i \| \theta_i - \theta_i' \|, \tag{10.6}$$

satisfied for all $\theta_i$ and $\theta_i', i = 1, 2, \ldots, K$. Note that the proximal step in Eq. (10.5) for $\theta_i$ is a (convex) quadratic optimization problem with the following closed-form solution:

$$\theta_i^{r+1} = z^{r+1} - \frac{\nabla l^{(i)}(z^{r+1}) + \beta_i^r}{\rho_i + L_i}. \tag{10.7}$$

As a consequence, the $(r+1)$th iteration of the pxADMM for solving Eq. (10.1) can be summarized as

$$z^{r+1} = \frac{1}{K} \sum_{i=1}^{K} \left( \theta_i^r + \frac{1}{\rho_i} \beta_i^r \right), \tag{10.8a}$$

$$\theta_i^{r+1} = z^{r+1} - \frac{\nabla l^{(i)}(z^{r+1}) + \beta_i^r}{\rho_i + L_i}, \tag{10.8b}$$

$$\beta_i^{r+1} = \beta_i^r + \rho_i(\theta_i^{r+1} - z^{r+1}). \tag{10.8c}$$

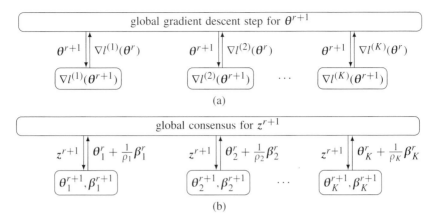

**Figure 10.3** Workflow of two existing distributed parameter optimization schemes: (a) PSGD [15] and (b) cADMM [13].

The pxADMM shares the same workflow with the cADMM as depicted in Fig. 10.3. The criteria for choosing $\rho_i$ and $L_i$ are given in [14], where the authors also proved that under mild conditions (a) $\theta_i^r$ converges to $z^r$ for all $i$ and (b) solution $(\{\theta_i^r\}, z^r, \{\beta_i^r\})$ converges to a stationary point of Eq. (10.1).

Principally, the pxADMM reduces the communication overhead in the same way as the cADMM does. However, the proximal step shown in Eq. (10.8b) leads to an approximated but closed-form solution of Eq. (10.3b) with considerably reduced computation cost. Although more iterations may be required by the pxADMM toward convergence, the overall computational time can be well reduced compared with the cADMM.

## PSGD

The parallel stochastic gradient descent (PSGD) based schemes are extremely popular owing to their capability of solving large-scale deep learning problems. They are widely used in the parameter server framework [15] with the following workflow. First, each local device at the bottom layer computes the gradient of the model parameters (called the gradient henceforth) by using a mini-batch of the full dataset. Second, the parameter server at the top layer aggregates all the gradients to update a set of globally shared parameters. Third, each device synchronizes its local estimates of the global parameters periodically via consensus. This parameter server framework is naturally compatible with the aforementioned MCC architecture.

The PSGD schemes exhibit the same convergence rate as the single-device mini-batch stochastic gradient descent (SGD) but at the cost of using a larger mini-batch. Moreover, they need to wait for the slowest learner to synchronize the update. Recent literatures has started to consider asynchronous PSGD methods, where the parameter server can update global parameters without waiting for all learners to finish. Their complexity depends on the number of iterations toward convergence and the size of the learning model. However, the local devices need to send the computed gradients to and obtain the global parameters from the parameter server in each gradient update step. Hence, it often requires more frequent communications between the local devices and the central server than the ADMM-based schemes.

### 10.3.2     Fully Distributed Learning Framework

Fully distributed learning is favoured when the underlying network does not allow centralized control, for example, in a fully distributed wireless sensor network or an ad hoc network. Its computation topology does not specify any central server but assumes that all distributed local devices are connected according to a (sometimes time-varying) communication graph. Therefore, specifying a proper communication protocol to enable efficient information exchange is crucial for fully distributed learning. Here, we would like to envision the trend of developing fully distributed variants of the parallel learning frameworks, such as the decentralized ADMM [16] and the decentralized PSGD [17]. Such variants are usually based on similar mathematical theories as their parallel counterparts, such that the data-driven wireless network may switch between the parallel and fully distributed computing mode without making a large change to system configurations. Hence, the distributed local devices will be able to rapidly adjust their collaboration scheme according to the actual network topology or computation demands, which is extremely appealing when flexibility and adaptability are of great importance.

### 10.3.3     Comparison

In practical applications, one needs to properly choose a scalable learning framework in accordance with the design specifications. For example, ADMM needs to perform localized parameter optimization, which requires the local devices to have higher computing power than PSGD. On the other hand, PSGD needs to perform frequent parameter consensus and is therefore more suitable to the applications where heavy communication cost is affordable. Notably, the fully distributed variants of the aforementioned scalable learning frameworks such as decentralized ADMM and decentralized PSGD are also under fast development. They specify no central device and are thus favored when the underlying network, such as a distributed wireless sensor network or an ad hoc network, does not allow centralized control.

## 10.4     Scalable Learning Models and Algorithms

While the scalable learning frameworks specify how distributed local devices communicate and coordinate with each other from a global view, the machine learning model and algorithms discussed in this section demonstrate how to extract desired patterns from a given dataset.

General machine learning enables each local device to predict its own future quantities, such as traffic variations and user movements, by discovering patterns from past data and exploiting them for better planning, which is the key to building an autonomous and adaptive data-driven wireless system. The current popularity of machine learning is mainly due to the revival of deep learning centered around DNNs models. However, traditional learning approaches based on DNNs are difficult (implicit and costly) to evaluate the modelling and prediction uncertainty, which largely hinders its application in critical intelligent systems. Therefore, one aim of this section is to draw public attention from the popular deep learning to Bayesian nonparametric learning and RL.

## 10.4.1    Bayesian Nonparametric Learning

Owing to the combination of big data, powerful computing facilities, and innovations on learning models and algorithms, we have witnessed the great success of deep learning (based on deep neural networks, convolutional neural networks, recurrent neural networks, etc.) in numerous applications, such as image recognition, computer vision, and natural language processing. However, traditional deep learning models are incapable of representing the prediction uncertainty in a natural and inexpensive way. A promising direction is to use Bayesian nonparametric learning models and algorithms, which are innately suited for representing the uncertainty of prediction and decision based on interpretable probabilistic models and small data [18].

**Models and Algorithms**

The golden era of Bayesian nonparametric learning traces back to the 1990s. In 1992, David J. C. MacKay published the seminal work on the Bayesian neural network [19], followed by Radford M. Neal's work on using sampling methods for inference [20], where he found that a Bayesian neural network with one infinitely long hidden layer and Gaussian priors on the neuron weights can be represented as a Gaussian process (GP).

According to [21], GP is defined as a collection of random variables, any finite number of which follow a Gaussian distribution. Mathematically, it is written as

$$f(\boldsymbol{x}) \sim \mathcal{GP}(m(\boldsymbol{x}), k(\boldsymbol{x}, \boldsymbol{x}'; \boldsymbol{\theta}_h)), \tag{10.9}$$

where $m(\boldsymbol{x})$ is the mean function, which is often set to zero in practice, especially when there is no prior knowledge about the underlying process, and $k(\boldsymbol{x}, \boldsymbol{x}'; \boldsymbol{\theta}_h)$ is the kernel function controlled by the parameters, $\boldsymbol{\theta}_h$.

Let us consider the GP regression model $y = f(\boldsymbol{x}) + e$, where $y \in \mathbb{R}$ is a continuous-valued, scalar output; the unknown function $f(\boldsymbol{x})\colon \mathbb{R}^d \mapsto \mathbb{R}$ is modelled as a zero-mean GP; and the noise $e$ is assumed to be Gaussian distributed with zero mean and variance $\sigma_e^2$. Moreover, the noise terms at different data points are assumed to be mutually independent. The set of all unknown GP model parameters is denoted by $\boldsymbol{\theta} := [\boldsymbol{\theta}_h^T, \sigma_e^2]^T$, and the dimension of $\boldsymbol{\theta}$ is assumed to be equal to $p$.

The joint prior distribution of the training output $\boldsymbol{y}$ and test output $\boldsymbol{y}_*$ can be compactly written as

$$\begin{bmatrix} y \\ y_* \end{bmatrix} \sim \mathcal{N} \left( \mathbf{0}, \begin{bmatrix} K(X,X) + \sigma_e^2 I_n, & K(X,X_*) \\ K(X_*,X), & K(X_*,X_*) + \sigma_e^2 I_{n_*} \end{bmatrix} \right), \tag{10.10}$$

where $K(X,X)$ is an $n \times n$ covariance matrix between the training inputs; $K(X,X_*)$ is an $n \times n_*$ covariance matrix between the training inputs and the test inputs; and $K(X_*,X_*)$ is an $n_* \times n_*$ covariance matrix between the test inputs. Here, we let $K(X,X)$ be a short term of $K(X,X;\theta_h)$ when parameter optimization is not the focus.

By applying some known results of the conditional Gaussian distribution, we can easily derive the posterior distribution as

$$p(y_*|\mathcal{D}, X_*; \theta_h) \sim \mathcal{N} \left( \bar{m}, \bar{V} \right), \tag{10.11}$$

where the posterior mean (vector) and posterior covariance (matrix) are, respectively,

$$\bar{m} = K(X_*,X) \left[ K(X,X) + \sigma_e^2 I_n \right]^{-1} y, \tag{10.12}$$

$$\bar{V} = K(X_*,X_*) + \sigma_e^2 I_{n_*} - K(X_*,X) \left[ K(X,X) + \sigma_e^2 I_n \right]^{-1} K(X,X_*). \tag{10.13}$$

Given an unseen input in the test dataset, this posterior mean gives a point prediction, while the posterior covariance gives the uncertainty region of the prediction. For clarity, we plot some sample functions generated from the prior and posterior distributions, respectively, in Fig. 10.4.

In addition to the introduced GP model, the development of Bayesian neural networks has been rapid in recent years [18]. In contrast to the traditional deep learning models, the Bayesian counterparts are rather flexible because they do not need to fix the network structure, concretely, the number of hidden layers and number of neurons a priori. Using nonparametric priors, these two numbers are estimated based on the posterior distributions.

The learning algorithms specifically refer to the model training and prediction/inference algorithms. Due to space limitations, we will constraint ourselves to GP models and scalable training and prediction algorithms to be introduced in the next subsection. We refer interested readers to [22] for general learning algorithms for Bayesian neural network type models.

## Good Interpretability of Kernel Functions

The kernel function determines the expressive power of the GP model to a large extent. To make a kernel function full of expressive power and automatically adaptive to the data, the following work can be adopted for wireless applications. In [23], a spectral mixture (SM) kernel was proposed to approximate the spectral density with a Gaussian mixture model in the frequency domain and transform it back into an universal stationary kernel. In [24], the authors modified the SM kernel to be a linear multiple low-rank subkernels with a favourable difference-of-convex optimization structure.

For data-driven wireless system, kernel design should consider the unique characteristics of the wireless data. Next, we showcase some representative kernel functions, whose parameters are full of physical meanings.

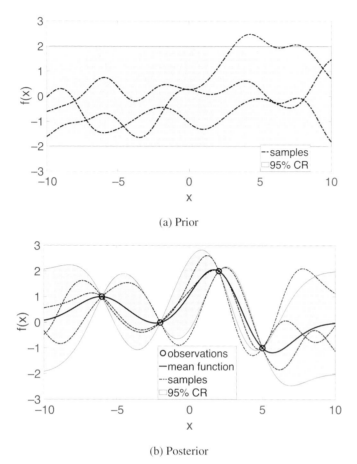

(a) Prior

(b) Posterior

**Figure 10.4** (a) Three sample functions randomly drawn from a GP prior with a specific squared-exponential kernel. (b) Three sample functions drawn from the posterior conditioned on the prior in (a) and four noisy observations indicated by small circles. The corresponding posterior mean function is depicted by the black curve. The grey shaded area represents the uncertainty region, namely, the 95 percent confidence region for both the prior and the posterior.

**Example 10.1** The *squared-exponential (SE) kernel* is a default kernel for both GP regression and classification. The corresponding kernel function is given by

$$k_{SE}(t, t') = \sigma_s^2 \exp\left[-\frac{(t - t')^2}{2l^2}\right], \tag{10.14}$$

which contains two parameters: the signal variance $\sigma_s^2$ representing the magnitude of functional fluctuation and the length scale $l$ determining how rapidly the underlying function may vary with time $t$.

**Example 10.2** The *automatic relevance determination (ARD) kernel* is another default GP kernel for multidimensional input $x = [x_1, x_2, \ldots, x_d] \in \mathbb{R}^d$. The corresponding kernel function is given by

$$k_{\text{ARD}}(x, x') = \sigma_s^2 \exp\left[-\sum_{i=1}^{d} \frac{(x_i - x_i')^2}{2l_i^2}\right], \qquad (10.15)$$

which contains $d + 1$ parameters, namely, the signal variance $\sigma_s^2$ and the lengthscales $l_i, i = 1, 2, \ldots, d$.

**Example 10.3** The *locally periodic kernel* is the valid kernel to represent periodicity in the data. The corresponding kernel function is given by

$$k_{\text{LP}}(t, t') = \sigma_s^2 \exp\left[-\frac{\sin^2\left(\frac{\pi(t-t')}{\lambda}\right)}{l_p^2}\right], \qquad (10.16)$$

which contains three parameters: the periodicity $\lambda$, lengthscale $l_p$, and signal variance $\sigma_s^2$.

**Example 10.4** The *SM kernel* is an optimal kernel proposed in [23]. The SM kernel was obtained by approximating the underlying spectral density with an $m$-mode Gaussian mixture in the frequency domain. The corresponding kernel function for the one-dimensional case is given by

$$k_{\text{SM}}(t, t') = \sum_{i=1}^{m} \alpha_i \exp\left[-2\pi^2\tau^2\sigma_i^2\right] \cos(2\pi\tau\mu_i), \qquad (10.17)$$

where $\tau := t' - t$. This kernel contains $3m$ parameters: the weights $\alpha_i$, frequency shift $\mu_i$, and variance component $\sigma_i$, for $i = 1, 2, \ldots, m$. After the kernel is trained, it will pinpoint the active frequency bands (with the magnitude $\alpha_i$, center frequency $\mu_i$, and bandwidth $\sigma_i$) of the given data.

---

In the following, we show some examples of selecting appropriate kernel functions according to the data patterns. As shown in Fig. 10.5, the wireless traffic in a real dataset demonstrates the following general patterns: (a) *weekly periodic pattern*, namely, the variation in accordance with weekdays and weekends; (b) *daily periodic pattern*, namely, the variation in accordance with weekdays and weekends; and (c) *deviations*, namely, the small-scale variation in addition to the above periodic trends. The first two patterns can be well captured by different periodic kernels, while the third pattern can be well captured by the SE kernel. Alternatively, one could use the optimal SM kernel if there is no prior knowledge about the data available and let the kernel adapt itself to the underlying pattern automatically through optimization.

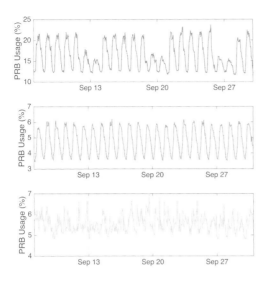

**Figure 10.5** The PRB usage curves of three base stations collected in three southern cities of China in 30 days. The data profile in the first panel reflects a typical office area, in which the traffic pattern shows a strong weekly periodic trend in accordance with weekdays and weekends. The data profile in the second panel reflects a typical residential area, in which the traffic pattern shows a strong daily trend with high demands in the daytime and low demands at night. The data profile in the third panel reflects a typical rural area, in which the traffic pattern is more or less random.

## Scalability

Next, we will introduce the scalable training and prediction algorithms for the GP models (called GP algorithms for short hereafter). The GP algorithms can be implemented in a distributed and principled manner, showing a perfect fit to the data-driven wireless communication system. Scalability occurs in both the training phase and the prediction/test phase. We elaborate on the two phases as follows.

**(1) Training phase.** To train the GP model parameters, one can resort to the classic ML-based estimation. Due to the Gaussian assumption on the noise, the log-likelihood function can be obtained in closed form as

$$l(X, y; \theta) := \log p(y; X, \theta)$$
$$= \log \mathcal{N}\big(y; m(X), K(X, X; \theta) + \sigma_e^2 I_n\big). \tag{10.18}$$

The GP model parameters can be equivalently optimized by minimizing the negative log-likelihood function (with the constant terms omitted):

$$l(X, y; \theta) = y^T C^{-1}(\theta) y + \log \det (C(\theta)), \tag{10.19}$$

where $C(\theta) \triangleq K(X, X; \theta_h) + \sigma_e^2 I_n$. This optimization problem is mostly solved using gradient descent-type methods such as L-BFGS-Newton or conjugate gradient [21], which requires computing the following closed-form partial derivatives for $i = 1, 2, \ldots, p$:

$$\frac{\partial l(\boldsymbol{\theta})}{\partial \theta_i} = \mathrm{tr}\left(\boldsymbol{C}^{-1}(\boldsymbol{\theta})\frac{\partial \boldsymbol{C}(\boldsymbol{\theta})}{\partial \theta_i}\right) - \boldsymbol{y}^T \boldsymbol{C}^{-1}(\boldsymbol{\theta})\frac{\partial \boldsymbol{C}(\boldsymbol{\theta})}{\partial \theta_i}\boldsymbol{C}^{-1}(\boldsymbol{\theta})\boldsymbol{y}. \qquad (10.20)$$

Using this ML estimation to train the GP model parameters requires $O(n^3)$ computational complexity due to the unavoidable matrix inverse and matrix determinant for general dense matrices.

To address the scalability issue of the standard GP, we need to first approximate the global objective in Eq. (10.18) by

$$l(\boldsymbol{X}, \boldsymbol{y}; \boldsymbol{\theta}) \approx \sum_{i=1}^{K} \log \mathcal{N}\left(\boldsymbol{y}_i; \boldsymbol{m}(\boldsymbol{X}_i), K(\boldsymbol{X}_i, \boldsymbol{X}_i; \boldsymbol{\theta}) + \sigma_e^2 \boldsymbol{I}\right), \qquad (10.21)$$

where $\mathcal{D}_i := \{\boldsymbol{X}_i, \boldsymbol{y}_i\}$, $i \in \{1, 2, \ldots, K\}$, is the dataset collected by the $i$th local device. Further applying either the cADMM- or pxADMM-based GP model parameter optimization schemes [5, 25] will solve the scalability issue of the model training process. The computational complexity can be reduced from $\mathcal{O}(n^3)$ for the standard GP to $\mathcal{O}(\frac{n^3}{K^3})$ for the proposed scalable GP, where $n$ is the number of training points and $K$ is the number of local devices. This computational complexity can be further reduced to $\mathcal{O}(\frac{n^2}{K^2})$ when the kernel matrix has a Toeplitz structure, which is often the case for uniformly sampled time series and spatio-temporal data.

**(2) Test/Prediction Phase.** After having trained the GP model parameters, we need to fuse the predictions from all local devices to obtain a global one. To this end, one can use the generalized product of experts (PoE) [26]. The generalized PoE model needs to introduce a set of fusion weight parameters, $\beta_i$, $i = 1, 2, \ldots, K$, to count the importance of the local predictions. The resulting PoE predictive distribution is

$$p(f_*|\boldsymbol{x}_*, \mathcal{D}) \approx \prod_{i=1}^{K} p_i^{\beta_i}(f_*|\boldsymbol{x}_*, \mathcal{D}^{(i)}), \qquad (10.22)$$

where $\beta_i$ is the weight for the $i$th local GP model, and the corresponding posterior mean and variance are, respectively,

$$\mu_* = \sigma_*^2 \sum_{i=1}^{K} \beta_i \sigma_i^{-2}(\boldsymbol{x}_*)\mu_i(\boldsymbol{x}_*), \qquad (10.23)$$

$$\sigma_*^2 = \left(\sum_{i=1}^{K} \beta_i \sigma_i^{-2}(\boldsymbol{x}_*)\right)^{-1}, \qquad (10.24)$$

where $\mu_i(\boldsymbol{x}_*)$ and $\sigma_i(\boldsymbol{x}_*)$ are the $i$th local predictive mean and variance evaluated at an unseen test point $\boldsymbol{x}_*$. The choice of $\beta_i$, $i = 1, 2, \ldots, K$, is vital to the accuracy of the global prediction.

In contrast to the existing fusion strategies that adopt empirical weights (e.g., directly using entropy as the weight [26]) we propose optimizing the weights via cross validation in [5], which can provide a reliable fusion strategy with concrete theoretical analysis. Meanwhile, we aim to achieve three desirable properties for the fusion process: (a) the predictions should be combined based on both the prior and the posterior information, which gives the combined model more generalization power; (b) the combination should follow a valid probabilistic model, which helps preserve the distinct GP properties (e.g., the posterior variance for prediction uncertainty evaluation); and (c) the combined prediction should be robust against bad local predictions.

The workflow of our proposed strategy is as follows. First, we divide the full training set into two parts, the training set and the validation set, where the validation set consists of the training points that are closer to the test set. To guarantee a robust performance for general regression tasks, we first optimize the prediction performance on the validation set and then use the optimized weights to combine all local predictions for a global one. Specifically, optimizing the prediction performance on the validation set can be formulated as the minimization of the prediction residuals:

$$\min_{\beta} \quad \sum_{m=1}^{M} (y_m - \tilde{y}_m)^2,$$

$$\text{s.t.} \quad \beta \in \Omega,$$

(10.25)

where $M$ is the size of the validation set, $\tilde{y}_m$ is the combined prediction on the validation point $m$, and $\Omega = \{\beta \in \mathbb{R}_+^K : \mathbf{1}^T \beta = 1\}$ restricts the weights $\beta$ to be in a probability simplex.

Based on the approximated global prediction given in Eqs. (10.22)–(10.24), we can derive

$$\tilde{y}_m = \arg\max_{\tilde{f}_m} \prod_{i=1}^{K} p_i^{\beta_i} \left( \tilde{f}_m | \mu_i(\mathbf{x}_m), \sigma_i(\mathbf{x}_m) \right)$$

$$= \frac{\sum_{i=1}^{K} \beta_i \sigma_i^{-2}(\mathbf{x}_m) \mu_i(\mathbf{x}_m)}{\sum_{i=1}^{K} \beta_i \sigma_i^{-2}(\mathbf{x}_m)}.$$

(10.26)

Therefore, the optimization problem proposed in Eq. (10.25) can be recast as

$$\min_{\beta} \quad f(\beta) = \sum_{m=1}^{M} \left( y_m - \frac{\sum_{i=1}^{K} a_i(\mathbf{x}_m) \beta_i}{\sum_{i=1}^{K} b_i(\mathbf{x}_m) \beta_i} \right)^2,$$

$$\text{s.t.} \quad \beta \in \Omega,$$

(10.27)

where $a_i(\mathbf{x}_m) := \sigma_i^{-2}(\mathbf{x}_m) \mu_i(\mathbf{x}_m)$ and $b_i(\mathbf{x}_m) := \sigma_i^{-2}(\mathbf{x}_m)$.

The convexity of the optimization problem in Eq. (10.27) depends on the size of the validation set. Specifically, when we only use a single point for validation (i.e., $M = 1$), this problem boils down to a convex problem, for which the global optimum can be obtained; when we use more than one point for validation (i.e., $M > 1$), the problem becomes nonconvex but can be solved efficiently using the mirror descent method. Detailed derivations and analyses of these results can be found in [5]. It is noteworthy that the former case with $M = 1$ is good for short-term prediction, whereas the latter case with $M > 1$ is better to use for relatively long-term prediction.

## Uncertainty

Compared with the traditional deep learning models (taking DNN as a concrete learning model in the sequel), the Bayesian nonparametric learning based on the GP model can better address the following uncertainty issues:

- *Functional uncertainty.* Learning tasks based on the DNN solely rely on its universal approximation property [27]. Having specified the network structure (often referring to its width and depth) and tuned the model parameters, the DNN turns out to be a deterministic function for representing the underlying regression mechanism. In contrast, the GP model assumes the underlying regression mechanism is a realization of a Gaussian stochastic process specified by the selected kernel, for example, an SM kernel [23]. In other words, model uncertainty has been considered prior to data collection. This has been demonstrated in the top subfigure of Fig. 10.4.
- *Prediction uncertainty.* Since the DNN performs a deterministic mapping, its prediction is a point estimate. In contrast, Bayesian learning based on the GP model provides a posterior distribution of the desired prediction with a natural uncertainty region, as shown in the bottom subfigure of Fig. 10.4. However, the effectiveness of the uncertainty region depends on the kernel selection. When more data points are collected, better understanding about this uncertainty region can be built.
- *Input uncertainty.* In contrast to the DNN, GP models can naturally handle input uncertainty. This is valuable to wireless applications as the model inputs are often subject to environmental noise due to wireless propagation. Since the GP model is a probabilistic model, the input uncertainty can be easily incorporated by assuming the training input $x$ to be a random variable with a known distribution $p(x)$. In [28], for instance, the mean function of GP with input uncertainty was obtained as

$$\tilde{m}(x) = \int m(x)p(x)dx, \tag{10.28}$$

and the kernel function was obtained as

$$\tilde{k}(x,x') = \int\int k(x,x')p(x)p(x')dx\,dx'. \tag{10.29}$$

The only difficulty lies in the evaluation of the two integrals. In general, they can be approximated by the Monte-Carlo integration [22, 29]. The remaining steps remain the same as the standard GP with clean input.

- *Output Uncertainty.* Due to the wireless propagation and imprecise measurement devices, the observed wireless measurements are often subject to noise aggregated from different sources. For example, as an important network performance indicator, the received-signal-strength contains noise due to fast fading, slow fading, non-line-of-sight bias, and measurement error. In general, such output uncertainty is difficult to remove completely but can be alleviated by smoothing the measurements either in time or in both time and space. Yet another method is to improve the precision of the measurement device by upgrading its hardware.

For instance, by using the massive multiple-in multiple-out (MIMO) and millimeter wave (mmWave) devices in 5G and beyond, the received signal strength is hoped to be more accurate than before in the 4G wireless system.

It is noteworthy that the first two uncertainties mentioned here belong to the class of model uncertainty, which is also referred to as *epistemic* uncertainty, whereas the last two uncertainties belong to the class of data uncertainty, which is also referred to as *aleatoric* uncertainty. Together, they lead to the predictive uncertainty, namely, the uncertainty level about our prediction [30].

### Wireless Applications

Before leaving this section, we present some representative wireless applications that are beneficial from the Bayesian nonparametric learning based on the GP model.

- First, it is particularly powerful for discovering complicated mechanisms, including 5G and vehicle to everything (V2X) channel impulse responses, multipath radio signal propagation, radio feature maps (such as the signal quality, uplink/downlink traffic, and wireless resources demand/supply) over time and space, indoor pedestrian motion, and so forth, in a more economical and cautious manner than the state-of-the-art DNN.
- Second, it is more favorable to use in terms of system identification, control, and integration. In particular, the GP algorithms can be combined with the traditional state-space model to accurately represent and rebuild complicated trajectories generated by human beings, autonomous vehicles, unmanned aerial vehicles (UAVs), and so on, and promising applications include super high-precision probabilistic fingerprinting and indoor/outdoor navigation based on wireless signals [31], among others.
- Third, it is also more natural to use for fusing multimodal data collected from different sensors, including raw sensory data from smart phone motion sensors, ultrasound data from UAV ultrasonic sensors, images and videos from surveillance camera, and so on, in different formats and with varying data qualities. Information dissemination and parameter inference can be carried out in a full probabilistic manner, using, for instance, message passing designed for the Bayesian network or Markov random field.

## 10.4.2 Reinforcement Learning

### Models and Algorithms

In RL, actions are taken to maximize the cumulative reward in an unknown environment [32]. Different from supervised learning where the training data are usually labelled, an RL agent needs to decide what to do solely based on the feedback (i.e., the reward signal without a label) when interacting with the environment. This interactive learning style endows RL with excellent adaptability to time-varying and unknown wireless environments. Moreover, RL naturally incorporates farsighted

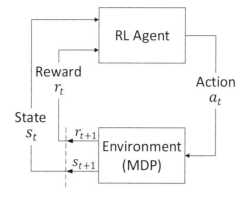

**Figure 10.6** Typical workflow of reinforcement learning.

decision-making capabilities by maximizing the cumulative reward over a long time horizon, which is desired by many data-driven wireless applications, such as path planning and content caching.

The dynamics in RL is usually modelled as a Markov decision process (MDP), which can be characterized by a state space $\mathcal{S}$, an action space $\mathcal{A}$, a reward function $r : \mathcal{S} \times \mathcal{A} \to \mathbb{R}^1$, and a stationary transition probability that satisfies the Markov property $p(s_{t+1}|s_1,a_1,\ldots,s_t,a_t) = p(s_{t+1}|s_t,a_t)$, where $s \in \mathcal{S}, a \in \mathcal{A}$, as shown in Fig. 10.6. At each state $s_t \in \mathcal{S}$, the RL agent selects an action $a_t \in \mathcal{A}$ by following a policy $\pi$ to interact with the environment and receives a reward $r(s_t,a_t)$; then, state $s_t$ moves on to $s_{t+1}$ to start the next round. The goal of RL is to obtain an optimal policy that maximizes the cumulative (discounted) rewards from the start state (i.e., $J(\pi) := \mathbb{E}\left[\sum_{k=1}^{\infty}\gamma^{k-1} \cdot r(s_k,a_k)\big|\pi\right]$), where $\gamma \in [0,1]$ is the discount factor.

Having introduced the model, we next move to the learning algorithms. Generally, most recently successful deep reinforcement learning (DRL) algorithms are developed based on *model-free* RL algorithms, which estimate a value function or a policy function directly from experience (i.e., from the trial and error interactions with the environment), as opposed to model-based RL algorithms, which directly estimate a model of the environment and make decisions by planning. Model-free RL algorithms can be classified into three categories: value-based algorithms, policy-based algorithms, and hybrid actor-critic algorithms.

*Value-based* RL algorithms estimate the expected future reward by iteratively updating a value function or a state-action-value function (also known as a Q-function). In particular, the value function for a given policy $\pi$ at state $s$ is defined to be the received long-term expected cumulative rewards starting at state $s$ and following policy $\pi$ thereafter [32]:

$$V^{\pi}(s) := \mathbb{E}^{\pi}\left[\sum_{t=0}^{\infty}\gamma^t r(s_t,a_t)\big|s_t = s\right]. \tag{10.30}$$

Similarly, the Q-function for a given policy $\pi$ when choosing action $\boldsymbol{a}$ at state $\boldsymbol{s}$ is defined as

$$Q^{\pi}(\boldsymbol{s}, \boldsymbol{a}) := \mathbb{E}^{\pi} \left[ \sum_{t=0}^{\infty} \gamma^t r(\boldsymbol{s}_t, \boldsymbol{a}_t) \big| \boldsymbol{s}_t = \boldsymbol{s}, \boldsymbol{a}_t = \boldsymbol{a} \right]. \qquad (10.31)$$

The Q-function can be updated via policy iterations based on the well-known Bellman equation [32]:

$$Q^{\pi}(\boldsymbol{s}_t, \boldsymbol{a}_t) = \mathbb{E}_{r_t, s_{t+1} \sim E} \left[ r(\boldsymbol{s}_t, \boldsymbol{a}_t) + \gamma \mathbb{E}_{a_{t+1} \sim \pi} \left[ Q^{\pi}(\boldsymbol{s}_{t+1}, \boldsymbol{a}_{t+1}) \right] \right]. \qquad (10.32)$$

For value-based algorithms, the optimal policy can be obtained by greedily selecting the action with the highest Q-value at each step.

Meanwhile, policy-based RL algorithms directly learn a parameterized policy to select optimal actions without consulting a value function. The policy function is usually updated based on the well-known policy gradient theorem [32]:

$$\nabla_{\theta} J(\pi_{\theta}) = \mathbb{E}_{s \sim \rho^{\pi}, a \sim \pi_{\theta}} \left[ \nabla_{\theta} \log \pi_{\theta}(a|s) Q^{\pi}(s, a) \right], \qquad (10.33)$$

where $\pi_{\theta}$ is the parameterized policy and $\rho^{\pi}$ is the state distribution depending on the policy. Note that the value function may still be needed to update the policy weights in policy-based RL algorithms but not for action selection.

The hybrid actor-critic algorithm combines the characteristic of value-based and policy-based algorithms. In particular, the critic maintains an estimate of the Q-function, while the actor updates the weights of a parameterized policy based on the estimated Q-function from the critic.

The popular DRL combines the deep learning technique and RL algorithms to greatly improve the generalization capability when solving complex tasks with high-dimensional state and action spaces. This revolution stems from the birth of deep Q-learning [33], which approximates the Q-function in Q-learning with a deep neural network (a.k.a., Q-network). The success of deep Q-learning depends on two innovative training techniques: (a) training the Q-network by storing/sampling experiences to/from a shuffled offline replay buffer to minimize the correlation among different samples and (b) training the Q-network with a target Q-network, which offers a consistent learning target. Many variants of deep Q-network (DQN) have been proposed to improve the learning performance. For example, Hasselt et al. proposed Double-Q networks to decouple the action selection from the value evaluation to alleviate the overoptimistic value estimates [34]; Wang et al. proposed the duelling DQN, which separately estimates the state-value function and state-dependent action advantage function to improve the learning rate and robustness [35]; and Schaul et al. proposed the prioritized experience replay, which samples the training batches according to the magnitude of their temporal-difference error to improve the learning efficiency [36].

However, these deep Q-learning based methods can only handle discrete and low-dimensional action spaces. Hence, another research line focuses on the combination of deep learning and actor-critic algorithm to adapt DRL to the continuous action domain. Most actor-critic based DRL algorithms approximate both value function and policy function with deep neural networks but employ different strategies to improve the learning efficiency. For example, the deep deterministic policy gradient (DDPG) algorithm [37] replaces the conventional stochastic policy gradient in actor-critic with deterministic policy gradient to learn the policy weights more efficiently; the asynchronous advantage actor-critic (A3C) algorithm [38] runs multiple actors in parallel on multiple instances of the environment to decorrelate the learning sample, which enables it to perform on-policy updates without using an off-line replay buffer; and the proximal policy optimization (PPO) algorithm [39] clips the policy changes per training step to improve the learning robustness.

## Scalability

Scalable DRL models for large-scale network management can be obtained in two ways. The first way is to develop a large DRL model to directly manage the entire network and accelerate the training process by performing parallel computing with multiple local devices based on the aforementioned PSGD framework. For example, the general reinforcement learning architecture named Gorila has been proposed to train the DRL models on massive parallel processes [40]. Each process in Gorila contains an actor that acts in its own copy of the environment, a replay memory to store historical experiences, and a learner that samples data from the replay buffer and computes gradients with respect to the model parameters. The local gradients are uploaded to the parameter server to update a set of global parameters. The global parameters are downloaded by each actor-learner at fixed intervals for consensus.

The second way is to partition a large-scale network into smaller pieces and handle them with DRL algorithms in a self-organized manner. In this case, each controller only needs to adapt its own DRL model to the local environment (i.e., a small portion of the entire network) to largely reduce the computation and communication overhead at each individual device, thereby increasing the system scalability. Our recent work [6] presents an instance of training the DDPG model in parallel and solving large-scale load balancing problems in a self-organized way. A general architecture is shown in Fig. 10.7.

In addition, many existing works on multiagent RL studied the scalability of RL algorithms under fully distributed settings [41–43]. Compared with the single-agent RL, multiagent RL usually considers that the reward functions of the agents may differ from each other and are private to each corresponding agent. This consideration makes the problem more challenging since each agent interacts with not only the environment but also the other agents. Related works on multiagent RL can be found in [44].

## Uncertainty

It is well known that RL can naturally address the system uncertainty since its learning goal is to adapt to a complex *unknown* environment through agent-environment interactions. Compared with the traditional deep learning, which has to split the uncertainty measurement and uncertainty-aware decision making into two separate processes, RL elegantly embeds uncertainty measurement by maximizing the cumulative long-term returns and is able to simultaneously perform uncertainty-aware decision making in

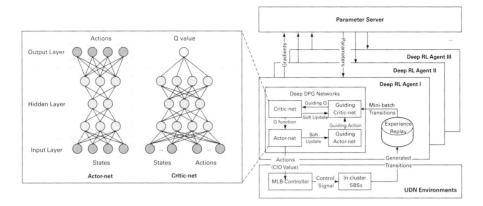

**Figure 10.7** A general scalable DRL architecture for large-scale network management.

a complete control loop. Hence, RL is a suitable and decent machine learning tool to solve control-related tasks in dynamic and uncertain environments.

The estimation of the expected rewards in RL can be considered a natural measurement of uncertainty on the long-term performance of a policy in the unknown environment. Similar to the discussion in Section 10.4.1, the uncertainty in RL can be categorized into aleatoric uncertainty and epistemic uncertainty [45].

- *Aleatoric uncertainty*: This uncertainty stems from the inherent stochastic nature of the environment. The estimation of aleatoric uncertainty can depict the true nature of the system and thus provide valuable information for risk-averse policy design [46, 47]. For example, many distributional RL algorithms were developed with such purpose, aiming at learning the entire distribution of the rewards instead of the expected distribution. The estimated reward density can assist in seeking large wins on rare occasions or avoiding a small chance of suffering a large loss [47].
- *Epistemic uncertainty*. This uncertainty arises from imperfect knowledge of the environment, which may be decreased with more information. The estimation of epistemic uncertainty can help improve the efficiency of exploring a new environment, such as by guiding the exploration on poorly understood states and actions [48, 49].

It is noteworthy that GP model and algorithm can be elegantly embedded into RL to give an explicit uncertainty measurement of the system dynamics. The result can be used to guide the policy learning process for performance enhancement. For example, Rasmussen et al. proposed to take the uncertainties measured from GP models to avoid slow policy evaluation iterations [50], and Saemundsson et al. proposed to use GP to estimate the long-term state evolution in a model-based RL setting in order to transfer knowledge from a set of training tasks to unseen but related tasks under the umbrella of metalearning [51].

### Wireless Applications

DRL can be directly employed for intelligent system control since the interactive learning process in RL forms a complete control loop. The representative applications include network operation/maintenance, resource management and security enhancement. Note that the context of MDP (such as state, action, and reward) should be properly specified in different applications. In particular, for network operation/maintenance such as handover management and user localization, DRL algorithms can be employed to decide the optimal actions of different operation/maintenance operations, such as handover and admission control [6, 52]. In contrast to traditional paradigms, which are developed under prior assumptions on the system (e.g., user mobility patterns and network topologies) the DRL does not require any prior knowledge about the underlying environment, thereby having a better capability to optimally configure the network under complex conditions. For network resource management such as beamforming control, channel assignment, and network caching, the RL actions can be specified as resource allocation operations, increasing or decreasing the transmitting power, and the RL reward can be specified according to the performance metrics or resource constraints, such as network throughput, communication latency, and quality of service (QoS). For network security enhancement, the DRL agent can be autonomously trained to recognize and avoid the network attacking attempts. For example, in jamming attacks, the attackers aim at sending jamming signals with high power to cause interference in the receivers. In this case, DRL can be employed to estimate the attacker's jamming policy and respond adaptively [53]. An extensive review of the DRL-relevant applications in wireless communications can be found in [54, 55].

## 10.5     Practical Use Cases

In this section, we provide a quantitative demonstration of how to take advantage of the presented scalable learning techniques for data-driven wireless networks. Specifically, we provide three case studies: (a) scalable GP-based wireless traffic prediction for base station (BS) on/off control, (b) scalable GP-based indoor motion modelling for data-driven target tracking, and (c) scalable RL-based load balancing for handover management.

### 10.5.1     Case I: Scalable GP-based Wireless Traffic Prediction

In the first use case, we show how to use the scalable GP model to predict traffic variations. Wireless traffic prediction can help evaluate the forthcoming network demand and supply, which is a key enabler for smart management in data-driven wireless networks, such as traffic-aware BS on/off control [5]. This is extremely valuable to 5G networks, in which the new-fashioned BSs are consuming a lot more energy than before.

In our early work [56], we first used the standard (centralized) GP-based Bayesian nonparametric learning, which achieved a prediction error as low as 3 percent in terms of the root-mean-squared-error (RMSE) computed with a real-world wireless traffic dataset. Therein, we compared four energy-saving schemes: (a) BS switch on/off based on the prophetic real traffic, which served as the performance lower bound; (b) BS switch on/off based on the predicted traffic; (c) BS switch on/off based on the current traffic; and (d) BSs that are always on. The energy saving utility ratio is defined as the ratio between the actual utility versus the utility in full service. It is shown that our traffic prediction helps improve the energy saving from 14.7 percent (for one-hour ahead prediction) to 24 percent (for eight-hour ahead prediction), and the outcomes almost reach the lower bound set by the ideal scheme.

Unfortunately, the computational complexity of the standard GP scales as $\mathcal{O}(n^3)$, where $n$ is the number of training samples, which makes it impractical for agile large-scale network managements. For example, predicting the next-hour traffic volume of one BS using 700 hourly recorded training samples takes approximately 16.8 seconds by the standard GP model.

To alleviate such constraint, in our recent work [5], we proposed a scalable GP model training algorithm based on the cADMM scheme to reduce the original computational complexity to $\mathcal{O}(\frac{n^3}{K^3})$ with acceptable performance loss. Recall that $K$ is the number of parallel computing units. For our GP model, we add up two periodic kernels with an SE kernel to represent the weekly periodic pattern, daily periodic pattern, and dynamic deviations observed from a set of real 4G traffic data, respectively.

Like before, we use 700 hourly recorded traffic samples (spanning approximately four weeks) to train the model and predict the next-hour traffic volume. As the results show in Table 10.1, the standard GP model requires approximately 16.8 seconds to complete the training (using one computing unit), while the scalable GP model requires as low as 0.1 seconds (using 16 computing units in parallel). In the online phase, we fuse the local predictions according to Eq. (10.22) with the weights $\beta_i$, $i = 1, 2, \ldots, K$, optimized using the cross-validation criterion introduced in Section 10.4. From Table 10.1, we also observed that the computational time required to fuse $K = 16$ local predictions is only 0.37 seconds.

Despite the scalable setup, the prediction performance of our proposed scalable GP model is fairly close to that of the standard (centralized) one and outperforms several state-of-the-art low-complexity competitors, including the robust Bayesian committee machine (rBCM) and subset-of-data (SOD) model, as shown in Fig. 10.8(a).

**Table 10.1.** Time consumption to train the scalable GP model and fuse the local predictions.

| Learning Phase | 1 Unit | 2 Units | 4 Units | 8 Units | 16 Units |
| --- | --- | --- | --- | --- | --- |
| Training | 16.8 s | 3.5 s | 1.1 s | 0.4 s | 0.1 s |
| Prediction | 0.04 s | 0.07 s | 0.13 s | 0.21 s | 0.37 s |
| Total | 16.84 s | 3.57 s | 1.23 s | 0.61 s | 0.47 s |

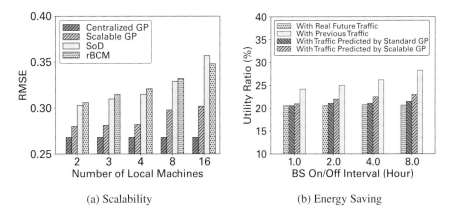

**Figure 10.8** Learning results of the scalable GP-based wireless traffic prediction for active BS on/off control in Case I.

Meanwhile, the energy saving utility ratio only degrades by approximately 2 percent for one-hour ahead prediction and 7 percent for eight-hour ahead prediction, as shown in Fig. 10.8(b).

## 10.5.2    Case II: Scalable GP-based Motion Modelling

In the second use case, we showcase the scalable GP-based indoor pedestrian motion modelling. Different from the first use case, the scalability is achieved through the cooperation of a batch of mobile users with on-device intelligence. This work can be considered a collaborative, data-driven model for trajectory learning, which is valuable for us to understand the behavior of pedestrians and predict their next movement.

The motion model can be written as $x_{t+1} = f(x_t) + e_t$, where vector $x_t = [x_t, y_t]^T$ contains the 2D position of a pedestrian at time instance $t$. For simplicity, we apply an individual GP for each dimension. Using the $x$-dimension as an example, we let

$$x_{t+1} = f_x(x_t) + e_t, \tag{10.34}$$

where $f_x(x_t)$ is modelled by a GP, whose kernel function $k_x(x_t, x_{t'})$ is selected to be the ARD kernel with the kernel parameters, $\sigma_{s,x}^2, l_{xx}, l_{xy}$. Here, we note that the ARD kernel may not be the optimal kernel. For optimality, the readers may refer to the SM kernel introduced in Section 10.4. This GP-based motion model constitutes an important component of the state space model for target tracking [57, 58].

To evaluate the performance, we collected a dataset in a real indoor office environment. This dataset contains more than 50 trajectories with approximately 25,000 samples. In the training phase, each of the three mobile users collected 15 trajectories. Similar to the first use case, the GP model is trained either using the cADMM or pxADMM introduced in Section 10.4 to approximate the standard (centralized) GP learning performance, and the computational complexity can be reduced to $\mathcal{O}(n^3/K^3)$, where $K$ is the number of collaborating mobile users. The configurations of the model

**Table 10.2.** Comparisons of two distributed ADMM schemes to tune the scalable GP model parameters.

|       | pxADMM-GP     | cADMM-GP       |
|-------|---------------|----------------|
| RMSE  | 0.1368 meter  | 0.1353 meter   |
| CT    | 714 seconds   | 10838 seconds  |

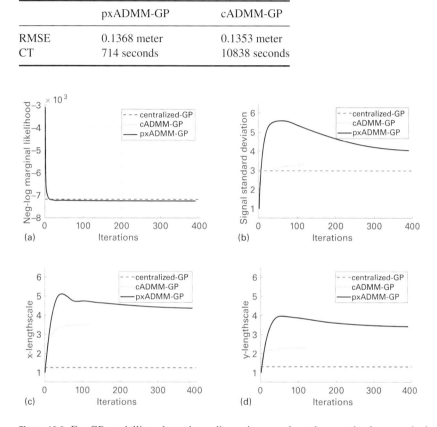

**Figure 10.9** For GP modelling along the $x$-dimension, we show the negative log-marginal likelihood functions (centralized formulation refers to Eq. (10.18), and distributed formulation refers to Eq. (10.21)) in (a); the ARD kernel parameters are estimated as a function of training iterations for the three input variables using pxADMM-GP and cADMM-GP in (b–d) for the signal variance $\sigma_s^2$, length-scale in $x$, and length-scale in $y$.

training algorithms are as follows: (a) pxADMM adopts regularization parameters $\rho_i = 500$ and $L_i = 5,000$, $\forall i$; and (b) cADMM adopts $\rho_i = 500$, for $i = 1, 2, 3$. We consider convergence when the difference in all optimization variables between two consequent iterations is within $10^{-3}$.

Due to space limitations, we only show the model training results for the $x$-dimension in Fig. 10.9. The two distributed ADMM schemes converge to different model parameter estimates compared with the global one. One reason is that the distributed schemes use different cost functions. Despite the difference in the parameter estimates, the corresponding negative log-marginal likelihood and the overall prediction RMSE results are fairly close, as shown in Table 10.2. We observed that the pxADMM scheme consumed the least computation time (CT) because the proximal step of the pxADMM is computationally less expensive than that of the exact optimization in the cADMM, although the latter often consumes fewer iterations toward a stationary point.

## 10.5.3   Case III: Scalable RL-based Load Balancing

In the third use case, we demonstrate a scalable RL-based load-balancing approach. Load balancing aims to automatically resolve the mismatch between network resource distribution and network traffic demand and is becoming increasingly important in data-driven wireless networks. Our recent work [6] presents a scalable DRL-based load-balancing framework based on the aforementioned MCC architecture and PSGD framework to handle the large-scale load-balancing problem in a scalable manner. The proposed framework dynamically groups the underlying cells into different clusters and perform in-cluster load-balancing with asynchronous parallel DRL. Each learning agent can autonomously accommodate its load-balancing policy to irregular network topologies and diversified user mobility patterns. The MDP of the DRL-based load-balancing model is defined as follows. The state includes the information of cell load distribution and user distribution. The action is the value of a handover parameter, which controls the user handover among adjacent cells. The reward signal is the inverse of the maximum load of all the cells, balancing the load distribution by alleviating the worst case.

The simulated scenario consists of multiple small base stations (SBSs) randomly distributed in a 300 m × 300 m area with 200 users randomly walking at the speed of 1 m/s to 10 m/s, each of which incurs a constant bit rate traffic demand. We compare the performance of (a) the centralized DRL model trained with a single device; (b) the scalable DRL model trained with three parallel devices; (c) the rule-based controller, which balances the load by executing predefined rules; (d) the Q-learning based controller, which does not employ DNNs for generalization; and (e) a plain baseline without performing any load-balancing operations. In particular, the scalable DRL is trained under the aforementioned PSGD framework, where multiple RL agents share the gradients for joint learning. We also use traditional

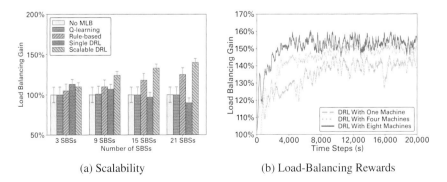

(a) Scalability                              (b) Load-Balancing Rewards

**Figure 10.10** Learning results of the scalable RL-based load balancing for handover management considered in Case III.

load balancing algorithms to generate high-quality training samples to guide the learning of our scalable DRL at the early stage. Figure 10.10(a) shows the load balancing performance for different numbers of SBSs, where the error bar reports one standard deviation as the uncertainty region to reflect the variance of the performance. Figure 10.10(b) reports the learning performance under 12 SBSs when using different numbers of local devices for parallel computing. The results show that the proposed scalable DRL-based method substantially outperforms all other methods in terms of load balancing.

## 10.6 Challenges and Future Directions

Despite the apparent opportunities, there are also challenges to apply scalable learning techniques to data-driven wireless networks. We summarize here a few promising yet challenging research directions.

First, it is well acknowledged that data collection is an indispensable step for data-driven management. However, the storage and transmission of wireless data requires a significant amount of storage and computational resources. This may put a huge burden on network entities and front-haul/back-haul links. Therefore, it is necessary to investigate data compression and recovery techniques in order to collect and transmit wireless data in a smart and efficient manner. Meanwhile, data collection also gives rise to the security and privacy concerns, which calls for the development of effective data encryption/decryption techniques.

Second, we believe that in the future, heavy computations will be distributed to edge devices and mobile devices to reduce the load of the central device as well as the latency. However, training an oversized model on such local devices with limited computation power is still impractical. Therefore, the development low-complexity or compact learning models under on-device constraints is in high demand.

Lastly, but most importantly, opening the black box of data-driven models to improve their interpretability is surely a major topic for future investigation. Such interpretability can help researchers understand how data-driven models work in order to design better scalable learning models and convince the customers. For example, the recently proposed GNN-based DRL [59] enables a mechanism for relational reasoning over structured representations. The learned internal representations can generalize to solve problems with more complex solutions than it had been trained within this task. Interpretability was also introduced in optimal deep kernel design for GP regression in [60].

## References

[1] X. Cheng et al., "Mobile big data: The fuel for data-driven wireless," *IEEE Internet of Things Journal*, vol. 4, no. 5, pp. 1489–1516, Oct. 2017.

[2] C. Jiang et al., "Machine learning paradigms for next-generation wireless networks," *IEEE Wireless Communications*, vol. 24, no. 2, pp. 98–105, Apr. 2017.

[3] W. Xu et al., "Data-cognition-empowered intelligent wireless networks: Data, utilities, cognition brain, and architecture," *IEEE Wireless Communications*, vol. 25, no. 1, pp. 56–63, Feb. 2018.

[4] A. Zappone, M. Di Renzo, and M. Debbah, "Wireless networks design in the era of deep learning: Model-based, AI-based, or both?" *IEEE Trans. on Communications*, vol. 67, no. 10, pp. 7331–7376, 2019.

[5] Y. Xu et al., "Wireless traffic prediction with scalable Gaussian process: Framework, algorithms, and verification," *IEEE Journal on Selected Areas in Communications*, vol. 37, no. 6, pp. 1291–1306, June 2019.

[6] Y. Xu et al., "Load balancing for ultradense networks: A deep reinforcement learning-based approach," *IEEE Internet of Things Journal*, vol. 6, no. 6, pp. 9399–9412, 2019.

[7] H. T. Dinh et al., "A survey of mobile cloud computing: architecture, applications, and approaches," *Wireless communications and mobile computing*, vol. 13, no. 18, pp. 1587–1611, Jan. 2013.

[8] N. Abbas et al., "Mobile edge computing: A survey," *IEEE Internet of Things Journal*, vol. 5, no. 1, pp. 450–465, Feb. 2018.

[9] W. Wang et al., "A survey on applications of model-free strategy learning in cognitive wireless networks," *IEEE Communications Surveys and Tutorials*, vol. 18, no. 3, pp. 1717–1757, 2016.

[10] S. Bi et al., "Wireless communications in the era of big data," *IEEE Communications Magazine*, vol. 53, no. 10, pp. 190–199, Oct. 2015.

[11] 3rd Generation Partnership Project (3GPP), "Service requirements for the 5G system," tech. specification (TS) 22.261, version 16.0.0, 2017.

[12] D. Kreutz et al., "Software-defined networking: A comprehensive survey," *Proceedings of the IEEE*, vol. 103, no. 1, pp. 14–76, Jan. 2015.

[13] S. Boyd et al., "Distributed optimization and statistical learning via the alternating direction method of multipliers," *Foundation Trends Machine Learning*, vol. 3, no. 1, pp. 1–122, Jan. 2011.

[14] M. Hong, Z.-Q. Luo, and M. Razaviyayn, "Convergence analysis of alternating direction method of multipliers for a family of nonconvex problems," *SIAM Journal on Optimization*, vol. 26, no. 1, pp. 337–364, Jan. 2016.

[15] M. Li et al., "Scaling distributed machine learning with the parameter server," in *Proc. USENIX Conf. on Operating Systems Design and Implementation*, pp. 583–598, 2014.

[16] N. S. Aybat et al., "Distributed linearized alternating direction method of multipliers for composite convex consensus optimization," *IEEE Trans. on Automatic Control*, vol. 63, no. 1, pp. 5–20, Jan. 2018.

[17] X. Lian et al., "Can decentralized algorithms outperform centralized algorithms? A case study for decentralized parallel stochastic gradient descent," in *Proc. Advances in Neural Information Processing Systems (NeurIPS)*, pp. 5336–5346, Dec. 2017.

[18] G. Zoubin, "Probabilistic machine learning and artificial intelligence," *Nature*, vol. 521, no. 1, pp. 452–459, May 2015.

[19] D. J. Mackay, "A practical Bayesian framework for backpropagation networks," *Neural Computation*, vol. 4, no. 3, pp. 448–472, May 1992.

[20] R. M. Neal, "Bayesian learning via stochastic dynamics," in *Proc. Advances in Neural Information Processing Systems (NeurIPS)*, pp. 475–482, Dec. 1993.

[21] C. E. Rasmussen and C. I. K. Williams, *Gaussian Processes for Machine Learning*, MIT Press, 2006.

[22] S. Theodoridis, *Machine Learning: A Bayesian and Optimization Perspective*, 2nd ed. Academic Press, 2020.

[23] A. G. Wilson and R. P. Adams, "Gaussian process kernels for pattern discovery and extrapolation," in *Proc. Int. Conf. on Machine Learning (ICML)*, pp. 1067–1075, July 2013.

[24] F. Yin et al., "Linear multiple low-rank kernel based stationary Gaussian processes regression for time series," *IEEE Trans. on Signal Processing*, vol. 68, pp. 5260–5275,, Sep. 2020.

[25] A. Xie et al., "Distributed Gaussian processes hyperparameter optimization for big data using proximal ADMM," *IEEE Signal Processing Letters*, vol. 26, no. 8, pp. 1197–1201, Aug. 2019.

[26] M. P. Deisenroth and J. W. Ng, "Distributed Gaussian processes," in *Proc. Int. Conf. on Machine Learning (ICML)*, pp. 1481–1490, Jul. 2015.

[27] K. Hornik, "Approximation capabilities of multilayer feedforward networks," *Neural Networks*, vol. 4, no. 2, pp. 251–257, 1991.

[28] A. Girard, "Approximate methods for propagation of uncertainty with Gaussian process model," Ph.D. dissertation, Univerity of Glasgow, 2004.

[29] C. Bishop, *Machine Learning and Pattern Recognition*. Springer, 2006.

[30] Y. Gal, "Uncertainty in deep learning," Ph.D. dissertation, University of Cambridge, 2016.

[31] F. Yin and F. Gunnarsson, "Distributed recursive Gaussian processes for RSS map applied to target tracking," *IEEE Journal of Selected Topics in Signal Processing*, vol. 11, no. 3, pp. 492–503, April 2017.

[32] R. S. Sutton and A. G. Barto, *Reinforcement Learning: An Introduction*. MIT press, 2018.

[33] V. Mnih et al., "Human-level control through deep reinforcement learning," *Nature*, vol. 518, no. 7540, pp. 529–533, Feb. 2015.

[34] H. Van Hasselt, A. Guez, and D. Silver, "Deep reinforcement learning with double Q-learning," in *Proc. AAAI Conf. on Artificial Intelligence*, pp. 2194–2200, Feb. 2016.

[35] Z. Wang et al., "Dueling network architectures for deep reinforcement learning," in *Proc. Int. Conf. on Machine Learning (ICML)*, pp. 1995–2003, June 2016.

[36] T. Schaul et al., "Prioritized experience replay," *arXiv preprint*, arXiv:1511.05952, Nov. 2015.

[37] T. P. Lillicrap et al., "Continuous control with deep reinforcement learning," in *Proc. Int. Conf. on Learning Representations (ICLR)*, pp. 1–14, May 2016.

[38] V. Mnih et al., "Asynchronous methods for deep reinforcement learning," in *Proc. Int. Conf. on Machine Learning (ICML)*, pp. 1928–1937, July 2016.

[39] J. Schulman et al., "Proximal policy optimization algorithms," in *Proc. Int. Conf. on Learning Representations (ICLR)*, pp. 1–8, May 2016.

[40] A. Nair et al., "Massively parallel methods for deep reinforcement learning," in Proc. Int. Conf. on Machine Learning (ICML), pp. 1–8, May 2015.

[41] K. Zhang et al., "Fully decentralized multi-agent reinforcement learning with networked agents," in *Proc. Int. Conf. on Machine Learning (ICML)*, pp. 9340–9371, July 2018.

[42] H.-T. Wai et al., "Multi-agent reinforcement learning via double averaging primal-dual optimization," in *Proc. Advances in Neural Information Processing Systems (NeurIPS)*, pp. 9649–9660, Dec. 2018.

[43] Y. Xu et al., "Voting-based multiagent reinforcement learning for intelligent IoT," in *IEEE Internet of Things Journal*, vol. 8, no. 4, pp. 2681–2693, Feb., 2021.

[44] L. Busoniu, R. Babuska, and B. De Schutter, "A comprehensive survey of multiagent reinforcement learning," *IEEE Trans. on Systems, Man, and Cybernetics*, vol. 38, no. 2, pp. 156–172, March 2008.

[45] W. R. Clements et al., "Estimating risk and uncertainty in deep reinforcement learning," *arXiv preprint*, arXiv:1905.09638, Feb. 2020.

[46] A. Tamar, D. D. Castro, and S. Mannor, "Learning the variance of the reward-to-go," *Journal of Machine Learning Research*, vol. 17, no. 13, pp. 1–36, Jan. 2016.

[47] W. Dabney et al., "Implicit quantile networks for distributional reinforcement learning," in *Proc. Int. Conf. on Machine Learning (ICML)*, pp. 1096–1105, July 2018.

[48] I. Osband et al., "Deep exploration via boot-strapped DQN," in *Proc. Advances in Neural Information Processing Systems (NIPS)*, pp. 4026–4034, Dec. 2016.

[49] A. Touati et al., "Randomized value functions via multiplicative normalizing flows," in *Proc. Int. Conf. on Uncertainty in Artificial Intelligence (UAI)*, pp. 422–432, June 2020.

[50] M. Kuss and C. E. Rasmussen, "Gaussian processes in reinforcement learning," in *Proc. Advances in Neural Information Processing systems (NIPS)*, pp. 751–758, Dec. 2004.

[51] S. Sæmundsson, K. Hofmann, and M. P. Deisenroth, "Meta reinforcement learning with latent variable Gaussian processes," in *Proc. Int. Conf. on Uncertainty in Artificial Intelligence (UAI)*, p. 642652, Aug. 2018.

[52] M. Chu et al., "Reinforcement learning-based multiaccess control and battery prediction with energy harvesting in iot systems," *IEEE Internet of Things Journal*, vol. 6, no. 2, pp. 2009–2020, April 2019.

[53] L. Xiao et al., "User-centric view of unmanned aerial vehicle transmission against smart attacks," *IEEE Trans. on Vehicular Technology*, vol. 67, no. 4, pp. 3420–3430, Dec. 2018.

[54] N. C. Luong et al., "Applications of deep reinforcement learning in communications and networking: A survey," *IEEE Communications Surveys and Tutorials*, vol. 21, no. 4, pp. 3133–3174, 2019.

[55] Z. Xiong et al., "Deep reinforcement learning for mobile 5G and beyond: Fundamentals, applications, and challenges," *IEEE Vehicular Technology Magazine*, vol. 14, no. 2, pp. 44–52, June 2019.

[56] Y. Xu et al., "High-accuracy wireless traffic prediction: A GP-based machine learning approach," in *Proc. IEEE Global Communications Conf. (GLOBECOM)*, pp. 1–6, 2017.

[57] R. Frigola, Y. Chen, and C. E. Rasmussen, "Variational Gaussian process state-space models," in *Proc. Advances in Neural Information Processing Systems (NeurIPS)*, pp. 3680–3688, 2014.

[58] Y. Zhao et al., "Cramer-Rao bounds for filtering based on Gaussian process state-space models," *IEEE Trans. on Signal Processing*, vol. 67, no. 23, pp. 5936–5951, Dec. 2019.

[59] P. W. Battaglia et al., "Relational inductive biases, deep learning, and graph networks," in *Proc. Int. Conf. on Learning Representations (ICLR)*, May 2019.

[60] Y. Dai et al., "An interpretable and sample efficient deep kernel for Gaussian process regression," in *Proc. Int. Conf. on Uncertainty in Artificial Intelligence (UAI)*, PMLR 124, pp. 759–768, 2020.

# 11 Capacity Estimation Using Machine Learning

Ziv Aharoni, Dor Tsur, Ziv Goldfeld, and Haim H. Permuter

## 11.1 Introduction

The capacity of a communication channel is the highest rate in which the transmitter can convey a message to a receiver with an arbitrary small probability of error. Characterizing the channel capacity is valuable for various reasons, from understanding the features of the considered communication channel or as a reference quantity for the design of codes.

Channel capacity is formulated as an optimization of a multiletter expression [1], or an optimization with respect to (w.r.t.) infinitely many variables. This poses a major practical difficulty since most optimization algorithms can only handle a finite number of variables. In some cases the multiletter expression can be reduced into a single-letter expression, or an optimization w.r.t. a finite number of variables. A canonical example where this happens is the discrete memoryless channel (DMC) whose capacity is expressed as the maximum of the single-letter input-output mutual information (MI) [2]. As such, the DMC capacity can be computed using convex optimization tools. Another case where the general capacity formula simplifies is the parallel additive white Gaussian noise (AWGN) channel subject to a common input average power constraint. For AWGN channels, capacity can be computed (and is achieved) by the water-filling algorithm [2].

For channels with memory, when the output depends on all previous inputs and outputs (rather than just the current input), solutions are available for some channels with discrete alphabet and for additive Gaussian noise (AGN) channels. For channels with discrete alphabets, the feedback capacity of finite state channels (FSCs) is be formulated as the optimal average reward of an infinite horizon Markov decision process (MDP). The average reward is then computed using dynamic programming (DP) algorithms; however, as the underlying MDP of channel capacity has continuous state and action spaces, DP algorithms are applied on a quantized representation of these spaces. This quantization causes the computational complexity to grow exponentially with the channel alphabet size. This makes DP effective only in the binary alphabet case. Evidently, this method was used to obtain various results on channels with binary alphabets [3–6]. As mentioned earlier, computable solutions are also available in the Gaussian case [2, 7–10], but these do not extent to other channel models. Therefore, the goal of this chapter is to provide an estimation algorithm of channel capacity that will perform well on discrete channels with nonbinary alphabet as well as for channels with continuous alphabets.

Our methods rely on representing channel capacity as an optimization of the directed information (DI) (between the channel input and output sequences) over the space input distributions. We focus on this representation since it unifies the capacity formulas for point-to-point channels with and without memory and with or without feedback (see Section 11.2.4). Since the DI formulation is a multiletter expression, we employ modern machine learning (ML) techniques to alleviate the practical difficulty of optimizing w.r.t. infinitely many variables. We distinguish between the cases of whether the channel transition kernel is known or not.[1] When it is known, the feedback capacity of FSCs is formulated as the average reward of an infinite horizon MDP. Hence the information rate of a fixed input distribution can be estimated (empirically) by averaging the rewards of the corresponding MDP. For Finding the capacity achieving input distribution, reinforcement learning (RL) algorithms[2] are used, in which NNs are use to approximate the value function and the policy of the underlying MDP. When the channel model is absent, we use NNs to directly approximate both the input distribution and the DI using only samples from the channel input and output. These approaches result in capacity estimation algorithms that are applicable for channels with memory. We describe the main ideas behind each method here, while deferring technical details to the respective sections.

Starting from the case of a known channel model, we present a method for estimating the feedback capacity of a FSC with memory using RL. In RL, one seeks to maximize cumulative rewards collected in a sequential decision-making environment. The computational efficiency of RL enables handling even MDPs with large state and action spaces. Leveraging this, RL is used to estimate the feedback capacity of FSCs with large alphabet size. The output of the RL algorithm sheds light on the properties of the optimal decision rule, which in our case, is the optimal input distribution of the channel. These insights can be converted into analytic, single-letter capacity expressions by solving corresponding lower and upper bounds. The efficiency of this method is demonstrated by analytically solving the feedback capacity of the well-known Ising channel for alphabet smaller and equal to eight. We also present a simple capacity-achieving coding scheme.

We then move to channels where the transition kernel is unknown. In this case, it is impossible to apply RL techniques as before due to the lack of a probabilistic model of the channel. Therefore, we leverage recent advance in ML to design a neural estimator of the DI from a finite sample of the input and output sequences of the channel (rather than rely on the MDP formulation to estimate the information rate

---

[1] The *channel transition kernel*, which we synonymize with *channel model* or just *channel*, refers to the conditional distribution of the current output given the current input as well as all past inputs and outputs, namely $P_{Y_i|Y^{i-1},X^i}$.

[2] The reader might consider approximate dynamic programming (ADP) algorithms as an alternative approach. That is since that Q-learning algorithms in RL are closely related to ADP. Nevertheless, policy optimization algorithms in RL were empirically more effective for channels with large alphabets and therefore RL is considered.

of a fixed input distribution). The DI neural estimator (DINE) uses the Donsker-
Varadhan variation form for KL divergences. DI is estimated by parametrizing the
DV function class by recurrent neural networks (RNNs) and approximating expected
values with sample means. Maximizing over the RNN parameters produces a provably consistent estimator of DI. With DINE at hand, the channel input distribution is
modeled by a generative RNN called the neural distribution transformer (NDT). The
NDT shapes a noise variable into samples drawn from the channel input distribution.
Passing these samples through the channel produces the input-output dataset needed
for DI estimation. We then run an optimization procedure that alternates between
estimating DI from the sample and maximizing the obtained value over the NDT
parameters (which is then sampled again, etc.). This method treats the channel as
a black box (solely to generate samples) and is applicable both when a feedback
link from the receiver to the transmitter is or is not present. Notably, the method
results not only in an estimate of the channel capacity, but also a generative model for
the input distribution that achieves this estimate. The capacity estimation algorithm
is demonstrated for scenarios with and without feedback by examining AGNs with
independent and identically distributed (i.i.d.) noise as well as for correlated noise.

## 11.2    Channel Capacity

The capacity of a point-to-point communication channel is given by input-output
mutual information (MI) maximized over all possible input distributions. In this section, an equivalent formulation of the capacity via DI is presented. As background
for subsequent sections, we first survey relations between MI and DI, while covering
basic properties of the latter.

### 11.2.1    Capacity of Discrete Memoryless Channel

A channel is *discrete* if the input alphabet $\mathcal{X}$ and output alphabet $\mathcal{Y}$ are finite. It is
said to be *memoryless* if the probability of the $i$th output symbol $Y_i$ given the $i$th input
symbol $X_i$ is conditionally independent of past inputs and outputs $\{(X_j, Y_j)\}_{j \leq i}$. As
established by Shannon in his landmark 1948 paper [11], the capacity omit of a DMC
is given by the following MI optimization objective

$$C := \sup_{P_X} I(X;Y), \tag{11.1}$$

where MI is defined by means of the KL divergence as

$$I(X;Y) := D_{\mathsf{KL}}\left(P_{X,Y} \| P_X \otimes P_Y\right), \tag{11.2}$$

with $P_X \otimes P_Y$ as the product of the marginal distributions, and $D_{\mathsf{KL}}(P\|Q) = \int \log \frac{P(x)}{Q(x)} \, dP(x)$. This single-letter characterization is a convex optimization problem
w.r.t. the input distribution $P_X$ [2].

## 11.2.2     Directed Information and Causal Conditioning

Originally proposed by Massey [12], DI quantifies the amount of information one stochastic process causally conveys about another. For two correlated processes $\mathbb{X} := \{X_i\}_{i=1}^{\infty}$ and $\mathbb{Y} := \{Y_i\}_{i=1}^{\infty}$, the DI from $X^n := (X_1, \ldots, X_n)$ to $Y^n := (Y_1, \ldots, Y_n)$ is

$$I\left(X^n \to Y^n\right) := \sum_{i=1}^{n} I\left(X^i; Y_i | Y^{i-1}\right), \tag{11.3}$$

where $I(A; B|C) := \int I(A; B|C = c) \, dP_C(c)$ is the conditional mutual information between $A$ and $B$ given $C$. A key property of DI is that it satisfies the law of information conservation [13]:

$$I\left(X^n \to Y^n\right) + I\left((0, Y^{n-1}) \to X^n\right) = I\left(X^n; Y^n\right), \tag{11.4}$$

where $(0, Y^{n-1})$ is the concatenation of a null value (e.g., the constant 0) with the $n-1$ first variables from $\mathbb{Y}$. Extending DI from finite-length sequences to the entire processes $\mathbb{X}$ and $\mathbb{Y}$, one obtains the *DI rate*

$$\mathcal{I}(\mathbb{X} \to \mathbb{Y}) := \lim_{n \to \infty} \frac{1}{n} I(X^n \to Y^n), \tag{11.5}$$

that is, the asymptotic per-letter DI. The DI rate exists whenever $\mathbb{X}$ and $\mathbb{Y}$ are jointly stationary.

Let $(X^n, Y^n)$ be a pair of continuous-valued, correlated random vectors jointly distributed according to $P_{X^n, Y^n}$, with probability density function (PDF) $p_{X^n, Y^n}$; we assume all marginal and conditional PDFs of $p_{X^n, Y^n}$ exist. Starting from the right-hand side (RHS) of Eq. (11.3), the DI from $X^n$ to $Y^n$ is given by

$$I(X^n \to Y^n) = h(Y^n) - h(Y^n \| X^n), \tag{11.6}$$

where $h(Y^n \| X^n) := \sum_{i=1}^{n} h(Y^i | X^i, Y^{i-1})$ is the differential entropy of $Y^n$ *causally conditioned* (CC) on $X^n$ [14]. Here $h(A|B)$ is the conditional differential entropy of $A$ given $B$, namely, for continuous variables $(A, B) \sim P_{A,B}$, we have $h(A|B) := \mathbb{E}_{P_{A,B}}\left[-\log p_{A|B}(A|B)\right]$, where $p_{A|B}$ is the conditional PDF. The term CC entropy reflects the conditioning on past and present values of the sequences only. DI can thus be interpreted as the reduction in uncertainty about $Y^n$ as a result of causally observing the elements of $X^n$.

CC entropy can be expressed similarly to (regular) conditional entropy as an expectation of a certain negative logarithm. To do so, we define the CC PDF of $Y^n$ given $X^n$ as

$$p_{Y^n \| X^n}\left(y^n \| x^n\right) := \prod_{i=1}^{n} p_{Y_i | Y^{i-1}, X^i}\left(y_i | y^{i-1}, x^i\right), \tag{11.7}$$

where $(x^n, y^n) \in \mathcal{X}^n \times \mathcal{Y}^n$, which again eliminates conditioning on future values.[3] It is straightforward to verify that $p_{Y^n \| X^n}(\cdot | x^n)$ is a valid PDF on $\mathcal{Y}^n$, for any $x^n \in \mathcal{X}^n$. The

---

[3] Compare $p_{Y^n \| X^n}$ to the conditional PDF $p_{X^n | Y^n}$, which factors as $p_{Y^n | X^n} = \prod_{i=1}^{n} p_{Y_i | Y^{i-1}, X^n}$.

induced CC probability distribution is denoted by $P_{Y^n \| X^n}$. CC can thus be interpreted as imposing the Markov relation $Y_i \leftrightarrow \left( Y^{i-1}, X^i \right) \leftrightarrow X_{i+1}^n$ on the original $(X^n, Y^n)$ pair. By definition of the CC PDF $p_{Y^n \| X^n}$, we obtain

$$
h\left( Y^n \| X^n \right) := \sum_{i=1}^{n} h\left( Y_i | Y^{i-1}, X^i \right) = \mathbb{E}_{P_{X^n, Y^n}}\left[ -\log p_{Y^n \| X^n}\left( Y^n \| X^n \right) \right]. \quad (11.8)
$$

### 11.2.3   Directed Information versus Mutual Information

Although intimately related, DI and MI differ in several key ways. While MI is symmetric, DI is generally not. Indeed, the MI $I(X^n; Y^n)$ is a measure of dependence between $X^n$ and $Y^n$, which is inherently a bidirectional relation. DI, on the other hand, disentangles the dependence between $X^n$ and $Y^n$ into, loosely speaking, directional information flows, as captured by the law of information conservation in Eq. (11.4). Consequently, while MI can be recovered from DI, the opposite is generally impossible. This implies that DI captures a finer granularity of dependence between the considered random sequences. The relation from Eq. (11.4) also implies the MI upper bounds DI, which is natural in light of the disentanglement of directional information interpretation.

---

**Example 11.1** We illustrate the relation and difference between MI and DI through the following binary example. Let $X^n$ be i.i.d. according to $\mathrm{Ber}(0.5)$ (Bernoulli Random Variable (RV) with $p = 0.5$) and define $Y_i = X_{i+1}$, for all $i = 1, \ldots, n$. Since the current $Y_i$ reveals information about future $X_j$'s (the opposite is not true), the causal information flow in this example is unidirectional from $Y^n$ to $X^n$. This is captured by DI since $I(X^n \to Y^n) = 0$ while $I\left( (0, Y^{n-1}) \to X^n \right) = n - 1$. MI, on the other hand, in ignorant to directionality and we have $I(X^n; Y^n) = n - 1$.

---

### 11.2.4   Capacity Characterization by the Directed Information

Similar to the capacity of a DMC, the capacity of channels with memory is also characterized by an optimization objective of an information measure. Specifically, the feedforward capacity (i.e., in the absence of feedback) is given in terms of the optimized average MI [1]

$$
C_{\mathrm{FF}} = \lim_{n \to \infty} \sup_{P_{X^n}} \frac{1}{n} I(X^n; Y^n), \quad (11.9)
$$

where $X^n, Y^n$ are the channel input and output sequences of length $n$, respectively. When feedback from the receiver to the transmitter is present, the capacity is given by [12]

$$
C_{\mathrm{FB}} = \lim_{n \to \infty} \sup_{P_{X^n \| Y^{n-1}}} \frac{1}{n} I(X^n \to Y^n), \quad (11.10)
$$

where $P_{X^n \| Y^{n-1}}$ is the distribution of $X^n$ CC on $Y^{n-1}$ (see [14], [15] for further details). As shown by Massey [12], if one chooses to ignore the available feedback link (i.e., optimize over $P_{X^n}$ rather than $P_{X^n \| Y^{n-1}}$), then Eqs. (11.9) and (11.10) coincide. Further restricting optimization to memoryless (product) input distributions, $P_{X^n} = P_X^{\otimes n}$, then Eq. (11.10) recovers Eq. (11.1). Thus, DI provides a unified framework for representing both feedforward and feedback capacity, whether the channel has memory or not.

### 11.2.5 Capacity Evaluation: Challenges and Overview of Proposed Techniques

Tools from convex optimization [2] and dynamic programming [4–6, 16, 17] are often used to solve the optimization problems defining $C_{FF}$ and $C_{FB}$. However, these tools are limited to single-letter capacity expressions [2], or for restricted scenarios (such as a binary channel with feedback) [4–6, 16, 17]. For the case where the channel has a continuous alphabet, solutions of the capacity are scarce and largely limited to the Gaussian case [2, 10]. In light of this, our goal is to develop a generic technique for (approximately) evaluating capacity under only minimal assumptions on the considered communication channel.

The capacity computation problem can by divided into two complementary tasks: evaluating the DI between two stochastic processes and optimizing it over the input processes distribution. In the following sections, we handle these tasks using ML techniques. We start from the DI optimization aspect and develop an algorithm based on RL. The algorithm requires knowledge of the channel model and availability of a feedback link. Having that, we consider the case where no channel model is available (and hence the aforementioned RL algorithm is no longer feasible). We develop an estimator of the DI rate between two stochastic processes, which enables evaluating the transmission rate for a given input distribution. The estimator is coupled with a generative modeling algorithm that enables optimizing over input distributions. This method only uses samples from the channel input and output processes and works whenever they are ergodic and stationary.

### 11.3 Capacity Estimation using Reinforcement Learning

Consider a communication channel with a known transition kernel. Under certain assumptions on the channel and the communication scenario, the capacity can be formulated as the average reward of an infinite-horizon MDP. This formulation has been used before for channels with a binary alphabet using DP algorithms. However, the size of the (quantized) state and action spaces of the underlying MDP grows exponentially with the channel alphabet size, which turns this method to be intractable for alphabets strictly larger than two. Leveraging the computational efficiency of RL for MDPs with large state and action spaces, we circumvent the limitations of DP and are able to compute the capacity of channels with large alphabets.

## 11.3.1 Feedback Capacity of Unifilar Finite-State Channels

A FSC is defined by the triplet $(\mathcal{X} \times \mathcal{S}, P_{Y, S'|X, S}, \mathcal{Y} \times \mathcal{S})$, where $X$ is the channel input, $Y$ is the channel output, $S$ is the channel state at the beginning of the transmission, and $S'$ is the channel state at the end of the transmission. Throughout this section it is assumed that the cardinalities $\mathcal{X}$, $\mathcal{Y}$, and $\mathcal{S}$ are finite. Furthermore, the channel is assumed to be a FSC; that is, for each time $t$, we have

$$P_{Y_t, S_t | X^t, S^{t-1}, Y^{t-1}} = P_{Y_t | X_t, S_{t-1}} P_{S_t | X_t, S_{t-1}, Y_t}. \tag{11.11}$$

The transition probabilities are specified by the conditional probability mass function (PMF) $p_{Y_t, S_t | X_t, S_{t-1}}(y_t, s_t | x_t, s_{t-1})$; when the subscripts of a PMF are uppercase version of the arguments, we omit them and write $p(y_t, s_t | x_t, s_{t-1})$. A FSC is called *unifilar* if the new channel state, $s_t$, is a time-invariant function $s_t = f_s(x_t, s_{t-1}, y_t)$. The feedback capacity of a strongly connected[4] unifilar FSC is stated next.

THEOREM 11.1  *[3, Thm 1] The feedback capacity of a strongly connected unifilar finite state channel, where the initial state $s_0$ is available for both encoder and decoder, is given by*

$$C_{\text{FB}} = \lim_{N \to \infty} \sup_{\{p_{X_t | S_{t-1}, Y^{t-1}}\}_{t=1}^N} \frac{1}{N} \sum_{i=1}^N I(X_i, S_{i-1}; Y_i | Y^{i-1}). \tag{11.12}$$

This optimization objective is a multiletter expression. Following [3], we formulate it as a MDP. Specifically, the feedback capacity can be viewed as the solution to an infinite horizon average reward MDP. The state is the probability vector $z_t := p_{S_t | Y^t}(\cdot | y^t)$, the action is the transition matrix $u_t := p_{X_t | S_{t-1}, Y^{t-1}}(\cdot | \cdot, y^{t-1})$, the instantaneous reward is $r_t := I(X_t, S_{t-1}; Y_t | Y^{t-1} = y^{t-1})$, and the next state is given by

$$z_t(z) \propto \sum_{x_t, s_{t-1}} z_{t-1}(s_{t-1}) u_t(x_t, s_{t-1}) p(y_t | x_t, s_{t-1}) \mathbb{1}[z = f_s(x_t, s_{t-1}, y_t)], \tag{11.13}$$

where $\mathbb{1}[\mathcal{A}]$ denotes the indicator of an event $\mathcal{A}$, and $z \in \{0, \dots, |\mathcal{X}| - 1\}$. The MDP formulation is summarized in Table 11.1.

**Table 11.1.** MDP formulation of the feedback capacity

| | |
|---|---|
| state | $p_{S_t | Y^t}(\cdot | y^t)$ |
| action | $p_{X_t | S_{t-1}, Y^{t-1}}(\cdot | \cdot, y^{t-1})$ |
| reward | $I(X_t, S_{t-1}; Y_t | Y^{t-1} = y^{t-1})$ |
| disturbance | $y_t$ |

---

[4] Strongly connected, as defined in [3, Def. 2], means that for any $s \in \mathcal{S}$, there exists an input distribution such that $s$ is reachable from any other state $s' \in \mathcal{S}$ with a finite amount of steps.

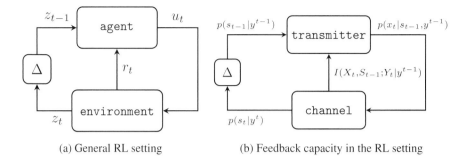

(a) General RL setting                    (b) Feedback capacity in the RL setting

**Figure 11.1**  A description of (a) the general RL setting and (b) the feedback capacity formulated in the RL setting.

## 11.3.2  The Reinforcement Learning Setting

The RL setting comprises an agent that interacts with a state-dependent environment whose input is an action, and the output is a state and a reward. Formally, at time $t$, the environment state is $z_{t-1}$, and an action $u_t \in \mathcal{U}$ is chosen by the agent. Then, a reward $r_t \in \mathcal{R}$ and a new state $z_t \in \mathcal{Z}$ are generated by the environment and are made available to the agent (Fig. 11.1). Denoting the corresponding RVs by uppercase letters with the appropriate subscripts, the environment is assumed to satisfy the Markov property

$$P_{R_t, Z_t | Z^{t-1}, U^t, R^{t-1}} = P_{R_t, Z_t | Z_{t-1}, U_t}. \tag{11.14}$$

The RHS is further assumed to be governed by a time-invariant distribution $P(r_t, z_t | z_{t-1}, u_t)$. The agent's *policy* is a sequence of actions $\pi := \{u_1, u_2, \dots\}$.

The objective of the agent is to choose a policy that yields maximal accumulated rewards across a predetermined horizon $h \in \mathbb{N}$. Here, we consider the *infinite-horizon average-reward* setting, where the agent-environment interaction lasts forever, and the goal of the agent is to maximize the average reward gained during the interaction. The average reward of the agent is thus given by

$$\rho(\pi) := \lim_{h \to \infty} \frac{1}{h} \sum_{t=1}^{h} \mathbb{E}_\pi[R_t | Z_0], \tag{11.15}$$

which depends on the initial state $Z_0$ and on the policy $\pi$.

The *differential return* of the agent is defined by

$$G_t := R_t - \rho(\pi) + R_{t+1} - \rho(\pi) + R_{t+1} - \rho(\pi) + \cdots. \tag{11.16}$$

Accordingly, the state-action value function $Q_\pi(z, u)$ is defined as

$$Q_\pi(z, u) = \mathbb{E}_\pi \left[ G_t | Z_{t-1} = z, U_t = u \right]. \tag{11.17}$$

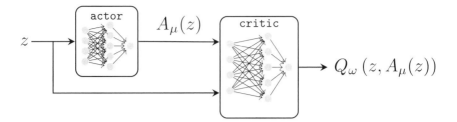

**Figure 11.2** Actor and critic networks. The actor network comprises a NN that maps the state $z$ to an action $A_\mu(z)$. The critic NN maps the tuple $(z, u)$ to an estimate of expected future cumulative rewards.

In words, $Q_\pi(z, u)$ is the expected rewards as a result of taking the action $u$ at state $z$ and thereafter following policy $\pi$. Using Eq. (11.14), one can write Eq. (11.17) as the sum of the immediate and future rewards:

$$Q_\pi(z, u) = \mathbb{E}\left[R_t | Z_{t-1} = z, U_t = u\right] - \rho(\pi) + \mathbb{E}_\pi\left[Q_\pi(Z_t, U_{t+1}) | Z_{t-1} = z, U_t = u\right].$$
(11.18)

This is the well-known Bellman equation [18], which is essential for estimating the function $Q_\pi$.

Given an estimate of the state-action value, the Bellman equation can be used to improve a given policy. Specifically, for each state $z \in \mathcal{Z}$, the current action $\pi(z)$ can be improved to the action $\pi'(z)$ by choosing

$$\pi'(z) = \arg\max_u Q_\pi(z, u).$$
(11.19)

The RL approach parametrizes the functions $Q_\pi(z, u)$ and $\pi(z)$ performs optimization over parameter space. The *actor* is defined by $A_\mu(z)$, a parametric model of $\pi(z)$, whose parameters are $\mu$. The *critic* is defined by $Q_\omega(z, u)$, a parametric model with parameters $\omega$ of the state-action value function that corresponds to the policy $A_\mu(z)$. In deep RL, the actor and critic are modeled by NNs, as shown in Fig. 11.2.

## 11.3.3    Formulating the Feedback Capacity as Reinforcement Learning

The main goal of this section is to relate the feedback capacity of unifilar FSC to the corresponding RL problem. Unlike the general RL setup, the obtained formulation has a fully known environment (which is a consequence of the known channel model). This enables us to employ two RL algorithms to compute the feedback capacity. The first is the deep deterministic policy gradient (DDPG) [19] algorithm with improvements based on the environment knowledge. The second algorithm is policy optimization through unfolding (POU) that exploits environment knowledge to optimize the DI.

The MDP formulation of the feedback capacity is used to convert the multi-letter expression from Theorem 11.1 into a RL problem (see Fig. 11.1). Recall that under this formulation the state, action, and reward are given by $z_{t-1} = p_{S_{t-1}|Y^{t-1}}(\cdot|y^{t-1})$, $u_t = p_{X_t|S_{t-1},Z_{t-1}}(\cdot|\cdot,z_{t-1})$, and $r_t = I(X_t,S_{t-1};Y_t|z_{t-1},u_t)$, respectively. The next state is given in Eq. (11.13). We introduce the shorthands

$$r_t := g(z_{t-1},u_t)$$
$$z_t := f(z_{t-1},u_t,w_t), \qquad (11.20)$$

where $g, f$ are the reward and next state function, respectively. Here, $w_t$ denotes the disturbance, which in our case is the channel output $y_t$.

## 11.3.4    Deep Deterministic Policy Gradient Algorithm

The DDPG algorithm [19] is a deep RL algorithm for deterministic policies and continuous state and action spaces. This corresponds to the MDP formulation of the feedback capacity. The training procedure comprises $M$ episodes, each containing $T$ sequential steps. A single step of the algorithm comprises two parallel *operations*: (a) collecting experience from the environment and (b) improving the actor and critic networks by training them over the accumulated data. We elaborate on these operations below.

**Operation 1: Collecting experience.** Given the current state $z_{t-1}$, the agent chooses an action $u_t$ according to an exploration policy. Here, the action is a probability distribution, and therefore exploration is performed by adding noise to the last hidden layer of the actor's network (as opposed to adding noise to the network output as done in [19]). This variation is introduced to account for the normalization constraint of the input distribution. We denote a noisy action at state $z_{t-1}$ by $A_\mu(z_{t-1};N_t)$, where $\{N_t\}$ is an i.i.d. Gaussian process with $N_t \sim \mathcal{N}(0,\sigma^2)$. After taking the action $A_\mu(z_{t-1};N_t)$, the agent observes the incurred reward $r_t$ and the next state $z_t$. Subsequently, the *transition* tuple

$$\tau = (z_{t-1},u_t,r_t,z_t)$$

is stored in a *replay buffer* (a bank of experience) that is used to improve the actor and critic networks in the second operation.

**Operation 2: Improving actor and critic networks.** We start by drawing uniformly at random $N$ transitions $\{\tau_i\}_{i=1}^N$ from the replay buffer. Then, for each transition, the target $b_i$ is computed based on the RHS of Eq. (11.18) as

$$b_i = r_i - \rho_{\text{MC}} + Q'_\omega(z_i, A'_\mu(z_i)), \quad i = 1,\ldots,N. \qquad (11.21)$$

The target is the sampled estimate of future rewards; from numerical considerations it is computed using a moving average of $Q_\omega, A_\mu$, which are the target networks,[5] $Q'_\omega, A'_\mu$. The term $\rho_{\text{MC}}$ is the estimate of the average reward, which is updated at the beginning of every episode by a Monte Carlo evaluation of $T_{\text{MC}}$ steps by $\frac{1}{T_{\text{MC}}}\sum_{t=0}^{T_{\text{MC}}-1} r_{t+1}$. Then, we minimize, over the parameters of the critic network $\omega$, the following objective:

---

[5] Target networks are used to make the algorithm numerically stable.

$$L(\omega) = \frac{1}{N} \sum_{i=1}^{N} \left[ Q_\omega \left( z_{i-1}, A_\mu(z_{i-1}) \right) - b_i \right]^2. \tag{11.22}$$

This update rule trains the critic to comply with the Bellman equation Eq. (11.18).

Afterwards, we train the actor to maximize the critic's estimate of future cumulative rewards. Accordingly, actor's parameters are updated as

$$\mu := \mu + \eta \frac{1}{N} \sum_{i=1}^{N} \nabla_a Q_\omega (z_{i-1}, a) \, |_{a=A_\mu(z_{i-1})} \nabla_\mu A_\mu (z_{i-1}), \tag{11.23}$$

where $\eta$ is a learning rate and the symbol $:=$ denotes the assignment operator. Finally, the agent updates its current state to be $z_t$ and moves to the next time step.

To conclude, the algorithm alternates between improving the critic's estimation of future cumulative rewards and training the actor to choose actions that maximize the critic's estimation. The algorithm is given in Algorithm 11.1 and its workflow is depicted in Fig. 11.3.

**Improvements**

We propose two improvements for the DDPG algorithm. First, we leverage the known environment to reduce the variance in estimating $Q_\pi$ by replacing sample means with expected values. Namely, instead of calculating $b_i$ as done in Eq. (11.21) we compute the expectation over all possible next states by

$$b_i = r_i - \rho_{\text{MC}} + \sum_{y \in \mathcal{Y}} p(y|z_{i-1}, u_i) Q'_\omega (z_i, A'_\mu(z_i)), \tag{11.24}$$

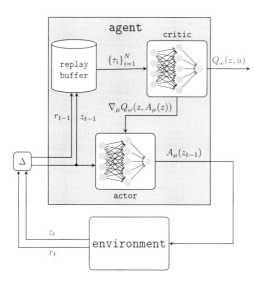

**Figure 11.3** DDPG algorithm workflow: At each time step $t$, the agent samples a transition from the environment using an noisy policy and stores the transition in the replay buffer. Simultaneously, $N$ past transitions $\{\tau_i\}_{i=1}^{N}$ are drawn from the replay buffer and used to update the critic and actor NN according to Eq. (11.22) and (11.23), respectively.

---

**Algorithm 11.1** DDPG algorithm for feedback capacity of unifilar FSC.

---

Initialize the critic $Q_\omega$ and actor $A_\mu$ networks with random weights $\omega$ and $\mu$, respectively

Initialize target networks $Q'_\omega$ and $A'_\mu$ with weights $\omega' \leftarrow \omega$, $\mu' \leftarrow \mu$

Initialize an empty replay buffer $R$

Initialize moving average parameter $\alpha$

**for** episode $= 1{:}M$ **do**

    Initialize a random process $\{N_t\}$ for action exploration

    Set $\rho_{\text{MC}} = \frac{1}{T_{\text{MC}}} \sum_{t=0}^{T_{\text{MC}}-1} r_{t+1}$ by a Monte Carlo evaluation of the average reward of $A_\mu$

    Randomize initial state $z_0$ from the $(|\mathcal{X}| - 1)$-simplex

    **for** step $= 1{:}T$ **do**

        Select noisy action $u_t = A_\mu(z_{t-1}; N_t)$

        Execute action $u_t$ and observe $(r_t, z_t)$

        Store transition $(z_{t-1}, u_t, r_t, z_t)$ in $R$

        Sample a random batch of $N$ transitions $\{(z_{i-1}, u_i, r_i, z_i)\}_{i=1}^N$ from $R$

        Set $b_i = r_i - \rho_{\text{MC}} + Q'_\omega(z_i, A'_\mu(z_i))$

        Update critic by minimizing the loss:

$$L(\omega) = \frac{1}{N} \sum_{i=1}^N \left[ Q_\omega\left(z_{i-1}, A_\mu(z_{i-1})\right) - b_i \right]^2$$

        Update the actor policy using the sampled policy gradient:

$$\frac{1}{N} \sum_{i=1}^N \nabla_a Q_\omega\left(z_{i-1}, a\right)\big|_{a=A_\mu(z_{i-1})} \nabla_\mu A_\mu\left(z_{i-1}\right)$$

        Update the target networks:

$$\omega' \leftarrow \alpha\omega + (1 - \alpha)\omega'$$

$$\mu' \leftarrow \alpha\mu + (1 - \alpha)\mu'$$

    **end for**

    **end for**

    **return** $\rho_{\text{MC}} = \frac{1}{T_{\text{MC}}} \sum_{t=0}^{T_{\text{MC}}-1} r_{t+1}$

---

where $z_i = f(z_{i-1}, u_i, y)$. This is possible since the disturbance (the channel output) has finite cardinality.

The second improvement is a variant of importance sampling [20]. This is essential since there are states that are rarely visited, and therefore the standard technique rarely uses them to improve the policy. For this purpose, we modify the replay buffer to store transitions as clusters. Each time a new transition arrives at the buffer, its max-norm distance with all cluster centers is calculated. The distance from the closest cluster is compared with a threshold (typically $\sim 0.1$). When the distance is smaller than the threshold, the transition is stored in the corresponding cluster; otherwise, a new cluster, which contains the new transition, is added. For sampling, instead of drawing transitions uniformly over the entire buffer, we first sample uniformly from the clusters, and then sample uniformly from within the sampled cluster. This modification increases the probability that *rare states* will be drawn from the replay buffer. This improves the value function estimates for these rare states and consequently yields better policies for them.

**Implementation**

We model $Q_\omega(z, u)$, $A_\mu(z)$ with two NNs, each of which are composed of three fully connected hidden layers of 300 units separated by a batch normalization layer. The actor network input is the state $z$, and its output is a matrix $A_\mu(z) \in \mathbb{R}^{|\mathcal{S}| \times |\mathcal{X}|}$ such that $A_\mu(z)\mathbf{1} = \mathbf{1}$. The critic network input is the tuple $(z, A_\mu(z))$, and its output is a estimate of the cumulative future rewards. In our experiments, we trained the networks for $M = 10^4$ episodes, each comprising $T = 500$ steps. The Monte Carlo evaluation length of average reward is $T_{\text{MC}} = 10^8$. For exploration, we added Gaussian noise with zero mean and variance $\sigma^2 = 0.05$ to the last layer of the actor network. The implementation details are available on Github.[6]

## 11.3.5    Policy Optimization by Unfolding

The POU algorithm utilizes the knowledge of the RL environment to optimize the policy without estimating the value function. That is, we optimize the average of consecutive rewards directly. This is done by using the reward function $g$ and the next state function $f$ (see Eq. (11.20)) to define a mapping between an initial MDP state and the average of the consecutive $n$ rewards. The mapping is finally used as an objective to optimize the policy.

Let us denote the policy-dependent reward function by

$$R_\mu(z) = g(z, A_\mu(z)).  \tag{11.25}$$

This function depends exclusively on $z$ since the policy $\mu$ is deterministic. Consequently, we define the average reward over $n$ consecutive time steps for an initial MDP state $z_0$ by

$$R_\mu^n(z_0) = \frac{1}{n} \sum_{t=1}^n \mathbb{E}\big[R_\mu(Z_{t-1})\big]$$

$$= \frac{1}{n} \left[ R_\mu(z_0) + \sum_{t=2}^n \mathbb{E}\big[R_\mu(Z_{t-1})\big] \right],  \tag{11.26}$$

---

[6] https://github.com/zivaharoni/capacity-rl

where $Z_t = f(Z_{t-1}, A_\mu(Z_{t-1}), W_t)$. This expectation is implicitly taken w.r.t. $P_{W_t | Z_{t-1}, A_\mu(Z_{t-1})}$. This is since the disturbance $W_t$ is conditionally independent of the past given the previous MDP state and the action $Z_{t-1}, A_\mu(Z_{t-1})$, respectively.

The choice of the interaction length parameter $n$ directly affects the performance of the optimized policy. As $n$ increases the policy is optimized over more rewards in future steps rather than over immediate rewards. For instance, choosing $n = 1$ translates to optimizing the immediate reward, which consequently yields a greedy policy. As shown later (see, e.g., Fig. 11.8), interactions over a relatively small $n$ (e.g., $n = 20$) are sufficient to achieve policies with long-term high performance. Smaller $n$ values are preferable from a computational standpoint. Indeed, the number of possible MDP states over $n$ interaction steps grows exponentially as $|\mathcal{Y}|^n$ (recall the disturbance in our case is the channel output $Y$).

To resolve this issue, the POU algorithm proposes a simple, yet efficient, method to *unfold* the interaction with the environment. Given a policy $A_\mu$ and an initial state $z_0$, we sample $n$ MDP states and rewards consecutively according the following law:

$$
\begin{aligned}
r_t &= R_\mu(z_{t-1}), \\
W_t &\sim P_{W|Z,U}(\cdot | Z = z_{t-1}, U = A_\mu(z_{t-1})), \\
z_t &= f(z_{t-1}, \mu(z_{t-1}), w_t),
\end{aligned}
\tag{11.27}
$$

where the disturbance $W_t$ is sampled conditioned on the previous MDP state and the action $A_\mu(Z_{t-1})$. Note that this law is dictated by the RL environment and the chosen policy and is independent of the planning horizon $n$. For a single time step $t$, the law in Eq. (11.27) describes a single step where the agent interacts with the environment, as shown in Fig. 11.4. The interaction with the environment for $n$ consecutive steps is shown in Fig. 11.5.

After applying Eq. (11.27) $n$ times, the disturbance sequence $(w_1, \ldots, w_{n-1}, w_n)$ is sampled, which gives rise to a deterministic, differentiable mapping between $z_0$ and the average reward. Thus, the gradient of Eq. (11.26) can be estimated by applying [21, Theorem 1], which yields the following unbiased estimate of the gradient:

$$
\nabla_\mu R_\mu^n(z_0) \approx \nabla_\mu \left[ \frac{1}{n} \sum_{t=1}^n \mathbb{E}\left[R_\mu(Z_{t-1}) - c\right] \right] = \frac{1}{n} \sum_{t=1}^n \mathbb{E}\left[ \frac{\partial}{\partial \mu} R_\mu(Z_{t-1}) \right]
$$
$$
+ \frac{1}{n} \sum_{t=2}^n \mathbb{E}\left[ \frac{\partial}{\partial \mu} \log\left(P_{W|Z,U}(W_t | Z_{t-1}, \mu(Z_{t-1}))\right) \left[ \sum_{s=t}^n R_\mu(Z_s) - c \right] \right],
\tag{11.28}
$$

where $c \in \mathbb{R}$ is a constant that is used as a baseline.

Then, we update the policy $A_\mu$ with the standard gradient ascent update as

$$
\mu := \mu + \eta \nabla_\mu R_\mu^n(z_0)
\tag{11.29}
$$

where $\eta$ is the step size. This procedure is repeated using the last state $z_n$ as the initial state of the next consecutive $n$ steps. See Algorithm 11.2 for details.

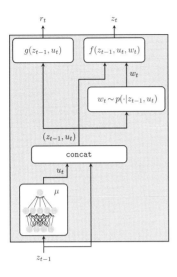

**Figure 11.4** A single step of the environment: The input of the block is the current RL state $z_{t-1}$ and the outputs are the immediate reward $r_t$ and the next sampled state $z_t$. Initially, the block uses the actor to construct the tuple $(z_{t-1}, u_t)$. Afterward, it samples the disturbance from $w_t \sim p(\cdot|z_{t-1}, u_t)$ and, finally, uses $g$ and $f$ to compute the reward and the next state, respectively.

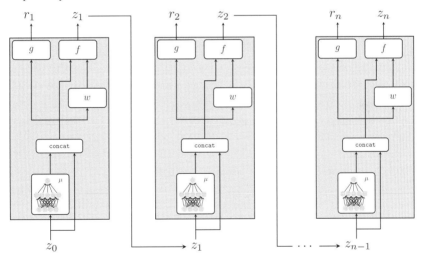

**Figure 11.5** The interaction with the channel unrolled across subsequent time steps. The weights of the actor network are shared across time steps.

## Implementation

The actor network is implemented exactly as described in Section 11.3.4. We train the actor for $M = 10^3$ episodes, each containing $T = 10^2$ consecutive $n$ blocks. Each block was chosen to be of length $n = 20$. The Monte Carlo evaluation length of average reward is $T_{\text{MC}} = 10^8$. For exploration, we use dropout [22] on the actor network throughout training. The implementation details are available on Github.[7]

---

[7] https://github.com/zivaharoni/capacity-rl-po

---

**Algorithm 11.2** POU algorithm for feedback capacity of unifilar FSC

Initialize actor $A_\mu$ with random weights $\mu$

Initialize learning rate $\eta$.

**for** episode = 1:M **do**

 Evaluate $\rho_{MC} = \frac{1}{T_{MC}} \sum_{t=0}^{T_{MC}-1} r_{t+1}$

 Sample $z_0$ uniformly from $(|\mathcal{X}| - 1)$-simplex

 **for** $t = 1 : T$ **do**

  Conditioned on $z_0$ and $A_\mu$, sample $(w_1, \ldots, w_{n-1})$ according to (11.27)

  Update the actor parameters using gradient ascent

$$\mu = \mu + \eta \left\{ \frac{1}{n} \sum_{t=1}^n \frac{\partial}{\partial \mu} R_\mu(z_{t-1}) + \frac{1}{n} \sum_{t=2}^n \frac{\partial}{\partial \mu} \right.$$
$$\left. \log\left(P_{W|Z,U}(w_t|z_{t-1}, \mu(z_{t-1}))\right) \left[ \sum_{s=t}^n \left(R_\mu(z_s) - \rho_{MC}\right) \right] \right\}$$

  Update the initial state $z_0 = z_n$

 **end for**

**end for**

**return** $\rho_{MC} = \frac{1}{T_{MC}} \sum_{t=0}^{T_{MC}-1} r_{t+1}$

---

## 11.3.6 Case Study: Ising Channel

We demonstrate the RL method developed in this section on the Ising channel [23]. Notably, beyond simply obtaining the numeral capacity value, the method enables an analytic capacity characterization. We start by a general description of how numerical results can be leveraged to obtain analytic ones. Then, we apply the ideas to the Ising channel example.

### Extracting Structure of Optimal Solutions

As part of the output of the RL algorithm, we obtain the actor, a parametric model of the input distribution of the channel. This network is used to obtain the structure of the solution via the following procedure. First, the actor is used for a Monte Carlo evaluation of the communication rate. During this evaluation, the MDP states and the channel outputs are recorded. These states are then clustered using common techniques, such as the k-means algorithm [24]. For instance, in Fig. 11.6 the MDP state histogram of the Ising channel with $|\mathcal{X}| = 3$ is shown, and it is evident that the estimated solution has only six discrete states.

Having the empirical results, the sequence of MDP states $\{Z_i\}_{i=1}^n$ ($n$ being the number of Monte Carlo iterations) is converted into a sequence of auxiliary RVs $\{Q_i\}_{i=1}^n$ with a discrete alphabet $\mathcal{Q}$, where each $q \in \mathcal{Q}$ corresponds to a node in a Q-graph [25]. The transitions between nodes are determined uniquely[8] by the channel outputs. The corresponding test distribution $P_{Y_i|Q_{i-1}} = T_{Y|Q}$ is estimated by counting

---

[8] The disturbance is the only randomness of the transition between RL states.

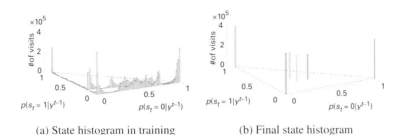

(a) State histogram in training       (b) Final state histogram

**Figure 11.6** State histogram of the policy as learnt by RL. The histogram is generated by a Monte Carlo evaluation of the policy: (a) histogram of the policy after 1,000 training iterations and (b) histogram of the policy after convergence.

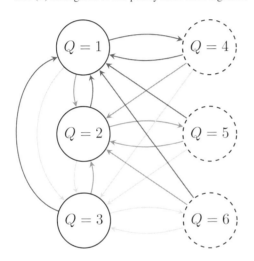

**Figure 11.7** Q-graph showing the transitions between states as a function of the channel's output. All edges going into states $Q = 1, Q = 2$, and $Q = 3$ correspond to $Y = 0, Y = 1$, and $Y = 2$, respectively. States with solid lines and dashed lines indicate whether the channel state is known or unknown to the decoder, respectively.

the channel output frequency at every Q-graph node. The Q-graph for $|\mathcal{X}| = 3$ is shown in Fig. 11.7.

## Upper Bound via Extracted Structure

The upper bound is derived by plugging the Q-graph and $T_{Y|Q}$ into the duality bound for unifilar FSCs with feedback, as presented in [26]. The duality bound is given in the following theorem.

THEOREM 11.2 *[26, Theorem 4] For any choice of Q-graph and test distribution $T_{Y|Q}$, the feedback capacity of a strongly connected unifilar FSC is bounded by*

$$C_{\text{FB}} \leq \lim_{n \to \infty} \max_{f(x^n \| y^n)} \max_{s_0, q_0} \frac{1}{n} \sum_{i=1}^{n} \mathbb{E}\left[ D_{\text{KL}}\left( P_{Y|X,S}\left( \cdot | x_i, S_{i-1} \right) \| T_{Y|Q}\left( \cdot | Q_{i-1} \right) \right) \right],$$

(11.30)

*where $f(x^n \| y^n) := \prod_{i=1}^{n} \mathbb{1}\left[ x_i = f_i\left( x^{i-1}, y^{i-1} \right) \right]$ stands for the causal conditioning of deterministic functions.*

**Table 11.2.** MDP formulation of the duality upper bound

| state | $(s_{t-1}, q_{t-1})$ |
|---|---|
| action | $x_t$ |
| reward | $D_{\mathsf{KL}}\left(P_{Y|X=x_t, S=s_{t-1}} \| T_{Y|Q=q_{t-1}}\right)$ |
| disturbance | $y_t$ |

The upper bound defines an infinite horizon average reward MDP, as described in Table 11.2. Unlike the MDP of the feedback capacity from Theorem 11.1, this MDP has finite state and action spaces. Therefore, it can be evaluated using standard DP algorithms, such as the value iteration algorithm [18]. The corresponding Bellman equation is

$$\rho + V(s,q) = \max_{x \in \mathcal{X}} D_{\mathsf{KL}}\left(P_{Y|X=x, S=s} \| T_{Y|Q=q}\right) + \sum_{y \in \mathcal{Y}} p(y|x,s) V\left(x, \phi(q, y)\right),$$

(11.31)

where $\rho$ is the average reward, $V(s,q)$ is the value function, and $\phi : \mathcal{Q} \times \mathcal{Y} \to \mathcal{Q}$ is the Q-graph transition function. Namely, $\phi(q, y)$ is the node reached from $Q = q$ when the channel output is $Y = y$.

Solving the Bellman equation numerically using the value iteration algorithm provides an estimate of the value function and the average reward, but most importantly, it produces a conjectured optimal policy

$$x(s,q) = \arg\max_{x \in \mathcal{X}} D_{\mathsf{KL}}\left(P_{Y|X=x, S=s} \| T_{Y|Q=q}\right) + \sum_{y \in \mathcal{Y}} p(y|x,s) V^*\left(x, \phi(q, y)\right),$$

(11.32)

where $V^*$ is the estimated optimal value function. Substituting $x(s,q)$ back into the Bellman equation converts it to a set of linear equations. By solving the system one obtains a conjectured optimal value function and average reward. To verify optimality for the conjectured values, it suffices to show they satisfy the fixed point condition in the Bellman equation.

The last step is to show that the obtained bound is tight. This is done by verifying two conditions. The first is that the model satisfies the Markov chain $Y^{i-1} \leftrightarrow Q_{i-1} \leftrightarrow Y_i$. In words, this means that there exists an input distribution that visits only the MDP states that formed the Q-graph, which yields an output distribution that satisfies $P_{Y_i|Y^{i-1}} = T_{Y|Q}$. The second condition is that the rate this distribution achieves coincides with the upper bound derived earlier. When both conditions hold, the bound is tight and the capacity characterization is completed.

## Bounds on the Ising Channel

We start from the case when the input alphabet satisfies $|\mathcal{X}| \leq 8$, in which case a complete characterization of the Ising channel feedback capacity is given. This analytic form is deduced from the numerical RL solution as described earlier.

---

**Algorithm 11.3** Capacity achieving coding scheme for $|\mathcal{X}| \leq 8$.

**Code construction and initialization:**

- Transform the $n$ uniform bits of the message into a stream of symbols (from $\mathcal{X}$) with the following statistics:

$$v_i = \begin{cases} v_{i-1} & \text{, w.p. } p \\ \text{Unif}[\mathcal{X} \setminus \{v_{i-1}\}] & \text{, w.p. } 1 - p, \end{cases} \tag{11.33}$$

with $v_0 = 0$. The mapping can be done using enumerative coding [27]
- Transmit a symbol twice to set the initial state of the channel $s_0$

---

**Encoder:**

Transmit $v_t$ and observe $y_t$

**if** $y_t = s_{t-1}$ **then**

    Re-transmit $v_t$

**end if**

---

**Decoder:**

Receive $y_t$

**if** $y_t \neq y_{t-1}$ **then**

    Store $y_t$ as an information symbol

**else**

    Ignore $y_t$ and store $y_{t+1}$ as a new information symbol

**end if**

---

THEOREM 11.3 (Feedback Capacity) *The feedback capacity of the Ising channel with $|\mathcal{X}| \leq 8$ is given by*

$$C_{\text{FB}}(\mathcal{X}) = \max_{p \in [0,1]} 2 \frac{H_2(p) + (1-p) \log (|\mathcal{X}|-1)}{p+3}. \tag{11.34}$$

*Equivalently, the feedback capacity can be expressed as*

$$C_{\text{FB}}(\mathcal{X}) = \frac{1}{2} \log \frac{1}{p}, \tag{11.35}$$

*where $p$ is the unique solution of $x^4 - ((|\mathcal{X}|-1)^4 + 4)x^3 + 6x^2 - 4x + 1 = 0$ on $[0,1]$.*

The proof of Theorem 11.3 is given [17]. The theorem only covers $|\mathcal{X}| \leq 8$ because while the derived upper and lower bounds (which we discuss later) hold in general, they turn to match only up to $|\mathcal{X}| = 8$.

For the lower bound, Algorithm 11.3 presents a simple achieving coding scheme. The scheme is applicable for any input alphabet size, but as stated next, it is optimal for $|\mathcal{X}| \leq 8$.

THEOREM 11.4 (Optimal coding scheme) *The coding scheme in Algorithm 11.3 achieves the capacity in Theorem 3 for $|\mathcal{X}| \leq 8$.*

In [17] it is shown that Algorithm 11.3 yields a zero-error code with a communication rate, when maximized over the parameter $p$, that equals the feedback capacity from Theorem 11.3.

For $|\mathcal{X}| > 8$, the structure of the analytic solution changes. Unlike in the $|\mathcal{X}| \le 8$ case, the Q-graph induced by the numerical results cannot be described with a finite set of nodes. Nevertheless, the numerical results dictate a suboptimal structure that induces an upper bound for $|\mathcal{X}| > 8$. The upper bound for $|\mathcal{X}| > 8$ is stated next.

THEOREM 11.5 (Upper bound for $|\mathcal{X}| > 8$)  *The feedback capacity of the Ising channel satisfies*

$$C_{FB}(\mathcal{X}) \le \frac{1}{2} \log \frac{|\mathcal{X}|}{p},$$

*where $p$ is the unique root of $x^2 - \left(2 + \frac{(|\mathcal{X}|-1)^2}{16|\mathcal{X}|}\right)x + 1$ in $[0, 1]$.*

Collecting the results, upper bound and lower bounds on the capacity of the Ising channel with any alphabet size are stated in the following theorem.

THEOREM 11.6 (Asymptotic performance)  *For any alphabet size $|\mathcal{X}| > 2$, the feedback capacity of the Ising channel satisfies*

$$C_{FB}(\mathcal{X}) \le \frac{3}{4} \log |\mathcal{X}|. \tag{11.36}$$

*Furthermore, there is a coding scheme that achieves the rate:*

$$R(\mathcal{X}) = \frac{3}{4} \log \frac{|\mathcal{X}|}{2}. \tag{11.37}$$

*Consequently, $\frac{3}{4} \log \frac{|\mathcal{X}|}{2} \le C_{FB}(\mathcal{X}) \le \frac{3}{4} \log |\mathcal{X}|$, whenever $|\mathcal{X}| > 2$.*

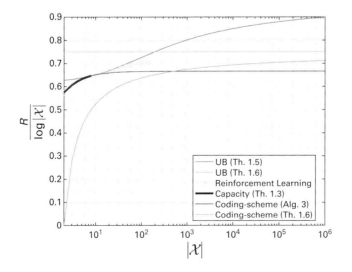

**Figure 11.8** Summary of the analytic bounds and the numerical results obtained by the RL simulations for varying alphabet sizes. The rates/bounds are normalized by $\log |\mathcal{X}|$.

The analytic and numerical results are summarized in Fig. 11.8, which shows the RL similation. We ran the RL algorithms up to alphabet size 150 due to a computational memory constraint. The bold black curve depicts the analytic capacity, as stated in Theorem 11.3 for $|\mathcal{X}| \leq 8$. One can see that the capacity achieving coding scheme coincides with the RL simulation for $|\mathcal{X}| \leq 8$. However, it converges to $\frac{2}{3} \log |\mathcal{X}|$ for large alphabets, while RL continues to improve. To back this observation up, the improved lower and upper bounds for the asymptotic case are shown. For large $|\mathcal{X}|$, both converge to $\frac{3}{4} \log |\mathcal{X}|$ with a constant difference of $\frac{3}{4}$. We also present the upper bound from Theorem 11.5, which outperforms the others for $8 < |\mathcal{X}| \leq 200$.

## 11.4 Directed Information Neural Estimation

Here we consider an unknown channel model. The only assumption is that the channel is stationary and ergodic. Since the channel model is not known, an MDP formulation of the DI is generally not available. Instead, advances in ML are utilized to estimate the DI rate of two stochastic processes using Donsker-Varadhan representation of KL-divergences. This results in an estimation algorithm called DINE for the estimation of the DI rate between two stationary and ergodic stochastic processes. This estimation is then used as a basis for the optimization over input distributions. Thus, the capacity estimation problem is addressed.

### 11.4.1 Related Work on DI

Previously known estimators of DI operate under rather restrictive assumptions on the data that are hard to verify (if not violated) in practice. Specifically, [28] estimated DI between *discrete-valued* processes based on universal probability assignments and context tree weighting (see [29–32] for some applications of that method). Continuous-valued processes, which are of central practical interest, were treated in [34] and [33] using $k$ nearest neighbors (kNN) estimation techniques. However, [33] assumed Markovian processes with short memory, while the method from [34] requires knowledge of the causal dependence graph between the processes.

DINE, on the other hand, accounts for any stationary and ergodic continuous-valued processes. The construction draws upon recent advances in mutual information estimation using deep neural networks (DNNs) [35, 36]. Specifically, [35] introduced the MI neural estimator (MINE), that estimates $I(X; Y)$ based on independently and identically distributed (i.i.d.) samples from $(X, Y)$ by optimizing its Donsker-Varadhan (DV) variational representation (as parametrized by a DNN). In the followup work [36], the original MINE design was altered (by incorporating a certain reference measure) with the goal of stabilizing and speeding up training. Although these estimators trivially account for pairwise i.i.d. stochastic process, they are incompatible for processes with memory.[9] DINE naturally extends the ideas from [35, 36] to this setting.

---

[9] This is unless we have many independent samples of the entire pairwise process, which is a highly impractical scenario.

## 11.4.2    Derivation

We devise a neural estimator of the DI rate between two jointly stationary and ergodic stochastic processes $\mathbb{X} := \{X_i\}_{i=1}^{\infty}$ and $\mathbb{Y} := \{Y_i\}_{i=1}^{\infty}$. Specifically, given a dataset $\mathcal{D}^{(n)} := (X^n, Y^n) \sim P_{X^n Y^n}$, we derive a parametrized empirical loss function that, when optimized, results in a provably consistent estimate of $\mathcal{I}(\mathbb{X} \to \mathbb{Y})$. To do so, we first express DI as a difference of certain KL divergence terms. These are then represented via the DV variational formula [37]. Lastly, the DV potentials are parametrized using RNNs and expected values are approximate by empirical means.

Recall that DI is given by

$$I\left(X^n \to Y^n\right) = h\left(Y^n\right) - h\left(Y^n \| X^n\right). \tag{11.38}$$

Assume $\mathcal{Y}$ is compact[10] and let $\widetilde{Y} \sim \mathsf{Unif}(\mathcal{Y}) =: P_{\widetilde{Y}}$ be independent of $\mathbb{X}$ and $\mathbb{Y}$. Using the uniform reference measure, we expand each entropy term as follows:

$$h(Y^n) = h_{\mathsf{CE}}\left(P_{Y^n}, P_{Y^{n-1}} \otimes P_{\widetilde{Y}}\right) - D_{\mathsf{KL}}\left(P_{Y^n} \big\| P_{Y^{n-1}} \otimes P_{\widetilde{Y}}\right), \tag{11.39}$$

where $h_{\mathsf{CE}}(P, Q) := \int -\log\left(Q(x)\right) \mathsf{d}P(x)$ and $D_{\mathsf{KL}}(P \| Q) := \int \log\left(\frac{\mathsf{d}P}{\mathsf{d}Q}(x)\right) \mathsf{d}P(x)$ are, respectively, the cross entropy (CE) and KL-divergence, while $\frac{\mathsf{d}P}{\mathsf{d}Q}$ is the Radon-Nikodym derivative of $P$ w.r.t. $Q$. Similarly we have

$$h(Y^n \| X^n) = h_{\mathsf{CE}}\left(P_{Y^n \| X^n}, P_{Y^{n-1} \| X^{n-1}} \otimes P_{\widetilde{Y}} \big| P_{X^n}\right)$$
$$- D_{\mathsf{KL}}\left(P_{Y^n \| X^n} \big\| P_{Y^{n-1} \| X^{n-1}} \otimes P_{\widetilde{Y}} \big| P_{X^n}\right), \tag{11.40}$$

where $h_{\mathsf{CE}}(P_{Y|X}, Q_{Y|X} | P_X) := \int h_{\mathsf{CE}}\left(P_{Y|X}(\cdot|x), Q_{Y|X}(\cdot|x)\right) \mathsf{d}P_X(x)$ is conditional CE, and $D_{\mathsf{KL}}(P_{Y|X} \| Q_{Y|X} | P_X) := \int D_{\mathsf{KL}}\left(P_{Y|X}(\cdot|x) \big\| Q_{Y|X}(\cdot|x)\right) \mathsf{d}P_X(x)$ is conditional KL divergence.

With some abuse of notation, let $\mathbb{X} := \{X_i\}_{i \in \mathbb{Z}}$ and $\mathbb{Y} := \{Y_i\}_{i \in \mathbb{Z}}$ be the two-sided extension of the considered processes (the underlying stationary and ergodic measure remains unchanged). Inserting Eqs. (11.39) and (11.40) into Eq. (11.38) and using stationarity we obtain

$$\mathcal{I}(\mathbb{X} \to \mathbb{Y}) = \lim_{n \to \infty} \left(D_{Y\|X}^{(n)} - D_Y^{(n)}\right), \tag{11.41}$$

where

$$D_{Y\|X}^{(n)} := D_{\mathsf{KL}}\left(P_{Y^0_{-(n-1)} \| X^0_{-(n-1)}} \big\| P_{Y^{-1}_{-(n-1)} \| X^{-1}_{-(n-1)}} \otimes P_{\widetilde{Y}} \big| P_{X^{-1}_{-(n-1)}}\right)$$

$$D_Y^{(n)} := D_{\mathsf{KL}}\left(P_{Y^0_{-(n-1)}} \big\| P_{Y^{-1}_{-(n-1)}} \otimes P_{\widetilde{Y}}\right). \tag{11.42}$$

To arrive at a variational form we make use of the DV theorem [37], as restated next.

---

[10] This is a technical assumption that arises due to the choice of a uniform reference measure. By changing the $P_{\widetilde{Y}}$ to, for example, Gaussian, this assumption is removed. We consider the uniform case for simplicity.

THEOREM 11.7 (DV representation) *For any two probability measures P and Q on* $\Omega$*, we have*

$$D_{\text{KL}}(P\|Q) = \sup_{\mathsf{T}:\Omega\to\mathbb{R}} \mathbb{E}_P[\mathsf{T}] - \log\mathbb{E}_Q[e^{\mathsf{T}}] \tag{11.43}$$

*where the supremum is over all measurable functions* $\mathsf{T}$ *with finite expectation.*

The optimal DV potentials for $D_Y^{(n)}$ and $D_{Y\|X}^{(n)}$ can be represented as open dynamical systems. As such, these potentials can be approximated to arbitrary precision by elements of the RNN function class [38].

DEFINITION 11.8 (RNN function class) Let $f : \mathbb{R}^d \to \mathbb{R}^d$ be Borel measurable and $\ell, k, T \in \mathbb{N}$. The RNN function class $\text{RNN}_{\ell,k}(f)$ is given by $\text{RNN}_{\ell,k}(f) = \bigcup_{m=1}^{\infty} \text{RNN}_{\ell,k}^{(m)}(f)$, where $\text{RNN}_{\ell,k}^{(m)}(f)$ comprises all dynamical systems parameterized with $m$ parameters whose input $x_n \in \mathbb{R}^\ell$, inner state $s_n \in \mathbb{R}^d$ and output $y_n \in \mathbb{R}^k$, are related through the following equations

$$s_{n+1} = f(As_n + Bx_n - b)$$
$$y_n = Cs_n, \tag{11.44}$$

$\forall n = 1, \dots, T$, where $A \in \mathbb{R}^{d\times d}$, $B \in \mathbb{R}^{d\times\ell}$, $C \in \mathbb{R}^{k\times d}$, $b \in \mathbb{R}^d$, and $f$ operates on vectors component-wise.

The DINE is thus given by

$$\widehat{\mathcal{I}}_n(\mathcal{D}_n) := \sup_{\mathsf{F}_2 \in \text{RNN}_{d_{xy},1}} \widehat{D}_{Y\|X}(\mathcal{D}_n, \mathsf{F}_2) \tag{11.45}$$

$$- \sup_{\mathsf{F}_1 \in \text{RNN}_{d_y,1}} \widehat{D}_Y(\mathcal{D}_n, \mathsf{F}_1) = \sup_{\mathsf{F}_2} \inf_{\mathsf{F}_1} \widehat{\mathcal{I}}_n(\mathcal{D}_n, \mathsf{F}_1, \mathsf{F}_2), \tag{11.46}$$

where, $d_{xy} = d_x + d_y$, and

$$\widehat{D}_Y(\mathcal{D}_n, \mathsf{F}_1) := \frac{1}{n}\sum_{i=1}^{n} \mathsf{F}_1\left(Y_i|Y^{i-1}\right) - \log\left(\frac{1}{n}\sum_{i=1}^{n} e^{\mathsf{F}_1(\widetilde{Y}_i|Y^{i-1})}\right) \tag{11.47a}$$

$$\widehat{D}_{Y\|X}(\mathcal{D}_n, \mathsf{F}_2) := \frac{1}{n}\sum_{i=1}^{n} \mathsf{F}_2\left(Y_i|Y^{i-1}X^i\right) - \log\left(\frac{1}{n}\sum_{i=1}^{n} e^{\mathsf{F}_2(\widetilde{Y}_i|Y^{i-1}X^i)}\right), \tag{11.47b}$$

and $\widehat{\mathcal{I}}_n(\mathcal{D}_n, \mathsf{F}_1, \mathsf{F}_2) := \widehat{D}_{Y\|X}(\mathcal{D}_n, \mathsf{F}_1) - \widehat{D}_Y(\mathcal{D}_n, \mathsf{F}_2)$. The optimization in Eq. (11.46) can be executed via gradient-ascent over the RNNs' parameters. We next state the consistency of DINE.

THEOREM 11.9 (Strong consistency) *Let* $\mathbb{X} = \{X_i\}_{i=1}^{\infty}$ *and* $\mathbb{Y} = \{Y_i\}_{i=1}^{\infty}$ *be jointly stationary ergodic stochastic processes, with an underlying probability measure* $\mathbb{P}$*. Denote by* $\mathcal{D}_n = \{(X_i, Y_i)\}_{i=1}^{n}$ *a sample of size n from the processes. Then, there exist RNNs* $\mathsf{F}_1 \in \text{RNN}_{d_y,1}(f), \mathsf{F}_2 \in \text{RNN}_{d_{xy},1}(f)$*, such that the DINE* $\widehat{\mathcal{I}}_n(\mathcal{D}_n, \mathsf{F}_1, \mathsf{F}_2)$ *is a strongly consistent estimator of* $\mathcal{I}(\mathbb{X} \to \mathbb{Y})$*, i.e., for any* $\epsilon > 0$ *there exists* $n_0 \in \mathbb{N}$*, such that for all* $n > n_0$*, we have*

$$\left|\widehat{\mathcal{I}}_n(\mathcal{D}_n, \mathsf{F}_1, \mathsf{F}_2) - \mathcal{I}(\mathbb{X} \to \mathbb{Y})\right| \le \epsilon \qquad \mathbb{P} - a.s. \tag{11.48}$$

The proof of Theorem 11.9 involves three main steps: (a) an *information-theoretic argument* that expresses the DI rate as a difference of KL divergence terms, followed by an application of the DV variational representation; (b) an *estimation step* that relies on a generalization of Birkhoff's ergodic theorem (due to Breiman [39] and Cover [40]) to approximate expected values (w.r.t. the underlying ergodic and stationary measure) by sample means; and (c) an *approximation* step, where a sequence of optimal DV potentials is approximated by elements of the RNN class.

*Remark 11.1* Despite the popularity of neural estimation, the theoretical performance guarantees for these estimators is still largely unknown. There are some results on the consistency of the original MINE method, and Theorem 11.9 provides an analogous consistency result for DINE. Going forward, it is of significant interest to derive nonasymptotic performance guarantees and sample complexity bounds. However, such results are not yet available, even for the original MINE method, yet alone for DINE method presented herein.

*Discriminator Interpretation*
The optimal DV potential in Eq. (11.43) is $\mathsf{T}^\star = \log\left(\frac{dP}{dQ}\right)$. This coincides with the log-likelihood ratio, which is the optimal test statistic for testing $H_0 : X \sim P$ against $H_1 : X \sim Q$. This observation motivated [41] to implement the conditional MINE of $I(X;Y)$ using a binary classifier between $P = P_{X,Y}$ and $Q = P_X \otimes P_Y$.

Adopting this perspective, the supremum achieving DV potentials for $D_Y^{(n)}$ and $D_{Y\|X}^{(n)}$ (see Eq. (11.42)), denoted by $\mathsf{T}_Y^\star$ and $\mathsf{T}_{Y\|X}^\star$, respectively, are interpreted as a discriminator that determines if the sample came from the conditional distribution or the uniform reference measure. Subtracting the discriminators' values, we obtain the optimal causal hypothesis test from [15]:

$$\mathsf{T}_Y^\star\left(Y_i, \widetilde{Y}_i\right) - \mathsf{T}_{Y\|X}^\star\left(X_i, Y_i, \widetilde{Y}_i\right) = \log\left(\frac{p_{Y_i|Y_{-\infty}^{i-1}}}{p_{Y_i|Y_{-\infty}^{i-1}X_{-\infty}^{i-1}}}\right).$$

Furthermore, by [15, Theorem 5], DI rate equals the asymptotic type-II error of this test.

## 11.4.3     Algorithm and Implementation

Calculating DINE entails optimizing over two RNNs $\mathsf{F}_i$, $i = 1, 2$ (see Eq. (11.47)). Each RNN comprises a *modified* long short-term memory (LSTM) layer and two fully connected networks (FCNs), with two hidden layers each. The architecture for $\widehat{D}_Y(\mathsf{F}_1, \mathcal{D}_n)$ is depicted in Fig. 11.9(a). We next describe the modified LSTM cell and afterward address the optimization process.

*Modified LSTM*
To capture the causal dependencies in $\mathcal{D}_n$, we adjust the structure of classic LSTM cell [42]. The modification is presented for $\widehat{D}_Y$; adapting to $\widehat{D}_{Y\|X}^{(n)}$ is straightforward.

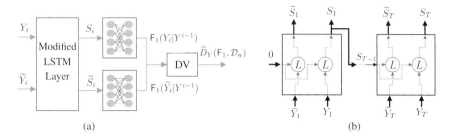

**Figure 11.9** (a) Block-diagram of $\widehat{D}_Y(\mathsf{F}_1, \mathcal{D}_B)$ estimate, and (b) unrolled modified LSTM.

The classic LSTM is an RNN that recursively computes a hidden state $S_i$ from its input $Y_i$ and the previous state $S_{i-1}$. We henceforth use the shorthand $S_i = L(Y_i, S_{i-1})$ for the relation between $S_i$ and $(Y_i, S_{i-1})$ defined by the LSTM. As DINE employs a reference sequence $\widetilde{Y}^n$, the modified LSTM collects hidden states for both $Y^n$ and $\widetilde{Y}^n$. At time $i = 1, \ldots, n$, the cell takes a pair $(Y_i, \widetilde{Y}_i)$ as input, and outputs two hidden states $S_i = L(Y_i, S_{i-1})$ and $\widetilde{S}_i = L(\widetilde{Y}_i, S_{i-1})$; only $S_i$ is passed for calculating the next state. The state sequence are then processed by the FCNs to obtain the elements of Eq. (11.47a). The calculation of the hidden states for $\widehat{D}_{Y\|X}$ in Eq. (11.47b) is performed analogously, by replacing $Y_i$ with $(X_i, Y_i)$. The modified LSTM cell is shown in Fig. 11.9(b).

*Algorithm*
The DINE algorithm computes $\widehat{I}_n(\mathcal{D}_n)$ by optimizing the parameters $\theta_1$ and $\theta_2$ of the RNNs $\mathsf{F}_1$ and $\mathsf{F}_2$, respectively. We divide the dataset into batches $\mathcal{D}_B := \left\{ \left( X^{iT}_{(i-1)T}, Y^{iT}_{(i-1)T} \right) \right\}_{i=1}^{B}$, each comprising $B$ sequences of length $T$, where $T$ is the number of time steps in the unrolled modified LSTM. We then sample the reference

---

**Algorithm 11.4** DI rate estimation.

**input:** Dataset $\mathcal{D}_n$.
**output:** $\widehat{I}_n(\mathcal{D}_n)$ DI rate estimate.

---

Initialize $\mathsf{F}_1, \mathsf{F}_2$ with parameters $\theta_1, \theta_2$.
**Step 1 – Optimization:**
**repeat**
Draw $\mathcal{D}_B$ batch & sample reference measure.
Compute $\widehat{D}_{Y\|X}(\mathsf{F}_2, \mathcal{D}_B)$, $\widehat{D}_Y(\mathsf{F}_1, \mathcal{D}_B)$.
Update networks parameters:
$$\theta_2 \leftarrow \theta_2 + \nabla \widehat{D}_{Y\|X}(\mathsf{F}_2, \mathcal{D}_B)$$
$$\theta_1 \leftarrow \theta_1 + \nabla \widehat{D}_Y(\mathsf{F}_1, \mathcal{D}_B)$$
**until** convergence
**Step 2 – Evaluation:**
Evaluate over $\mathcal{D}_n$ and subtract losses to obtain $\widehat{I}_n(\mathcal{D}_n, \mathsf{F}_1, \mathsf{F}_2)$

measure[11] and feed the sequences through the DINE architecture to obtain the DV potentials $F_1$ and $F_2$. The weights of the FCNs within each RNN are shared since we wish to produce the same DV potential acting on different inputs. We draw several reference samples for each element of $\mathcal{D}_B$ and average the resulting RNN outputs to reduce noise and stabilize the training process. See Algorithm 11.4 for the full list of steps.

### 11.4.4    Applications and Experiments

We tested DINE on both synthetic and real-world stock market data. We trained on a 1080Ti NVIDIA GPU, using a batch size of 50, LSTM time unroll factor of 50 and the Adam optimizer [43] with initial learning rate $10^{-4}$.

*Synthetic Data*
Consider the linear Gaussian hidden Markov model (HMM), which is a popular model for speech processing [44, 45], computational biology [46, 47] and numerous other fields [48–50]. In this setting, the processes $\mathbb{X}$ and $\mathbb{Y}$ are defined through the recursive relation:

$$X_i = \alpha X_{i-1} + N_{1,i}$$
$$Y_i = \beta X_i + N_{2,i} \tag{11.49}$$

where $\{N_{1,i}\}_{i\in\mathbb{N}}$ and $\{N_{2,i}\}_{i\in\mathbb{N}}$ are i.i.d. centered Gaussian processes with variances $\sigma_1^2$ and $\sigma_2^2$, respectively. Although no closed expression form $\mathcal{I}(\mathbb{X} \to \mathbb{Y})$ is known, upper and lower bounds can be derived. Since $\mathbb{Y}$ is an HMM with unknown parameters, it

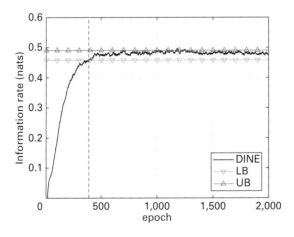

**Figure 11.10**  DINE convergence for linear Gaussian HMM. The solid black line shows the DINE, and the other lines show the upper and lower bounds on the DI rate. The vertical dashed line marks the epoch when the estimate reaches values between the bounds.

---

[11]  In practice, we sample uniformly from the smallest $d$-dimensional bounding hypercube of $Y^n$.

falls outside the frameworks from [33, 34], so these methods do not apply here. DINE, on the contrary, is applicable and its estimate the DI rate is illustrated in Fig. 11.10. The figure shows the convergence of DINE toward $\mathcal{I}(\mathbb{X} \to \mathbb{Y})$ and the aforementioned bounds, for $\alpha = \beta = \sigma_1^2 = \sigma_2^2 = 1$.

## 11.5     Capacity Estimation without a Channel Model

Calculating the capacity (with or without feedback) of channels with memory and continuous alphabets is a challenging task. It requires optimizing the DI rate over all channel input distributions. The objective is a multiletter expression, whose analytic solution is only known for a few specific cases. When no analytic solution is present or the channel model is unknown, there is no unified framework for calculating or even approximating capacity. Herein, a capacity estimation algorithm [51] is presented, that treats the channel as a black box, both when feedback is or is not present. The algorithm has two main ingredients: (a) a neural distribution transformer (NDT) model that shapes a noise variable into the channel input distribution, which we are able to sample, and (b) the DINE that estimates the communication rate of the current NDT model. These models are trained by an alternating maximization procedure to both estimate the channel capacity and obtain an NDT for the optimal input distribution. The method is demonstrated on the moving average additive Gaussian noise channel, where it is shown that both the capacity and feedback capacity are estimated without knowledge of the channel transition kernel. The proposed estimation framework opens the door to a myriad of capacity approximation results for continuous alphabet channels that were inaccessible until now.

### 11.5.1     Neural Distribution Transformer

DINE accounts for one of the two tasks involved in estimating capacity, and it estimates the objective of Eq. (11.10). It then remains to optimize this objective over input distributions. To that end, we design a generative recurrent model, termed the NDT, characterized by a neural network $F_X$ to approximate the channel input distributions. This is similar in flavor to generators used in generative adversarial networks [52]. For this purpose, we present the NDT model that represents a general input distribution of the channel. At each iteration $i = 1, \ldots, n$ the NDT maps an i.i.d noise vector $N^i$ to a channel input variable $X_i$. When feedback is present the NDT maps $(N^i, Y^{i-1}) \longmapsto X_i$. Thus, NDT is represented by an RNN with parameters $\mu$ as shown in Fig. 11.11.

### 11.5.2     Capacity Estimation

The capacity estimation algorithm trains DINE and NDT models together via an alternating optimization procedure (i.e., fixing the parameters of one model while training

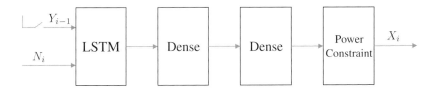

**Figure 11.11** The NDT. The noise and past channel output (if feedback is applied) are fed into an NN. The last layer performs normalization to obey the input constraint (e.g., power constraint) if needed.

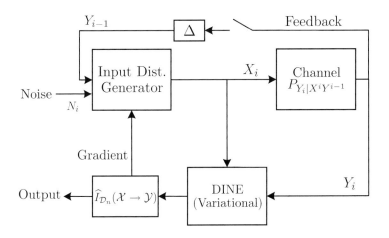

**Figure 11.12** The overall capacity estimator: NDT generates samples that are fed into the channel. DINE uses these samples to improve its estimation of the communication rate. DINE then supplies gradient for the optimization of NDT.

the other). DINE estimates the communication rate of a fixed NDT input distribution, and the NDT is trained to increase its rate w.r.t. fixed DINE model. Proceeding until convergence, this results in the capacity estimate, as well as an NDT generative model for the achieving input distribution. We demonstrate our method on the MA(1)-AGN channel. Both $C_{FF}$ and $C_{FB}$ are estimated using the same algorithm, using the channel as a black box to solely generate samples. The estimation results are compared with the analytic solution to show the effectiveness of the proposed approach.

As shown in Fig. 11.12, the NDT model is fed with i.i.d. noise and its output is the samples $X^n$. These samples are fed into the channel to generate outputs. Then, DINE uses $(X^n, Y^n)$ to produce the estimate $\widehat{\mathcal{I}}_n(\mathcal{D}_n)$. To estimate capacity, DINE and NDT models are trained together. The training scheme, as shown in Algorithm 11.5, is a variant of alternated maximization procedure. This procedure iterates between updating the DINE networks parameters $\theta_1, \theta_2$ and the NDT parameters $\theta_X$, each time keeping one of the models fixed. At the end of training a long Monte Carlo evaluation of $\sim 10^6$ samples of the DI rate. Applying this algorithm to channels with memory estimates their capacity without any specific knowledge of the channel underlying

---

**Algorithm 11.5** Capacity estimation.

---

**input:** Continuous channel, feedback indicator
**output:** $\widehat{\mathcal{I}}(\mathcal{D}_n, \mathsf{F}_1\mathsf{F}_2, \mathsf{F}_X)$, estimated capacity.

---

Initialize $\mathsf{F}_1, \mathsf{F}_2, \mathsf{F}_X$ with parameters $\theta_1, \theta_2, \theta_X$
**if** feedback indicator **then**
    Add feedback to NDT
**end if**
**repeat**
    **Step 1: Train DINE model**
    Generate B sequences of length T of i.i.d random noise
    Compute $\mathcal{D}_B = \{(x_i^T, y_i^T)\}_{i=1}^B$ with NDT and channel
    Compute $\widehat{D}_{Y\|X}(\theta_{Y\|X}, \mathcal{D}_B), \widehat{D}_Y(\theta_Y, \mathcal{D}_B)$
    Update DINE parameters:
        $\theta_1 \leftarrow \theta_1 + \nabla \widehat{D}_Y(\mathcal{D}_B, \mathsf{F}_1, \mathsf{F}_X)$
        $\theta_2 \leftarrow \theta_2 + \nabla \widehat{D}_{Y\|X}(\mathcal{D}_B, \mathsf{F}_2, \mathsf{F}_X)$
    **Step 2: Train NDT**
    Generate B sequences of length T of i.i.d random noise
    Compute $\mathcal{D}_B = \{(x_i^T, y_i^T)\}_{i=1}^B$ with NDT and channel
    compute the objective:
        $\widehat{I}(\mathcal{D}_n, \mathsf{F}_1\mathsf{F}_2, \mathsf{F}_X) = \widehat{D}_{Y\|X}(\mathcal{D}_B, \mathsf{F}_2, \mathsf{F}_X) - \widehat{D}_Y(\mathcal{D}_B, \mathsf{F}_1, \mathsf{F}_X)$
    Update NDT parameters:
        $\theta_X \leftarrow \theta_X + \nabla_{\theta_X} \widehat{I}(\mathcal{D}_B, \mathsf{F}_1\mathsf{F}_2, \mathsf{F}_X)$
**until** convergence
Monte Carlo evaluation of $\widehat{\mathcal{I}}(\mathcal{D}_n, \mathsf{F}_1\mathsf{F}_2, \mathsf{F}_X))$
**return** $\widehat{\mathcal{I}}(\mathcal{D}_n, \mathsf{F}_1\mathsf{F}_2, \mathsf{F}_X)$

---

distribution. The implementation is available on Github.[12] Next, we demonstrate the effectiveness of this algorithm on continuous alphabet channels.

## 11.5.3   Numerical Results

This section demonstrates the empirical effectiveness of Algorithm 11.5 on the capacity estimation of continuous channels with memory. For that purpose, it is tested on memoryless channels and on channels with memory on both cases where a feedback link is or is not available. Even though Algorithm 11.5 is applicable for any ergodic and stationary channel, it necessary to provide experiments on channels whose analytic solution is already known in order to show the correctness of the algorithm. Accordingly, the AWGN channel is chosen as an instance of a memoryless channel, and a first-order MA-AGN channel is chosen as an instance of a channel with memory.

---

[12] https://github.com/zivaharoni/capacity-estimator-via-dine

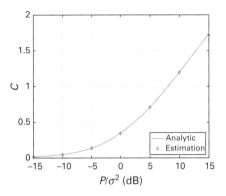

**Figure 11.13** Estimation of AWGN channel capacity for various SNR values

*AWGN Channel*

The power constrained AWGN channel is considered. This is an instance of a memory-less, continuous-alphabet channel for which an analytic capacity expression is known. The channel model is

$$Y_i = X_i + Z_i, \quad i \in \mathbb{N}, \tag{11.50}$$

where $Z_i \sim \mathcal{N}(0, \sigma^2)$ are i.i.d RVs, and $X_i$ is the channel input sequence bound to the power constraint $\mathbb{E}[X_i^2] \leq P$. The capacity of this channel is given by $C = \frac{1}{2} \log\left(1 + \frac{P}{\sigma^2}\right)$. In our implementation, we chose $\sigma^2 = 1$ and estimated capacity for a range of $P$ values. The numerical results are compared to the analytic solution in Fig. 11.13, where a clear correspondence is seen.

*Gaussian MA(1) Channel*

We consider both the feedback ($C_{\text{FB}}$) and the feedforward ($C_{\text{FB}}$) capacity of the MA(1) Gaussian channel. The model here is

$$Z_i = \alpha U_{i-1} + U_i,$$
$$Y_i = X_i + Z_i, \tag{11.51}$$

where $U_i \sim \mathcal{N}(0, 1)$ are i.i.d., $X_i$ is the channel input sequence bound to the power constraint $\mathbb{E}[X_i^2] \leq P$, and $Y_i$ is the channel output.

The feedforward capacity of the MA(1) Gaussian channel with input power constraint can be obtained via the water-filing algorithm [2]. This is the benchmark against which we compare the quality of the $C_{\text{FF}}$ estimate produced by Algorithm 11.5. Results are shown in Fig. 11.14(a).

Computing the feedback capacity of the ARMA(k) Gaussian channel can be formulated as a dynamic programming, which is then solved via an iterative algorithm [8]. For the particular case of Eq. (11.51), $C_{\text{FB}}$ is given by $-\log(x_0)$, where $x_0$ is a solution to a fourth-order polynomial equation. The estimates for $C_{\text{FB}}$ produced by Algorithm 11.5 are compared to the analytic solutions in Fig. 11.14(b).

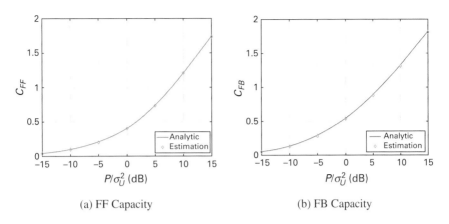

(a) FF Capacity                    (b) FB Capacity

**Figure 11.14** Capacity estimation over MA(1)-AGN channel. In (a) feedback is not present, and in (b) feedback is present.

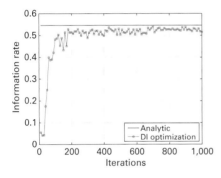

**Figure 11.15** Optimization progress of DI rate of Algorithm 11.5 for the feedback setting with $P = 1$. The information rates were estimated by a Monte Carlo evaluation with $10^5$ samples.

## References

[1] R. G. Gallager, *Information Theory and Reliable Communication*. Wiley, 1968.

[2] T. M. Cover and J. A. Thomas, *Elements of Information Theory*, 2nd ed. Wiley, 2006.

[3] H. H. Permuter et al., "Capacity of the trapdoor channel with feedback," *IEEE Trans. on Information Theory*, vol. 54, no. 7, pp. 3150–3165, 2009.

[4] O. Elishco and H. H. Permuter, "Capacity and coding for the Ising channel with feedback," *IEEE Trans. on Information Theory*, vol. 60, no. 9, pp. 5138–5149, 2014.

[5] A. Sharov and R. M. Roth, "On the capacity of generalized Ising channels," *IEEE Trans. on Information Theory*, 2016.

[6] O. Sabag, H. Permuter, and N. Kashyap, "The feedback capacity of the binary erasure channel with a no-consecutive-ones input constraint," *IEEE Trans. on Information Theory*, vol. 62, no. 1, pp. 8–22, 2016.

[7] S. Yang, A. Kavcic, and S. Tatikonda, "Feedback capacity of stationary sources over Gaussian intersymbol interference channels," in *GLOBECOM*, 2006.

[8] S. Yang, A. Kavcic, and S. Tatikonda, "On the feedback capacity of power-constrained gaussian noise channels with memory," *IEEE Trans. on Information Theory*, vol. 53, no. 3, pp. 929–954, 2007.

[9] S. Tatikonda and S. Mitter, "The capacity of channels with feedback," *IEEE Trans. on Information Theory*, vol. 55, no. 1, pp. 323–349, 2009.

[10] W. Hirt and J. L. Massey, "Capacity of the discrete-time gaussian channel with intersymbol interference," *IEEE Trans. on Information Theory*, vol. 34, no. 3, pp. 38–38, 1988.

[11] C. E. Shannon, "A mathematical theory of communication," *The Bell System Technical Journal*, vol. 27, no. 3, pp. 379–423, 1948.

[12] J. Massey, "Causality, feedback and directed information," *Proc. Int. Symp. Information Theory Applications (ISITA-90)*, pp. 303–305, 1990.

[13] J. L. Massey and P. C. Massey, "Conservation of mutual and directed information," in *Proc. Int. Symp. on Information Theory (ISIT)*, pp. 157–158, 2005.

[14] G. Kramer, *Directed Information for Channels with Feedback*. Citeseer, 1998.

[15] H. H. Permuter, Y.-H. Kim, and T. Weissman, "Interpretations of directed information in portfolio theory, data compression, and hypothesis testing," *IEEE Trans. on Information Theory*, vol. 57, no. 6, pp. 3248–3259, 2011.

[16] O. Sabag, H. H. Permuter, and N. Kashyap, "Feedback capacity and coding for the BIBO channel with a no-repeated-ones input constraint," *IEEE Trans. on Information Theory*, vol. 64, no. 7, pp. 4940–4961, 2018.

[17] Z. Aharoni, O. Sabag, and H. H. Permuter, "Computing the feedback capacity of finite state channels using reinforcement learning," in *IEEE Int. Symp. on Information Theory (ISIT)*, 2019.

[18] R. Bellman, "A Markovian decision process," *Indiana University Mathematics Journal*, vol. 6, pp. 679–684, 1957.

[19] T. P. Lillicrap et al., "Continuous control with deep reinforcement learning," *arXiv preprint*, arXiv:1509.02971, 2015.

[20] W. K. Hasting, "Monte Carlo sampling methods using Markov chains and their applications," *Biometrika*, vol. 57, no. 1, pp. 97–109, 1970.

[21] J. Schulman, N. Heess, T. Weber, and P. Abbeel, "Gradient estimation using stochastic computation graphs," *Advances in Neural Information Processing Systems*, vol. 28, 2015.

[22] N. Srivastava et al., "Dropout: A simple way to prevent neural networks from overfitting," *Journal of Machine Learning Research*, vol. 15, no. 56, pp. 1929–1958, 2014.

[23] E. Ising, "Beitrag zur theorie des ferromagnetismus," *Zeitschrift für Physik*, vol. 31, no. 1, pp. 253–258, 1925.

[24] S. Lloyd, "Least squares quantization in PCM," *IEEE Trans. on Information Theory*, vol. 28, no. 2, pp. 129–137, 1982.

[25] O. Sabag, H. H. Permuter, and H. D. Pfister, "A single-letter upper bound on the feedback capacity of unifilar finite-state channels," *IEEE Trans. on Information Theory*, vol. 63, no. 3, pp. 1392–1409, 2017.

[26] O. Sabag and H. H. Permuter, "The duality upper bound for unifilar finite-state channels with feedback," in *Proc. Int. Zurich Seminar on Information and Communication (IZS 2020)*, pp. 68–72, 2020.

[27] T. Cover, "Enumerative source encoding," *IEEE Trans. on Information Theory*, vol. 19, no. 1, pp. 73–77, 1973.

[28] J. Jiao et al., "Universal estimation of directed information," *IEEE Trans. on Information Theory*, vol. 59, no. 10, pp. 6220–6242, 2013.

[29] J. Etesami and N. Kiyavash, "Measuring causal relationships in dynamical systems through recovery of functional dependencies," *IEEE Trans. on Signal and Information Processing over Networks*, vol. 3, no. 4, pp. 650–659, 2016.

[30] Z. Wang et al., "Causality analysis of fmri data based on the directed information theory framework," *IEEE Trans. on Biomedical Engineering*, vol. 63, no. 5, pp. 1002–1015, 2015.

[31] R. Malladi et al., "Identifying seizure onset zone from the causal connectivity inferred using directed information," *IEEE Journal of Selected Topics in Signal Processing*, vol. 10, no. 7, pp. 1267–1283, 2016.

[32] B. Oselio, S. Liu, and A. Hero, "Multi-layer relevance networks," in *Proc. IEEE 19th Int. Workshop on Signal Processing Advances in Wireless Communications (SPAWC)*, pp. 1–5, 2018.

[33] Y. Murin, "k-NN estimation of directed information," *arXiv preprint*, arXiv:1711.08516, 2017.

[34] A. Rahimzamani et al., "Estimators for multivariate information measures in general probability spaces," in *Advances in Neural Information Processing Systems 31*, S. Bengio et al., eds., pp. 8664–8675, 2018.

[35] M. I. Belghazi et al., "MINE: mutual information neural estimation," *arXiv preprint*, arXiv:1801.04062, 2018.

[36] C. Chung et al., "Neural entropic estimation: A faster path to mutual information estimation," *arXiv preprint*, arXiv:1905.12957, 2019.

[37] M. D. Donsker and S. S. Varadhan, "Asymptotic evaluation of certain Markov process expectations for large time. iv," *Communications on Pure and Applied Mathematics*, vol. 36, no. 2, pp. 183–212, 1983.

[38] A. M. Schäfer and H. G. Zimmermann, "Recurrent neural networks are universal approximators," in *Proc. Int. Conf. on Artificial Neural Networks*, pp. 632–640, 2006.

[39] L. Breiman, "The individual ergodic theorem of information theory," *The Annals of Mathematical Statistics*, vol. 28, no. 3, pp. 809–811, 1957.

[40] P. H. Algoet and T. M. Cover, "A sandwich proof of the Shannon-McMillan-Breiman theorem," *The Annals of Probability*, pp. 899–909, 1988.

[41] S. Mukherjee, H. Asnani, and S. Kannan, "CCMI: Classifier based conditional mutual information estimation," *arXiv preprint*, arXiv:1906.01824, 2019.

[42] S. Hochreiter and J. Schmidhuber, "Long short-term memory," *Neural Computation*, vol. 9, no. 8, pp. 1735–1780, 1997.

[43] D. P. Kingma and J. Ba, "Adam: A method for stochastic optimization," *arXiv preprint*, arXiv:1412.6980, 2014.

[44] A. Varga and R. Moore, "Hidden Markov model decomposition of speech and noise," in *Proc. Int. Conf. on Acoustics, Speech, and Signal Processing*, pp. 845–848, 1990.

[45] B. Schuller, G. Rigoll, and M. Lang, "Hidden Markov model-based speech emotion recognition," in *Proc. IEEE Int. Conf. on Acoustics, Speech, and Signal Processing (ICASSP)*, vol. 2, pp. II–1, 2003.

[46] A. Krogh et al., "Predicting transmembrane protein topology with a hidden Markov model: Application to complete genomes," *Journal of Molecular Biology*, vol. 305, no. 3, pp. 567–580, 2001.

[47] E. L. Sonnhammer et al., "A hidden Markov model for predicting transmembrane helices in protein sequences," *ISMB*, vol. 6, pp. 175–182, 1998.

[48] A. Srivastava et al., "Credit card fraud detection using hidden Markov model," *IEEE Trans. on Dependable and Secure Computing*, vol. 5, no. 1, pp. 37–48, 2008.

[49] J. Li, A. Najmi, and R. M. Gray, "Image classification by a two-dimensional hidden Markov model," *IEEE Trans. on Signal Processing*, vol. 48, no. 2, pp. 517–533, 2000.

[50] M. R. Hassan and B. Nath, "Stock market forecasting using hidden markov model: a new approach," in *Proc. 5th Int. Conf. on Intelligent Systems Design and Applications (ISDA)*, pp. 192–196, 2005.

[51] Z. Aharoni et al., "Capacity of continuous channels with memory via directed information neural estimator," in *IEEE Int. Symp. on Information Theory (ISIT)*, pp. 2014–2019, 2020.

[52] I. Goodfellow et al., "Generative adversarial nets," in *Advances in Neural Information Processing Systems*, pp. 2672–2680, 2014.

# Part II

## Wireless Networks for Machine Learning

# 12 Collaborative Learning over Wireless Networks: An Introductory Overview

Emre Ozfatura, Deniz Gündüz, and H. Vincent Poor

## 12.1 Introduction

The number of devices connected to the Internet has already surpassed 1 billion. With the increasing proliferation of mobile devices, the amount of data collected and transmitted over wireless networks is growing at an exponential rate. Along with this growth in data traffic, its content is also rapidly changing thanks to new applications, such as autonomous systems, factory automation, and wearable technologies, referred to as the Internet of Things (IoT) paradigm in general. While the current data traffic is dominated by video and voice content that are transmitted mainly for user consumption, data generated by IoT devices is intended for machine analysis and inference in order to incorporate intelligence into the underlying IoT applications. The current approach to IoT intelligence is to offload all the relevant data to a cloud server and train a powerful machine learning (ML) model using all the available data and processing power. However, such a centralized solution is not applicable in many ML applications due to the significant communication latency it would introduce. Moreover, IoT devices are typically limited in power and bandwidth, and the communication links become a major bottleneck as the volume of collected data increases. For example, autonomous cars are expected to generate 5–20 terabytes of data each day. The increasing data volume becomes particularly challenging when the information density of the collected data is low; that is, large volumes of data need to be offloaded with only limited relevant information for the underlying learning task. Centralized approaches also incite privacy concerns, and users are increasingly averse to sharing their data with third parties even in return of certain utility. Privacy concerns can be particularly deterring for personal data, which is often the case with IoT devices, even if it may not be obvious at first.

The solution to all these limitations is to bring the intelligence to the network edge by enabling wireless devices to implement learning algorithms in a distributed and collaborative manner [1, 2]. The two core components behind the recent success of ML algorithms are massive datasets and computational power, which allow training highly complex models. Both of these are abundant at the network edge, but in a highly distributed manner. Hence, a main challenge in edge intelligence is to design learning algorithms that can exploit these distributed resources in a seamless and efficient manner.

In this chapter, we mainly focus on collaborative training across wireless devices. Training a ML model is equivalent to solving an optimization problem, and many distributed optimization algorithms have been developed over the last decades. These distributed ML algorithms provide data locality; that is, a joint model can be trained collaboratively while the data available at each participating device remains local. This addresses, to some extend, the privacy concern. They also provide computational scalability as they allow exploiting computational resources distributed across many edge devices. However, in practice, this does not directly lead to a linear gain in the overall learning speed with the number of devices. This is partly due to the communication bottleneck limiting the overall computation speed. Additionally, wireless devices are highly heterogeneous in their computational capabilities, and both their computation speed and communication rate can be highly time-varying due to physical factors. Therefore, distributed learning algorithms, particularly those to be implemented at the wireless network edge, must be carefully designed, taking into account the impact of time-varying communication network as well as the heterogeneous and stochastic computation capabilities of devices.

In recent years, many solutions have been proposed in the ML literature to reduce the communication load of distributed ML algorithms to mitigate the communication bottleneck; however, these solutions typically aim to reduce the amount of information that needs to be exchanged between the computing servers, assuming rate-limited perfect communication links, and they do not take into account the physical characteristics of the communication medium. However, such an abstraction of the communication channel is far from reality, particularly in the case of wireless links. Converting the wireless channel into a reliable bit pipe may not even be possible (e.g., in the case of fading channels) or may introduce substantial complexity and delays into the system and increase the energy cost. Moreover, such an approach is based on an inherent separation between the design of ML algorithms and the communication protocols enabling the exchange of messages between devices.

Our goal in this chapter is to show that the speed and final performance of distributed learning techniques can be significantly improved by taking the wireless channel characteristics into account. We will provide several examples and point to some of the recent literature for further details. We will mainly focus on the joint optimization of the parameters of the underlying communication protocols (e.g., resource allocation, scheduling, etc.) together with the underlying distributed learning framework. An alternative approach is to design the learning and communication algorithms jointly, completely stepping out of the current digital communication framework [1, 3, 4].

In the remainder of the chapter, we first introduce the general collaborative learning problem. We then present centralized and decentralized approaches to collaborative learning and highlight federated learning (FL) as a popular distributed learning framework. Afterward we will overview some of the techniques that can be used to reduce the communication load of distributed learning. We will then present device scheduling and resource allocation algorithms for distributed learning over wireless networks. We should emphasize that the goal of this chapter is not to provide a comprehensive survey of all existing works, but to provide an introduction to the challenges and typical solution approaches.

## 12.2    Collaborative Learning

Many parameterized ML problems can be modeled as the following *stochastic optimization problem*:

$$\min_{\boldsymbol{\theta} \in \mathbb{R}^d} \mathbb{E}_{\zeta \sim \mathcal{P}} F(\boldsymbol{\theta}, \zeta), \tag{12.1}$$

where $\boldsymbol{\theta} \in \mathbb{R}^d$ denotes the model parameters to be optimized, $\zeta$ is the random data sample with distribution $\mathcal{P}$, and $F$ is the problem specific stochastic loss function. Typically, the underlying distribution $\mathcal{P}$ is not known, and instead we have access to a dataset $\mathcal{D}$ sampled from this distribution. Accordingly, we minimize the empirical loss function $\mathbb{E}_{\zeta \sim \mathcal{D}} F(\boldsymbol{\theta}, \zeta)$. This problem is often solved iteratively using stochastic gradient descent (SGD), where the model is updated at each iteration along the direction of the gradient estimate, computed using a subset of the dataset. That is, at iteration $t$, we update the model $\boldsymbol{\theta}_t$ as follows:

$$\boldsymbol{\theta}_{t+1} = \boldsymbol{\theta}_t - \eta_t \cdot \mathbf{g}_t, \tag{12.2}$$

where

$$\mathbf{g}_t = \nabla_{\boldsymbol{\theta}_t} F(\boldsymbol{\theta}_t, \zeta_t), \tag{12.3}$$

$\zeta_t$ is a random sample from the dataset $\mathcal{D}$, and $\eta_t$ is the learning rate at iteration $t$. Note that $\mathbf{g}_t$ is an unbiased estimate of the full gradient of $\mathbb{E}_{\zeta \sim \mathcal{D}} F(\boldsymbol{\theta}, \zeta)$. SGD is guaranteed to converge to the optimal solution under certain conditions on function $F$ (e.g., convex and smooth), and it is particularly beneficial in practice when the dataset is large, and hence, a full gradient computation at each iteration is costly.

In the distributed setting, we consider $N$ devices denoted by set $\mathcal{S}$, each with its own local dataset, denoted by $\mathcal{D}_i$. The stochastic optimization problem we gave earlier can be solved in a distributed manner over these $N$ devices by rewriting the minimization problem in Eq. (12.1) as

$$\min_{\boldsymbol{\theta} \in \mathbb{R}^d} f(\boldsymbol{\theta}) = \frac{1}{N} \sum_{i=1}^{N} \underbrace{\mathbb{E}_{\zeta \sim \mathcal{D}_i} F(\boldsymbol{\theta}, \zeta)}_{:=f_i(\boldsymbol{\theta})}, \tag{12.4}$$

which is an average of $N$ stochastic functions. The main idea behind collaborative learning is to solve the minimization problem in Eq. (12.4) in a distributed manner, such that each device tries to minimize its local loss function $f_i(\boldsymbol{\theta})$, while seeking a consensus with other devices on the global model $\boldsymbol{\theta}$. The consensus is facilitated either with the help of a central entity called a parameter server (PS) or by each device communicating with only a limited number of neighboring devices. Hence, based on the consensus framework, collaborative learning methods can be divided into two classes, *centralized* and *decentralized*. The centralized approach has received significantly more attention in the literature due to its simplicity in both the implementation and the analysis. Before going into the details of the solution techniques, we briefly introduce the distributed learning problem in both cases and highlight the main differences.

## 12.2.1   Centralized Learning

We classify a distributed learning framework as centralized when the learning framework is orchestrated by a PS. To be more precise, in the centralized framework each user/worker performs local computation under the supervision of the PS, and these computations are then utilized at the PS to solve the optimization problem. One of the most popular centralized strategies, particularly for training deep neural network (DNN) architectures, is *parallel synchronous stochastic gradient descent (PSSGD)*.

We recall that the objective in Eq. (12.4) is written as the sum of $N$ functions; accordingly, the PSSGD works as follows: at the beginning of iteration $t$, each device pulls the current global model $\boldsymbol{\theta}_t$ from the PS and computes the local gradient estimate to minimize its own loss function $f_i(\boldsymbol{\theta}_t)$:

$$\mathbf{g}_{i,t} = \nabla_{\boldsymbol{\theta}} F(\boldsymbol{\theta}_t, \zeta_{i,t}), \tag{12.5}$$

where $\zeta_{i,t}$ is the data randomly sampled from the local dataset $\mathcal{D}_i$ in the $t$th iteration. We note here that the devices can also use mini-batches with multiple samples rather than a single data sample to compute the local gradient estimate. Once each device completes the computation of the local gradient estimate, the results are sent to the PS to be aggregated and used to update the model parameter $\boldsymbol{\theta}$ as follows:

$$\boldsymbol{\theta}_{t+1} = \boldsymbol{\theta}_t - \eta_t \cdot \frac{1}{|\mathcal{S}|} \sum_{i \in \mathcal{S}} \mathbf{g}_{i,t}, \tag{12.6}$$

where $\eta_t$ is the learning rate at iteration $t$.

The PSSGD algorithm is presented in Algorithm 12.1. Here we note that, the algorithm is referred to as synchronous since the PS waits for all the devices to send their local gradient estimates before the model update, and at each iteration $t$, all the devices use the same global model to compute their local gradient estimates. This synchronization may induce delays when the computation speed of some workers

---

**Algorithm 12.1** Parallel synchronous stochastic gradient descent (PSSGD).

1: **for** $t = 1, 2, \ldots$ **do**
2:     **for** $i \in \mathcal{S}$ **do** in parallel
3:         Pull recent model $\boldsymbol{\theta}_t$ from PS
4:         Compute and send local SGD: $\mathbf{g}_{i,t} = \nabla_{\boldsymbol{\theta}} F(\boldsymbol{\theta}_t, \zeta_{i,t})$
5:     **Model update**:
6:     $G_t = \frac{1}{|\mathcal{S}|} \sum_{i \in \mathcal{S}} \mathbf{g}_{i,t}$
7:     $\boldsymbol{\theta}_{t+1} = \boldsymbol{\theta}_t - \eta_t \cdot G_t$

is slower than the others; hence, asynchronous variations of the centralized learning strategy have also been studied in the literature [5–7].

## 12.2.2    Decentralized Learning

Centralized learning is based on the star topology; that is, a central/ edge server, which can be an access point or a base station in the wireless context, orchestrates the devices participating in the collaborative training process. While this may be preferable in the presence of a base station that can communicate directly with all the devices in the coverage area, it may increase the burden on the cell-edge devices, and the multiple access channel from the devices to a single central server may become a bottleneck and limit the number of devices that can participate in the learning process at each iteration. Moreover, the star topology leads to a single point of failure and increases the privacy and security risks. The alternative is fully decentralized learning, where the devices directly communicate with each other in a device-to-device fashion. Decentralized learning has been extensively studied in the literature [8–12].

The main alteration in the decentralized scenario takes place in the consensus step, in which the devices combine their local models with those of their neighbors according to a certain rule that is driven by the network topology. Therefore, compared to the star topology used in centralized learning, the models at different devices are no longer fully synchronized after each global averaging step, introducing additional noise in the framework. Formally speaking, each device $i$ seeks a consensus with the neighboring devices according to the following combining strategy:

$$\tilde{\boldsymbol{\theta}}_{i,t} = \sum_{j \in \mathcal{S}} \mathbf{W}_{ij} \boldsymbol{\theta}_{j,t}, \tag{12.7}$$

where $\boldsymbol{\theta}_{j,t}$ denotes the model at device $j$ at iteration $t$, and $\mathbf{W}$ is a weight matrix. We set $\mathbf{W}_{ij} = 0$ if there is no connection between device $i$ and device $j$. Under certain assumptions on $\mathbf{W}$, such as being *doubly stochastic*, convergence of the decentralized framework has been shown, and its convergence speed is driven by the second largest *eigenvalue* of $\mathbf{W}$. The decentralized learning algorithm is presented in Algorithm 12.2.

The weight matrix $\mathbf{W}$ can be constructed based on the network topology, which can be represented by a graph $\mathcal{G}$ in several ways. A common approach is to use the *graph Laplacian*, such that $\mathbf{W}$ is written as

$$\mathbf{W} = \mathbf{I} - \frac{1}{d_{\max} + 1}(\mathbf{D} - \mathbf{A}), \tag{12.8}$$

---

**Algorithm 12.2** Decentralized learning.

---

**for** $t = 1, \ldots, T$ **do**

    For $i \in \mathcal{S}$ in parallel $\mathbf{g}_{i,t} = \nabla_{\boldsymbol{\theta}} \mathcal{L}_i(\boldsymbol{\theta}_{i,t}, \zeta_{i,t})$

    Consensus step: $\tilde{\boldsymbol{\theta}}_{i,t} = \sum_{j \in \mathcal{S}} \mathbf{W}_{ij} \boldsymbol{\theta}_{j,t}$

    Update local model: $\boldsymbol{\theta}_{i,t} = \tilde{\boldsymbol{\theta}}_{i,t} + \eta(t)\mathbf{g}_{i,t}$

---

where $\mathbf{A}$ is the adjacency matrix of the graph $\mathcal{G}$ representing the connectivity between the devices, $\mathbf{D}$ is a diagonal matrix whose entry in position $D_{ii}$ is the degree of the node corresponding to device $d_i$ in $\mathcal{G}$, and $d_{\max}$ is the maximum node degree. One can easily check that weight matrix designed according to Eq. (12.8) will be symmetric and doubly stochastic. For further details on the network topology and the convergence behavior readers can refer to [13, 14] for the implementation of decentralized learning over a wireless network.

## 12.3     Communication Efficient Distributed ML

One of the main challenges in distributed learning is the communication bottleneck due to the large size of the trained models. Numerous communication efficient learning strategies have been proposed in the ML literature to reduce the number of bits exchanged between the devices and the PS per global iteration. We classify these approaches into three main groups – namely *sparsification*, *quantization*, and *local updates* – and present each of these approaches in detail in this section. We would like to highlight that these strategies are independent of the communication medium and the communication protocol employed to exchange model updates or gradient estimates between the devices and the PS, as they mainly focus on reducing the size of the messages exchanged. Therefore, these techniques can be incorporated into the resource allocation and device selection policies that will be presented in Section 12.4 when optimizing learning algorithms over wireless networks.

### 12.3.1     Sparsification

The objective of sparse SGD is to transform the $d$-dimensional gradient estimate $\mathbf{g}$ in Eq. (12.3) to its sparse representation $\tilde{\mathbf{g}}$, where the nonzero elements of $\tilde{\mathbf{g}}$ are equal to the corresponding elements of $\mathbf{g}$. Sparsification can be considered as applying a $d$-dimensional mask vector $\mathbf{M} \in \{0, 1\}^d$ on $\mathbf{g}$; that is,

$$\tilde{\mathbf{g}} = \mathbf{M} \otimes \mathbf{g}, \qquad (12.9)$$

where $\otimes$ denotes element-wise multiplication. We denote the *sparsification level* of this mask by

$$\phi \triangleq \frac{||\mathbf{M}||_1}{d} \ll 1 . \qquad (12.10)$$

It has been shown that during the training of complex DNN architectures, such as ResNet [15] or VGG [16], use of sparse SGD with $\phi \in [0.01, 0.001]$ provides a significant reduction in the communication load with minimal or no loss in the generalization performance [17–25]. Next, we give a brief overview of some of the common sparsification strategies used in practice.

**Random Sparsification**

In random sparsification, introduced in [18], each element of vector $\mathbf{g}$ is set to zero independently with a prescribed probability. The $i$th element of the sparsified vector $\tilde{\mathbf{g}}^{\text{rand}}$, for $i = 1, \ldots, d$, is obtained by

$$\tilde{\mathbf{g}}_i^{\text{rand}} = \begin{cases} \frac{\mathbf{g}_i}{\mathbf{p}_i}, & \text{with probability } \mathbf{p}_i, \\ 0, & \text{with probability } 1 - \mathbf{p}_i. \end{cases}$$

This is an unbiased operator since

$$\mathbb{E}[\tilde{\mathbf{g}}_i^{\text{rand}}] = \mathbf{p}_i \times \frac{\mathbf{g}_i}{\mathbf{p}_i} = \mathbf{g}_i, \tag{12.11}$$

and its variance is bounded by

$$\mathbb{E}\left[\sum_{i=1}^{d} \left(\tilde{\mathbf{g}}_i^{\text{rand}}\right)^2\right] = \sum_{i=1}^{d} \frac{\mathbf{g}_i^2}{\mathbf{p}_i}. \tag{12.12}$$

The key design issue behind random gradient sparsification is to minimize the number of nonzero values while controlling the variance, equivalently, solving the following optimization problem:

$$\textbf{P1:} \quad \min_{p_1, \ldots, p_d} \sum_{i=1}^{d} \mathbf{p}_i$$

$$\text{subject to: } \sum_{i=1}^{d} \frac{\mathbf{g}_i^2}{\mathbf{p}_i} \leq (1 + \epsilon) \sum_{i=1}^{d} \mathbf{g}_i^2, \tag{12.13}$$

for some given $\epsilon > 0$. The solution of this optimization problem is given by [18]

$$\mathbf{p}_i = \min(\lambda \cdot \mathbf{g}_i, 1), \tag{12.14}$$

where $\lambda$ is a parameter chosen based on $\epsilon$ and $\mathbf{g}$.

**Synchronous Sparse Parameter Averaging**

In this strategy, at each iteration, only a subset of the parameters are averaged to reduce the communication load. We call it synchronous since all the devices use an identical sparsification mask $\mathbf{M}_t$ at iteration $t$. The sparse parameter averaging strategy is executed in two steps [19]: In the first step, device $i$ updates its model based on the local gradient estimate

$$\boldsymbol{\theta}_{i,t+1/2} = \boldsymbol{\theta}_{i,t} - \eta_t \cdot \nabla_{\boldsymbol{\theta}} F(\boldsymbol{\theta}_{i,t}, \zeta_{i,t}). \tag{12.15}$$

Then, every device masks its local model with $\mathbf{M}_t$ and shares the masked parameters with the PS for aggregation. Accordingly, $\boldsymbol{\theta}_{i,t+1}$ evolves over the iterations in the following way:

$$\boldsymbol{\theta}_{i,t+1} = \frac{1}{N} \sum_{n=1}^{N} \boldsymbol{\theta}_{n,t+1/2} \otimes \mathbf{M}_t. \tag{12.16}$$

The convergence of the synchronous sparse parameter averaging strategy can be shown under the constraint that the sampling period of each parameter should not exceed some predefined value $\tau_{max}$; that is,

$$|| \otimes_{t=j}^{k} (1 - \mathbf{M}_t)||_1 = 0, \tag{12.17}$$

for any $k, j > 0 : k - j \geq \tau_{max}$.

## Top-$K$, Rand-$K$, and $R$-top-$K$ Sparsification

Top-$K$ sparsification is one of the most commonly used strategies for gradient sparsification. In the top-$K$ sparsification strategy each device constructs a sparsification mask $\mathbf{M}_{i,t}$ by identifying the indices of the largest $K$ values in $|\mathbf{g}_{i,t}|$ [20, 22, 26]. We use $S_{top}(\cdot, K)$ do denote this operation:

$$\mathbf{M}_{i,t} = S_{top}(|\mathbf{g}_{i,t}|, K). \tag{12.18}$$

Further details on the convergence of top-$K$ sparsification can be found in [22, 26].

The rand-$K$ sparsification strategy [22] selects the sparsification mask $\mathbf{M}_{i,t}$ randomly from the set of masks with sparsification level $K$:

$$\mathcal{M} = \left\{ \mathbf{M} : \{0,1\}^d, ||\mathbf{M}||_1 = K \right\}. \tag{12.19}$$

We note that, unlike the random sparsification strategy in [18], rand-$K$ and top-$K$ do not result in an unbiased compression strategy. In the case of rand-$K$, scaling $\mathbf{M}_{i,t}$ with $d/K$ can be used to obtain unbiased compressed gradient estimates; however, such correction also scales the variance of the compression scheme and thus may not be desirable in practice [22]. In general, top-$K$ sparsification strategy has been shown to perform better compared to rand-$K$ in practice in terms of the test accuracy and convergence speed. However, top-$K$ sparsification comes with the additional complexity due to the sorting of the elements of $|\mathbf{g}_{i,t}|$. It also requires increased communication load since the indices of the nonzero elements of each mask vector $\mathbf{M}_{i,t}$ should also be transmitted. This may not be a requirement for rand-$K$ if a pseudo-random generator with a common seed is used so that all the devices can generate the same mask.

We also note that, when employed for distributed training of DNN architectures, these sparse communication strategies can be applied to each layer of the network separately, since it is observed that different layers have different tolerance to sparsification of their weights. We refer the readers to [17] for further information on gradient sparsification. Finally, a hybrid approach, called $R$-top-$K$, is considered in [23]. In this scheme, each device first identifies the largest $R$ values in $|\mathbf{g}_{i,t}|$ and then selects $K$ of them randomly. This hybrid approach results in better compression as well as less bias.

We also want to remark that the use of top-$K$ sparsification locally (that is, having a separate mask for each worker) has the drawback that the sparsification is achieved only in the uplink direction, whereas the global model update from the PS to the devices will not be sparse. Moreover, the mismatch between the sparsity patterns across the workers limits the use of more efficient communication protocols, such as `all-reduce`. This problem is addressed in [27], where a majority voting based strategy is proposed to seek a consensus among the workers to form a common sparsification mask.

### Error Accumulation

The idea behind the error accumulation is to compensate in each iteration for the compression error that has been introduced in the previous iteration by simply adding the previous compression error to the current gradient estimate. It has been shown that [20, 22, 26] error accumulation improves the convergence performance. When error accumulation is employed, the sparsification step at the devices is rewritten as

$$\widetilde{\mathbf{g}}_{n,t} = \mathbf{M}_{n,t} \otimes (\mathbf{g}_{n,t} + \mathbf{e}_{n,t}), \tag{12.20}$$

where $\mathbf{e}_{n,t}$ is the accumulated error from the previous iteration. Following gradient sparsification, the compression error is updated as follows:

$$\mathbf{e}_{n,t+1} = (1 - \mathbf{M}_{n,t}) \otimes (\mathbf{g}_{n,t} + \mathbf{e}_{n,t}). \tag{12.21}$$

The overall sparse SGD framework with error accumulation is illustrated in Algorithm 12.3. We remark that gradient sparsification can be employed both in the device-to-PS (uplink) and PS-to-device (downlink) directions to improve the communication efficiency further. In Algorithm 12.3, sparsification at the PS side

---

**Algorithm 12.3** Sparse SGD with error accumulation.

---

1: **for** $t = 1, \ldots, T$ **do**
2:     **Device side:**
3:     **for** $n = 1, \ldots, N$ **do** in parallel
4:       Receive $\widetilde{\mathbf{G}}_{t-1}$ from PS
5:       **Update model:** $\boldsymbol{\theta}_t = \boldsymbol{\theta}_{t-1} + \eta_t \widetilde{\mathbf{G}}_{t-1}$
6:       **Compute SGD:** $\mathbf{g}_{n,t} = \nabla_{\boldsymbol{\theta}} F(\boldsymbol{\theta}_t, \zeta_{n,t})$
7:       **Sparsification with error accumulation:**
8:       $\widetilde{\mathbf{g}}_{n,t} = \mathbf{M}_{n,t} \otimes (\mathbf{g}_{n,t} + \mathbf{e}_{n,t})$
9:       **Update the error:**
10:      $\mathbf{e}_{n,t+1} = (1 - \mathbf{M}_{n,t}) \otimes (\mathbf{g}_{n,t} + \mathbf{e}_{n,t})$
11:      Send $\widetilde{\mathbf{g}}_{n,t}$ to PS
12:     **Sparse communication PS side:**
13:     **Aggregate local sparse gradient:**
14:     $\widetilde{\mathbf{G}}_t = \sum_{n \in [N]} \widetilde{\mathbf{g}}_{n,t}$
15:     send $\widetilde{\mathbf{G}}_t$ to all devices
16:     **Aggregate local sparse gradients with error accumulation:**
17:     $\widetilde{\mathbf{G}}_t = \mathbf{M}_{n,t} \otimes (\sum_{n \in [N]} \widetilde{\mathbf{g}}_{n,t} + \mathbf{e}_t)$
18:     Send $\widetilde{\mathbf{G}}_t$ to all devices
19:     **Update error:**
20:     $\mathbf{e}_{t+1} = (1 - \mathbf{M}_t) \otimes (\widetilde{\mathbf{G}}_{n,t} + \mathbf{e}_{n,t})$

---

is applied in lines 16–20. A further analysis on this double sparsification strategy can be found in [28]. In general, unbiased compression schemes are preferred for distributed learning as they lend themselves to theoretical convergence analysis; however, unbiased sparsification methods typically suffer from large variation, making them less desirable for practical applications [22].

The key benefit of the error accumulation mechanism is to make a biased sparsification method an efficient compression strategy for distributed learning, backed with a theoretical convergence analysis. It is shown in [22] that a gradient sparsification scheme with error accumulation converges if it is in the form of a *k-contraction operator*:

DEFINITION 12.1   For a parameter $0 < k \leq d$, a *k-contraction operator* is a (possibly randomized) operator comp : $\mathbb{R}^d \mapsto \mathbb{R}^d$ that satisfies the contraction property

$$\mathbb{E}\left[||\mathbf{x} - \text{comp}(\mathbf{x})||^2\right] \leq \left(1 - \frac{k}{d}\right)||\mathbf{x}||^2, \quad \forall \mathbf{x} \in \mathbb{R}^d. \tag{12.22}$$

In [22], it is also shown that both the rand-$K$ and top-$K$ sparsification schemes are indeed $K$-contraction operators. The notion of contraction operation is further generalized in [29].

## Sparse Representation

When a sparsity mask is employed to reduce the communication load in distributed learning, the mask also needs to be communicated to the PS together with the nonzero entries. In general, $\log_2 d$ bits are needed to convey the position of each nonzero element of a sparse masking vector. However, this number can be reduced by employing the following coding scheme. Let $\mathbf{v}$ be a sparse vector with sparsification level $\phi$. Assume that $\mathbf{v}$ is divided into equal-length blocks of size $1/\phi$. Position of a nonzero value within a particular block can be represented by only $\log_2(1/\phi)$ bits. In order to identify each block we use an additional bit in the following way: we use 0 to skip to the next block, and 1 to indicate that the next $\log_2(1/\phi)$ bits represent the location within the current block. Hence, on average $\log_2(1/\phi) + 1$ bits are needed to represent the location of each nonzero value, and a total of $\phi d$ bits to determine the block indices.

In the decoding part, given the encoded binary vector $\mathbf{v}_{loc}$ for the position of the nonzero values in $\mathbf{v}$, the PS starts reading the bits from the first index and checks whether it is 0 or 1. If it is 1, then the next $\log_2(1/\phi)$ bits are used to recover the position of the nonzero value in the current block. If the index is 0, then the PS increases the *block index*, which tracks the current block, by one, and moves on to the next index. The overall decoding procedure is illustrated in Algorithm 12.4.

For example, consider a sparse vector of size $d = 24$ and $\phi = 1/8$; that is, there are only three nonzero values, and let those values be at indices 1, 5, and 17. Since $\phi = 1/8$, the vector is divided into three blocks, each of size eight, and the position

---

**Algorithm 12.4** Sparse position decoding.

1: Initialize: $pointer = 0$
2: **while** $pointer < length(\mathbf{v}_{loc})$ **do**
3:     **if** $\mathbf{v}_{loc}(pointer) = 0$ **then**
4:         $blockindex = blockindex + 1$
5:         $pointer = pointer + 1$
6:     **else**
7:         Recover $Intra\,Block\,Position$:
8:         Read the next $\log_2(1/\phi)$ bits
9:         Recover the location of a non-zero value:
10:         $(1/\phi) \cdot blockindex + Intra\,Block\,Position$
11:         $pointer = pointer + \log_2(1/\phi) + 1$

---

of nonzero values within each block can be represented by 3 bits (i.e., $000, 100, 000$); hence, overall the sparse representation can be written as a bit stream of

$$100011000\underline{0}10000, \tag{12.23}$$

where the bits in normal font represent the position within the block, bold bits indicate that the following three bits refer to the position within the current block, and the underlined bits represent the end of a block. We note that other methods such *Elias coding* [30] and *Golomb coding* [31] can also be used for an efficient representation of the sparse mask vector.

## 12.3.2    Quantization

Even after applying a sparsifying transform as in Eq. (12.9), we inherently assume that the exact values of the local updates, $\tilde{\mathbf{g}}$, are communicated from the devices to the PS. In practice, the communication links from the devices to the PS are finite capacity, and hence, these real-valued vectors must be quantized. In the standard setting, floating point precision with 32 bits is assumed to represent each real value, so 32 bits are required to represent each element of the local gradient estimate. Quantization techniques aim to represent each element of the gradient estimates with as few bits as possible to reduce the communication load [30, 32–40]. In the most extreme case, only the sign of each element is sent, using only a single bit per dimension, to achieve up to a 32 times reduction in the communication load [34, 38–40].

**Stochastic Uniform Quantization**

In stochastic uniform quantization strategy [30, 32], denoted by $\mathcal{Q}_s(\cdot)$, we uniformly allocate representation points as in uniform quantization, but instead of mapping each number to the nearest representation point, we probabilistically assign them to one of the two nearest points. We first normalize and take the absolute values of the elements of the given vector $\mathbf{u}$: $\mathbf{u} \in \mathbb{R}^d$:

$$\tilde{\mathbf{u}} = \frac{|\mathbf{u}|}{||\mathbf{u}||}. \tag{12.24}$$

Then, consider dividing the interval $[0, 1]$ into $L$ equal length sub-intervals $I_1, \ldots, I_L$, where $I_l = [I_l^{lower}, I_l^{upper})$. In the last step, each normalized value $\tilde{\mathbf{u}}_i$ is randomly mapped to one of the boundary points of the corresponding interval according to $q_{rand}(\cdot)$ and scaled by the sign of $\mathbf{u}_i$. Let $\tilde{\mathbf{u}}_i \in I_l$; we have

$$q_{rand}(\tilde{\mathbf{u}}_i) = \begin{cases} I_l^{lower}, & \text{with probability } (I_l^{upper} - \tilde{\mathbf{u}}_i)L \\ I_l^{upper}, & \text{with probability } (\tilde{\mathbf{u}}_i - I_l^{lower})L. \end{cases}$$

Hence, overall, the quantized representation of $\mathbf{u}$ can be written as

$$\mathcal{Q}_s(\mathbf{u})_i = \text{sign}(\mathbf{u}_i) q_{rand}(\tilde{\mathbf{u}}_i) ||\mathbf{u}||. \tag{12.25}$$

In [33], stochastic quantization framework is combined with the error accumulation mechanism.

### Ternary Gradient Quantization

Similarly to stochastic gradient quantization, ternary gradient quantization is another unbiased compression strategy, which can be considered as a mixture of random sparsification and stochastic quantization [39]. It is called ternary since the quantized values can take only three values:

$$Q_{tern}(\mathbf{g})_i \in \{-1, 0, +1\}. \tag{12.26}$$

It can be represented as follows:

$$Q_{tern}(\mathbf{g}) = g_{max} \cdot \text{sign}(\mathbf{g}) \otimes \mathbf{b} \tag{12.27}$$

where

$$g_{max} = ||\mathbf{g}||_\infty, \tag{12.28}$$

and $\mathbf{b}$ is a random binary vector of independent Bernoulli variables, where

$$\mathbf{b}_i = \begin{cases} 1, & \text{with probability } \frac{|\mathbf{g}_i|}{g_{max}}, \\ 0, & \text{with probability } 1 - \frac{|\mathbf{g}_i|}{g_{max}}. \end{cases}$$

### SignSGD and 1-bit Quantization

One of the earliest schemes that has been employed in distributed learning to reduce the communication load is the 1-bit quantization strategy [34, 35], where each device quantizes its gradient using a single threshold value so that only a single bit is required to represent each element.

A similar approach that uses only one bit for the gradient values is *SignSGD* [36], where each device only sends the sign of the gradient values. Equivalently, in SignSGD the distributed learning framework works as a majority vote since the sum of the signs indicates the result of a majority vote for the gradient direction. The SignSGD algorithm is presented in Algorithm 12.5. We note that, thanks to the

---

**Algorithm 12.5** SignSGD.

1: **for** $t = 1, \ldots, T$ **do**
2:     **Device side:**
3:     **for** $n = 1, \ldots, N$ **do** in parallel
4:         **Update model:** $\boldsymbol{\theta}_t = \boldsymbol{\theta}_{t-1} + \eta \times \text{sign}\left(\sum_{n \in [N]} \text{sign}(\mathbf{g}_{n,t})\right)$
5:         **Compute SGD:** $\mathbf{g}_{n,t} = \nabla_{\boldsymbol{\theta}} F(\boldsymbol{\theta}_t, \zeta_{n,t})$
6:         Send $\text{sign}(\mathbf{g}_{n,t})$ to PS
7:     **Sparse communication PS side:**
8:     Send $\text{sign}\left(\sum_{n \in [N]} \text{sign}(\mathbf{g}_{n,t})\right)$ to devices

---

reduced dimensionality, SignSGD strategy also provides certain robustness against adversarial attacks [37].

### Scaled Sign Operator

Another quantization scheme is the scaled sign operator, which has a similar structure to ternary quantization, although it is not an unbiased compression scheme. The scaled sign operator sends the sign of the gradient values by scaling them with the mean value of the gradient:

$$\mathcal{Q}(\mathbf{g}) = \frac{||\mathbf{g}||_1}{d} \text{sign}(\mathbf{g}), \tag{12.29}$$

where $d$ is the dimension of vector $\mathbf{g}$. Although the scaled sign operator is biased, it is a $\delta$-approximate compressor; that is, it satisfies the following inequality:

$$||\mathcal{Q}(\mathbf{x}) - \mathbf{x}||_2^2 \leq 1 - \delta \cdot ||\mathbf{x}||_2^2. \tag{12.30}$$

In [38], it is shown that when a quantization scheme is a $\delta$-approximate compressor, then the quantized SGD framework converges when the quantization scheme is employed together with error accumulation.

Block-wise implementation of the scaled sign operator is considered in [39], where the gradient vector $\mathbf{g}$, or the momentum term if momentum SGD is used as an optimizer, is divided into $L$ disjoint smaller blocks $\{\mathbf{g}_1, \ldots, \mathbf{g}_L\}$, and the scaled sign operator is applied to each block separately. This way, variations on the mean gradient values of different layers can be incorporated to the compression procedure to further reduce the quantization error. We also note that in [39] quantization is employed both in the uplink and downlink directions. An efficient compression scheme for transmission in the PS-to-device direction is presented in [41].

## 12.3.3 Local Iteration

Local SGD, also known as *H-step averaging*, aims to reduce the frequency of global model updates at the PS and, hence, the communication of local model updates from the devices to the PS [26, 42–47]. In order to reduce the frequency of global aggregation, devices update their models locally for $H$ consecutive SGD iterations based on their local gradient estimates, before sending their local model updates to the PS. The global model evolves over communication rounds as follows:

$$\theta_{t+1} = \theta_t + \frac{1}{N} \sum_{n=1}^{N} \underbrace{\sum_{h=1}^{H} -\eta_t \cdot \mathbf{g}_{n,t}^h}_{\Delta \theta_{n,t}}, \tag{12.31}$$

where

$$\mathbf{g}_{n,t}^h = \nabla_\theta F(\theta_{n,t}^h, \zeta_{n,h}), \tag{12.32}$$

and

$$\theta_{n,t}^h = \theta_{n,t}^{h-1} - \eta_t \cdot \mathbf{g}_{n,t}^h, \tag{12.33}$$

where we set $\theta_{n,t}^0 = \theta_t$.

In this section, we presented three groups of strategies to reduce the communication load in distributed learning. We would like to highlight that these strategies can be combined to reduce the communication load further [31, 48]. In Algorithm 12.6, a general communication efficient distributed learning framework is presented where $C$ can be one of the compression operators previously explained or a combination of these schemes. We note that the lines 15–17 in Algorithm 12.6 are used instead of lines 13 and 14 when the compression strategy is also employed in the downlink direction.

---

**Algorithm 12.6** Compressed local SGD with error accumulation.

---

1: **for** $t = 1, \ldots, T$ **do**
2:     **Device side:**
3:     **for** $n = 1, \ldots, N$ **do** in parallel
4:         Receive $\widetilde{\Delta}_{t-1}$ from PS
5:         **Update model:** $\theta_{n,t} = \theta_{n,t-1} + \widetilde{\Delta}_{t-1}$
6:         Initialize $\theta_{n,t}^0 = \theta_{n,t}$
7:         **Perform local SGD for $H$ iteartions:**
8:         $\Delta_{n,t} = \theta_{n,t}^H - \theta_{n,t}^0$
9:         $\widetilde{\Delta}_{n,t} = C(\Delta_{n,t} + \mathbf{e}_{n,t})$
10:        Send $\widetilde{\Delta}_{n,t}$ to PS
11:        $\mathbf{e}_{n,t} = \Delta_{n,t} - \widetilde{\Delta}_{n,t}$
12:     **Sparse communication PS side:**
13:     $\widetilde{\Delta}_t = \sum_{n\in[N]} \widetilde{\Delta}_{n,t}$
14:     Send $\widetilde{\Delta}_t$ to all devices
15:     $\widetilde{\Delta}_t = C(\sum_{n\in[N]} \widetilde{\Delta}_{n,t} + \mathbf{e}_t)$
16:     Send $\widetilde{\Delta}_t$ to all devices
17:     $\mathbf{e}_{t+1} = \sum_{n\in[N]} \widetilde{\Delta}_{n,t} + \mathbf{e}_t - \widetilde{\Delta}_t$

## 12.3.4    Federated Learning (FL)

FL is a centralized distributed learning strategy, which was introduced to solve the stochastic optimization problem given in Eq. (12.4) in a distributed manner without sharing local datasets in order to offer a certain level of privacy to users. This is particularly relevant when sensitive data, such as personal medical records, are used to train a model, such as for digital healthcare or remote diagnosis applications [49, 50].

From the implementation perspective, FL framework differs from other centralized learning strategies since it aims to practice collaborative learning on a large scale. Therefore, practical implementations of FL require additional mechanisms to make them communication efficient. The first mechanism used in FL framework is the use of local SGD strategy to reduce the communication frequency between the PS and the devices. Second, to prevent overload at the PS, at each iteration $t$, only a subset of the active devices in the network, denoted by $\mathcal{S}_t \subseteq \mathcal{S}$, are selected to participate the learning process. The devices can be sampled randomly, or by some more advanced selection schemes as we will address later.

In a broad sense, FL employs distributed SGD, and each SGD iteration consists of three main steps: *device sampling/selection*, *local computation*, and *consensus/aggregation*. Once the PS samples the devices $\mathcal{S}_t \subseteq \mathcal{S}$ for the $t$th iteration, each device first pulls the latest global model $\boldsymbol{\theta}_{t-1}$ from the PS and then locally minimizes its own loss function $f_n(\boldsymbol{\theta}_t)$ with an SGD update:

$$\boldsymbol{\theta}_{n,t}^h = \boldsymbol{\theta}_{n,t}^{h-1} - \eta_t \cdot \mathbf{g}_{n,t}^h \qquad (12.34)$$

for $h = 1, \ldots, H$, where we set $\boldsymbol{\theta}_{i,t}^0 = \boldsymbol{\theta}_{t-1}$,

$$\mathbf{g}_{n,t}^h = \nabla_{\boldsymbol{\theta}} F(\boldsymbol{\theta}_{n,t}^h, \zeta_{n,h}), \qquad (12.35)$$

and $\zeta_{n,h}$ is the data sampled from dataset $\mathcal{D}_n$ in the $h$th local iteration.

In each iteration, each device carries out $H$ SGD steps before the aggregation at the PS. Once each device finishes its local updates, it sends the latest local model $\boldsymbol{\theta}_{n,t}^H$ to the PS for aggregation. The common approach for the aggregation/consensus in FL is to simply take the average of the participating devices' local models:

$$\boldsymbol{\theta}_t = \frac{1}{|\mathcal{S}_t|} \sum_{i \in \mathcal{S}_t} \boldsymbol{\theta}_{i,t}^H. \qquad (12.36)$$

However, we remark here that more advanced aggregation approaches have also been introduced recently in [51–53]. The generic FL framework iterating over these three steps, called as *federated averaging (FedAVG)*, is provided in Algorithm 12.7. We note that the particular scenario with full participation, $\mathcal{S}_t = \mathcal{S}$ and $H = 1$, is referred to as *federated SGD (FedSGD)*.

### Accelerated FL

There have been many efforts to accelerate FL convergence. A classical technique for accelerating SGD is the momentum method [54], which accumulates a velocity vector in the directions of persistent reduction in the objective function across iterations.

---

**Algorithm 12.7** Federated averaging (FedAVG) Federated Learning (FL).

1: **for** $t = 1, 2, \ldots$ **do**

2:     Choose a subset of devices $\mathcal{S}_t \subseteq \mathcal{S}$

3:     **for** $n \in \mathcal{S}_t$ **do** in parallel

4:       Pull $\boldsymbol{\theta}_{t-1}$ from PS: $\boldsymbol{\theta}_{n,t}^0 = \boldsymbol{\theta}_{t-1}$

5:       **for** $h = 1, \ldots, H$ **do**

6:         Compute SGD: $\mathbf{g}_{n,t}^h = \nabla_{\boldsymbol{\theta}} F(\boldsymbol{\theta}_{n,t}^{h-1}, \zeta_{n,h})$

7:         Update model: $\boldsymbol{\theta}_{n,t}^h = \boldsymbol{\theta}_{n,t}^{h-1} - \eta_t \mathbf{g}_{n,t}^h$

8:       Push $\boldsymbol{\theta}_{n,t}^H$

9:     **Federated averaging**: $\boldsymbol{\theta}_t = \frac{1}{|\mathcal{S}_t|} \sum_{i \in \mathcal{S}_t} \boldsymbol{\theta}_{n,t}^H$

---

**Algorithm 12.8** SlowMo framework.

1: **for** $t = 1, \ldots, T$ **do**

2:     **Local iteration:**

3:     Sample participating devices $\mathcal{S}_t \subseteq \mathcal{S}$

4:     **for** $i \in \mathcal{S}_t$ **do** in parallel

5:       $\boldsymbol{\theta}_{i,t}^0 = \boldsymbol{\theta}_{t-1}$

6:       **for** $h = 1, \ldots, H$ **do** local update:

7:         Compute SGD: $\mathbf{g}_{i,t}^h = \nabla_{\boldsymbol{\theta}} F(\boldsymbol{\theta}_{i,t}^{h-1}, \zeta_{i,h})$

8:         Update model: $\boldsymbol{\theta}_{i,t}^h = \boldsymbol{\theta}_{i,t}^{h-1} - \eta_t \mathbf{g}_{i,t}^h$

9:     **Communication phase:**

10:     **for** $i \in \mathcal{S}_t$ **do**

11:       Send to PS: $\boldsymbol{\Delta}_{i,t} \triangleq \boldsymbol{\theta}_{i,t}^H - \boldsymbol{\theta}_{t-1}$

12:     **Compute pseudo gradient:**

13:     $\bar{\mathbf{G}}_t = \frac{1}{|\mathcal{S}_t|} \frac{1}{\eta_t} \sum_{i \in \mathcal{S}_t} \boldsymbol{\Delta}_{i,t}$

14:     **Compute pseudo momentum:**

15:     $\mathbf{m}_{t+1} = \beta \mathbf{m}_t + \bar{\mathbf{G}}_t$

16:     **Model update:**

17:     $\boldsymbol{\theta}_{t+1} = \boldsymbol{\theta}_t - \alpha \eta_t \mathbf{m}_{t+1}$

---

State-of-the-art results for most ML tasks incorporate momentum. Momentum can also be incorporated into distributed SGD in various ways. *Global momentum* can be used at the PS when updating the global model, the devices can apply *local momentum* during their local update steps, or a combination of the two can also be used [43].

We can treat the FL framework as a combination of two loops, where in the inner loop devices update their local models, and in the outer loop PS updates the global model. In principle, it is possible to accelerate the convergence of FL by using different optimizers in either of these loops. One such strategy is the SlowMo framework [55], which is summarized in Algorithm 12.8. The core idea behind the SlowMo framework is to utilize the local model updates from the inner loop to compute a *pseudo gradient* at the PS, which is then used by the PS as the optimizer for the outer loop to update the global model. Formally speaking, each device scheduled for the $t$th iteration of the FL algorithm performs $H$ local updates as before and sends the accumulated model difference $\Delta_{n,t} \triangleq \theta_{i,t}^H - \theta_{t-1}$ to the PS (line 10 of Algorithm 12.8). PS utilizes the average of the model updates as the pseudo gradient and updates the momentum of the outer loop accordingly (line 14 of Algorithm 12.8), which is then used to update the model (line 16 of Algorithm 12.8). We note that SlowMo uses *momentum SGD* at the PS, but this approach can also be extended to other optimizers [56].

## 12.4    Device Selection and Resource Allocation in Distributed Learning Over Wireless Networks

A fundamental design problem in centralized collaborative learning is to find the optimal number of uplink devices and the optimal frequency of global model updates, or equivalently, the number of local iterations in between consecutive global updates, to seek a balance between the accuracy of the model update and the communication cost in order to achieve the best convergence performance based on the wall clock time. Collaborative learning across wireless devices introduces new challenges as the communication bottleneck becomes even more stringent due to the limited bandwidth available for model updates, varying nature of the channel quality across users and iterations, and the energy-limited nature of most wireless devices. Hence, the optimal collaborative learning design across wireless networks must take into account the channel conditions of the devices and optimize the wireless networking parameters together with the parameters of the collaborative learning algorithm [57–60].

The two most common device selection mechanisms employed in the FL framework are *random scheduling*, where devices participating in each iteration are randomly sampled, and *round-robin scheduling*, where the devices are divided into groups and the groups are scheduled according to a predefined order throughout the learning process. However, these scheduling mechanisms do not take into account the channel condition of each individual device, and since the global model update requires certain synchronization across selected devices at each iteration, devices with weaker channel conditions may become the bottleneck, resulting in significant delays.

To overcome the aforementioned issue, device selection mechanism in FL should be modified by taking into account the channel conditions of the devices. Let $L_{i,t}^{\text{comm}}$ and $L_{i,t}^{\text{comp}}$ be the communication and computation latency of the $i$th device at iteration $t$, respectively. We have $L_{i,t}^{\text{comm}} = d / R_{i,t}$, where $d$ is the size of the local update in bits, and $R_{i,t}$ (bits per unit time) is the transmission rate of the $i$th device at iteration $t$. The transmission rate of device $i$, $R_{i,t}$, is a function of its transmission power $P_{i,t}$, channel state $h_{i,t}$, and the resource allocation policy $\Phi_t$, which decides on the time/frequency resources allocated to each device at each iteration. From the latency perspective, collaborative learning in a wireless network setup can be formulated as a joint device selection ($\mathbf{\Pi}_t$) and power/resource allocation ($\mathbf{P}_t, \mathbf{\Phi}_t$) problem, where the objective can be written, generically, in the following form:

$$\min_{\mathbf{P}_t, \mathbf{\Pi}_t, \mathbf{\Phi}_t | \mathbf{h}_t} \max_{i \in \mathbf{\Pi}_t} \underbrace{\frac{d}{R_{i,t}(\mathbf{P}_t, \mathbf{h}_t, \mathbf{\Phi}_t)}}_{L_{i,t}^{\mathrm{comm}}} + L_{i,t}^{\mathrm{comp}}, \qquad (12.37)$$

where $\mathbf{\Pi}_t \subseteq \mathcal{S}$ denotes the set of scheduled devices at iteration $t$. We note that the objective in Eq. (12.37) depicts a general form, while in a practical scenario we may assume fixed transmission power, in which case the objective function reduces to joint device selection and resource allocation problem, and it can be further simplified to only a device selection problem by assuming a uniform, or a fixed resource allocation strategy. We can also impose a constraint on the number of scheduled devices at each round (i.e., $|\mathbf{\Pi}_t| = K, \forall t$) for some $0 < K \leq |\mathcal{S}|$, or leave it as an optimization parameter.

In the FL context, it is possible to establish a connection between the number of scheduled devices and the number of communication rounds to achieve a certain level of final accuracy. However, this assumes that the devices are scheduled independently and with identical probability, such that, although one device may be preferred over another at a certain iteration $t$, devices participate in the model updates equally in the long run in order not to introduce any bias in the overall process. However, in the wireless setting, scheduling statistics may not be identical for devices, such as when the access point acting as the PS is not equally distant to the devices. Such heterogeneity requires further constraints on communication efficient device selection strategies for FL. To clarify this point, we present an example that highlights why purely channel-dependent scheduling strategies are not sufficient in the FL framework.

In this example, we consider training a CNN architecture in a federated manner with the help of a single-antenna base station. We assume there are 100 single-antenna edge devices uniformly distributed around the base station within a radius of 500 meters, and at each iteration 20 devices participate in the global model update. In the simulation, we consider both large-scale path loss and small-scale Rayleigh fading with unit power and consider transmission at the corresponding channel capacity assuming a random channel realization at each iteration. We assume both BS and devices transmits with a fixed power of 15 dbm and 10 dbm, respectively, and we set the bandwidth to $B = 2 \times 10^7$ and the noise power to $N_0 = -204$ dbW/Hz.

We compare random scheduling with channel-aware scheduling, which schedules the devices in order to minimize the latency as in Eq. (12.37). We plot the test accuracy of the learned model throughout the learning process in Fig. 12.1. We observe that the channel-aware strategy performs better initially, quickly learning a reasonably good model, but its performance drops sharply after a certain number of iterations and eventually records a large generalization gap compared to the simple random scheduling strategy. This behavior is due to the biased updates since some devices, those closer to the access point, participate in the federated averaging updates more frequently since they have better channel states on average.

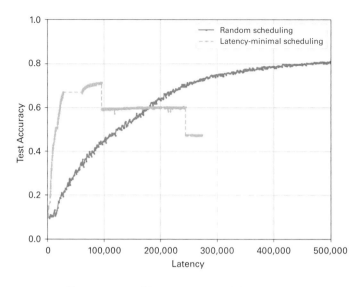

**Figure 12.1** Test accuracy with respect to overall latency (seconds) analyzed over 1,000 communication rounds for collaborative learning of classification on the CIFAR-10 dataset using a CNN architecture. In latency-minimal scheduling, we schedule the users with the best channel states at each iteration.

## Age-Aware Device Scheduling

As evident from this example, since only a certain portion of the devices are selected for model update in FL, the selection strategy may cause some devices to participate for frequently than others, resulting in biased model updates. Hence, it is important to regulate the frequency of devices' participation in the learning process to prevent any dominance of particular set of devices on the model update. To this end, in [58], an *age* based regulation strategy is introduced, where the term age is used as a metric to represent the number of iterations since the last scheduling of a device. Hence, age of the $i$th device, denoted by $a_i$, evolves over the iterations in the following way:

$$a_i(t) = \begin{cases} 0, & \text{if } i \in \mathcal{S}_t \\ a_i(t) + 1, & \text{otherwise.} \end{cases}$$

One can observe that the age metric in this context measures the *staleness* due to the lack of participation. To this end, the device scheduling strategy can be designed to minimize the impact of the *aggregate staleness* at each communication round. For this reason, [58] uses the following parameterized function, which guarantees a certain fairness across the devices according to the staleness of their updates:

$$f_\alpha(x) = \begin{cases} \frac{x^{1-\alpha}}{1-\alpha}, & \text{if } \alpha \neq 1 \\ \log(1 + x), & \alpha = 1. \end{cases}$$

Hence, under certain latency constraint on the communication round, the overall optimization problem can be rewritten as

$$\textbf{P2:} \qquad \min_{\mathbf{P}_t, \mathbf{\Pi}_t, \mathbf{\Phi}_t} \sum_{i \in \mathcal{S}_t} f(a_{i,t})$$

$$\text{subject to: } R_{i,t}(\mathbf{P}_t, \mathbf{\Phi}_t) \geq R_{min}, \quad \forall i \in \mathcal{S}_t, \tag{12.38}$$

$$\mathbf{P}_t \in \mathcal{P}, \mathbf{\Phi}_t \in \mathcal{F}, \forall t, \tag{12.39}$$

where $\mathcal{P}$ and $\mathcal{F}$ denote the feasible power and resource allocation sets, respectively. The first constraint in Eq. (12.38) is introduced to ensure that the duration of each communication round does not exceed a predefined latency constraint $L_{max} = \frac{d}{R_{min}}$.

The problem **P2** is fairly general. Consider now a more specific scenario, where we have a set of orthogonal channel resources, i.e., subchannels, denoted by $\mathcal{W}$ that we allocate to the scheduled devices at each iteration. In this scenario, the feasible sets for power and resource allocation in Eq. (12.39) can be specified to guarantee constraints on the peak and/or average power and the total channel resources, including practical requirements, such as allocating only a single device for each available channel resource. Let $\mathcal{W}_{i,t}$ denote the subset of channel resources allocated to the $i$th device in the $t$th iteration under the resource allocation function $\mathbf{\Phi}_t$.

Using Shannon capacity as the transmission rate, which would serve as an upper bound on the rate that can be achieved in practice, and assuming unit-variance additive Gaussian noise at each subchannel, with power allocation $\mathbf{P}_t$ and resource allocation $\mathbf{\Phi}_t$, the transmission rate of the $i$th device is given by

$$R_{i,t}(\mathbf{P}_t, \mathbf{\Phi}_t) = \sum_{n \in \mathcal{W}_{i,t}} \frac{1}{2} \log(1 + G_{i,t,n} P_{i,t,n}), \tag{12.40}$$

where $G_{i,t,n}$ denotes the channel gain experienced by the $i$th device in the $t$th iteration over subchannel $n$, and $P_{i,t,n}$ is the power allocated to this subchannel.

We can impose a power constraint on each device, such that the total power consumption of device $i$ does not exceed $P_{\max}$:

$$\sum_{n \in \mathcal{W}_{i,t}} P_{i,n} \leq P_{\max}. \tag{12.41}$$

For resource allocation, we can require orthogonal transmission to prevent any interference between the devices. Hence, $\mathcal{F}$ is the set of all possible orthogonal resource allocation policies; that is,

$$\mathcal{W}_{i,t} \subseteq \mathcal{W}, \forall i, t, \quad \text{and} \quad \mathcal{W}_{i,t} \cap \mathcal{W}_{j,t} = \emptyset, \forall i, j \in \mathcal{S}_t. \tag{12.42}$$

Under orthogonal channel allocation, the optimization problem illustrated in **P2** is a combinatorial one; hence, in [58], a greedy heuristic is designed to solve it. The greedy algorithm consists of two phases, executed subsequently. Given the set of available subchannels $\widetilde{W}$, the first algorithm solves a resource and power allocation problem for for each device that has not been scheduled yet; that is, for $i \in \mathcal{S} \setminus \mathcal{S}_t$, where initially $\mathcal{S}_t = \emptyset$, such that the number of channels required for each device is minimized. This is equivalent to solving the following optimization problem for each device $i \in \mathcal{S} \setminus \mathcal{S}_t$:

$$\textbf{P3:} \quad \min_{\mathbf{P}_{i,t}, \mathbf{\Phi}_{i,t}} |\mathcal{W}_{i,t}|$$

$$\text{subject to: } R_{i,t}(\mathbf{P}_{i,t}, \mathbf{\Phi}_{i,t}) \geq R_{min}, \tag{12.43}$$

$$\sum_{n \in \mathcal{W}_{i,t}} P_{i,t,n} \leq P_{\max}. \tag{12.44}$$

The results of **P3** are fed to the second algorithm, which checks the ratio between the impact of the staleness and the number of required subchannels for the model update:

$$\frac{f_\alpha(a_{i,t})}{|\mathcal{W}_{i,t}|}, \tag{12.45}$$

and the device with the highest ratio is added to the set of scheduled devices, and the subchannels identified by **P3** for that device are removed from the set of available subchannels

$$\mathcal{S}_t \leftarrow \mathcal{S}_t \cup \{i\} \text{ and } \widetilde{W} \leftarrow \widetilde{W} \setminus \mathcal{W}_{i,t}. \tag{12.46}$$

Both algorithms are executed subsequently until the remaining subchannels are not sufficient to add any other device to the set $\mathcal{S}_t$, which means there is no feasible solution to the optimization problem **P3**.

### Device Scheduling with Fixed Power Allocation

In general, it is hard to provide a tractable solution to the optimization problem in Eq. (12.37). Instead, one can assume fixed power transmission and focus only on the impact of device scheduling, considering many PSs in parallel potentially interfering with each other. We can then replace the communication latency term with the *update success probability* to characterize the convergence performance [59]. To be more precise, consider multiple PSs distributed according Poisson point process (PPP) with parameter $\lambda$, which represents the density of the PSs. It is assumed that, in each cluster formed around a PS, there are $N$ devices distributed uniformly, and $K < N$ subchannels are available to be allocated to the devices at each iteration. The channel between the devices and the corresponding PS is assumed to behave according to the *block-fading propagation* model, in which the channel state remains fixed within a communication block and changes in an independent and identically distributed (i.i.d.) fashion from one transmission block to the next.

According to the block-fading propagation model together with the large-scale path-loss model, the received signal-to-interference-plus-noise ratio (SINR) at the PS for device $n$ can be written as

$$\gamma_{n,t} = \frac{p h_n d_n^{-\alpha}}{\sum_{c \in \mathcal{C}_{outer}} p h_c d_c^{-\alpha} + \sigma^2}, \tag{12.47}$$

where $p$ is the fixed transmit power for devices, $d_n$ is the distance of device $n$ to the PS, $\sigma^2$ is the variance of the additive noise term, $\alpha$ is the path-loss exponent, $h_n$ is the small scale fading coefficient, and $\mathcal{C}_{outer}$ is the set of interfering devices belonging to other clusters. Under the given SINR model, a transmission from device $n$ to the PS is considered successful if $\gamma_{n,t} > \gamma^\star$ for some predefined threshold $\gamma^\star$. Accordingly, the update success probability for device $n$, denoted by $U_n$, is defined as the probability of being scheduled for the model update and completing the transmission successfully:

$$U_n = \mathbb{P}(\gamma_{n,t} > \gamma^\star, n \in \mathcal{S}_t). \tag{12.48}$$

Assuming i.i.d. channel statistics across devices within the same clusters and law of large numbers the update success probability can be written as

$$U_n = \lim_{t \to \infty} \frac{1}{Nt} \sum_{\tau=0}^{t} \sum_{n=1}^{N} \mathbb{1}_{\{n \in \mathcal{S}_\tau, \gamma_{n,\tau} > \gamma^\star\}}. \tag{12.49}$$

It has been shown in [59] that the number of communication rounds to achieve a certain accuracy level can be written as a function of $U_n$. More precisely, it is shown to be proportional to $\frac{1}{\log(1-U_n)}$. The key design question is to find the scheduling policy $\Pi$ that maximizes the update success probability in order to achieve a better convergence result. Three different scheduling policies are investigated in [59] for this purpose: random scheduling (RS), round robin scheduling (RR), and proportional fair scheduling (PF).

RS selects $K$ devices randomly for the model update. Under RS, $U_n$ can be approximated as

$$U_n \approx \frac{K/N}{1 + V(\gamma^\star, \alpha)}, \tag{12.50}$$

where

$$V(\gamma^\star, \alpha) = \frac{\sigma^2 \gamma^\star \lambda^{1-\frac{\alpha}{2}}}{P 2^{\alpha-2}} (\gamma^\star)^{\alpha/2} \int_{u=0}^{\infty} \frac{1 - \epsilon^{-\frac{12}{5\pi}(\gamma^\star)^{\alpha/2} u}}{1 + u^{\alpha/2}}. \tag{12.51}$$

Accordingly the minimum number of required iterations under RS, $T_{RS}$, is proportional to

$$\frac{1}{\log\left(1 - \frac{K/N}{1+V(\gamma^\star,\alpha)}\right)}. \tag{12.52}$$

RR policy, on the other hand, divides the devices into $G = K/N$ groups and sequentially schedules one group at a time. Compared to RS, the RR policy introduces some level of fairness by giving a chance to each device to contribute to the model update periodically. Under the RR policy, the update success probability can be written as

$$U_n \approx \begin{cases} \frac{1}{1+V(\gamma^\star,\alpha)}, & \text{if scheduled} \\ 0 & \text{otherwise,} \end{cases} \tag{12.53}$$

and accordingly the minimum number of required iterations under RR policy, $T_{RR}$ is proportional to

$$\frac{K/N}{\log\left(1 - \frac{1}{1+V(\gamma^\star,\alpha)}\right)}. \tag{12.54}$$

Finally, the PF policy selects the devices opportunistically based on the channel statistics. It sorts all the devices according to the ratio, $\frac{\tilde{\rho}_i}{\bar{\rho}_i}$, where $\tilde{\rho}_i$ and $\bar{\rho}_i$ are the instantaneous signal-to-noise ratio (SNR) and time-averaged SNR, respectively. The opportunistic behaviour behind the PF policy helps to increase the update success probability, which can be written as

$$U_n \approx \sum_{i=1}^{N-K+1} \binom{N-K+1}{i} \frac{(-1)^{i+1} N/K}{1 + V(i\gamma^\star, \alpha)}, \tag{12.55}$$

and the minimum number of required iterations under the RR policy, $T_{PF}$, is proportional to

$$\frac{1}{\log \left( 1 - \sum_{i=1}^{K-N+1} \binom{K-N+1}{i} \frac{(-1)^{i+1} K/N}{1+V(i\gamma^\star, \alpha)} \right)}. \tag{12.56}$$

To compare the performance of these three scheduling strategies (RR, RS, and PF), a classification problem on MNIST data consisting of handwritten numerals, using a convolutional neural network (CNN) architecture has been considered in [59]. In the simulations, two different scenarios with high and low SINR threshold regimes with $\gamma^\star = 20dB$ and $\gamma^\star = -25dB$, respectively, have been analyzed. The results indicate that, in the high SINR threshold regime, PF significantly outperforms the RR policy; while PF achieves 94 percent average test accuracy, RR gets stuck around 50 percent. This is mainly because PF achieves further successful global aggregation rounds compared to RR. In the low SINR threshold regime, all three scheduling policies exhibit similar performance since the chance of successfully participating in global aggregation increases with reduced threshold. Further details on the numerical experiments can be found in [59].

In Eq. (12.37), we assumed a fixed number of scheduled devices at each iteration and tried to minimize the latency of getting updates from so many devices. Alternatively, we can impose a deadline on each iteration and try to receive updates from as many devices as possible at each iteration [61]. Assuming that the devices transmit one by one using all the available channel resources, the device selection problem can be formulated as follows:

**P4:** $\quad \max |\mathcal{S}_t|$

$\qquad$ subject to: $\mathcal{S}_t \subseteq \bar{\mathcal{S}}_t, \quad \forall t,$ $\hfill (12.57)$

$$\sum_{i=1}^{|\mathcal{S}_t|} \left( L_{s_i,t}^{\text{comm}} + \max \left( \sum_{i=1}^{|\mathcal{S}_t|} L_{s_i,t}^{\text{comm}}, L_{s_i,t}^{\text{comp}} \right) \right) \leq T_{max}, \tag{12.58}$$

where the objective is to schedule the maximum number devices from among the randomly sampled set $\bar{\mathcal{S}}_t$, while the overall latency is bounded by some predefined $T_{max}$. We want to highlight that, in this formulation, the device selection strategy also incorporates the scheduling decision for the model updates, such that the devices in set $\mathcal{S}_t = \{s_1, \ldots, s_{|\mathcal{S}_t|}\}$ are ordered. This ordering also affects the latency since, as illustrated in Eq. (12.58), the computation time of the $s_i$th device can be overlapped with the cumulative latency of the previously scheduled devices. In general, **P4** is a combinatorial problem, so a greedy strategy is proposed in [61], similar to those used for the knapsack problem. This strategy iteratively adds to set $\mathcal{S}_t$ the device that introduces the minimum additional delay, until the cumulative latency reaches $T_{max}$.

Latency-aware device selection problem with a fairness constraint is also studied in [57], where the optimal device selection problem is analyzed as a multi-armed bandit (MAB) problem.

### Update-Aware Device Scheduling

In all the policies discussed so far, the devices are scheduled according to the channel conditions, with the additional consideration of the age or the fairness in order to make sure that each device is scheduled at some minimal frequency. An alternative approach to these policies is considered in [62], where the updates of the devices are also taken into account when selecting the devices to be scheduled at each iteration.

Assume that we schedule a fixed number of $K$ devices at each iteration. Unlike the previous formulation, here we fix both the latency and the number of devices at each iteration, but instead change the update resolution sent from each device, which was set to $d$ bits in Eq. (12.37). Once the scheduled devices are decided, the channel resources are allocated between these devices such that each device can approximately send the same number of bits to the PS.

If we schedule the devices with the best channel states, the scheduled devices will be able to transmit their model updates with the highest fidelity; however, not all model updates may have the same utility for the SGD process. For example, if the data at one of the devices is already in accordance with the current model, this device will not want to change the model and, hence, will send a very small or zero model update to the PS, which can be ignored without much loss in the optimization. To enable such an "update-aware" scheduling policy, [62] proposes to schedule the devices taking into account the $l_2$ norm of their model updates.

Four different scheduling policies are proposed and compared in [62]: the *best channel* (BC) policy schedules devices soleley based on their channel conditions; that is, selects the $K$ devices with the best channel conditions. The *best $l_2$-norm* (BN2) strategy schedules the $K$ devices with the best channel conditions. On the other hand, *best channel-best $l_2$-norm* (BC-BN2) strategy first chooses $K_c$ devices according to their channel conditions and then chooses $K$ out of these $K_c$ devices, depending on the $l_2$ norm of their updates. This guarantees scheduling users with both good channel states and significant model updates. Note however that the real quality of the update transmitted to the PS depends also on the available channel resources, which determines how much the device needs to quantize its model update. For example, a device might have an update with a relatively large $l_2$ norm; however, this might reduce drastically after quantization if the channel of this device is relatively weak. To take into account this reduction in the update quality due to quantization, the *best $l_2$-norm-channel* (BN2-C) strategy chooses the final devices based on the $l_2$ norm of the updates they would send if they were the sole transmitter.

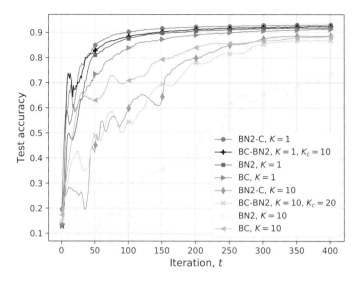

**Figure 12.2** Performance of different scheduling policies with $N = 40$ devices.

We compare the test accuracy of these four schemes in Fig. 12.2 considering the training of a neural network model for MNIST image classification. We considering 40 devices, each with 1,000 randomly chosen MNIST samples. We can see from this plot that scheduling only a single device at each iteration achieves the best performance for all four schemes. We can also see that scheduling solely based on the channel gains or on the update significance has a slower convergence speed, and converges to a lower test accuracy. It is clear that the best performance is achieved when both the channel gain and the update significance is taken into account when scheduling the devices.

## 12.4.1    Hierarchical Edge Learning

In general, to prevent excessive communication load and congestion at the PS only a small portion of devices are scheduled at each round to upload their locally trained models. On the other hand, scheduling only a small number of $K$ devices at each round increases the required number of communication rounds to achieve a target accuracy level. In order to reduce the communication load while keeping the number of communication rounds fixed, orchestration of the edge learning framework can be modified such that multiple PSs are employed to utilize parallel aggregation in the network under the orchestration of a main coordinator PS [63, 64].

The generic structure of the *hierarchical FL (HFL)* framework is provided in Algorithm 12.9. In HFL, devices in the network are grouped into $L$ clusters and the devices within the same cluster are assigned to the same PS. Therefore, intracluster federated averaging is executed in parallel (lines 2–10 in Algorithm 12.9). Following $H$ rounds of intracluster averaging, PSs communicate with each other through the main PS to seek a consensus on the global model (line 13 in Algorithm 12.9).

---

**Algorithm 12.9** Hierarchical federated learning (HFL).

1: **for** $t = 1, 2, \ldots$ **do**
2:     **for** $l = 1, \ldots, L$ **do** in parallel
3:       **for** $i \in C_l$ **do** in parallel
4:        Pull $\boldsymbol{\theta}_{l,t-1}$ from $PS_l$: $\boldsymbol{\theta}^0_{i,t-1} = \boldsymbol{\theta}_{l,t-1}$
5:        **for** $h = 1, \ldots, H_{local}$ **do**
6:          Compute SGD: $\mathbf{g}_{i,h} = \nabla_{\boldsymbol{\theta}} \mathcal{L}_n(\boldsymbol{\theta}^h_{i,t-1}, \zeta_{i,h})$
7:          Update model: $\boldsymbol{\theta}^h_{i,t-1} = \boldsymbol{\theta}^{h-1}_{i,t-1} - \eta_t \mathbf{g}_{i,h}$
8:       Push $\boldsymbol{\theta}^{H_{local}}_{i,t-1}$ to $PS_l$
9:     **Intra-cluster averaging**:
10:     $\boldsymbol{\theta}_{l,t} = \frac{1}{|\mathcal{C}_l|} \sum_{i \in \mathcal{C}_l} \boldsymbol{\theta}^{H_{local}}_{i,t-1}$
11:    **if** $t | H$ **then**:
12:     **Inter-cluster Averaging**
13:     $\boldsymbol{\theta}_t = \frac{1}{L} \sum_{l=1}^{L} \boldsymbol{\theta}_{l,t}$

---

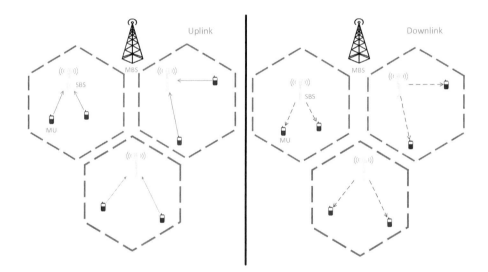

**Figure 12.3** Intracluster consensus.

In the wireless setting, this hierarchical edge learning strategy can be implemented through small-cell base stations (SBS) such that the clusters are formed by assigning devices to SBSs and the SBSs are orchestrated by a macro-cell base station (MBS), as illustrated in Figs. 12.3 and 12.4. In this scenario, we refer to the devices as mobile users (MUs).

To illustrate the benefits of the hierarchical FL strategy, we consider 28 MUs uniformly distributed in a circular area with radius 750 meters. We consider hexagonal clusters, where the diameter of a circle inscribed in each of them is 500 meters. We consider in total seven clusters with hexagonal shape, and the SBSs are located

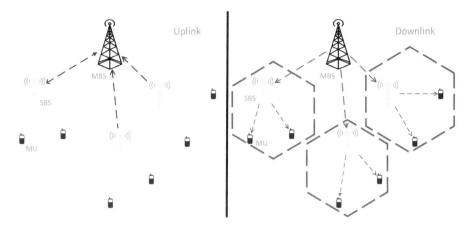

**Figure 12.4** Intercluster consensus.

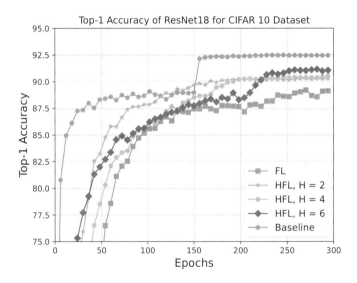

**Figure 12.5** Test accuracy through iterations.

at the center of the hexagons. We further assume that the fronthaul links between the MBS and the SBSs are 100 times faster than the uplink and downlink connections between the MUs and the SBSs. We assume 600 subcarriers with subcarrier spacing of 30 KHz. The maximum transmit powers of the MBS, SBSs, and the MUs are set to 20 W, 6.3 W, and 0.2 W, respectively.

For the simulations, we consider the image classification problem over the CIFAR-10 dataset with 10 different image classes and collaboratively train the ResNet18 architecture. The dataset is divided disjointly among the devices in an i.i.d. manner. We set the batch size for training to 64 and the initial learning rate to 0.25, which is reduced by a factor of 10 at the end of the 150th epoch and 225th epoch. Finally, we apply 99, 90, 90, and 90 percent sparsification for communication from the MUs to

**Table 12.1.** Test accuracy results for
different strategies.

| | |
|---|---|
| Baseline | $92.48 \pm 0.13$ |
| FL | $89.23 \pm 0.42$ |
| HFL, $H = 2$ | $90.27 \pm 0.11$ |
| HFL, $H = 4$ | $90.474 \pm 0.20$ |
| HFL, $H = 6$ | $91.03 \pm 0.19$ |

the SBSs, SBSs to MUs, SBSs to the MBS, and the MBS to SBSs, respectively. We set $H_{local} = 1$ in Algorithm 12.9, and use the *momentum* optimizer for the local updates.

The convergence results for the HFL framework for parameters $H = 2, 4, 6$, as well as the centralized FL and the baseline method corresponding to centralized single machine training are illustrated in Fig. 12.5, while the final test accuracy results are presented in Table 12.1. We observe that HFL achieves a higher final accuracy as well as faster convergence compared to conventional FL. Additionally, the hierarchical FL strategy provides five to seven times speed up in the latency required to reach these accuracy levels compared to FL orchestrated directly by the MBS.

## 12.5    Conclusion

In this chapter, we presented a brief introduction to distributed/collaborative learning, particularly targeting wireless edge applications. We first introduced general communication reduction methods such as sparsification, quantization, and local iterations. We then provided an overview of device scheduling and resource allocation strategies for distributed learning over wireless networks. We particularly highlighted the fact that both the scheduling and the resource allocation strategy can be substantially different from conventional solutions for wireless network optimization, which typically aims at maximizing the throughput, or minimizing the delay. In distributed learning, the goal is to solve the underlying optimization problem as fast and as accurately as possible, and we often employ iterative algorithms, such as SGD. This means that, in addition to receiving accurate solutions from the participating devices, we also need to guarantee that each device is scheduled with some minimal regularity. Additionally, selecting the devices depending on the significance of their updates on the solution can increase the convergence to the optimal solution.

There are many other factors that can be considered in formulating the optimal scheduling scheme, such as the energy consumption at the devices [65] or impact of the compression in the PS-to-device links [41]. We also remark that we focused exclusively on digital schemes in this chapter, where the devices first compress their model updates into a finite number of bits and transmit these bits to the PS over orthogonal channels. An alternative approach is to transmit all the model updates from the devices simultaneously in an uncoded/analog manner, which allows exploiting the signal superposition property of the wireless channel for over-the-air computation [3, 4].

# References

[1] D. Gündüz et al., "Communicate to learn at the edge," *IEEE Communications Magazine*, vol. 58, no. 12, pp. 14–19, 2020.

[2] J. Park et al., "Wireless network intelligence at the edge," *Proc. IEEE*, vol. 107, no. 11, pp. 2204–2239, 2019.

[3] M. Mohammadi Amiri and D. Gündüz, "Machine learning at the wireless edge: Distributed stochastic gradient descent over-the-air," *IEEE Trans. on Signal Processing*, vol. 68, pp. 2155–2169, 2020.

[4] M. M. Amiri and D. Gündüz, "Federated learning over wireless fading channels," *IEEE Trans. on Wireless Communications*, vol. 19, no. 5, pp. 3546–3557, 2020.

[5] W. Zhang et al., "Staleness-aware async-SGD for distributed deep learning," in *Proc. 25th Int. Joint Conf. on Artificial Intelligence (IJCAI)*, pp. 2350–2356, 2016.

[6] X. Lian et al., "Asynchronous parallel stochastic gradient for nonconvex optimization," in *Advances in Neural Information Processing Systems*, vol. 28, C. Cortes et al., eds., Curran Associates, Inc., pp. 2737–2745, 2015.

[7] S. Zheng et al., "Asynchronous stochastic gradient descent with delay compensation," in *Proc. Machine Learning Research*, vol. 70, D. Precup and Y. W. Teh, eds., pp. 4120–4129, 2017.

[8] X. Lian et al., "Can decentralized algorithms outperform centralized algorithms? A case study for decentralized parallel stochastic gradient descent," in *Advances in Neural Information Processing Systems 30*, pp. 5330–5340, 2017.

[9] H. Tang et al., "d2: Decentralized training over decentralized data," in *Proc. 35th Int. Conf. on Machine Learning*, J. Dy and A. Krause, eds., pp. 4848–4856, 2018.

[10] Z. Jiang et al., "Collaborative deep learning in fixed topology networks," in *Advances in Neural Information Processing Systems 30*. Curran Associates, Inc., pp. 5904–5914, 2017.

[11] K. Yuan, Q. Ling, and W. Yin, "On the convergence of decentralized gradient descent," *SIAM Journal on Optimization*, 2016.

[12] J. Zeng and W. Yin, "On nonconvex decentralized gradient descent," *IEEE Trans. on Signal Processing*, vol. 66, no. 11, pp. 2834–2848, June 2018.

[13] J. C. Duchi, A. Agarwal, and M. J. Wainwright, "Dual averaging for distributed optimization: Convergence analysis and network scaling," *IEEE Trans. on Automatic Control*, vol. 57, no. 3, pp. 592–606, 2012.

[14] E. Ozfatura, S. Rini, and D. Gündüz, "Decentralized SGD with over-the-air computation," *arXiv preprint*, arXiv 2003.04216, 2020.

[15] K. He et al., "Deep residual learning for image recognition," in *IEEE Conf. on Computer Vision and Pattern Recognition (CVPR)*, pp. 770–778, 2016.

[16] K. Simonyan and A. Zisserman, "Very deep convolutional networks for large-scale image recognition," in *Int. Conf. on Learning Representations*, 2015.

[17] N. F. Eghlidi and M. Jaggi, "Sparse communication for training deep networks," *arXiv preprint*, arXiv 2009.09271, 2020.

[18] J. Wangni et al., "Gradient sparsification for communication-efficient distributed optimization," in *Advances in Neural Information Processing Systems 31*, S. Bengio et al., eds. Curran Associates, Inc., pp. 1305–1315, 2018.

[19] P. Jiang and G. Agrawal, "A linear speedup analysis of distributed deep learning with sparse and quantized communication," in *Advances in Neural Information Processing Systems 31*, S. Bengio et al., eds. Curran Associates, Inc., pp. 2529–2540, 2018.

[20] A. F. Aji and K. Heafield, "Sparse communication for distributed gradient descent," in *Proc. Conf. on Empirical Methods in Natural Language Processing*, pp. 440–445, 2017.

[21] D. Alistarh et al., "The convergence of sparsified gradient methods," in *Advances in Neural Information Processing Systems 31*, S. Bengio et al., eds. Curran Associates, Inc., pp. 5976–5986, 2018.

[22] S. U. Stich et al., "Sparsified SGD with memory," in *Advances in Neural Information Processing Systems 31*, S. Bengio et al., eds. Curran Associates, Inc., pp. 4448–4459, 2018.

[23] L. P. Barnes et al., "rtop-k: A statistical estimation approach to distributed SGD," *arXiv preprint*, arXiv 2005.10761, 2020.

[24] Y. Lin et al., "Deep gradient compression: Reducing the communication bandwidth for distributed training," in *Int. Conf. on Learning Representations*, 2018.

[25] S. Shi et al., "A distributed synchronous SGD algorithm with global top-k sparsification for low bandwidth networks," in *IEEE 39th Int. Conf. on Distributed Computing Systems (ICDCS)*, pp. 2238–2247, 2019.

[26] H. Yu, R. Jin, and S. Yang, "On the linear speedup analysis of communication efficient momentum SGD for distributed non-convex optimization," in *Proc. 36th Int. Conf. on Machine Learning*, K. Chaudhuri and R. Salakhutdinov, eds., pp. 7184–7193, 2019.

[27] K. Ozfatura, E. Ozfatura, and D. Gunduz, "Distributed sparse SGD with majority voting," *arXiv preprint*, arXiv 2011.06495, 2020.

[28] H. Tang et al., "`DoubleSqueeze`: Parallel stochastic gradient descent with double-pass error-compensated compression," in *Proc. 36th Int. Conf. on Machine Learning*, K. Chaudhuri and R. Salakhutdinov, eds., pp. 6155–6165, 2019.

[29] S. P. Karimireddy et al., "Error feedback fixes SignSGD and other gradient compression schemes," in *Proc. 36th Int. Conf. on Machine Learning*, pp. 3252–3261, 2019.

[30] H. Zhang et al., "ZipML: Training linear models with end-to-end low precision, and a little bit of deep learning," in *Proc. Machine Learning Research*, vol. 70, D. Precup and Y. W. Teh, eds., pp. 4035–4043, 2017.

[31] F. Sattler et al., "Robust and communication-efficient federated learning from non-i.i.d. data," *IEEE Trans. on Neural Networks and Learning Systems*, pp. 1–14, 2019.

[32] D. Alistarh et al., "QSGD: Communication-efficient SGD via gradient quantization and encoding," in *Advances in Neural Information Processing Systems 30*, I. S. Guyon et al., eds. Curran Associates, Inc., pp. 1709–1720, 2017.

[33] J. Wu et al., "Error compensated quantized SGD and its applications to large-scale distributed optimization," in *Proc. 35th Int. Conf. on Machine Learning*, vol. 80, J. Dy and A. Krause, eds., pp. 5325–5333, 2018.

[34] F. Seide et al., "1-bit stochastic gradient descent and application to data-parallel distributed training of speech DNNs," in *Interspeech*, 2014.

[35] N. Strom, "Scalable distributed DNN training using commodity GPU cloud computing," in *INTERSPEECH*, 2015.

[36] J. Bernstein et al., "signSGD: Compressed optimisation for non-convex problems," in *Proc. 35th Int. Conf. on Machine Learning*, J. Dy and A. Krause, eds., pp. 560–569, 2018.

[37] J. Bernstein et al., "signSGD with majority vote is communication efficient and fault tolerant," in *Int. Conf. on Learning Representations*, 2019.

[38] S. P. Karimireddy et al., "Error feedback fixes SignSGD and other gradient compression schemes," in *Proc. 36th Int. Conf. on Machine Learning*, K. Chaudhuri and R. Salakhutdinov, eds., pp. 3252–3261, 2019.

[39] S. Zheng, Z. Huang, and J. Kwok, "Communication-efficient distributed blockwise momentum SGD with error-feedback," in *Advances in Neural Information Processing Systems 32*. Curran Associates, Inc., pp. 11450–11460, 2019.

[40] W. Wen et al., "Terngrad: Ternary gradients to reduce communication in distributed deep learning," in *Advances in Neural Information Processing Systems 30*, I. S. Guyon et al., eds. Curran Associates, Inc., pp. 1509–1519, 2017.

[41] M. M. Amiri, D. Gündüz, S. R. Kulkarni, and H. V. Poor, Convergence of federated learning over a noisy downlink, IEEE Transactions on Wireless Communications, vol. 20, no. 6, pp. 3643–3658, Jun. 2021.

[42] S. U. Stich, "Local SGD converges fast and communicates little," in *International Conference on Learning Representations*, 2019.

[43] T. Lin et al., "Don't use large mini-batches, use local SGD," in *Int. Conf. on Learning Representations*, 2020.

[44] F. Zhou and G. Cong, "On the convergence properties of a k-step averaging stochastic gradient descent algorithm for nonconvex optimization," in *Proc. 27th Int. Joint Conf. on Artificial Intelligence (IJCAI)*, pp. 3219–3227, 2018.

[45] J. Wang and G. Joshi, "Cooperative SGD: A unified framework for the design and analysis of communication-efficient SGD algorithms," *CoRR*, vol. abs/1808.07576, 2018.

[46] H. Yu, S. Yang, and S. Zhu, "Parallel restarted SGD with faster convergence and less communication: Demystifying why model averaging works for deep learning," *arXiv preprint*, arXiv 1807.06629, 2018.

[47] F. Haddadpour et al., "Local SGD with periodic averaging: Tighter analysis and adaptive synchronization," in *Advances in Neural Information Processing Systems 32*, Curran Associates Inc., pp. 11082–11094, 2019.

[48] D. Basu et al., "Qsparse-local-SGD: Distributed SGD with quantization, sparsification and local computations," in *Advances in Neural Information Processing Systems 32*, Curran Associates Inc., pp. 14695–14706, 2019.

[49] W. Li et al., "Privacy-preserving federated brain tumour segmentation," in *Machine Learning in Medical Imaging*, H.-I. Suk et al., eds. Springer International Publishing, 2019, pp. 133–141.

[50] N. Rieke et al., "The future of digital health with federated learning," *arXiv preprint*, arXiv 2003.08119, 2020.

[51] H. Wang et al., "Federated learning with matched averaging," in *Int. Conf. on Learning Representations*, 2020.

[52] T. Lin et al., "Ensemble distillation for robust model fusion in federated learning," *arXiv preprint*, arXiv 2006.07242, 2020.

[53] E. Jeong et al., "Communication-efficient on-device machine learning: Federated distillation and augmentation under non-iid private data," *arXiv preprint*, arXiv 1811.11479, 2018.

[54] B. T. Polyak, "Some methods of speeding up the convergence of iteration methods," *USSR Computational Mathematics and Mathematical Physics*, vol. 4, no. 5, pp. 1–17, Dec. 1964.

[55] J. Wang et al., "Slowmo: Improving communication-efficient distributed SGD with slow momentum," in *International Conference on Learning Representations*, 2020.

[56] S. Reddi et al., "Adaptive federated optimization," *arXiv preprint*, arXiv 2003.00295, 2020.

[57] W. Xia et al., "Multi-armed bandit based client scheduling for federated learning," *IEEE Trans. on Wireless Communications*, pp. 1–1, 2020.

[58] H. H. Yang et al., "Age-based scheduling policy for federated learning in mobile edge networks," in *IEEE Int. Conf. on Acoustics, Speech and Signal Processing (ICASSP)*, pp. 8743–8747, 2020.

[59] H. H. Yang et al., "Scheduling policies for federated learning in wireless networks," *IEEE Trans. on Communications*, vol. 68, no. 1, pp. 317–333, 2020.

[60] W. Shi, S. Zhou, and Z. Niu, "Device scheduling with fast convergence for wireless federated learning," in *IEEE Int. Conf. on Communications (ICC)*, pp. 1–6, 2020.

[61] T. Nishio and R. Yonetani, "Client selection for federated learning with heterogeneous resources in mobile edge," in *IEEE Int. Conf. on Communications (ICC)*, pp. 1–7, 2019.

[62] M. M. Amiri, D. Gunduz, S. Kulkarni, and H. V. Poor, Convergence of update aware device scheduling for federated learning at the wireless edge , IEEE Transactions on Wireless Communications, vol. 20, no. 6, pp. 3643–3658, Jun. 2021.

[63] M. S. H. Abad et al., "Hierarchical federated learning across heterogeneous cellular networks," in *IEEE Int. Conf. on Acoustics, Speech and Signal Processing (ICASSP)*, 2020, pp. 8866–8870.

[64] L. Liu et al., "Client-edge-cloud hierarchical federated learning," in *IEEE Int. Conf. on Communications (ICC)*, pp. 1–6, 2020.

[65] Q. Zeng et al., "Energy-efficient resource management for federated edge learning with CPU-GPU heterogeneous computing," *arXiv preprint*, arXiv 2007.07122, 2020.

# 13 Optimized Federated Learning in Wireless Networks with Constrained Resources

Shiqiang Wang, Tiffany Tuor, and Kin K. Leung

## 13.1 Introduction

When compared with existing networks, future wireless networks are expected to provide a much higher data rate and lower latency to support advanced communication services and new user applications. In particular, these represent the main advances for the 5G and beyond 5G networks. In addition to communications among human users, other forms of communications such as machine-to-machine, Internet of Things (IoT), vehicle-to-vehicle, and sensors are going to be increasingly popular. All of these machines and devices attached to the edge of wireless networks will produce and consume huge volumes of different kinds of data, such as high-quality images and videos that utilize a great deal of communication resources for transmission.

Machine learning (ML) techniques have been widely applied to a variety of applications, ranging from speech, image and video processing, surveillance, to system monitoring and control, and network resource management, to name a few. By nature, ML is used to process and extract useful information and knowledge from a large quantity of data generated by or collected from various sources. On top of many existing types of devices, consider self-driving vehicles mounted with camera and sensors as one type of mobile device in the near future. Clearly, they will become new data sources for ML. The availability of additional data naturally will lead to new ML applications, which will also result in new challenges for wireless networks.

To enable ML applications over wireless networks, radio bandwidth and transmission power may not be the only network resources in limited supply. Furthermore, computational capability and speed, on-board memory storage, energy and power, etc. may also be resources of concern on the wireless clients that can significantly affect the overall quality of service for ML applications. Existing wireless research has been focusing on optimizing the use of limited communication resources such

The research presented in this chapter was partly sponsored by the US Army Research Laboratory and the UK Ministry of Defence under agreement number W911NF-16-3-0001. The views and conclusions contained in this document are those of the authors and should not be interpreted as representing the official policies, either expressed or implied, of the US Army Research Laboratory, the US government, the UK Ministry of Defence, or the UK government. The US and UK governments are authorized to reproduce and distribute reprints for government purposes notwithstanding any copyright notation hereon.

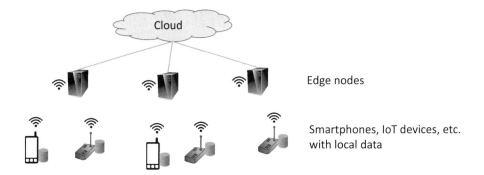

**Figure 13.1** Model training at the edge.

as bandwidth and transmission power among competing users or applications, but we have not adequately considered the other factors and issues that can be crucially important to the overall performance of ML services over wireless networks.

## 13.1.1    Model Training at the Wireless Edge

A major benefit of the vast amount of data generated by modern applications is that they can be used to train new ML models, which can in turn enable new applications and functionalities. However, as most of such data are generated on the device where the application runs, it is often infeasible to transmit all the on-device information to a central cloud, due to the large amount of data located on billions of end devices and privacy regulations such as the General Data Protection Regulation (GDPR).[1]

To make use of locally collected data, new ML techniques have been developed in recent years that are broadly known as *edge learning* [1, 2]. As illustrated in Fig. 13.1, in a 5G and beyond system with mobile edge computing (MEC), edge nodes located at base stations or elsewhere in close proximity to users act as processing units with a high-bandwidth and low-latency connection to end devices. There are two different ways to enable model training in this context: (a) Each end device shares a small portion of its data with its nearest edge node or the cloud, in the form of a *coreset* [3], so that the edge node or cloud can train models using this subset of data. (b) The devices, edge nodes, or both collaboratively train a model without sharing their raw data, which is known as *federated learning* (FL) [4, 5]. Both approaches avoid sending all the data from sources to a central place. Hence, they are more communication-efficient and privacy-preserving compared to traditional ML techniques that require central data storage. They can coexist in a single system. For instance, each end device may transmit a coreset of its data to the nearest edge node; then, multiple edge nodes can collaboratively train a model using FL.

---

[1]  http://data.consilium.europa.eu/doc/document/ST-9565-2015-INIT/en/pdf

## 13.1.2    Adaptive Federated Learning (FL)

FL is particularly challenging at the resource-constrained wireless edge, since it includes multiple rounds of computation and communication among the participating clients and the server. Here and throughout this chapter, we refer to the FL participants with local datasets as *clients*, which can be either an end device or an edge node; the FL *server* is a logical entity acting as an aggregator, which does not have access to any data and is often located in the cloud, but can be located on an edge node or end device as well. In a typical FL system, data is generated or collected by the distributed clients, which have some processing capability and disk space to store the data. After processing, each local client sends its learned model parameters through wireless links to the server for further processing and aggregation. For good overviews of FL, see [6–8]. From an implementation's perspective, some recent developments have built FL prototypes using embedded systems with wireless connections [9–11].

Evidently, the amount of radio resource consumption depends on how often the clients transmit their model parameters. Even without transmitting the model parameters, local processing can continue on the clients, which consumes energy and may also lead to divergent results from the perspective of the central controller (aggregator). Therefore, consumption and usage of communication, computation, and energy resources form a complicated relationship and trade-off that determines the overall quality of the FL process. We present an *adaptive federated learning* technique in later sections that dynamically adapts the FL algorithm to make the best use of available resources [9].

## 13.1.3    Outline

The rest of this chapter is organized as follows. We describe the definitions and algorithms of FL in Section 13.2. In Sections 13.3 and 13.4, we present a formulation and solution of the adaptive FL problem with resource constraints, respectively. Some experimentation results are given in Section 13.5. Section 13.6 includes a summary and some further discussions.

## 13.2    Definitions and Algorithms of Federated Learning

## 13.2.1    Mathematical Definition of Federated Learning

In general, ML systems include trainable models. These models map the input data to the desired output, as illustrated in Fig. 13.2. Each model has its computational logic and a set of trainable parameters. During the *training* phase, the model parameters are learned using a training dataset with input data samples and their ground-truth labels.[2] Afterward, this model is applied to predict the labels of new data with unknown labels

---

[2] This is assuming supervised learning. Unsupervised learning works in a similar manner, but the training data is unlabeled.

**Table 13.1.** Per-sample loss functions of exemplar models.

| Model | Per-sample loss function $l(\mathbf{w}, \mathbf{x}, y)$ |
|---|---|
| Linear regression | $\frac{1}{2}\|y - \mathbf{w}^{\mathrm{T}}\mathbf{x}\|^2$ |
| SVM (smooth loss) | $\frac{\lambda}{2}\|\mathbf{w}\|^2 + \frac{1}{2}\max\left\{0; 1 - y\mathbf{w}^{\mathrm{T}}\mathbf{x}\right\}^2$ ($\lambda \geq 0$ is a constant) |
| K-means ($Q$ clusters) | $\frac{1}{2}\min_q \|\mathbf{x} - \mathbf{w}_{(q)}\|^2$ where $\mathbf{w} := [\mathbf{w}_{(1)}^{\mathrm{T}}, \ldots, \mathbf{w}_{(q)}^{\mathrm{T}}, \ldots, \mathbf{w}_{(Q)}^{\mathrm{T}}]^{\mathrm{T}}$ |
| Convolutional neural network | Cross-entropy on neural network output, see [15] |

**Figure 13.2** Machine learning model.

during the *inference* phase. We focus on model training in this chapter. For resource-efficient inference, we refer interested readers to [12–14].

Let the function $g(\mathbf{w}, \mathbf{x})$ denote the model with a trainable parameter vector $\mathbf{w}$. For a given input data sample (image in Fig. 13.2) $\mathbf{x}$, the model gives a predicted output (label) $\hat{y} := g(\mathbf{w}, \mathbf{x})$, where ":=" denotes "defined to be equal to." Based on this, a per-sample loss can be defined by $l(\hat{y}, y) = l(g(\mathbf{w}, \mathbf{x}), y)$, where $y$ is the ground-truth output (label) of $\mathbf{x}$ that is available in the training data but not available during the inference phase. The loss function $l(\hat{y}, y)$ can be a common error function, such as mean square error, cross-entropy, etc. [15]. In many cases, the loss function can include additional terms related to the model parameter $\mathbf{w}$, which are not included in $g(\cdot, \cdot)$, such as a regularization term. Hence, it can be more convenient to write the loss function as related to $\mathbf{w}$ and $\mathbf{x}$ directly, instead of relating them through $g(\cdot, \cdot)$, in the form of $l(\mathbf{w}, \mathbf{x}, y)$. An example of per-sample loss functions $l(\mathbf{w}, \mathbf{x}, y)$ of some well-known ML models is given in Table 13.1.

As the training data includes multiple samples, it is the average of the per-sample losses of all data samples in the training dataset that we care about. In the federated setting, assume that there are $N$ clients in total. Each client $i \in \{1, 2, \ldots, N\}$ has its local training dataset denoted by $\mathcal{S}_i$, which includes data samples $\{\mathbf{x}, y\} \in \mathcal{S}_i$ that capture both the input $\mathbf{x}$ and its desired output/label $y$. Based on this, we can define the *local average loss*[3] of client $i$ as

$$f_i(\mathbf{w}) := \frac{1}{|\mathcal{S}_i|} \sum_{\{\mathbf{x}, y\} \in \mathcal{S}_i} l(\mathbf{w}, \mathbf{x}, y), \qquad (13.1)$$

---

[3] The local and global average losses are also known as local and global empirical risks in the literature. We use the term "average loss" in this chapter, and we omit the word "average" when it is unambiguous from the context.

where $|\cdot|$ denotes the cardinality (i.e., number of elements) in the set. We can further define the *global average loss* as

$$f(\mathbf{w}) := \sum_{i=1}^{N} p_i f_i(\mathbf{w}), \tag{13.2}$$

where $p_i$ is a weighting coefficient, such that $\sum_{i=1}^{N} p_i = 1$. We assume that $\mathcal{S}_i \cap \mathcal{S}_j = \emptyset$ for $i \neq j$ and discuss a few possible ways of choosing $p_i$.

- If we choose $p_i = \frac{|\mathcal{S}_i|}{\left|\cup_{j=1}^{N} \mathcal{S}_j\right|}$, we have

$$f(\mathbf{w}) = \frac{1}{\left|\cup_{j=1}^{N} \mathcal{S}_j\right|} \sum_{\{\mathbf{x},y\} \in \cup_{j=1}^{N} \mathcal{S}_j} l(\mathbf{w}, \mathbf{x}, y),$$

  according to Eqs. (13.1) and (13.2), which is equal to the average loss in the case of having all the local datasets (i.e., $\cup_{j=1}^{N} \mathcal{S}_j$) at a central place. With this definition of $p_i$, all *data samples* have equal importance, and a client with a smaller number of data samples than others will be less important during the FL process.
- If we choose $p_i = \frac{1}{N}$, all *clients* have equal importance. For example, a client with only one data sample is equally important as another client with 100 data samples. Equivalently, this means that the data sample in the former client is more important than each data sample in the latter client.
- Choosing $p_i$ somewhere between $\frac{|\mathcal{S}_i|}{\left|\cup_{j=1}^{N} \mathcal{S}_j\right|}$ and $\frac{1}{N}$ strikes a balance between the two cases. In general, $p_i$ is chosen by the practitioner defining the FL problem, and its choice can control the trade-off between sample bias and client bias.

### The Learning Problem
With the definitions thus far, we can write the FL problem as a minimization of the global loss. We would like to find the optimal parameter vector $\mathbf{w}^*$ that minimizes $f(\mathbf{w})$:

$$\mathbf{w}^* := \arg \min_{\mathbf{w}} f(\mathbf{w}). \tag{13.3}$$

From Eq. (13.3), we can see that ML and FL problems are essentially optimization problems. The challenge in solving Eq. (13.3) is twofold: (a) due to the complexity of ML models and the fact that the loss functions are data-dependent, it is usually impossible to obtain a closed-form solution. (b) In FL systems, the global loss $f(\mathbf{w})$ is not directly observable since it is defined on distributed datasets at clients, only the local loss $f_i(\mathbf{w})$ is observable at each client $i$.

Due to these challenges, ML problems are usually solved using gradient descent approaches [15]. In FL, a distributed version of gradient descent is often used so that data remain at the clients and are not shared with the (central) server, which we discuss next.

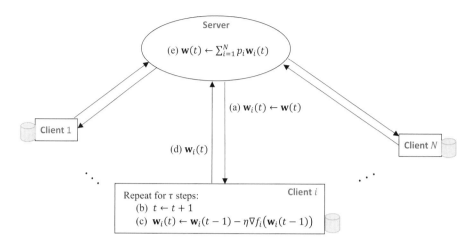

**Figure 13.3** Federated averaging (FedAvg).

## 13.2.2  Federated Averaging (FedAvg)

A typical way of solving Eq. (13.3) in the FL setting is to use the *federated averaging* (FedAvg) algorithm [4]. FedAvg includes multiple *rounds*, where each round includes *local iteration* steps at each client followed by a *parameter aggregation* step that involves both clients and the server. The steps of FedAvg are illustrated in Fig. 13.3 [16].

In particular, each round starts with the server sending its current parameter vector $\mathbf{w}(t)$ to all the clients (Step (a) in Fig. 13.3), where $t$ is the iteration index, starting at $t = 0$ for the parameter vector at initialization. Upon receiving $\mathbf{w}(t)$ from the server, each client $i$ sets its local parameter vector $\mathbf{w}_i(t)$ to be equal to $\mathbf{w}(t)$. Then, every client $i$ performs $\tau$ steps of gradient descent using its local loss function (defined on its local dataset, as in Eq. (13.1)). Each (out of $\tau$) gradient descent step updates the local parameter vector $\mathbf{w}_i$ according to

$$\mathbf{w}_i(t) = \mathbf{w}_i(t-1) - \eta \nabla f_i(\mathbf{w}_i(t-1)), \tag{13.4}$$

as shown in Steps (b) and (c) in Fig. 13.3, where $\eta > 0$ is the gradient descent step size. Then, the resulting parameter vector from each client is sent to the server, and the server computes the aggregated parameter using

$$\mathbf{w}(t) = \sum_{i=1}^{N} p_i \mathbf{w}_i(t), \tag{13.5}$$

as in Steps (d) and (e) in Fig. 13.3.

This process repeats until a certain stopping condition is reached, which can be either a prespecified maximum number of rounds, a given resource budget, or when the decrease in the loss function is smaller than some threshold. Assume that training stops after $K$ rounds. For any $\tau$, there are then $T = K\tau$ local iterations (counted on any single client) before stopping. The final model parameter can be either $\mathbf{w}(T)$ obtained after aggregation in the last round or

$$\mathbf{w}^{\mathrm{f}} := \arg \min_{k \in \{1, 2, \dots, K\}} f(\mathbf{w}(k\tau)), \qquad (13.6)$$

that is the parameter that gives the smallest global loss among all the rounds. The former is often used in practice, whereas the latter is useful for certain kinds of theoretical analysis, as we will see later. Note that $f(\mathbf{w}(k\tau))$ can be computed if each client $i$ sends its local loss $f_i(\mathbf{w}(k\tau))$ to the server.

The rationale behind this approach is that when $\tau = 1$ (i.e., parameter aggregation is performed after every local iteration), FedAvg is equivalent to gradient descent in the centralized setting. The reason is that, as gradient computation is linear, the aggregation ($p_i$-weighted average) of Eq. (13.4) is equivalent to performing gradient descent on the global loss $f(\cdot)$, since we always have $\mathbf{w}_i(t) = \mathbf{w}(t)$ when $\tau = 1$ and $f(\cdot)$ is defined as the same $p_i$-weighted average of the local losses $f_i(\cdot)$ (see Eqs. (13.2) and (13.5)). This equivalence does not hold when $\tau > 1$ though. Hence, as we will see in later sections, we can optimize for $\tau$ to reach a good trade-off between the communication overhead and learning convergence.

We note that the FedAvg procedure presented earlier exchanges the full parameter vector between clients and the server at the end of each round. This standard FedAvg process can be extended to exchanging a sparse gradient after every local iteration (i.e., $\tau = 1$) [17] or exchanging a compressed model that can be obtained after pruning the original model [10]. In general, when $\tau > 1$, the gradient computed in a single iteration does not represent the change in parameter vector across all iterations within the same federated learning round; hence, the parameter vector or its difference between the start and end of the round would need to be communicated, instead of the gradient alone.

In cases where the training dataset is very large, such as for deep learning, stochastic gradient descent (SGD) is often used in Eq. (13.4) in place of deterministic gradient descent (DGD) [4, 16]. We focus on DGD in this chapter for the ease of analysis. Empirically, the approach that we present here also works for SGD, as we will discuss in Section 13.5.

## 13.2.3    Other Model Fusion Algorithms

Besides FedAvg, other model fusion algorithms have also been developed in recent years. For example, FedProx extends FedAvg to better support FL with heterogeneous data and devices by adding a proximal term to the objective function [18]. Bayesian approaches for efficient model fusion in FL settings was also developed [19]. In general, these extensions are based on the basic FedAvg algorithm or at least its main concept of alternating between local computation and model fusion in each round. We mainly focus on FedAvg in the resource-constrained FL setting in the remainder of this chapter.

## 13.3    Resource-Constrained Federated Learning

Based on our experiences from the 2G to 4G cellular networks and anticipation of the 5G and beyond networks, whenever communication bandwidth and capabilities are enhanced from one network generation to the next, they will always trigger new applications, services, and demands that subsequently utilize the newly available bandwidth and capabilities. As a result, unlike wired networks that are often characterized by stable and adequate capacity, communication resources in wireless networks are often limited when faced with ever increasing traffic demands from users and data-intensive applications. Some of the reasons for such highly constrained environments are due to the sharing of limited, expensive radio bandwidth as well as the mutual interference among adjacent transmissions that can significantly reduce the overall capacity of wireless networks.

To address such challenges, the purpose of this section is to formulate the FL process as an optimization problem that maximizes the learning quality (i.e., in terms of minimizing the defined loss function) subject to the amount of different types of system resources available for the learning process. The resources can be in the form of energy, elapsed time, monetary cost, etc., which are incurred during both communication and computation. Because the system usually has many other tasks running simultaneously, it is important to constrain the resource consumption of FL tasks to a certain limit (or budget), so that they do not interfere with other higher-priority tasks. This is particularly necessary at the resource-constrained wireless edge, where the amount of available resources is substantially lower than datacenter environments.

For a given resource budget, a natural question is how to perform FL so that the global loss function is minimized while not exceeding the resource budget. For the FedAvg algorithm presented in Section 13.2.2, one possible way is to control the number of local iterations in each round (i.e., the value of $\tau$) to find the best trade-off between communication and computation resource consumption. As a case study, we scope on the adjustment of $\tau$ in the following. There are other ways to balance various trade-offs in the FL process, such as using dynamic client selection [20, 21], sending sparsified and quantized gradients [17, 22], etc., which are beyond the scope of this chapter.

Consider $M$ types of resources, where a resource type can correspond to time, energy, communication bandwidth, monetary cost, etc., or linear combinations (e.g., sum) of them. For every $m \in \{1, 2, \ldots, M\}$, let $c_m$ denote the amount of type-$m$ resource consumed in each local iteration involving all clients; similarly, let $a_m$ denote the amount of type-$m$ resource consumed in each parameter aggregation step. For each resource type $m$, we assume that we are given a *budget* of $B_m$. We can then formulate the overall resource-constrained FL problem as follows:

$$\min_{\tau, K \in \{1,2,3,\ldots\}} \quad f(\mathbf{w}^{\mathrm{f}}) \tag{13.7}$$

$$\text{s.t.} \quad (K\tau + 1)c_m + (K + 1)a_m \leq B_m, \quad \forall m \in \{1, \ldots, M\},$$

where we recall the notations $\mathbf{w}^{\mathrm{f}}$, $\tau$, and $K$ from Section 13.2.2. Depending on how the final model is chosen by the FL algorithm, we can use either $\mathbf{w}^{\mathrm{f}}$ or $\mathbf{w}(T)$ in the objective function. We use $\mathbf{w}^f$ here to match with the convergence analysis presented later. Note $\mathbf{w}^f$ is still obtained using FedAvg, and the resource budget constraint is similar to a stopping condition. To compute $\mathbf{w}^f$ (see Eq. (13.6)), each client needs to compute $f_i(\mathbf{w}(T))$ and transmit the result to the server, based on which the global loss $f(\mathbf{w}(T))$ is computed, which is why there is an additional local iteration and parameter aggregation in the resource constraint in Eq. (13.7).

Different from the ML objective in Eq. (13.3), the focus of Eq. (13.7) is to optimize for $\tau$ and $K$ so that the global loss of the final result obtained from FedAvg is minimized. To do so, one usually needs to find the relationship between the objective $f(\mathbf{w}^f)$ and $\tau$ and $K$. However, this is difficult because it depends on how FedAvg converges over time, for which only upper bounds can be usually obtained. The bound should also reflect the impact of different $\tau$ and $K$, which adds another challenge beyond the analysis of standard gradient descent approaches. In the next section, we present a convergence upper bound and use this result as an approximation to solve Eq. (13.7).

## 13.4 Solution

Compared to distributed ML in datacenters, a challenge in FL systems is that the data at different clients are usually non-identically distributed (non-i.i.d.). This invalidates a lot of convergence results for distributed ML that were derived for independent and identically distributed (i.i.d.) data. In this section, we first analyze the convergence bound of FedAvg with non-i.i.d. data and a given value of $\tau$ (i.e., the number of local iterations in each round), based on which we present an approximately optimal solution to Eq. (13.7). We will see that the optimal solution depends on the "non-i.i.d.-ness" of data distribution, as well as several other interesting characteristics.

A summary of main notations used throughout this chapter is given in Table 13.2. These notations will be frequently reused in this section.

### 13.4.1 Convergence Analysis

In the following, we derive an upper bound of $f(\mathbf{w}^f) - f(\mathbf{w}^*)$, where $\mathbf{w}^f$ is found using FedAvg.

**Auxiliary Procedure with Centralized Gradient Descent**

To derive the bound, we define an auxiliary procedure that performs gradient descent in the centralized setting (i.e., with all the data available at a central place) within each FL round. This auxiliary procedure is not performed in the actual system. It is only used as a way to derive the gap between the federated case and the centralized (ideal) case at the end of each round. As each round $k \in \{1, 2, \ldots, K\}$ corresponds to the interval of iterations $[(k-1)\tau, k\tau]$, we use $[k]$ to denote this interval in short: $[k] := [(k-1)\tau, k\tau]$. An illustration of these intervals and the centralized gradient descent procedure described next is given in Fig. 13.4.

**Table 13.2.** Summary of main notations.

| | |
|---|---|
| $N$ | Total number of clients |
| $f_i(\mathbf{w})$ | Local loss at client $i$ |
| $f(\mathbf{w})$ | Global loss |
| $p_i$ | Weighting coefficient in global loss definition and parameter aggregation |
| $\mathbf{w}^*$ | Optimal parameter vector that minimizes the global loss $f(\mathbf{w})$ |
| $t$ | Iteration index |
| $\mathbf{w}_i(t)$ | Local parameter vector in iteration $t$ at client $i$ |
| $\mathbf{w}(t)$ | Global parameter vector in iteration $t$ |
| | (after parameter aggregation if it is the last iteration in round) |
| $\mathbf{w}^{\mathrm{f}}$ | Final parameter vector at the end of FL process |
| $\eta$ | Step size of gradient descent |
| $\tau$ | Number of local iterations in each round |
| $T$ | Total number of local iterations (seen at a single client) |
| $K$ | Total number of parameter aggregations, equal to $T/\tau$ |
| $m\ (M)$ | Resource type index (total number of resource types) |
| $B_m$ | Budget of type-$m$ resource |
| $c_m$ | Type-$m$ resource consumption in one local iteration |
| $a_m$ | Type-$m$ resource consumption in one parameter aggregation |
| $\rho$ | Lipschitz coefficient of $f_i(\mathbf{w})$ ($\forall i$) and $f(\mathbf{w})$ |
| $\beta$ | Smoothness coefficient of $f_i(\mathbf{w})$ ($\forall i$) and $f(\mathbf{w})$ |
| $\delta$ | Gradient divergence |
| $h(\tau)$ | Gap between the model parameter vector obtained from |
| | centralized gradient descent and FL – function defined in (13.12) |
| $\varphi$ | Control parameter – constant defined in Lemma 13.5 |
| $G(\tau)$ | Control objective – function defined in (13.17) |
| $\tau^*$ | Optimal $\tau$ that minimizes $G(\tau)$ |

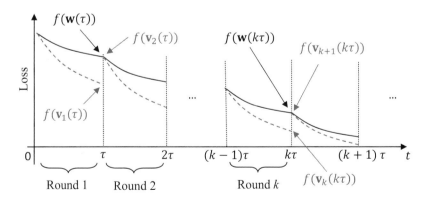

**Figure 13.4** Auxiliary centralized gradient procedure within each round.

For every interval $[k]$, $\mathbf{v}_k(t)$ denotes the auxiliary parameter vector obtained using gradient decent in the centralized setting, according to

$$\mathbf{v}_k(t) = \mathbf{v}_k(t-1) - \eta \nabla f(\mathbf{v}_k(t-1)), \tag{13.8}$$

where $\mathbf{v}_k(t)$ is defined only for $t \in [k]$, for any $k \in \{1, 2, \ldots, K\}$.

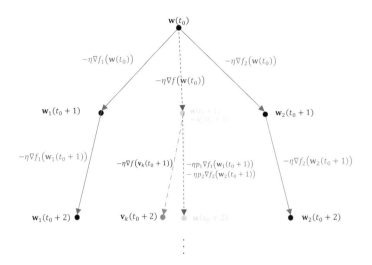

**Figure 13.5** Example showing the gap between $\mathbf{w}(t)$ and $\mathbf{v}_k(t)$ for an FL system with two clients, where $t_0 := (k-1)\tau$.

At the beginning of each round $k$, the value of $\mathbf{v}_k(t)$ is set to be equal to the parameter vector obtained from FedAvg. In other words, we define

$$\mathbf{v}_k((k-1)\tau) := \mathbf{w}((k-1)\tau), \tag{13.9}$$

for any $k \in \{1, 2, \ldots, K\}$. Note that $\mathbf{w}((k-1)\tau)$ is the parameter vector obtained by FedAvg at the end of the previous round $k-1$ after parameter aggregation (or initialization if $k=1$).

The key difference between $\mathbf{w}(t)$ and $\mathbf{v}_k(t)$ is as follows. The vector $\mathbf{w}(t)$ is obtained by FedAvg that only includes one parameter aggregation step at the end of round $k$, after iteration $t = k\tau$; there is no aggregation in any other iteration $t \in ((k-1)\tau, k\tau)$ where the gradient descent continues on the local parameter vector, according to Eq. (13.4). However, the values of $\mathbf{w}(t)$ for $t \in ((k-1)\tau, k\tau)$ are still defined according to Eq. (13.5), although they are not observable by the system. In contrast, $\mathbf{v}_k(t)$ is obtained by assuming parameter aggregation is performed after every iteration $t$, so that gradient descent is conducted on the global parameter in each iteration, which is equivalent to the centralized setting.

An example showing the progression of $\mathbf{w}(t)$ and $\mathbf{v}_k(t)$ for an FL system with two clients (i.e., $N = 2$) is given in Fig. 13.5. To simplify the description, we define $t_0 := (k-1)\tau$ in the following. We discuss the first and second local iterations in each round.

- At the beginning of any round $k$, the parameter vectors at all clients are equal, (i.e., $\mathbf{w}_i(t_0) = \mathbf{w}(t_0)$ for any client $i \in \{1, 2, \ldots, N\}$), because they are obtained after parameter aggregation in the previous round. According to the definition of the global loss in Eq. (13.2) and the linearity of gradient operator, we have

$\nabla f(\mathbf{w}) = \sum_{i=1}^{N} p_i \nabla f_i(\mathbf{w})$. Combining with the global parameter definition in Eq. (13.5), we have $\mathbf{w}(t_0 + 1) = \mathbf{v}_k(t_0 + 1)$ after the first local iteration, which means that the global parameter vector obtained from FL is equal to that obtained from centralized gradient descent after one local iteration. However, when $\tau > 1$, the value of $\mathbf{w}(t_0 + 1)$ is not observable by the system, because it would only be observable if a parameter aggregation step is performed.

- In the second local iteration $t = t_0 + 2$, the gradient at each client $i$ is computed on the local parameter obtained from the first local iteration $t = t_0 + 1$. Because the local parameters $\mathbf{w}_1(t_0 + 1)$ and $\mathbf{w}_2(t_0 + 1)$ are generally different from their $p_i$-weighted averaged value $\mathbf{w}(t_0 + 1)$, the gradient used in the centralized setting is now different from the "gradient" in the FL setting, yielding $\mathbf{w}(t_0 + 2) \neq \mathbf{v}_k(t_0 + 2)$, as shown in Fig. 13.5.

The third and further iterations are similar to the second iteration. Then, the gap between $\mathbf{w}(t)$ and $\mathbf{v}_k(t)$ quantifies how much FL falls short of the performance of centralized gradient descent.

With these definitions, we can derive the convergence bound of FedAvg in two steps. The first step derives an upper bound of the gap between $\mathbf{w}(k\tau)$ and $\mathbf{v}_k(k\tau)$ obtained at the end of each round $k$. The second step combines this upper bound with the convergence of $\mathbf{v}_k(k\tau)$ to obtain the convergence bound with respect to $\mathbf{w}^f$.

## Definitions and Assumptions

We introduce some definitions and assumptions that are required for the convergence result presented later.

DEFINITION 13.1    For any client $i$ and parameter vector $\mathbf{w}$, we define the local *gradient divergence* $\delta_i$ as an upper bound of $\|\nabla f_i(\mathbf{w}) - \nabla f(\mathbf{w})\|$:

$$\|\nabla f_i(\mathbf{w}) - \nabla f(\mathbf{w})\| \leq \delta_i. \tag{13.10}$$

We also define the global gradient divergence as $\delta := \sum_{i=1}^{N} p_i \delta_i$.

The gradient divergence captures the degree of non-i.i.d. across client data. In the extreme case when all clients have exactly the same data, we always have $f_i(\mathbf{w}) = \nabla f(\mathbf{w})$ and the gradient divergence is zero.

We also make the following assumption for our analysis, which is relatively standard in various convergence analysis work.

ASSUMPTION 13.2    We assume that the following holds for the local loss function $f_i(\mathbf{w})$ for all clients $i$:

1. $f_i(\mathbf{w})$ is convex.
2. $f_i(\mathbf{w})$ is $\rho$-Lipschitz; that is, $\|f_i(\mathbf{w}) - f_i(\mathbf{w}')\| \leq \rho\|\mathbf{w} - \mathbf{w}'\|$ for any $\mathbf{w}$ and $\mathbf{w}'$.
3. $f_i(\mathbf{w})$ is $\beta$-smooth; that is, $\|\nabla f_i(\mathbf{w}) - \nabla f_i(\mathbf{w}')\| \leq \beta \|\mathbf{w} - \mathbf{w}'\|$ for any $\mathbf{w}$ and $\mathbf{w}'$.

When the components of $\mathbf{w}$ and $\mathbf{w}'$ are within finite upper and lower bounds, which is usually the case in practice, the conditions in Assumption 13.2 hold for the linear regression and support vector machine (SVM) losses defined in Table 13.1. For other loss function definitions that are nonconvex, our experimental results in Section 13.5 show that the control algorithm derived from our convergence bound still works well in an empirical sense.

When Assumption 13.2 holds, the same conditions also hold for the global loss $f(\mathbf{w})$, as shown in the following lemma.

LEMMA 13.3 *The global loss $f(\mathbf{w})$ is convex, $\rho$-Lipschitz, and $\beta$-smooth, where the Lipschitzness and smoothness definitions are the same as in Assumption 13.2 by replacing $f_i(\cdot)$ with $f(\cdot)$.*

The proof is straightforward from the definition of $f(\mathbf{w})$ and triangle inequality.

## Main Results

In the following, we present the main convergence results from [9] and explain their intuitions. For detailed proofs, see [9] and its extended version on arXiv.[4]

THEOREM 13.4 *When Assumption 13.2 holds, for any interval $[k]$ and $t \in [k]$, we have*

$$\|\mathbf{w}(t) - \mathbf{v}_k(t)\| \le h(t - (k-1)\tau), \qquad (13.11)$$

*where*

$$h(x) := \frac{\delta}{\beta}\left((\eta\beta + 1)^x - 1\right) - \eta\delta x, \qquad (13.12)$$

*for any nonnegative integer $x$.*

*Furthermore, as $f(\cdot)$ is $\rho$-Lipschitz, we also have*

$$f(\mathbf{w}(t)) - f(\mathbf{v}_k(t)) \le \rho \|\mathbf{w}(t) - \mathbf{v}_k(t)\| \le \rho h(t - (k-1)\tau). \qquad (13.13)$$

The main idea for proving this theorem is to construct a geometric progression using the gradient divergence bound and the bounds in Assumption 13.2.

For nontrivial gradient descent and loss definitions, we have $\eta > 0$ and $\beta > 0$. Hence, from Bernoulli's inequality, we always have $(\eta\beta + 1)^x \ge \eta\beta x + 1$, which implies $h(x) \ge 0$ for any integer $x \ge 0$. We can easily see that $h(0) = h(1) = 0$, which aligns with our example in Fig. 13.5 and its related discussion, showing that the gap between $\mathbf{w}(t)$ and $\mathbf{v}_k(t)$ is zero after the first local iteration in each round. Obviously, this gap is also zero before the first local iteration, due to the definition of $\mathbf{v}_k(t)$ in Eq. (13.9). According to the definition of the interval $[k]$, when $\tau = 1$, we always have $t - (k-1)\tau \in \{0, 1\}$ for any $t$ such that $\mathbf{v}_k(t)$ is defined. Thus, FedAvg is equivalent to centralized gradient descent when $\tau = 1$, as also informally explained in Section 13.2.2.

Based on Theorem 13.4, we can first characterize the convergence of centralized gradient descent within each round $k$ and derive an upper bound related to $f(\mathbf{v}_k(t))$. Then, we can combine this with the upper bound of $f(\mathbf{w}(t)) - f(\mathbf{v}_k(t))$ given in Theorem 13.4 (see Fig. 13.4). This gives us the following lemma.

---

[4] https://arxiv.org/abs/1804.05271

LEMMA 13.5    *When all of the following are satisfied:*

1. $\eta \leq \frac{1}{\beta}$
2. $\eta\varphi - \frac{\rho h(\tau)}{\tau\varepsilon^2} > 0$
3. $f(\mathbf{v}_k(k\tau)) - f(\mathbf{w}^*) \geq \varepsilon$ *for all k*
4. $f(\mathbf{w}(T)) - f(\mathbf{w}^*) \geq \varepsilon$

*for some $\varepsilon > 0$, where we define $\varphi := \omega\left(1 - \frac{\beta\eta}{2}\right)$ and $\omega := \min_k \frac{1}{\|\mathbf{v}_k((k-1)\tau) - \mathbf{w}^*\|^2}$, then the convergence upper bound of FedAvg after T local iterations (seen at a single client) is given by*

$$f(\mathbf{w}(T)) - f(\mathbf{w}^*) \leq \frac{1}{T\left(\eta\varphi - \frac{\rho h(\tau)}{\tau\varepsilon^2}\right)}. \tag{13.14}$$

The bound in Lemma 13.5 depends on a prespecified parameter $\epsilon$ and is an intermediate result. The dependence on $\epsilon$ can be removed by solving for an $\epsilon$ value using the upper bound in Eq. (13.14) and the lower bounds in Conditions 3 and 4 of the lemma, which gives the following final result.

THEOREM 13.6    *When Assumption 13.2 holds and the gradient descent step size $\eta \leq \frac{1}{\beta}$, we have*

$$f(\mathbf{w}^f) - f(\mathbf{w}^*) \leq \frac{1}{2\eta\varphi T} + \sqrt{\frac{1}{4\eta^2\varphi^2 T^2} + \frac{\rho h(\tau)}{\eta\varphi\tau}} + \rho h(\tau), \tag{13.15}$$

*where $\mathbf{w}^f$ is defined according to Eq. (13.6).*

We make the following observations from Theorem 13.6.

- When $\tau = 1$, we have $h(\tau) = 0$ according to its definition in Eq. (13.12). In this case, the optimality gap $f(\mathbf{w}^f) - f(\mathbf{w}^*)$ approaches zero as $T \to \infty$.
- When $\tau > 1$, we generally have $h(\tau) > 0$. In this case, $f(\mathbf{w}^f) - f(\mathbf{w}^*)$ is nonzero even if $T \to \infty$, which means we have a nonzero optimality gap even if we have abundant resource.

These observations show that if we have an infinite resource budget, it is always beneficial to choose $\tau = 1$. However, for a finite resource budget, it may be better to choose a larger value of $\tau$ because it may yield a smaller $f(\mathbf{w}^f)$ when constrained on the resource budget, stopping at some finite $T$.

We further note that $h(\tau)$ is proportional to the gradient divergence $\delta$, as defined in Eq. (13.12). Thus, for $\tau > 1$, the optimality gap increases with the gradient divergence, which suggests that $\tau$ should be smaller if the data distribution at clients is more non-i.i.d., causing a larger gradient divergence. This is intuitive because the more the clients are different, the faster their parameters diverge, and hence they need to be aggregated more frequently. It also aligns with the experimental findings of our control algorithm that tries to find the best $\tau$, which we will see later.

### Other Convergence Results

The convergence result we have presented is based on [9]. Other results have also been derived in subsequent work. In particular, it was shown that the optimality gap can also approach zero when $\tau > 1$, if the step size $\eta$ either decays over time or is defined as a decreasing function of the total number of iterations $T$, where the latter requires knowing $T$ at the start of the algorithm [23, 24]. Intuitively, this is because using a small gradient descent step size has a similar effect as decreasing $\tau$ while keeping the same step size.

In addition, while our result in Theorem 13.6 only holds for the DGD case, the work in [23] has provided a convergence bound for FedAvg with SGD updates. A similar procedure for finding the optimal $\tau$ as what we present next should work for other convergence bounds too. For consistency, we use the result in Theorem 13.6 in our following analysis and confirm that our method also works with SGD empirically in Section 13.5.

## 13.4.2    Approximate Solution

Based on the convergence result in Theorem 13.6, we present an approximate solution to Eq. (13.7) in the following. Noting that the optimal parameter vector $\mathbf{w}^*$ is fixed (but unknown) for any predefined global loss $f(\mathbf{w})$, the optimal global loss $f(\mathbf{w}^*)$ is a constant. Hence, the minimization of $f(\mathbf{w}^f)$ in Eq. (13.7) is equivalent to minimizing $f(\mathbf{w}^f) - f(\mathbf{w}^*)$, for which we use the upper bound in Eq. (13.15) as an approximation; that is, $f(\mathbf{w}^f) - f(\mathbf{w}^*) \approx \frac{1}{2\eta\varphi T} + \sqrt{\frac{1}{4\eta^2\varphi^2 T^2} + \frac{\rho h(\tau)}{\eta\varphi\tau}} + \rho h(\tau)$.

Using this upper-bound approximation and after rearranging the constraint in Eq. (13.7), we can rewrite Eq. (13.7) as the following:

$$\min_{\tau, K \in \{1,2,3,\ldots\}} \quad \frac{1}{2\eta\varphi K\tau} + \sqrt{\frac{1}{4\eta^2\varphi^2 K^2\tau^2} + \frac{\rho h(\tau)}{\eta\varphi\tau}} + \rho h(\tau) \qquad (13.16)$$

$$\text{s.t.} \quad K \le \frac{B'_m}{c_m\tau + a_m}, \quad \forall m \in \{1,\ldots,M\},$$

where $B'_m := B_m - a_m - c_m$.

The objective function in Eq. (13.16) decreases with $K$. Hence, for any $\tau$, the optimal choice of $K$ is $\left\lfloor \min_m \frac{B'_m}{c_m\tau + a_m} \right\rfloor$, which is the largest $K$ that satisfies the constraint in Eq. (13.16), where $\lfloor \cdot \rfloor$ stands for rounding down to integer (i.e., floor operation). For simplicity, we further approximate by ignoring the floor operation and substitute $K\tau \approx \min_m \frac{B'_m\tau}{c_m\tau + a_m} = 1 / \max_m \frac{c_m\tau + a_m}{B'_m\tau}$ into the objective of Eq. (13.16), which gives us

$$G(\tau) := \frac{\max_m \frac{c_m\tau + a_m}{B'_m\tau}}{2\eta\varphi} + \sqrt{\frac{\left(\max_m \frac{c_m\tau + a_m}{B'_m\tau}\right)^2}{4\eta^2\varphi^2} + \frac{\rho h(\tau)}{\eta\varphi\tau}} + \rho h(\tau). \qquad (13.17)$$

Now, the (approximately) optimal $\tau$ can be computed by

$$\tau^* := \arg \min_{\tau \in \{1, 2, 3, \dots\}} G(\tau), \tag{13.18}$$

based on which we also obtain the (approximately) optimal total number of rounds $K$ as

$$K^* = \left\lfloor \min_m \frac{B'_m}{c_m \tau^* + a_m} \right\rfloor. \tag{13.19}$$

Hence, instead of solving Eq. (13.7) directly, which is mathematically intractable, we solve Eq. (13.18) as an approximation to Eq. (13.7). Our experimental results in the next section show that our approximate solution is close to the actual optimal solution on various real datasets and models.

**Solution Properties**

Although $G(\tau)$ has a complicated form as shown in Eq. (13.17), its optimal solution $\tau^*$ in Eq. (13.18) has some nice properties as shown in the following.

THEOREM 13.7    When $\eta \leq \frac{1}{\beta}$, $\rho > 0$, $\beta > 0$, $\delta > 0$, we have $\lim_{B_{\min} \to \infty} \tau^* = 1$, where $B_{\min} := \min_m B_m$.

THEOREM 13.8    When $\eta \leq \frac{1}{\beta}$, $\rho > 0$, $\beta > 0$, $\delta > 0$, there exists a finite value $\tau_0$ that only depends on $\eta$, $\beta$, $\rho$, $\delta$, $\varphi$, $c_m$, $a_m$, $B'_m$ ($\forall m$), such that $\tau^* \leq \tau_0$, where

$$\tau_0 := \max \left\{ \max_m \frac{a_m B'_v - a_v B'_m}{c_v B'_m - c_m B'_v}; \ \frac{\varphi(2 + \eta\beta)}{2\rho\delta} \left( \frac{2c_v a_v}{C_2} + \frac{2a_v^2}{C_2} \right); \right.$$

$$\left. \frac{1}{\rho\delta\eta \log D} \left( \frac{a_v}{C_1} + \rho\eta\delta \right) - \frac{1}{\eta\beta}; \ \frac{1}{\eta\beta} + \frac{1}{2} \right\},$$

in which index $v := \arg \max_{m \in V} \frac{a_m}{B'_m}$ (set $V := \arg \max_m \frac{c_m}{B'_m}$), $D := \eta\beta + 1$, $C_1 := 2\eta\varphi B'_v$, $C_2 := 4\eta^2\varphi^2 B'^2_v$. Here, for convenience, we allow $\arg \max$ to return either a set and an arbitrary value in that set, and we also define $\frac{0}{0} := 0$.

We also note that $\tau_0 \geq \frac{1}{\eta\beta} + \frac{1}{2} > 1$ since $0 < \eta\beta \leq 1$.

These two theorems can be proven using some analytical manipulation of the function $G(\tau)$. For further details, see [9]. Their implications are discussed as follows.

Theorem 13.7 shows that when we have abundant resources (of all types), the optimal solution is $\tau^* = 1$. Since zero optimality gap can only be guaranteed when $\tau = 1$, as discussed in Section 13.4.1, this shows that our approximate solution guarantees convergence to the optimal global loss when the resource budget goes to infinity.

The main observation from Theorem 13.8 is that the optimal solution $\tau^*$ is upper bounded by a constant $\tau_0$, although the expression of this constant appears nontrivial. This allows us to find $\tau^*$ using a linear search on $G(\tau)$, computing $G(\tau)$ for $\tau = 1$, $\tau = 2$, $\tau = 3$, and so on, up to some maximum value of $\tau$, which is useful because $G(\tau)$ includes both exponential and polynomial terms and thus we cannot express its optimal solution analytically.

### Other Approaches

Besides the approach presented earlier, a similar method of optimizing $\tau$ using convergence bound approximation was described in [25], which does not support non-i.i.d. data distribution though. For communication-efficient FL in general, other optimization techniques such as online learning [17] and reinforcement learning [21] may also be applied, which may be either more or less preferred over our convergence bound approximation approach depending on specific problem settings.

## 13.5 Implementation and Experimental Validation

### 13.5.1 Implementation

#### Estimation of Parameters in $G(\tau)$

In a practical FL system, the computation of $\tau^*$ from $G(\tau)$ needs to be incorporated into the FedAvg procedure (or any other FL algorithm). A challenge is that $G(\tau)$ has many parameters that are unknown to the system, including $\rho, \beta, \delta, \varphi$ (see Eq. (13.17) and the definition of $h(\tau)$ in Eq. (13.12)). Other resource-related parameters $c_m$ and $a_m$ (for all $m$) also need to be estimated during the real-time operation of the system.

In an actual system implementation, the Lipschitz and smoothness coefficients $\rho$ and $\beta$ can be estimated based on the change of the loss or its gradient for different parameter vectors $\mathbf{w}$ and $\mathbf{w}'$, as defined in Assumption 13.2. A natural way to do so is to estimate them during the parameter aggregation step, where each client $i$ knows its local parameter vector $\mathbf{w}_i(t)$ and the aggregated parameter vector $\mathbf{w}(t)$ received from the server, so they can be considered as $\mathbf{w}$ and $\mathbf{w}'$, respectively.

The gradient divergence $\delta$ can be estimated by having each client sending the gradient of its local loss function to the server, so that the server can compute the distance between the local and global gradients to estimate $\delta$, as in Definition 13.1. To avoid additional overhead of sending the gradients in addition to the model parameters, the gradient may also be estimated by the server using the change in model parameters in one FL round divided by the step size $\eta$ and the number of per-round iterations $\tau$, which is equal to the average gradient seen in this entire round.

The quantity $\varphi$ is difficult to estimate, because it depends on the maximum distance between $\mathbf{v}_k$ and $\mathbf{w}^*$, as defined in Lemma 13.5, where this distance is largely unknown because the true optimal solution $\mathbf{w}^*$ is unknown. In practice, $\varphi$ can be considered as a tuneable hyperparameter that may be model-specific. From our experiments, we found that our algorithm's performance is not very sensitive to the choice of $\varphi$.

When using SGD, the estimations we have discussed need to be based on losses computed on the same mini-batch of data to avoid noise in mini-batch sampling causing significant error in the estimation.

#### Protocol

The overall method of optimized FL with adaptive $\tau$ is shown in Fig. 13.6, which integrates the standard FedAvg procedure, estimation of parameters in $G(\tau)$, monitoring

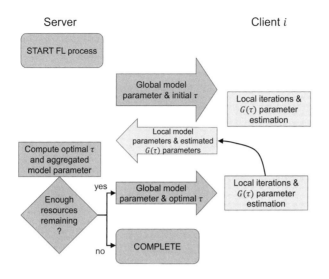

**Figure 13.6** Protocol of FedAvg with adaptive $\tau$.

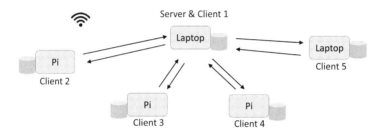

**Figure 13.7** Experiment setup.

of consumed resources, and checking against the resource budget. The resource budget is seen as an input to the system, which can be specified by a system administrator or by a separate job scheduling mechanism that is beyond the scope of this discussion.

## 13.5.2     Experiments

We now present an experimental validation of the optimized FL approach presented earlier. We consider a networked prototype system with five clients, including three Raspberry Pi devices and two laptop computers, which are interconnected using an office Wi-Fi. We also consider a simulated environment based on measurements from the prototype system. Each client has its own dataset on which local iterations are performed. One of the laptop clients also acts as the server simultaneously, which is possible because the server in FL is simply a logical entity that can colocate with a client. This experiment setup is illustrated in Fig. 13.7.

We consider the *elapsed time* as our resource type of interest in the experiments, for example, to reflect a timeliness requirement of obtaining the trained model, which can be important for time-critical tasks. The elapsed time captures the resource overhead caused by both communication and computation in the FL process. A maximum time budget is specified for each FL task. The resource consumption (elapsed time) $c$ and $a$ for local iteration and parameter aggregation, respectively, are set using real measurements from the prototype system, where we omit the subscript $m$ because we only consider a single resource type here. For the simulated system, they are randomly generated by a Gaussian distribution with statistics obtained from the prototype system.

We compare our approach against the following three baseline approaches:

- Gradient descent in the centralized setting with all training data available at a central location [26];
- FedAvg with a fixed and prespecified number of local iterations $\tau$ in each round [4]; and
- Synchronous distributed gradient descent, which is equivalent to setting $\tau = 1$ [27].

For fairness reasons, the estimation of resource consumption is implemented for all these baselines and the FL procedure terminates when the resource budget is reached.

We consider four different models, including SVM, linear regression, K-means, and deep convolutional neural network (CNN). These models are trained using five different datasets. Their combinations are summarized as follows.

- SVM is trained on the MNIST dataset [28], which contains images of handwritten digits. The problem is converted into a binary classification problem, where the SVM predicts whether a digit is even or odd. The SVM uses a smooth loss function as defined in Table 13.1.
- Linear regression is trained using the energy dataset [29], which records 19,735 measurements taken at different sensors, and the goal is to forecast the energy usage of appliances.
- K-means is performed on the user knowledge modeling dataset [30], where each data sample is a user described by five attributes representing its interaction with a web-based adaptive learning environment. The goal is to classify users into four different groups according to their knowledge levels.
- CNN is trained on the original MNIST (O-MNIST) [28], Fashion MNIST (F-MNIST) [31], and CIFAR-10 [32] datasets. For each dataset, a CNN is trained to classify images into 10 different classes. The CNN has nine layers in the following architecture: $5 \times 5 \times 32$ Convolutional $\rightarrow 2 \times 2$ MaxPool $\rightarrow$ Local Response Normalization $\rightarrow 5 \times 5 \times 32$ Convolutional $\rightarrow$ Local Response Normalization $\rightarrow 2 \times 2$ MaxPool $\rightarrow \gamma \times 256$ Fully connected $\rightarrow 256 \times 10$ Fully connected $\rightarrow$ Softmax, where $\gamma$ depends on the size of the input image and we have $\gamma = 1,568$ for O-MNIST and F-MNIST and $\gamma = 2,048$ for CIFAR-10.

The experiments also include four different ways of distributing data across clients.

- In Case *i.i.d.*, data samples are randomly assigned to each client, which results in an i.i.d. data distribution.
- In Case *non-i.i.d.*, different clients have data with different labels (classes), yielding a non-i.i.d. data distribution across clients. When there are more clients than labels, data with the same label may be partitioned across multiple clients.
- In Case *equal*, the full dataset is replicated at each node.
- In Case *combined*, half of the clients have data distributed in a similar way as Case i.i.d. and the remaining half in a similar way as Case non-i.i.d.

For the energy dataset [29] that does not have explicit labels, the data distribution is made using an unsupervised clustering approach.

Both deterministic and stochastic gradient descent approaches (i.e., DGD and SGD) are considered. The maximum value of $\tau$ in linear search to find the optimal $G(\tau)$ is set to 100. The hyperparameter $\varphi$ is set to 0.025 for SVM, linear regression, and K-means, and $5 \times 10^{-5}$ for CNN. The gradient descent step size $\eta$ is 0.01. We set the total time budget $B$ as 15 seconds.

Fig. 13.8 shows results of our experiments in an FL system with five clients, where the results are averaged over 15 independent experiments. SVM, linear regression, and K-means models were trained on the prototype system described earlier. CNN was trained in a simulated environment due to computational resource constraint of Raspberry Pis.

The global loss of the optimal FL approach with adaptive $\tau$ (i.e., proposed) is compared to the baseline approaches. For SVM and CNN models, accuracy is also compared. In Fig. 13.8, our approach is represented using a single marker (symbol) for each case. The single marker is the average value of $\tau^*$ and its corresponding loss/accuracy obtained from the 15 independent experiments. The centralized case, which has also a single value, is shown by a flat line to make the comparison easier.

The results show that $\tau^*$ found by our optimized FL approach is always close to the (empirically) optimal $\tau$ for all kinds of data distributions and all models. The results also underline the importance of having an adaptive $\tau$, as we can see that the (empirically) optimal $\tau$ varies for different cases and models. In general, $\tau^*$ is large for i.i.d. and equal cases and small for non-i.i.d. and combined cases, which is intuitive because the parameters need to be aggregated more frequently if the client data are more different. This also aligns with our intuition discussed in Section 13.4.1.

In some cases, FL performs better than the centralized approach, because for a given time budget, FL makes use of the computational resources of all clients, whereas the centralized approach was run only on a single Raspberry Pi client. Case i.i.d. outperforms Case equal when the DGD optimizer is used, because the amount of data to process in Case equal is larger. Consequently, the time of each local iteration in Case equal is larger than that of Case i.i.d.

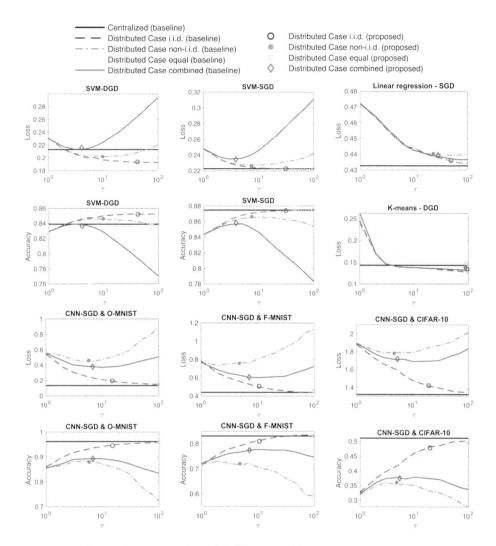

**Figure 13.8** Loss and accuracy values with different models and datasets.

## 13.6    Conclusion

In this chapter, we described a resource-optimized FL technique, where the near-minimum model training loss and near-maximum model accuracy is attained by optimizing the number of local iterations ($\tau$) in each FL round. Effectively, the choice of $\tau$ controls the trade-off between computation and communication. As each local iteration (computation) and parameter aggregation (primarily communication) step may consume different amounts and types of resources and have different effects on model training, finding the best trade-off between them allows us to make the best use of the available resources. We have presented an algorithm for achieving this goal, which uses a convergence upper bound to approximate the objective. Experimental results have demonstrated the usefulness of the optimized FL algorithm and shed further insights on the impact of $\tau$ on model loss/accuracy with various data distributions across clients.

The area of optimized FL with resource constraints extends beyond the optimization of $\tau$. Other aspects of FL can be optimized as well, such as which clients should participate in each FL round (client selection), which model parameters should be transmitted during parameter aggregation (model pruning, compression, quantization, and sparsification), how to couple FL with wireless channel assignments, etc. While some progress has been made toward these directions in recent years, many outstanding questions still exist. For example, the behavior of an FL system that integrates all these efficient communication and computation techniques would be worth studying from both theoretical and empirical (experimental) perspectives.

In addition, most existing research on FL is based on theoretical analysis and data-driven simulations, where many challenges in practical systems may not be adequately captured. To enable efficient and reliable FL in 5G systems and beyond, comprehensive studies that implement FL in actual cellular systems with real users would be necessary. Such studies could start with an FL implementation on a 5G testbed deployed in a campus environment, for instance. The results obtained from this testbed could provide important insights that enable a larger-scale deployment afterward.

## References

[1]  J. Park et al., "Wireless network intelligence at the edge," in *Proc. IEEE*, vol. 107, no. 11, pp. 2204–2239, 2019.

[2]  G. Zhu et al., "Toward an intelligent edge: wireless communication meets machine learning," *IEEE Communications Magazine*, vol. 58, no. 1, pp. 19–25, 2020.

[3]  H. Lu et al., "Robust coreset construction for distributed machine learning," *IEEE Journal on Selected Areas in Communications*, vol. 38, no. 10, pp. 2400–2417, Oct. 2020.

[4]  H. B. McMahan et al., "Communication-efficient learning of deep networks from decentralized data," in *AISTATS*, 2017.

[5]  M. Chen et al., "A joint learning and communications framework for federated learning over wireless networks," *IEEE Transactions on Wireless Communications*, 2020.

[6]  T. Li et al., "Federated learning: Challenges, methods, and future directions," *IEEE Signal Processing Magazine*, vol. 37, no. 3, pp. 50–60, 2020.

[7]  Q. Yang et al., "Federated machine learning: Concept and applications," *ACM Transactions on Intelligent Systems and Technology (TIST)*, vol. 10, no. 2, p. 12, 2019.

[8]  P. Kairouz et al., "Advances and open problems in federated learning," *arXiv preprint*, arXiv:1912.04977, 2019.

[9] S. Wang et al., "Adaptive federated learning in resource constrained edge computing systems," *IEEE Journal on Selected Areas in Communications*, vol. 37, no. 6, pp. 1205–1221, Jun. 2019.

[10] Y. Jiang et al., "Model pruning enables efficient federated learning on edge devices," *arXiv preprint*, arXiv:1909.12326, 2019.

[11] Z. Xu et al., "Elfish: Resource-aware federated learning on heterogeneous edge devices," *arXiv preprint*, arXiv:1912.01684, 2019.

[12] T. Bolukbasi et al., "Adaptive neural networks for efficient inference," in *Proc. 34th Int. Conf. on Machine Learning*, pp. 527–536, 2017.

[13] Z. Wu et al., "Blockdrop: Dynamic inference paths in residual networks," in *Proc. IEEE Conf. on Computer Vision and Pattern Recognition*, pp. 8817–8826, 2018.

[14] M. Saeed Shafiee, M. Javad Shafiee, and A. Wong, "Dynamic representations toward efficient inference on deep neural networks by decision gates," in *Proc. IEEE Conf. on Computer Vision and Pattern Recognition Workshops*, 2019.

[15] I. Goodfellow, Y. Bengio, and A. Courville, *Deep Learning*. MIT Press, 2016.

[16] T. Tuor et al., "Distributed machine learning in coalition environments: Overview of techniques," in *21st Int. Conf. on Information Fusion (FUSION)*, pp. 814–821, 2018.

[17] P. Han, S. Wang, and K. K. Leung, "Adaptive gradient sparsification for efficient federated learning: An online learning approach," in *IEEE International Conference on Distributed Computing Systems (ICDCS)*, 2020.

[18] T. Li et al., "Federated optimization in heterogeneous networks," in *Conf. on Machine Learning and Systems (MLSys)*, 2020.

[19] M. Yurochkin et al., "Bayesian nonparametric federated learning of neural networks," in *Proc. of Machine Learning Research (PMLR)*, vol. 97, pp. 7252–7261, Jun. 2019.

[20] Y. Jin et al., "Resource-efficient and convergence-preserving online participant selection in federated learning," in *IEEE Int. Conf. on Distributed Computing Systems (ICDCS)*, 2020.

[21] H. Wang et al., "Optimizing federated learning on non-iid data with reinforcement learning," in *IEEE Conf. on Computer Communications (INFOCOM)*, pp. 1698–1707, 2020.

[22] J. Konečnỳ et al., "Federated learning: Strategies for improving communication efficiency," *arXiv preprint*, arXiv:1610.05492, 2016.

[23] H. Yu, S. Yang, and S. Zhu, "Parallel restarted SGD with faster convergence and less communication: Demystifying why model averaging works for deep learning," in *Proc. AAAI Conf. on Artificial Intelligence*, vol. 33, 2019, pp. 5693–5700.

[24] X. Li et al., "On the convergence of fedavg on non-iid data," in *Int. Conf. on Learning Representations*, 2020.

[25] J. Wang and G. Joshi, "Adaptive communication strategies to achieve the best error-runtime trade-off in local-update SGD," in *Systems and Machine Learning (SysML) Conf.*, 2019.

[26] S. Shalev-Shwartz and S. Ben-David, *Understanding Machine Learning: From Theory to Algorithms*. Cambridge University Press, 2014.

[27] J. Chen et al., "Revisiting distributed synchronous SGD," *arXiv preprint*, arXiv:1604.00981, 2016.

[28] Y. LeCun et al., "Gradient-based learning applied to document recognition," in *Proc. IEEE*, vol. 86, no. 11, pp. 2278–2324, 1998.

[29] L. M. Candanedo, V. Feldheim, and D. Deramaix, "Data driven prediction models of energy use of appliances in a low-energy house," *Energy and buildings*, vol. 140, pp. 81–97, 2017.

[30] H. T. Kahraman, S. Sagiroglu, and I. Colak, "The development of intuitive knowledge classifier and the modeling of domain dependent data," *Knowledge-Based Systems*, vol. 37, pp. 283–295, 2013.

[31] H. Xiao, K. Rasul, and R. Vollgraf, "Fashion-mnist: a novel image dataset for benchmarking machine learning algorithms," *arXiv preprint*, arXiv:1708.07747, 2017.

[32] A. Krizhevsky et al., "Learning multiple layers of features from tiny images," Tech. report, Citeseer, 2009.

# 14    Quantized Federated Learning

Nir Shlezinger, Mingzhe Chen, Yonina C. Eldar, H. Vincent Poor,
and Shuguang Cui

## 14.1    Introduction

Recent years have witnessed unprecedented success of machine learning methods in a broad range of applications [1]. These systems utilize highly parameterized models, such as deep neural networks, trained using a massive amount of labeled data samples. In many applications, samples are available at remote users, such as smartphones and other edge devices, and the common strategy is to gather these samples at a computationally powerful server, where the model is trained [2]. Often, datasets, such as images and text messages, contain private information, and thus the user may not be willing to share them with the server. Furthermore, sharing massive datasets can result in a substantial burden on the communication links between the edge devices and the server. To allow centralized training without data sharing, federated learning (FL) was proposed in [3] as a method combining distributed training with central aggregation. This novel method of learning has been the focus of growing research attention over the last few years [4]. FL exploits the increased computational capabilities of modern edge devices to train a model on the users' side, while the server orchestrates these local training procedures and, in addition, periodically synchronizes the local models into a global one.

FL is trained by an iterative process [5]. In particular, at each FL iteration, the edge devices train a local model using their (possibly) private data and transmit the updated model to the central server. The server aggregates the received updates into a single global model and sends its parameters back to the edge devices [6]. Therefore, to implement FL, edge devices only need to exchange their trained model parameters, which avoids the need to share their data, thereby preserving privacy. However, the repeated exchange of updated models between the users and the server given the large number of model parameters involves massive transmissions over throughput-limited communication channels. This challenge is particularly relevant for FL carried out over wireless networks, such as when the users are wireless edge devices. In addition to overloading the communication infrastructure, these repeated transmissions imply that the time required to tune the global model not only depends on the number of training iterations, but also depends on the delay induced by transmitting the model updates at each FL iteration [7]. Hence, this communication bottleneck may affect the training time of global models trained via FL, which in turn may degrade their resulting accuracy. This motivates the design of schemes in which the purpose is to limit the communication overhead due to the repeated transmissions of updated model parameters in the distributed training procedure.

Various methods have been proposed in the literature to tackle the communication bottleneck induced by the repeated model updates in FL. The works [8–17] focused on FL over wireless channels and reduced the communication by optimizing the allocation of the channel resources, such as bandwidth, among the participating users as well as limiting the amount of participating devices while scheduling when each user takes part in the overall training procedure. An additional related strategy treats the model aggregation in FL as a form of over-the-air computation [18–20]. Here, the users exploit the full resources of the wireless channel to convey their model updates at high throughput, and the resulting interference is exploited as part of the aggregation stage at the server side. These communication-oriented strategies are designed for scenarios in which the participating users communicate over the same wireless media and are thus concerned with the division of the channel resources among the users.

An alternative approach to reduce the communication overhead, which holds also when the users do not share the same wireless channel, is to reduce the volume of the model updates conveyed at each FL iteration. Such strategies do not focus on the communication channel and how to transmit over it, but rather on what is being transmitted. As a result, they can commonly be combined with the aforementioned communication-oriented strategies. One way to limit the volume of conveyed parameters is to have each user transmit only part of its model updates; that is, they implement dimensionality reduction by sparsifying or subsampling [6, 21–25]. An alternative approach is to discretize the model updates, such that each parameter is expressed using a small number of bits, as proposed in [26–31]. More generally, the compression of the model updates can be viewed as the conversion of a high-dimensional vector, where entries take continuous values, into a set of bits communicated over the wireless channel. Such formulations are commonly studied in the fields of quantization theory and source coding. This motivates the formulation of such compression methods for FL from a quantization perspective, which is the purpose of this chapter.

The goal of this chapter is to present a unified FL framework utilizing quantization theory, which generalizes many of the previously proposed FL compression methods. The purpose of the unified framework is to facilitate the comparison and the understanding of the differences between existing schemes in a systematic manner, as well as identify quantization theoretic tools that are particularly relevant for FL. We first introduce the basic concepts of FL and quantization theory in Section 14.2. We conclude this section by identifying the unique requirements and characteristics of FL, which affect the design of compression and quantization methods. Based on these requirements, we present in Section 14.3 quantization theory tools that are relevant for the problem at hand in a gradual and systemic manner: We begin with the basic concept of scalar quantization and identify the need for a probabilistic design. Then, we introduce the notion of subtractive dithering as a means of reducing the distortion induced by discretization and explain why it is specifically relevant in FL scenarios, as it can be established by having a random seed shared by the server and each user. Next, we discuss how the distortion can be further reduced by jointly discretizing multiple parameters (i.e., switching from scalar to vector quantization) in a universal fashion, without violating the basic requirements and constraints of FL. Finally, we demonstrate how quantization can be combined with lossy source coding, which provides further performance benefits from the underlying nonuniformity and sparsity of the digital representations. The resulting compression method combining all of these aforementioned quantization concepts is referred to as universal vector quantization for federated learning (UVeQFed). In Section 14.4 we analyze the performance measures of UVeQFed, including the resulting distortion induced by quantization as well as the convergence profile of models trained using UVeQFed combined with conventional federated averaging. Our performance analysis begins by considering the distributed learning of a model using a smooth convex objective measure. The analysis demonstrates that proper usage of quantization tools result in achieving a similar asymptotic convergence profile as that of FL with uncompressed model updates, without communication constraints. Next, we present an experimental study that evaluates the performance of such quantized FL models when training neural networks with nonsynthetic datasets. These numerical results illustrate the added value of each of the studied quantization tools, demonstrating that both subtractive dithering and universal vector quantizers achieves more accurate recovery of model updates in each FL iteration for the same number of bits. Furthermore, the reduced distortion is translated into improved convergence with the MNIST and CIFAR-10 datasets. We conclude this chapter with a summary of the unified UVeQFed framework and its performance in Section 14.5.

## 14.2       Preliminaries and System Model

Here we formulate the system model of quantized FL. We first review some basics in quantization theory in Section 14.2.1. Then, in Section 14.2.2 we formulate the conventional FL setup, which operates without the need to quantize the model updates. Finally, we show how the throughput constraints of uplink wireless communications (i.e., the transmissions from the users to the server) gives rise to the need for quantized FL, which is formulated in Section 14.2.3.

## 14.2.1       Preliminaries in Quantization Theory

We begin by briefly reviewing the standard quantization setup and state the definition of a quantizer:

DEFINITION 14.1 (Quantizer)    A quantizer $Q_R^L(\cdot)$ with $R$ bits, input size $L$, input alphabet $\mathcal{X}$, and output alphabet $\hat{\mathcal{X}}$, consists of
1. An encoder $e: \mathcal{X}^n \mapsto \{0, \ldots, 2^R - 1\} \triangleq \mathcal{U}$ that maps the input into a discrete index.
2. A decoder $d: \mathcal{U} \mapsto \hat{\mathcal{X}}^L$ that maps each $j \in \mathcal{U}$ into a codeword $\boldsymbol{q}_j \in \hat{\mathcal{X}}^L$.

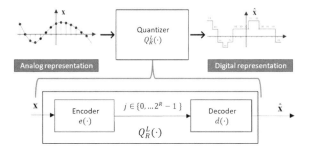

**Figure 14.1** Quantization operation.

We write the output of the quantizer with input $\boldsymbol{x} \in \mathcal{X}^L$ as

$$\hat{\boldsymbol{x}} = d\left(e\left(\boldsymbol{x}\right)\right) \triangleq Q_R^L\left(\boldsymbol{x}\right). \tag{14.1}$$

*Scalar quantizers* operate on a scalar input (i.e., $L = 1$ and $\mathcal{X}$ is a scalar space), while *vector quantizers* have a multivariate input. An illustration of a quantization system is depicted in Fig. 14.1.

The basic problem in quantization theory is to design a $Q_R^L\left(\cdot\right)$ quantizer in order to minimize some distortion measure $\delta : \mathcal{X}^L \times \hat{\mathcal{X}}^L \mapsto \mathcal{R}^+$ between its input and its output. The performance of a quantizer is characterized using its quantization rate $\frac{R}{L}$ and the expected distortion $\mathbb{E}\{\delta\left(\boldsymbol{x}, \hat{\boldsymbol{x}}\right)\}$. A common distortion measure is the mean-squared error (MSE): $\delta\left(\boldsymbol{x}, \hat{\boldsymbol{x}}\right) \triangleq \|\boldsymbol{x} - \hat{\boldsymbol{x}}\|^2$.

Characterizing the optimal quantizer, the one that minimizes the distortion for a given quantization rate, and its trade-off between distortion and rate is in general a difficult task. Optimal quantizers are thus typically studied assuming either high quantization rate (i.e., $\frac{R}{m} \to \infty$, see, e.g., [32]) or asymptotically large inputs, namely, $L \to \infty$, via rate-distortion theory [33, Ch. 10]. One of the fundamental results in quantization theory is that vector quantizers are superior to scalar quantizers in terms of their rate-distortion trade-off. For example, for large quantization rate, even for independent and identically distributed (i.i.d.) inputs, vector quantization outperforms scalar quantization, with a distortion gap of 4.35 dB for Gaussian inputs with the MSE distortion [34, Ch. 23.2].

## 14.2.2    Preliminaries in FL

### FL System Model

In this section we describe the conventional FL framework proposed in [3]. Here, a centralized server is training a model consisting of $m$ parameters based on labeled samples available at a set of $K$ remote users. The model is trained to minimize a loss function $\ell(\cdot; \cdot)$. Letting $\{\boldsymbol{x}_i^{(k)}, \boldsymbol{y}_i^{(k)}\}_{i=1}^{n_k}$ be the set of $n_k$ labeled training samples available at the $k$th user, $k \in \{1, \ldots, K\} \triangleq \mathcal{K}$, FL aims at recovering the $m \times 1$ weights vector $\boldsymbol{w}^o$ satisfying

$$w^{\circ} = \arg\min_{w} \left\{ F(w) \triangleq \sum_{k=1}^{K} \alpha_k F_k(w) \right\}. \qquad (14.2)$$

Here, the weighting average coefficients $\{\alpha_k\}$ are nonnegative satisfying $\sum \alpha_k = 1$, and the local objective functions are defined as the empirical average over the corresponding training set:

$$F_k(w) \equiv F_k\left(w; \{x_i^{(k)}, y_i^{(k)}\}_{i=1}^{n_k}\right) \triangleq \frac{1}{n_k} \sum_{i=1}^{n_k} \ell\left(w; (x_i^{(k)}, y_i^{(k)})\right). \qquad (14.3)$$

**Federated Averaging**

*Federated averaging* [3] aims at recovering $w^{\circ}$ using iterative subsequent updates. In each update of time instance $t$, the server shares its current model, represented by the vector $w_t \in \mathcal{R}^m$, with the users. The $k$th user, $k \in \mathcal{K}$, uses its set of $n_k$ labeled training samples to retrain the model $w_t$ over $\tau$ time instances into an updated model $\tilde{w}_{t+\tau}^{(k)} \in \mathcal{R}^m$. Commonly, $\tilde{w}_{t+\tau}^{(k)}$ is obtained by $\tau$ stochastic gradient descent (SGD) steps applied to $w_t$, executed over the local dataset:

$$\tilde{w}_{t+1}^{(k)} = \tilde{w}_t^{(k)} - \eta_t \nabla F_k^{i_t^{(k)}}\left(\tilde{w}_t^{(k)}\right), \qquad (14.4)$$

where $i_t^{(k)}$ is a sample index chosen uniformly from the local data of the $k$th user at time $t$. When the local updates are carried out via Eq. (14.4), federated averaging specializes the local SGD method [35], and the terms are often used interchangeably in the literature.

Having updated the model weights, the $k$th user conveys its model update, denoted as $h_{t+\tau}^{(k)} \triangleq \tilde{w}_{t+\tau}^{(k)} - w_t$, to the server. The server synchronizes the global model by averaging the model updates via

$$w_{t+\tau} = w_t + \sum_{k=1}^{K} \alpha_k h_{t+\tau}^{(k)} = \sum_{k=1}^{K} \alpha_k \tilde{w}_{t+\tau}^{(k)}. \qquad (14.5)$$

By repeating this procedure over multiple iterations, the resulting global model can be shown to converge to $w^{\circ}$ under various objective functions [35–37]. The number of local SGD iterations can be any positive integer. For $\tau = 1$, the model updates $\{h_{t+\tau}^{(k)}\}$ are the scaled stochastic gradients, and thus the local SGD method effectively implements mini-batch SGD. While such a setting results in a learning scheme that is simpler to analyze and is less sensitive to data heterogeneity compared to using large values of $\tau$, it requires much more communications between the participating entities [38] and may give rise to privacy concerns [39].

## 14.2.3    FL with Quantization Constraints

The federated averaging method relies on the ability of the users to repeatedly convey their model updates to the server without errors. This implicitly requires the users and the server to communicate over ideal links of infinite throughput. Since upload speeds are typically more limited compared to download speeds [40], the users needs to communicate a finite-bit quantized representation of their model update, resulting in the quantized FL setup model we now describe.

### Quantized FL System Model

The number of model parameters $m$ can be very large, particularly when training highly parameterized deep neural networks (DNNs). The requirement to limit the volume of data conveyed over the uplink channel implies that the model updates should be quantized prior to its transmission. Following the formulation of the quantization problem in Section 14.2.1, this implies that each user should encode the model update into a digital representation consisting of a finite number of bits (a codeword), and the server has to decode each of these codewords describing the local model into a global model update. The $k$th model update $\boldsymbol{h}_{t+\tau}^{(k)}$ is therefore encoded into a digital codeword of $R_k$ bits denoted as $u_t^{(k)} \in \{0, \ldots, 2^{R_k} - 1\} \triangleq \mathcal{U}_k$, using an encoding function with input $\boldsymbol{h}_{t+\tau}^{(k)}$:

$$e_{t+\tau}^{(k)} : \mathcal{R}^m \mapsto \mathcal{U}_k. \tag{14.6}$$

The uplink channel is modeled as a bit-constrained link with a transmission rate that does not exceed Shannon capacity; that is, each $R_k$ bit codeword is recovered by the server without errors, as commonly assumed in the FL literature [6, 21–23, 25–31, 41]. The server uses the received codewords $\{u_{t+\tau}^{(k)}\}_{k=1}^K$ to reconstruct $\hat{\boldsymbol{h}}_{t+\tau} \in \mathcal{R}^m$, obtained via a joint decoding function

$$d_{t+\tau} : \mathcal{U}_1 \times \ldots \times \mathcal{U}_K \mapsto \mathcal{R}^m. \tag{14.7}$$

The recovered $\hat{\boldsymbol{h}}_{t+\tau}$ is an estimate of the weighted average $\sum_{k=1}^K \alpha_k \boldsymbol{h}_{t+\tau}^{(k)}$. Finally, the global model $\boldsymbol{w}_{t+\tau}$ is updated via

$$\boldsymbol{w}_{t+\tau} = \boldsymbol{w}_t + \hat{\boldsymbol{h}}_{t+\tau}. \tag{14.8}$$

An illustration of this FL procedure is depicted in Fig. 14.2. Clearly, if the number of allowed bits is sufficiently large, the distance $\|\hat{\boldsymbol{h}}_{t+\tau} - \sum_{k=1}^K \alpha_k \boldsymbol{h}_{t+\tau}^{(k)}\|^2$ can be made arbitrarily small, allowing the server to update the global model as the desired weighted average Eq. (14.5).

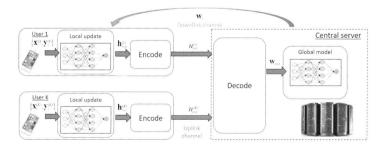

**Figure 14.2** Federated learning with bit-rate constraints.

In the presence of a limited bit budget (i.e., small values of $\{R_k\}$) distortion is induced, which can severely degrade the ability of the server to update its model. This motivates the need for efficient quantization methods for FL, which are to be designed under the unique requirements of FL that we now describe.

## Quantization Requirements

Tackling the design of quantized FL from a purely information theoretic perspective (i.e., as a lossy source code [33, Ch. 10]) inevitably results in utilizing complex vector quantizers. In particular, in order to approach the optimal tradeoff between number of bits and quantization distortion, one must jointly map $L$ samples together, where $L$ should be an arbitrarily large number, by creating a partition of the $L$-dimensional hyperspace that depends on the distribution of the model updates. Furthermore, the distributed nature of the FL setups implies that the reconstruction accuracy can be further improved by utilizing infinite-dimensional vector quantization as part of a distributed lossy source coding scheme [42, Ch. 11], such as Wyner-Ziv coding [43]. However, these coding techniques tend to be computationally complex and require each of the nodes participating in the procedure to have accurate knowledge of the joint distribution of the complete set of continuous-amplitude values to be discretized, which is not likely to be available in practice.

Therefore, to study quantization schemes while faithfully representing FL setups, one has to account for the following requirements and assumptions:

A1. All users share the same encoding function, denoted as $e_t^{(k)}(\cdot) = e_t(\cdot)$ for each $k \in \mathcal{K}$. This requirement, which was also considered in [6], significantly simplifies FL implementation.

A2. No a-priori knowledge or distribution of $h_{t+\tau}^{(k)}$ is assumed.

A3. As in [6], the users and the server share a source of common randomness. This is achieved by, for example, letting the server share with each user a random seed along with the weights. Once a different seed is conveyed to each user, it can be used to obtain a dedicated source of common randomness shared by the server and each of the users for the entire FL procedure.

These requirements, and particularly A1–A2, are stated to ensure feasibility of the quantization scheme for FL applications. Ideally, a compression mechanism should exploit knowledge about the distribution of its input, and different distributions would yield different encoding mechanisms. However, although it is desirable to design a single encoding mechanism that can be used by all devices, the fact that prior statistical knowledge about the distribution of the model updates is likely to be unavailable, particularly when training deep models, motivates the requirements A1–A2. In fact, these conditions give rise to the need for a *universal quantization* approach, namely, a scheme that operates reliably regardless of the distribution of the model updates and without prior knowledge of this distribution.

## 14.3     Quantization for FL

Next, we detail various mechanisms for quantizing the model updates conveyed over the uplink channel in FL. We begin with the common method of probabilistic scalar quantization in Section 14.3.1. Then in Sections 14.3.2 and 14.3.3 we show how distortion can be reduced while accounting for the FL requirements A1–A3 by introducing subtractive dithering and vector quantization, respectively. Finally, we present a unified formulation for quantized FL based on the aforementioned techniques combined with lossless source coding and overload prevention in Section 14.3.4.

### 14.3.1    Probabilistic Scalar Quantization

The most simple and straight forward approach to discretize the model updates is to utilize quantizers. Here, a scalar quantizer $Q_R^1(\cdot)$ is set to some fixed partition of the real line, and each user encodes its model update $h_{t+\tau}^{(k)}$ by applying $Q_R^1(\cdot)$ to it entry-wise. Arguably the most common scalar quantization rule is the uniform mapping, which for a given support $\gamma > 0$ and quantization step size $\Delta = \frac{2\gamma}{2^R}$ is given by

$$Q_R^1(x) = \begin{cases} \Delta\left(\lfloor \frac{x}{\Delta} \rfloor + \frac{1}{2}\right), & \text{for } |x| < \gamma \\ \text{sign}(x)\left(\gamma - \frac{\Delta}{2}\right), & \text{else,} \end{cases} \tag{14.9}$$

where $\lfloor \cdot \rfloor$ denotes rounding to the next smaller integer and $\text{sign}(\cdot)$ is the signum function. The overall number of bits used for representing the model update is thus $mR$, which adjusts the resolution of the quantizers to yield a total of $R_k$ bits to describe the model update. The server then uses the quantized model update to compute the aggregated global model by averaging these digital representations into Eq. (14.8). In its coarsest form, the quantizer represents each entry using a single bit, as in, for example, signSGD [26, 31]. An illustration of this simplistic quantization scheme is given in Fig. 14.3.

The main drawback of the scalar quantization described in Eq. (14.9) follows from the fact that its distortion is a deterministic function of its input. To see this, consider for example the case in which the quantizer $Q_R^1(\cdot)$ implements rounding to the nearest integer, and all the users compute the first entry of their model updates as 1.51. In such a case, all the users will encode this entry as the integer value 2, and thus the first

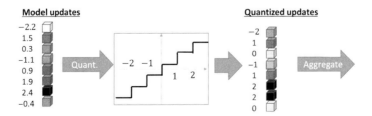

**Figure 14.3** Model scalar quantization.

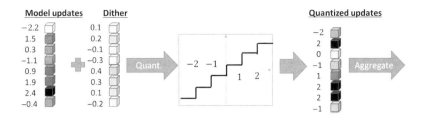

**Figure 14.4** Model probabilistic scalar quantization.

entry of $\hat{\boldsymbol{h}}_{t+\tau}$ in Eq. (14.8) also equals 2, resulting in a possibly notable error in the aggregated model. This motivates the usage of *probabilistic quantization*, where the distortion is a random quantity. Considering again the previous example, if instead of having each user encode 1.51 into 2, the users would have a 51 percent probability of encoding it 2 and a 49 percent probability of encoding it as 1. The aggregated model is expected to converge to the desired update value of 1.51 as the number of users $K$ grows by the law of large numbers.

Various forms of probabilistic quantization for distributed learning have been proposed in the literature [41]. These include using one-bit sign quantizers [26], ternary quantization [27], uniform quantization [28], and nonuniform quantization [29]. Probabilistic scalar quantization can be treated as a form of *dithered quantization* [44, 45], where the continuous-amplitude quantity is corrupted by some additive noise, referred to as dither signal, which is typically uniformly distributed over its corresponding decision region. Consequently, the quantized updates of the $k$th user are given by applying $Q_R^1(\cdot)$ to $\boldsymbol{h}_{t+\tau}^{(k)} + \boldsymbol{z}_{t+\tau}^{(k)}$ element-wise, where $\boldsymbol{z}_{t+\tau}^{(k)}$ denotes the dither signal. When the quantization mapping consists of the uniform partition in Eq. (14.9), as used in the quantized SGD (QSGD) algorithm [28], this operation specializes nonsubtractive dithered quantization [45], and the entries of the dither signal $\boldsymbol{z}_{t+\tau}^{(k)}$ are typically i.i.d. and uniformly distributed over $[-\Delta/2, \Delta/2]$ [45]. An illustration of this continuous-to-discrete mapping is depicted in Fig. 14.4

Probabilistic quantization overcomes errors of a deterministic nature, as discussed earlier in the context of conventional scalar quantization. However, the addition of the dither signal also increases the distortion in representing each model update in discrete form [46]. In particular, while probabilistic quantization reduces the effect of the distortion induced on the aggregated $\hat{\boldsymbol{h}}_{t+\tau}$ in Eq. (14.8) compared to conventional scalar quantization, it results in the discrete representation of each individual update $\boldsymbol{h}_{t+\tau}^{(k)}$ being less accurate. This behavior is also observed when comparing the quantized updates in Figs. 14.3 and 14.4. This excess distortion can be reduced by utilizing subtractive dithering strategies, as detailed in the following section.

## 14.3.2     Subtractive Dithered Scalar Quantization

Subtractive dithered quantization extends probabilistic quantization by introducing an additional decoding step, rather than directly using the discrete codewords as the

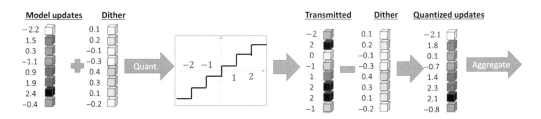

**Figure 14.5**  Model subtractive dithered scalar quantization.

compressed digital representation. The fact that each of the users can share a source of local randomness with the server by assumption A3 implies that the server can generate the realization of the dither signal $z_{t+\tau}^{(k)}$. Consequently, instead of using the element-wise quantized version of $h_{t+\tau}^{(k)} + z_{t+\tau}^{(k)}$ received from the $k$th user, denoted here as $Q(h_{t+\tau}^{(k)} + z_{t+\tau}^{(k)})$, the user sets its representation of $h_{t+\tau}^{(k)}$ to be $Q(h_{t+\tau}^{(k)} + z_{t+\tau}^{(k)}) - z_{t+\tau}^{(k)}$. An illustration of this procedure is depicted in Fig. 14.5.

The subtraction of the dither signal upon decoding reduces its excess distortion in a manner that does not depend on the distribution of the continuous-amplitude value [47]. As such, it is also referred to as *universal scalar quantization*. In particular, for uniform quantizers of the form Eq. (14.9) where the input lies within the support $[-\gamma, \gamma]$, it holds that the effect of the quantization can be rigorously modeled as an additive noise term whose entries are uniformly distributed over $[-\Delta/2, \Delta/2]$, regardless of the values of the realization and the distribution of the model updates $h_{t+\tau}^{(k)}$ [45]. This characterization implies that the excess distortion due to dithering is mitigated for each model update individually, while the overall distortion is further reduced in aggregation, as federated averaging results in this additive noise term effectively approaching its mean value of zero by the law of large numbers.

### 14.3.3    Subtractive Dithered Vector Quantization

While scalar quantizers are simple to implement, they process each sample of the model updates using the same continuous-to-discrete mapping. Consequently, scalar quantizers are known to be inferior to vector quantizers, which jointly map a set of $L > 1$ samples into a single digital codeword, in terms of their achievable distortion for a given number of bits. In this section, we detail how the concept of subtractive dithered quantization discussed in the previous section can be extended to vector quantizers, as illustrated in Fig. 14.6. The extension of universal quantization via subtractive dithering to multivariate samples reviewed here is based on lattice quantization [48]. To formulate the notion of such dithered vector quantizers, we first briefly review lattice quantization, after which we discuss its usage for FL uplink compression.

**Lattice Quantization**

Let $L$ be a fixed positive integer, referred to henceforth as the lattice dimension, and let $G$ be a nonsingular $L \times L$ matrix, which denotes the lattice generator matrix.

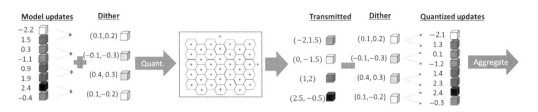

**Figure 14.6** Model subtractive dithered vector quantization.

For simplicity, we assume that $M \triangleq \frac{m}{L}$ is an integer, where $m$ is the number of model parameters, although the scheme can also be applied when this does not hold by replacing $M$ with $\lceil M \rceil$. Next, we use $\mathcal{L}$ to denote the lattice, which is the set of points in $\mathcal{R}^L$ that can be written as an integer linear combination of the columns of $\boldsymbol{G}$:

$$\mathcal{L} \triangleq \{\boldsymbol{x} = \boldsymbol{Gl} : \boldsymbol{l} \in \mathcal{Z}^L\}. \tag{14.10}$$

A lattice quantizer $Q_{\mathcal{L}}(\cdot)$ maps each $\boldsymbol{x} \in \mathcal{R}^L$ to its nearest lattice point (i.e., $Q_{\mathcal{L}}(\boldsymbol{x}) = \boldsymbol{l}_x$ where $\boldsymbol{l}_x \in \mathcal{L}$ if $\|\boldsymbol{x} - \boldsymbol{l}_x\| \le \|\boldsymbol{x} - \boldsymbol{l}\|$ for every $\boldsymbol{l} \in \mathcal{L}$). Finally, let $\mathcal{P}_0$ be the basic lattice cell [49], or the set of points in $\mathcal{R}^L$ that are closer to $\boldsymbol{0}$ than to any other lattice point:

$$\mathcal{P}_0 \triangleq \{\boldsymbol{x} \in \mathcal{R}^L : \|\boldsymbol{x}\| < \|\boldsymbol{x} - \boldsymbol{p}\|, \forall \boldsymbol{p} \in \mathcal{L}/\{\boldsymbol{0}\}\}. \tag{14.11}$$

For example, when $\boldsymbol{G} = \Delta \cdot \boldsymbol{I}_L$ for some $\Delta > 0$, then $\mathcal{L}$ is the square lattice, for which $\mathcal{P}_0$ is the set of vectors $\boldsymbol{x} \in \mathcal{R}^L$ that have a $\ell_\infty$ norm that is not larger than $\frac{\Delta}{2}$. For this setting, $Q_{\mathcal{L}}(\cdot)$ implements entry-wise scalar uniform quantization with spacing $\Delta$ [34, Ch. 23]. This is also the case when $L = 1$, for which $Q_{\mathcal{L}}(\cdot)$ specializes scalar uniform quantization with spacing dictated by the (scalar) $\boldsymbol{G}$.

### Subtractive Dithered Lattice Quantization

Lattice quantizers can be exploited to realize universal vector quantization. Such mappings jointly quantize multiple samples, thus benefiting from the improved rate-distortion trade-off of vector quantizers, while operating in a manner that is invariant of the distribution of the continuous-amplitude inputs [48].

To formulate this operation, we note that in order to apply an $L$-dimensional lattice quantizer to $\boldsymbol{h}_{t+\tau}^{(k)}$, it must first be divided into $M$ distinct $L \times 1$ vectors, denoted $\{\bar{\boldsymbol{h}}_i^{(k)}\}_{i=1}^M$. In order to quantize each subvector $\bar{\boldsymbol{h}}_i^{(k)}$, it is first corrupted with a dither vector $\boldsymbol{z}_i^{(k)}$, which is here uniformly distributed over the basic lattice cell $\mathcal{P}_0$ and then quantized using a lattice quantizer $Q_{\mathcal{L}}(\cdot)$. On the decoder side, the representation of $\bar{\boldsymbol{h}}_i^{(k)}$ is obtained by subtracting $\boldsymbol{z}_i^{(k)}$ from the discrete $Q_{\mathcal{L}}(\bar{\boldsymbol{h}}_i^{(k)} + \boldsymbol{z}_i^{(k)})$. An example of this procedure for $L = 2$ is illustrated in Fig. 14.7.

## 14.3.4   Unified Formulation

The aforementioned strategies give rise to a unified framework for model updates quantization in FL. Before formulating the resulting encoding and decoding mappings

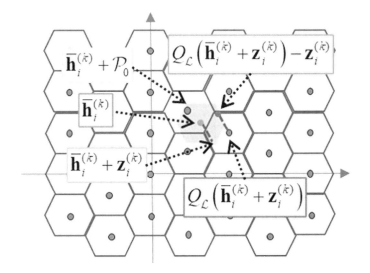

**Figure 14.7** Subtractive dithered lattice quantization.

in a systematic manner, we note that two additional considerations of quantization and compression should be accounted for in the formulation. These are the need to prevent overloading of the quantizers and the ability to further compress the discretized representations via lossless source coding.

## Overloading Prevention

Quantizers are typically required to operate within their dynamic range, namely, when the input lies in the support of the quantizer. For uniform scalar quantizers as in Eq. (14.9), this implies that the magnitude of the input must not be larger than $\gamma$. The same holds for multivariate lattice quantizers; as an infinite number of lattice regions are required to span $\mathcal{R}^L$, the input must be known to lie in some $L$-dimensional ball in order to utilize a finite number of lattice points, which in turn implies a finite number of different discrete representations. In particular, the desired statistical properties that arise from the usage of dithering (i.e., that the distortion is uncorrelated with the input and can thus be reduced by averaging) hold when the input is known to lie within the quantizer support [45].

In order to guarantee that the quantizer is not overloaded, namely, that each continuous-amplitude value lies in the quantizer support, the model updates vector $\boldsymbol{h}_{t+\tau}^{(k)}$ can be scaled by $\zeta \|\boldsymbol{h}_{t+\tau}^{(k)}\|$ for some parameter $\zeta > 0$. This setting guarantees that the elements $\{\bar{\boldsymbol{h}}_i^{(k)}\}_{i=1}^M$ of which the compressed $\boldsymbol{h}_{t+\tau}^{(k)}$ is comprised all reside inside the $L$-dimensional ball with radius $\zeta^{-1}$. The number of lattice points is not larger than $\frac{\pi^{L/2}}{\zeta^L \Gamma(1+L/2)\det(\boldsymbol{G})}$ [50, Ch. 2], where $\Gamma(\cdot)$ is the Gamma function. Note that the scalar quantity $\zeta \|\boldsymbol{h}_{t+\tau}^{(k)}\|$ depends on the vector $\boldsymbol{h}_{t+\tau}^{(k)}$ and must thus be quantized with high resolution and conveyed to the server, to be accounted for in the decoding process. The overhead in accurately quantizing the single scalar quantity $\zeta \|\boldsymbol{h}_{t+\tau}^{(k)}\|$

is typically negligible compared to the number of bits required to convey the set of vectors $\{\bar{\boldsymbol{h}}_i^{(k)}\}_{i=1}^M$, hardly affecting the overall quantization rate.

## Lossless Source Coding

As defined in Section 14.2.1, quantizers can output a finite number of different codewords. The amount of codewords dictates the number of bits used $R$, as each codeword can be mapped into a different combination of $R$ bits, and conveyed in digital form over the rate-constrained uplink channel. However, these digital representations are in general not uniformly distributed. In particular, model updates are often approximately sparse, which is the property exploited in sparsification-based compression schemes [21, 23]. Consequently, the codeword corresponding to (almost) zero values is likely to be assigned more often than other codewords by the quantizer.

This property implies that the discrete output of the quantizer, the vector $Q(\boldsymbol{h}_{t+\tau}^{(k)} + \boldsymbol{z}_{t+\tau}^{(k)})$, can be further compressed by lossless source coding. Various lossless source coding schemes, including arithmetic, Lempel-Ziv, and Elias codes, are capable of compressing a vector of discrete symbols into an amount of bits approaching the most compressed representation, dictated by the entropy of the vector [33, Ch. 13]. When the distribution of the discrete vector $Q(\boldsymbol{h}_{t+\tau}^{(k)} + \boldsymbol{z}_{t+\tau}^{(k)})$ is notably different from being uniform, as is commonly the case in quantized FL, the incorporation of such entropy coding can substantially reduce the number of bits in a lossless manner, without inducing additional distortion.

## Encoder-Decoder Formulation

Based on these considerations of overloading prevention and the potential of entropy coding, we now formulate a unified quantized FL strategy coined UVeQFed. UVeQFed generalizes the quantization strategies for FL detailed in Sections 14.3.1 through 14.3.3, where the usage of scalar versus vector quantizers is dictated by the selection of the dimension parameter $L$: When $L = 1$ UVeQFed implements scalar quantization, while with $L > 1$ it results in subtractive dithered lattice quantization. Specifically, UVeQFed consists of the following encoding and decoding mappings:

**Encoder:** The encoding function $e_{t+\tau}(\cdot)$ in Eq. (14.6) for each user includes the following steps:

E1. **Normalize and partition:** The $k$th user scales $\boldsymbol{h}_{t+\tau}^{(k)}$ by $\zeta\|\boldsymbol{h}_{t+\tau}^{(k)}\|$ for some $\zeta > 0$, and divides the result into $M$ distinct $L \times 1$ vectors, denoted $\{\bar{\boldsymbol{h}}_i^{(k)}\}_{i=1}^M$. The scalar quantity $\zeta\|\boldsymbol{h}_{t+\tau}^{(k)}\|$ is quantized separately from $\{\bar{\boldsymbol{h}}_i^{(k)}\}_{i=1}^M$ with high resolution, for example, using a uniform scalar quantizer with at least 12 bits, such that it can be recovered at the decoder with negligible distortion.

E2. **Dithering:** The encoder utilizes the source of common randomness, such as a shared seed, to generate the set of $L \times 1$ dither vectors $\{\boldsymbol{z}_i^{(k)}\}_{i=1}^M$, which are randomized in an i.i.d. fashion, independently of $\boldsymbol{h}_{t+\tau}^{(k)}$, from a uniform distribution over $\mathcal{P}_0$.

E3. **Quantization:** The vectors $\{\bar{h}_i^{(k)}\}_{i=1}^M$ are discretized by adding the dither vectors and applying lattice/uniform quantization, namely, by computing $\{Q_{\mathcal{L}}(\bar{h}_i^{(k)} + z_i^{(k)})\}$.

E4. **Entropy coding:** The discrete values $\{Q_{\mathcal{L}}(\bar{h}_i^{(k)} + z_i^{(k)})\}$ are encoded into a digital codeword $u_{t+\tau}^{(k)}$ in a lossless manner.

**Decoder:** The decoding mapping $d_{t+\tau}(\cdot)$ implements the following:

D1. **Entropy decoding:** The server first decodes each digital codeword $u_{t+\tau}^{(k)}$ into the discrete value $\{Q_{\mathcal{L}}(\bar{h}_i^{(k)} + z_i^{(k)})\}$. Since the encoding is carried out using a lossless source code, the discrete values are recovered without any errors.

D2. **Dither subtraction:** Using the source of common randomness, the server generates the dither vectors $\{z_i^{(k)}\}$, which can be carried out rapidly and at low complexity using random number generators as the dither vectors obey a uniform distribution. The server then subtracts the corresponding vector by computing $\{Q_{\mathcal{L}}(\bar{h}_i^{(k)} + z_i^{(k)}) - z_i^{(k)}\}$.

D3. **Collecting and scaling:** The values computed in the previous step are collected into an $m \times 1$ vector $\hat{h}_{t+\tau}^{(k)}$ using the inverse operation of the partitioning and normalization in Step E1.

D4. **Model recovery:** The recovered matrices are combined into an updated model based on Eq. (14.8). Namely,

$$w_{t+\tau} = w_t + \sum_{k=1}^{K} \alpha_k \hat{h}_{t+\tau}^{(k)}. \tag{14.12}$$

A block diagram of the proposed scheme is depicted in Fig. 14.8. The use of subtractive dithered quantization in Steps E2–E3 and D2 allow obtaining a digital representation that is relatively close to the true quantity, without relying on prior knowledge of its distribution. For nonsubtractive dithering, as in conventional probabilistic quantization such as QSGD [28], one must skip the dither subtraction Step D2, as illustrated in Fig. 14.8. For clarity, we use the term UVeQFed to refer to the encoder-decoder pair consisting of all the detailed steps E1–D4, with subtractive dithered quantization. The joint decoding aspect of these schemes is introduced in the final model recovery Step D4. The remaining encoding-decoding procedure, Steps E1–D3, is carried out independently for each user.

UVeQFed has several clear advantages. First, while it is based on information theoretic arguments, the resulting architecture is rather simple to implement. In particular, both subtractive dithered quantization as well as entropy coding are concrete and established methods that can be realized with relatively low complexity and feasible hardware requirements. Increasing the lattice dimension $L$ reduces the distortion, thus leading to more improved trained models, at the cost of increased complexity in the quantization step E3. The numerical study presented in Section 14.4.2 demonstrates that the accuracy of the trained model can be notably improved by using two-dimensional lattices compared to utilizing scalar quantizers, setting $L = 2$ instead of

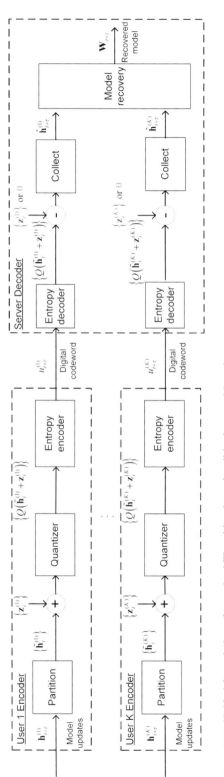

**Figure 14.8** Unified formulation of quantized FL, with subtractive dithering or without it.

$L = 1$. The source of common randomness needed for generating the dither vectors can be obtained by sharing a common seed between the server and users, as discussed in the statement of requirement A3. The statistical characterization of the quantization error of such quantizers does not depend on the distribution of the model updates. This analytical tractability allows us to rigorously show that combining UVeQFed with federated averaging mitigates the quantization error, which we show in the following section.

## 14.4  Performance Analysis

In this section we study and compare the performance of different quantization strategies that arise from the unified formulation presented in the previous section. We first present the theoretical performance of UVeQFed in terms of its resulting distortion and FL convergence for convex objectives. Then, we numerically compare the FL performance with different quantization strategies using both synthetic and nonsynthetic datasets.

### 14.4.1  Theoretical Performance

Here, we theoretically characterize the performance of UVeQFed, namely, the encoder-decoder pair detailed in Section 14.3.4. The characterization holds for both uniform scalar quantizers as well as lattice vector quantizers, depending on the setting of the parameter $L$.

**Assumptions**
We first introduce the assumptions on the objective functions $F_k(\cdot)$ in light of which the theoretical analysis of the resulting distortion and the convergence of local SGD with UVeQFed is carried out in the sequel. In particular, our theoretical performance characterization utilizes the following assumptions:

AS1.  The expected squared $\ell_2$ norm of the random vector $\nabla F_k^i(w)$, representing the stochastic gradient evaluated at $w$, is bounded by some $\xi_k^2 > 0$ for all $w \in \mathcal{R}^m$.

AS2.  The local objective functions $\{F_k(\cdot)\}$ are all $\rho_s$-smooth, namely, for all $v_1, v_2 \in \mathcal{R}^m$ it holds that

$$F_k(v_1) - F_k(v_2) \le (v_1 - v_2)^T \nabla F_k(v_2) + \frac{1}{2}\rho_s \|v_1 - v_2\|^2.$$

AS3.  The local objective functions $\{F_k(\cdot)\}$ are all $\rho_c$-strongly convex, namely, for all $v_1, v_2 \in \mathcal{R}^m$ it holds that

$$F_k(v_1) - F_k(v_2) \ge (v_1 - v_2)^T \nabla F_k(v_2) + \frac{1}{2}\rho_c \|v_1 - v_2\|^2.$$

Assumption AS1 on the stochastic gradients is often employed in distributed learning studies [35, 36, 51] Assumptions AS2–AS3 are also commonly used in FL convergence studies [35, 36] and hold for a broad range of objective functions used in FL systems, including $\ell_2$-norm regularized linear regression and logistic regression [36].

## Distortion Analysis

We begin our performance analysis by characterizing the distortion induced by quantization. The need to represent the model updates $\boldsymbol{h}_{t+\tau}^{(k)}$ using a finite number of bits inherently induces some distortion; that is, the recovered vector is $\hat{\boldsymbol{h}}_{t+\tau}^{(k)} = \boldsymbol{h}_{t+\tau}^{(k)} + \boldsymbol{\epsilon}_{t+\tau}^{(k)}$. The error in representing $\zeta \|\boldsymbol{h}_{t+\tau}^{(k)}\|$ is assumed to be negligible. For example, the normalized quantization error is on the order of $10^{-7}$ for 12 bit quantization of a scalar value and decreases exponentially with each additional bit [34, Ch. 23].

Let $\bar{\sigma}_{\mathcal{L}}^2$ be the normalized second-order lattice moment, defined as $\bar{\sigma}_{\mathcal{L}}^2 \triangleq \int_{\mathcal{P}_0} \|\boldsymbol{x}\|^2 d\boldsymbol{x} / \int_{\mathcal{P}_0} d\boldsymbol{x}$ [52]. For uniform scalar quantizers Eq. (14.9), this quantity equals $\Delta^2/3$, the second-order moment of a uniform distribution over the quantizer support. The moments of the quantization error $\boldsymbol{\epsilon}_{t+\tau}^{(k)}$ satisfy the following theorem, which is based on the properties of nonoverloaded subtractive dithered quantizers [49]:

THEOREM 14.2 *The quantization error vector $\boldsymbol{\epsilon}_{t+\tau}^{(k)}$ has zero-mean entries and satisfies*

$$\mathbb{E}\left\{ \left\| \boldsymbol{\epsilon}_{t+\tau}^{(k)} \right\|^2 \middle| \boldsymbol{h}_{t+\tau}^{(k)} \right\} = \zeta^2 \|\boldsymbol{h}_{t+\tau}^{(k)}\|^2 M \bar{\sigma}_{\mathcal{L}}^2. \tag{14.13}$$

Theorem 14.2 characterizes the distortion in quantizing the model updates using UVeQFed. Due to the usage of vector quantizers, the dependence of the expected error norm on the number of bits is not explicit in Eq. (14.13), but rather encapsulated in the lattice moment $\bar{\sigma}_{\mathcal{L}}^2$. To observe that Eq. (14.13) indeed represents lower distortion compared to previous FL quantization schemes, we note that even when scalar quantizers are used, when $L = 1$ for which $\frac{1}{L}\bar{\sigma}_{\mathcal{L}}^2$ is known to be largest [52], the resulting quantization is reduced by a factor of two compared to conventional probabilistic scalar quantizers due to the subtraction of the dither upon decoding in Step D2 [45, Thms. 1–2].

We next bound the distance between the desired model $\boldsymbol{w}_{t+\tau}^{\mathrm{des}}$, which is given by $\boldsymbol{w}_{t+\tau}^{\mathrm{des}} = \sum_{k=1}^{K} \alpha_k \tilde{\boldsymbol{w}}_{t+\tau}^{(k)}$ in Eq. (14.8), and the recovered one $\boldsymbol{w}_{t+\tau}$, as stated in the following theorem:

THEOREM 14.3 [53, Thm. 2] *When AS1 holds, the mean-squared distance between $\boldsymbol{w}_{t+\tau}$ and $\boldsymbol{w}_{t+\tau}^{\mathrm{des}}$ satisfies*

$$\mathbb{E}\left\{ \left\| \boldsymbol{w}_{t+\tau} - \boldsymbol{w}_{t+\tau}^{\mathrm{des}} \right\|^2 \right\} \leq M \zeta^2 \bar{\sigma}_{\mathcal{L}}^2 \tau \left( \sum_{t'=t}^{t+\tau-1} \eta_{t'}^2 \right) \sum_{k=1}^{K} \alpha_k^2 \xi_k^2. \tag{14.14}$$

Theorem 14.3 implies that the recovered model can be made arbitrarily close to the desired one by increasing $K$, namely, the number of users. For example, when $\alpha_k = 1/K$ (i.e., conventional averaging), it follows from Theorem 14.3 that the mean-squared error in the weights decreases as $1/K$. In particular, if $\max_k \alpha_k$ decreases with $K$, which essentially means that the updated model is not based only on a small part of the participating users, then the distortion vanishes in the aggregation process. Furthermore, when the step size $\eta_t$ gradually decreases, which is known to contribute to the convergence of FL [36], it follows from Theorem 14.3 that the distortion decreases accordingly, further mitigating its effect as the FL iterations progress.

### Convergence Analysis

We next study the convergence of FL with UVeQFed. We do not restrict the labeled data of each of the users to be generated from an identical distribution; that is, we consider a statistically heterogeneous scenario, thus faithfully representing FL setups [4, 54]. Such heterogeneity is in line with requirement A2, which does not impose any specific distribution structure on the underlying statistics of the training data. Following [36], we define the heterogeneity gap as

$$\psi \triangleq F(\boldsymbol{w}^{\mathrm{o}}) - \sum_{k=1}^{K} \alpha_k \min_{\boldsymbol{w}} F_k(\boldsymbol{w}). \tag{14.15}$$

The value of $\psi$ quantifies the degree of heterogeneity. If the training data originates from the same distribution, then $\psi$ tends to zero as the training size grows. However, for heterogeneous data, its value is positive. The convergence of UVeQFed with federated averaging is characterized in the following theorem:

THEOREM 14.4 [53, Thm. 3]  *Set $\gamma = \tau \max(1, 4\rho_s/\rho_c)$ and consider a UVeQFed setup satisfying AS1–AS3. Under this setting, local SGD with step size $\eta_t = \frac{\tau}{\rho_c(t+\gamma)}$ for each $t \in \mathcal{N}$ satisfies*

$$\mathbb{E}\{F(\boldsymbol{w}_t)\} - F(\boldsymbol{w}^{\mathrm{o}}) \leq \frac{\rho_s}{2(t+\gamma)} \max\left(\frac{\rho_c^2 + \tau^2 b}{\tau \rho_c}, \gamma \|\boldsymbol{w}_0 - \boldsymbol{w}^{\mathrm{o}}\|^2\right), \tag{14.16}$$

*where*

$$b \triangleq \left(1 + 4M\zeta^2\bar{\sigma}_{\mathcal{L}}^2\tau^2\right) \sum_{k=1}^{K} \alpha_k^2\xi_k^2 + 6\rho_s\psi + 8(\tau-1)^2 \sum_{k=1}^{K} \alpha_k\xi_k^2.$$

Theorem 14.4 implies that UVeQFed with local SGD (i.e., conventional federated averaging) converges at a rate of $\mathcal{O}(1/t)$. This is the same order of convergence as FL without quantization constraints for i.i.d. [35] as well as heterogeneous data [36, 55]. A similar order of convergence was also reported for previous probabilistic quantization schemes which typically considered i.i.d. data [28, Thm. 3.2]. While it is difficult to identify the convergence gains of UVeQFed over previously proposed FL quantizers by comparing Theorem 14.4 to their corresponding convergence bounds, in the following section we empirically demonstrate that UVeQFed converges to more accurate global models compared to FL with probabilistic scalar quantizers, when trained using i.i.d. as well as heterogeneous datasets.

## 14.4.2    Numerical Study

In this section we numerically evaluate UVeQFed. We first compare the quantization error induced by UVeQFed to competing methods utilized in FL. Then, we numerically demonstrate how the reduced distortion is translated in FL performance gains using both the MNIST and CIFAR-10 datasets.

### Quantization Distortion

We begin by focusing only on the compression method, studying its accuracy using synthetic data. We evaluate the distortion induced in quantization of UVeQFed operating with a two-dimensional hexagonial lattice (i.e., $L = 2$ and $G = [2, 0; 1, 1/\sqrt{3}]$ [56]) as well as with scalar quantizers, namely, $L = 1$ and $G = 1$. The normalization coefficient is set to $\zeta = \frac{2 + \tilde{R}/5}{\sqrt{M}}$, where $\tilde{R}$ is the quantization rate, or the ratio of the number of bits to the number of model updates entries $m$. The distortion of UVeQFed is compared to QSGD [28], which can be treated as UVeQFed without dither subtraction in Step D2 and with scalar quantizers (i.e., $L = 1$). In addition, we evaluate the corresponding distortion achieved when using uniform quantizers with random unitary rotation [6] and subsampling by random masks followed by uniform three-bit quantizers [6]. All the simulated quantization methods operate with the same quantization rate (i.e., the same overall number of bits).

Let $H$ be a $128 \times 128$ matrix with Gaussian i.i.d. entries, and let $\Sigma$ be a $128 \times 128$ matrix with entries that are given by $(\Sigma)_{i,j} = e^{-0.2|i-j|}$, representing an exponentially decaying correlation. In Figs. 14.9 and 14.10 we depict the per-entry squared-error in quantizing $H$ and $\Sigma H \Sigma^T$, representing independent and correlated data, respectively, versus the quantization rate $\tilde{R}$. The distortion is averaged over 100 independent realizations of $H$. To meet the bit-rate constraint when using lattice quantizers, we scaled $G$ such that the resulting codewords use less than $128^2 \tilde{R}$ bits. For the scalar quantizers and subsampling-based scheme, the rate determines the quantization resolution and the subsampling ratio, respectively.

We observe in Figs. 14.9–14.10 that implementing the complete encoding-decoding steps E1–D4 allows UVeQFed to achieve a more accurate digital representation compared to previously proposed methods. It is also observed that UVeQFed with vec-

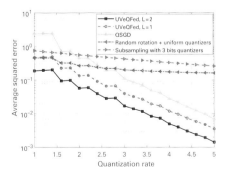

**Figure 14.9** Quantization distortion, i.i.d. data.

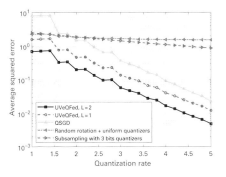

**Figure 14.10** Quantization distortion, correlated data.

tor quantization outperforms its scalar counterpart and that the gain is more notable when the quantized entries are correlated. This demonstrates the improved accuracy of jointly encoding multiple samples via vector quantization as well as the ability of UVeQFed to exploit statistical correlation in a universal manner by using fixed lattice-based quantization regions that do not depend on the underlying distribution.

### Convergence Performance

Next, we demonstrate that the reduced distortion that follows from the combination of subtractive dithering and vector quantization in UVeQFed also translates into FL performance gains. To that aim, we evaluate UVeQFed for training neural networks using the MNIST and CIFAR-10 datasets, and compare its performance to that achievable using QSGD.

For MNIST, we use a fully connected network with a single hidden layer of 50 neurons and an intermediate sigmoid activation. Each of the $K = 15$ users has 1,000 training samples, which are distributed sequentially among the users (i.e., the first user has the first 1,000 samples in the dataset, and so on), resulting in an uneven heterogeneous division of the labels of the users. The users update their weights using gradient descent, where federated averaging is carried out on each iteration. The resulting accuracy versus the number of iterations is depicted in Figs. 14.11 and 14.12 for quantization rates $\tilde{R} = 2$ and $\tilde{R} = 4$, respectively.

For CIFAR-10, we train the deep convolutional neural network architecture used in [57], where trainable parameters constitute three convolution layers and two fully connected layers. Here, we consider two methods for distributing the 50,000 training images of CIFAR-10 among the $K = 10$ users: An i.i.d. division, where each user has the same number of samples from each of the 10 labels, and a heterogeneous division, in which at least 25 percent of the samples of each user correspond to a single distinct label. Each user completes a single epoch of SGD with a mini-batch size of 60 before the models are aggregated. The resulting accuracy versus the number of epochs is depicted in Figs. 14.13 and 14.14 for quantization rates $\tilde{R} = 2$ and $\tilde{R} = 4$, respectively.

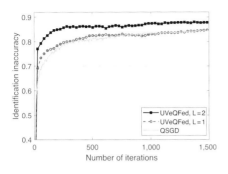

**Figure 14.11**  Convergence profile, MNIST, $\tilde{R} = 2$.

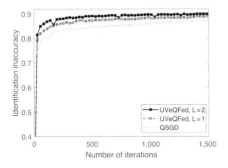

**Figure 14.12**  Convergence profile, MNIST, $\tilde{R} = 4$.

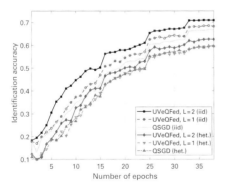

**Figure 14.13**  Convergence profile, CIFAR-10, $\tilde{R} = 2$.

We observe in Figs. 14.11–14.14 that UVeQFed with vector quantizer, $L = 2$, results in convergence to the most accurate model for all the considered scenarios. The gains are more dominant for $\tilde{R} = 2$, implying that the usage of UVeQFed with multidimensional lattices can notably improve the performance over low rate channels. Particularly, we observe in Figs. 14.13 and 14.14 that similar gains of UVeQFed are noted for both i.i.d. as well as heterogeneous setups, while the heterogeneous division of the data degrades the accuracy of all considered schemes

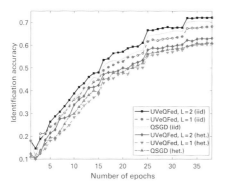

**Figure 14.14** Convergence profile, CIFAR-10, $\tilde{R} = 4$.

compared to the i.i.d division. It is also observed that UVeQFed with scalar quantizers, $L = 1$, achieves improved convergence compared to QSGD for most considered setups, which stems from its reduced distortion.

The results presented in this section demonstrate that the theoretical benefits of UVeQFed, which rigorously hold under AS1–AS3, translate into improved convergence when operating under rate constraints with nonsynthetic data. They also demonstrate how introducing the state-of-the-art quantization theoretic concepts of subtractive dithering and universal joint vector quantization contributes to both reducing the quantization distortion as well as improving FL convergence.

## 14.5      Conclusion

In the emerging machine learning paradigm of FL, the task of training highly parameterized models becomes the joint effort of multiple remote users, orchestrated by the server in a centralized manner. This distributed learning operation, which brings forth various gains in terms of privacy, also gives rise to a multitude of challenges. One of these challenges stems from the reliance of FL on the ability of the participating entities to reliably and repeatedly communicate over channels whose throughput is typically constrained, motivating the users to convey their model updates to the server in a compressed manner with minimal distortion.

In this chapter we considered the problem of model update compression from a quantization theory perspective. We first identified the unique characteristics of FL in light of that quantization methods for FL should be designed. Then, we presented established concepts in quantization theory that facilitate uplink compression in FL. These include (a) the usage of probabilistic (dithered) quantization to allow the distortion to be reduced in federated averaging, (b) the integration of subtractive dithering to reduce the distortion by exploiting a random seed shared by the server and each user, (c) the extension to vector quantizers in a universal manner to further improve the rate-distortion trade-off, and (d) the combination of lossy quantization with lossless source coding to exploit nonuniformity and sparsity of the digital representations. These concepts are summarized in a concrete and systematic FL quantization scheme, referred to as UVeQFed. We analyzed UVeQFed, proving that its error term is mitigated by federated averaging. We also characterized its convergence profile, showing that its asymptotic decay rate is the same as an unquantized local SGD. Our numerical study demonstrates that UVeQFed achieves more accurate recovery of model updates in each FL iteration compared to previously proposed schemes for the same number of bits, and that its reduced distortion is translated into improved convergence with the MNIST and CIFAR-10 datasets.

## References

[1] Y. LeCun, Y. Bengio, and G. Hinton, "Deep learning," *Nature*, vol. 521, no. 7553, p. 436, 2015.

[2] J. Chen and X. Ran, "Deep learning with edge computing: A review," *Proc. IEEE*, vol. 107, no. 8, pp. 1655–1674, 2019.

[3] H. B. McMahan et al., "Communication-efficient learning of deep networks from decentralized data," *arXiv preprint*, arXiv:1602.05629, 2016.

[4] P. Kairouz et al., "Advances and open problems in federated learning," *arXiv preprint*, arXiv:1912.04977, 2019.

[5] J. Konečný et al., "Federated optimization: Distributed machine learning for on-device intelligence," *arXiv preprint*, arXiv:1610.02527, Oct. 2016.

[6] J. Konečný et al., "Federated learning: Strategies for improving communication efficiency," *arXiv preprint*, arXiv:1610.05492, 2016.

[7] M. Chen et al., "Wireless communications for collaborative federated learning in the internet of things," *arXiv preprint*, arXiv:2006.02499, 2020.

[8] N. H. Tran et al. "Federated learning over wireless networks: Optimization model design and analysis," in *Proc. IEEE Conf. on Computer Communications*, Apr. 2019.

[9] M. Chen et al., "Convergence time optimization for federated learning over wireless networks," *arXiv preprint*, arXiv:2001.07845, 2020.

[10] T. T. Vu et al., "Cell-free massive MIMO for wireless federated learning," *IEEE Trans. on Wireless Communications*, vol. 19, no. 10, pp. 6377–6392, Oct. 2020.

[11] Z. Yang et al., "Energy efficient federated learning over wireless communication networks," *arXiv preprint*, arXiv:1911.02417, 2019.

[12] G. Zhu et al., "Toward an intelligent edge: Wireless communication meets machine learning," *IEEE Communications Magazine*, vol. 58, no. 1, pp. 19–25, Jan. 2020.

[13] S. Wang et al., "Adaptive federated learning in resource constrained edge computing systems," *IEEE Journal on Selected Areas in Communications*, vol. 37, no. 6, pp. 1205–1221, June 2019.

[14] M. Chen et al., "A joint learning and communications framework for federated learning over wireless networks," *IEEE Trans. on Wireless Communications*, vol. 20, no. 1, pp 269–283, Jan. 2020.

[15] J. Ren, G. Yu, and G. Ding, "Accelerating DNN training in wireless federated edge learning system," *arXiv preprint*, arXiv:1905.09712, 2019.

[16] Q. Zeng et al., "Energy-efficient resource management for federated edge learning with CPU-GPU heterogeneous computing," *arXiv preprint*, arXiv:2007.07122, 2020.

[17] R. Jin, X. He, and H. Dai, "On the design of communication efficient federated learning over wireless networks," *arXiv preprint*, arXiv:2004.07351, 2020.

[18] M. M. Amiri and D. Gunduz, "Machine learning at the wireless edge: Distributed stochastic gradient descent over-the-air," in *Proc. IEEE Int. Symp. Information Theory (ISIT)*, July 2019.

[19] G. Zhu et al., "One-bit over-the-air aggregation for communication-efficient federated edge learning: Design and convergence analysis," *arXiv preprint*, arXiv:2001.05713, 2020.

[20] T. Sery et al., "COTAF: Convergent over-the-air federated learning," in *Proc. IEEE Global Communications Conf. (GLOBECOM)*, 2020.

[21] Y. Lin et al., "Deep gradient compression: Reducing the communication bandwidth for distributed training," *arXiv preprint*, arXiv:1712.01887, 2017.

[22] C. Hardy, E. Le Merrer, and B. Sericola, "Distributed deep learning on edge-devices: feasibility via adaptive compression," in *Proc. IEEE Int. Symp. on Network Computing and Applications (NCA)*, 2017.

[23] A. F. Aji and K. Heafield, "Sparse communication for distributed gradient descent," *arXiv preprint*, arXiv:1704.05021, 2017.

[24] D. Alistarh et al., "The convergence of sparsified gradient methods," in *Neural Information Processing Systems*, pp. 5973–5983, 2018.

[25] P. Han, S. Wang, and K. K. Leung, "Adaptive gradient sparsification for efficient federated learning: An online learning approach," *arXiv preprint*, arXiv:2001.04756, 2020.

[26] J. Bernstein et al., "SignSGD: Compressed optimisation for non-convex problems," *arXiv preprint*, arXiv:1802.04434, 2018.

[27] W. Wen et al., "Terngrad: Ternary gradients to reduce communication in distributed deep learning," in *Neural Information Processing Systems*, pp. 1509–1519, 2017.

[28] D. Alistarh et al., "QSGD: Communication-efficient SGD via gradient quantization and encoding," in *Neural Information Processing Systems*, pp. 1709–1720, 2017.

[29] S. Horvath et al., "Natural compression for distributed deep learning," *arXiv preprint*, arXiv:1905.10988, 2019.

[30] A. Reisizadeh et al., "FedPAQ: A communication-efficient federated learning method with periodic averaging and quantization," *arXiv preprint*, arXiv:1909.13014, 2019.

[31] S. P. Karimireddy et al., "Error feedback fixes signSGD and other gradient compression schemes," in *Int. Conf. on Machine Learning*, 2019, pp. 3252–3261.

[32] R. M. Gray and D. L. Neuhoff, "Quantization," *IEEE Trans. on Information Theory*, vol. 44, no. 6, pp. 2325–2383, 1998.

[33] T. M. Cover and J. A. Thomas, *Elements of Information Theory*. John Wiley & Sons, 2012.

[34] Y. Polyanskiy and Y. Wu, "Lecture notes on information theory," *Lecture Notes for 6.441 (MIT), ECE563 (University of Illinois Urbana-Champaign), and STAT 664 (Yale)*, 2012–2017.

[35] S. U. Stich, "Local SGD converges fast and communicates little," *arXiv preprint*, arXiv:1805.09767, 2018.

[36] X. Li et al., "On the convergence of fedavg on non-iid data," *arXiv preprint* arXiv:1907.02189, 2019.

[37] B. Woodworth et al., "Is local SGD better than minibatch SGD?" *arXiv preprint*, arXiv:2002.07839, 2020.

[38] J. Zhang et al., "Parallel SGD: When does averaging help?" *arXiv preprint*, arXiv:1606.07365, 2016.

[39] L. Zhu, Z. Liu, and S. Han, "Deep leakage from gradients," in *Advances in Neural Information Processing Systems*, pp. 14774–14784, 2019.

[40] Speedtest.net, "Speedtest United States market report," tech. report; www.speedtest.net/reports/united-states/.

[41] S. Horváth et al., "Stochastic distributed learning with gradient quantization and variance reduction," *arXiv preprint*, arXiv:1904.05115, 2019.

[42] A. El Gamal and Y.-H. Kim, *Network Information Theory*. Cambridge University Press, 2011.

[43] A. Wyner and J. Ziv, "The rate-distortion function for source coding with side information at the decoder," *IEEE Trans. on Information Theory*, vol. 22, no. 1, pp. 1–10, 1976.

[44] S. P. Lipshitz, R. A. Wannamaker, and J. Vanderkooy, "Quantization and dither: A theoretical survey," *Journal of the Audio Engineering Society*, vol. 40, no. 5, pp. 355–375, 1992.

[45] R. M. Gray and T. G. Stockham, "Dithered quantizers," *IEEE Trans. on Information Theory*, vol. 39, no. 3, pp. 805–812, 1993.

[46] B. Widrow, I. Kollar, and M.-C. Liu, "Statistical theory of quantization," *IEEE Trans. on Instrumentation and Measurement*, vol. 45, no. 2, pp. 353–361, 1996.

[47] J. Ziv, "On universal quantization," *IEEE Trans. on Information Theory*, vol. 31, no. 3, pp. 344–347, 1985.

[48] R. Zamir and M. Feder, "On universal quantization by randomized uniform/lattice quantizers," *IEEE Trans. on Information Theory*, vol. 38, no. 2, pp. 428–436, 1992.

[49] R. Zamir and M. Feder, "On lattice quantization noise," *IEEE Trans. on Information Theory*, vol. 42, no. 4, pp. 1152–1159, 1996.

[50] J. H. Conway and N. J. A. Sloane, *Sphere Packings, Lattices and Groups*, Springer Science & Business Media, 2013.

[51] Y. Zhang, J. C. Duchi, and M. J. Wainwright, "Communication-efficient algorithms for statistical optimization," *Journal of Machine Learning Research*, vol. 14, no. 1, pp. 3321–3363, 2013.

[52] J. Conway and N. Sloane, "Voronoi regions of lattices, second moments of polytopes, and quantization," *IEEE Trans. on Information Theory*, vol. 28, no. 2, pp. 211–226, 1982.

[53] N. Shlezinger et al., "UVeQFed: Universal vector quantization for federated learning," *IEEE Trans. on Signal Processing*, vol. 69, pp. 500–514, Dec. 2020.

[54] T. Li et al., "Federated learning: Challenges, methods, and future directions," *IEEE Signal Processing Magazine*, vol. 37, no. 3, pp. 50–60, May 2020.

[55] A. Koloskova et al., "A unified theory of decentralized SGD with changing topology and local updates," *arXiv preprint*, arXiv:2003.10422, 2020.

[56] A. Kirac and P. Vaidyanathan, "Results on lattice vector quantization with dithering," *IEEE Trans. on Circuits and Systems—Part II: Analog and Digital Signal Processing*, vol. 43, no. 12, pp. 811–826, 1996.

[57] MathWorks Deep Learning Toolbox Team, "Deep learning tutorial series," *MATLAB Central File Exchange*, 2020.

# 15 Over-the-Air Computation for Distributed Learning over Wireless Networks

Mohammad Mohammadi Amiri and Deniz Gündüz

## 15.1 Introduction

Future wireless networks will need to support traffic for a wide range of applications, from autonomous vehicles and intelligent robots with wireless connections to e-health, smart cities, and virtual/augmented reality. However, there is an important distinction between the traffic generated by such Internet of Things (IoT) applications and content-generated traffic that dominates communication networks today: most IoT applications aim at making inferences from the data collected by IoT devices at the network edge; that is, rather than the data itself, the destination node is interested in learning some statistics from the data, for example, for state estimation or anomaly detection, or use the data to take actions in a control loop, such as an autonomous driving car avoiding a person on the road or a health-monitoring device sending an alarm in the event of a health emergency.

The conventional approach to IoT edge intelligence is to offload the collected data to an edge or a cloud server, where the computing or learning tasks are carried out. However, with the increasing amount of information collected by edge devices, this centralized approach is becoming increasingly demanding for the communication network, and alternative distributed learning approaches are being considered [1]. Distributed learning also provides a certain level of privacy to the clients, as they can carry out learning without offloading their sensitive datasets to a central server [2].

Distributed learning over wireless edge networks poses many interesting research challenges [3–9]. In distributed learning, the objective is not necessarily aligned with maximizing the throughput of individual users. Since we would like to increase the data diversity in learning, we often want to allocate communication resources as fairly as possible among the devices. Moreover, since learning is typically carried out in many iterative steps, the accuracy of individual iterations may not be critical. As a consequence, devices typically operate at different trade-off points compared to conventional network operations, and new resource allocation and network optimization techniques need to be employed [10–13].

Interference management is another fundamental aspect of network design that has been challenged in distributed learning applications. In the current network architectures, transmissions from different devices are treated as independent data streams; therefore, in order to increase the reliability of each transmission, we try to minimize interference among different transmissions as much as possible. Interference can be mitigated by orthogonalizing transmitting devices in time, code, or frequency. On the other hand, when the goal is to carry out computations on distributed data, it is possible to exploit the signal superposition property of the wireless medium rather than mitigate it. This is called over-the-air computation (OAC), and it allows us to *exploit the wireless medium not only to convey information, but also to carry out computations directly on data.* Turning the wireless channel into a computing medium allows us to bring the intelligence to the network edge and to seamlessly combine learning and communication tasks.

In this chapter, we show how OAC can be used for federated training of a shared machine learning model across multiple edge devices each with its own local dataset. Before going into the details of how OAC can be exploited for distributed learning in wireless systems, we first present the federated learning (FL) paradigm and its implementation at the wireless network edge, called federated edge learning (FEEL).

### 15.1.1    FEEL with OAC

In FL, multiple devices collaborate to train a joint model orchestrated by a parameter server (PS) [2]. The main motivation in FL is to train the joint model without sharing the local datasets due to privacy concerns. This can be achieved by distributed stochastic gradient descent (SGD), where the global model is iteratively updated based on local model updates from the devices, which are computed in a distributed fashion based on their local datasets. Therefore, the devices only share their model updates with the parameter server, and the datasets are localized. To further reduce the communication load of FL, devices can apply multiple SGD iterations on the global model before sharing their updates, which reduces the communication frequency from the devices to the PS. Also, in typical FL scenarios, it is assumed that a large number of devices contribute to the collaborative training process; however, only a subset of those devices participate in each iteration.

FEEL refers to the application of FL at the wireless edge, where the device-to-PS and PS-to-device communication take place over a shared wireless medium [5, 9]; equivalently, it is assumed that the devices are in geographical proximity of each other. This may be the case, for example, when IoT devices within a macro cell are orchestrated by a macro base station or by a Wi-Fi access point to jointly train a model. Therefore, in a FEEL scenario, participating devices must share the available wireless resources when transmitting their model updates to the PS at each iteration. Note that a collaborative training process in the context of FEEL can be preferable not only due to privacy concerns, but also to the limited channel bandwidth, particularly when the datasets are relatively big.

An important observation in the FEEL scenario is that the goal of the PS at each iteration is to compute the average of the model updates from the devices. The PS is not interested in the individual model updates, and indeed, from a privacy perspective, it is even more advantageous if the PS can recover the model average without being able to learn the individual updates of the devices. In this sense, FEEL is an ideal framework to apply OAC for the transmission of model updates from the devices over the multiple access channel (MAC) to the PS. At each iteration of the training process, participating devices can transmit their local updates in an uncoded fashion such that the transmitted waveforms are superposed over the wireless channel, allowing the PS to directly recover the sum of the model updates.

There are several challenges in the implementation of OAC in the context of FEEL: First, it will require all the transmitting devices to be synchronized in time or frequency to make sure that the model updates are aligned for superposition. This can be achieved by the timing advance procedure in 4G and 5G systems, where the synchronization error will depend on the resources dedicated to it. While we expect that the benefits of OAC for FEEL will continue to hold even in the presence of some synchronization errors, this has not yet been quantified. Second, the correct summation of the model updates at the PS requires them to be received at the same power level. This requires precoding at the transmitters and, hence, knowledge of channel state information (CSI). As we will see later in this chapter, lack of CSI can be compensated with the introduction of multiple antennas at the PS. The third challenge in applying OAC in FEEL is the large bandwidth requirement. When digital communication techniques are employed, model updates can be quantized to a lower number of bits, which can then be transmitted to the PS over the available bandwidth using an appropriate modulation and coding scheme [4, 14]. However, when the model update is transmitted in an uncoded fashion, this will require as many channel resources as the model dimension, which can be large in modern deep learning architectures. For example, architectures used for machine vision applications typically have millions of parameters; the well-known AlexNet, ResNet50, and VGG16 architectures have 50, 26, and 138 million parameters, respectively. Models for natural language processing applications typically have much larger networks, with even billions of parameters. Therefore, training such large networks with uncoded transmissions would require very large bandwidth, often not available at the network edge. In the following sections, we will present a solution to this problem through sparsification and random projection, which allows the devices to significantly reduce the bandwidth requirement without sacrificing the performance much. Note also that OAC already has inherent bandwidth efficiency due to using the entire channel resources by all the devices instead of sharing them.

The fundamental limits of OAC have been studied in the information theory literature [15, 16]. It has also been considered as a method to increase network capacity by allowing intermediate nodes to decode functions of messages rather than individual messages [17–19]. Computation over a wireless MAC was first studied in [20]. Employing OAC for efficient edge learning has received significant recent attention. This idea was introduced in [4] for a Gaussian MAC, while a wireless MAC was considered in [21]. In [22], the authors focus on the device selection and receive beamforming design considering a multiantenna PS. The idea of bandwidth compression was applied in [23] to FEEL over wireless fading channels. Multiple antenna PS was also considered in [24, 25], but without CSI at the devices. In [26] the authors studied OAC when devices are limited to using a 4-QAM constellation to transmit the model updates. The results showed that the benefits are available even under this limitation and, hence, can be exploited in conventional systems, which are often limited to finite constellation due to hardware design. The benefits of OAC were combined with federated distillation in [27]. Device scheduling in over-the-air FEEL was studied in [28] using energy availability.

In the rest of this chapter, we first introduce the general system model for FEEL over wireless fading channels. We then present both the digital transmission of model updates and the uncoded transmission with OAC. We then present the so-called blind FEEL, which employs OAC without any CSI at the transmitting devices.

## 15.2    System Model

We consider $M$ devices, each with its local dataset, training a global model $\boldsymbol{\theta} \in \mathbb{R}^d$ collaboratively with the help of a remote PS, as depicted in Fig. 15.1. The goal is to minimize a loss function

$$F(\boldsymbol{\theta}) = \frac{1}{M} \sum_{m=1}^{M} F_m(\boldsymbol{\theta}), \tag{15.1}$$

where

$$F_m(\boldsymbol{\theta}) = \frac{1}{B_m} \sum_{\boldsymbol{u} \in \mathcal{B}_m} f(\boldsymbol{\theta}, \boldsymbol{u}), \quad m \in [M], \tag{15.2}$$

is the local loss function at device $m$, which is minimized with respect to the global model $\boldsymbol{\theta}$ having access to local dataset $\mathcal{B}_m$ with size $B_m = |\mathcal{B}_m|$, and $f$ is an empirical loss function.

During iteration $t$, the PS shares the global model, denoted by $\boldsymbol{\theta}(t)$, with the devices. After receiving $\boldsymbol{\theta}(t)$, devices perform $\tau$-step SGD to minimize their local loss functions. The $i$th step SGD at device $m$ corresponds to the following update:

$$\boldsymbol{\theta}_m^{i+1}(t) = \boldsymbol{\theta}_m^i(t) - \eta_{lm}^i(t) \nabla F_m\left(\boldsymbol{\theta}_m^i(t), \xi_m^i(t)\right), \quad i \in [\tau], m \in [M], \tag{15.3}$$

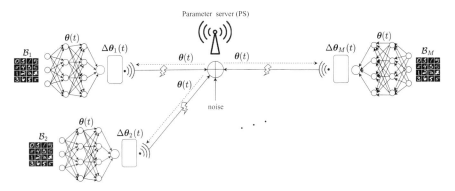

**Figure 15.1**  FEEL system model at iteration $t$.

where $\boldsymbol{\theta}_m^1(t) = \boldsymbol{\theta}(t)$, $\eta_m^i(t)$ represents the learning rate, and $\nabla F_m\big(\boldsymbol{\theta}_m^i(t), \xi_m^i(t)\big)$ is an unbiased estimate of $\nabla F_m\big(\boldsymbol{\theta}_m^i(t)\big)$ according to a local mini-batch data $\xi_m^i(t)$ selected randomly from the local dataset $\mathcal{B}_m$; that is,

$$\mathbb{E}_\xi\big[\nabla F_m\big(\boldsymbol{\theta}_m^i(t), \xi_m^i(t)\big)\big] = \nabla F_m\big(\boldsymbol{\theta}_m^i(t)\big), \quad \forall i \in [\tau], \forall m \in [M], \forall t, \qquad (15.4)$$

where $\mathbb{E}_\xi$ represents expectation with respect to the mini-batch selection randomness. After the $\tau$th SGD step, device $m$ computes the local model update $\Delta\boldsymbol{\theta}_m(t) = \boldsymbol{\theta}_m^{\tau+1}(t) - \boldsymbol{\theta}(t)$, for $m \in [M]$, which is shared with the PS. Assuming that the PS can receive the local model updates accurately from the devices, it updates the global model as follows:

$$\boldsymbol{\theta}(t+1) = \boldsymbol{\theta}(t) + \Delta\boldsymbol{\theta}(t), \qquad (15.5)$$

where we have defined

$$\Delta\boldsymbol{\theta}(t) \triangleq \frac{1}{M} \sum_{m=1}^M \Delta\boldsymbol{\theta}_m(t). \qquad (15.6)$$

We model the transmission from the devices to the PS as a wireless fading MAC with $s$ subchannels, where we assume that $s \leq d$ (for practical purposes we are particularly interested in $s \ll d$ regime), and each communication round takes place over $N$ time slots. We let $\boldsymbol{x}_m^n(t) = [x_{m,1}^n(t), \ldots, x_{m,s}^n(t)]^T \in \mathbb{C}^s$ denote the channel input at device $m$ at the $n$th time slot during iteration $t$, $n \in [N]$, $m \in [M]$. We also denote the channel gains from device $m$ to the PS and the additive noise at the PS at the $n$th time slot during iteration $t$ by $\boldsymbol{h}_m^n(t) = [h_{m,1}^n(t), \ldots, h_{m,s}^n(t)]^T \in \mathbb{C}^s$ and $\boldsymbol{z}^n(t) \in \mathbb{C}^s$ with each entry independent and identically distributed (i.i.d.) according to $\mathcal{CN}(0, \sigma^2)$ and $\mathcal{CN}(0, 1)$, respectively, $n \in [N]$, $m \in [M]$. The received signal at the PS during the $n$th time slot of iteration $t$ is given by

$$\boldsymbol{y}^n(t) = \sum_{m=1}^M \boldsymbol{h}_m^n(t) \circ \boldsymbol{x}_m^n(t) + \boldsymbol{z}^n(t), \quad n \in [N]. \qquad (15.7)$$

We assume that CSI is available at the devices. Given a total number of $T$ iterations, we impose an average transmit power constraint $\bar{P}$ at each device; that is,

$$\frac{1}{NT} \sum_{t=1}^T \sum_{n=1}^N \mathbb{E}\big[||\boldsymbol{x}_m^n(t)||_2^2\big] \leq \bar{P}, \quad \forall m \in [M], \qquad (15.8)$$

where the expectation is with respect to the randomness of the channel gains.

The underlying wireless fading MAC from the devices to the PS dictates that the PS can only receive a noisy version of $\Delta\boldsymbol{\theta}(t)$, denoted by $\Delta\widehat{\boldsymbol{\theta}}(t)$, which is used to update the global model according to

$$\boldsymbol{\theta}(t+1) = \boldsymbol{\theta}(t) + \Delta\widehat{\boldsymbol{\theta}}(t). \qquad (15.9)$$

We will consider both the conventional digital approach and the uncoded transmission with OAC for the transmission of the local model updates from the devices to the PS. With the digital approach, devices perform quantization followed by channel coding to transmit the quantized local model updates over the MAC. Alternatively, with OAC, devices transmit their local model updates in an uncoded manner without employing any channel coding in order to benefit from the signal superposition property of the wireless medium.

## 15.3    Digital FEEL

We first consider digital transmission of the local model updates from the devices to the PS, where we refer to this approach as digital FEEL. For the digital approach, we consider $N = 1$; that is, each communication round takes place over a single time slot. For simplicity, we drop the dependency of the variables on the time slot index $n$.

The goal is to deliver the local model updates from the devices to the PS as accurately as possible. Due to the resource sharing nature of digital transmission over the MAC, when all the devices participate in training, each device receives a limited amount of resources and may send a coarse description of its local model update over the MAC. We instead schedule the devices opportunistically, where at each iteration, only a subset of the devices are selected to send their local model updates over the MAC (refer to [29] for various device scheduling techniques).

### 15.3.1    Upper Bound on Capacity

We consider scheduling a single device and allocating all the resources to it. In particular, having CSI at the devices, we schedule a device with the largest value of $\sum_{i=1}^{s} |h_{m,i}(t)|^2$, $m \in [M]$, during iteration $t$. Accordingly, the index of the participating device at iteration $t$ is given by:

$$m^*(t) = \arg \max_{m \in [M]} \left\{ \sum_{i=1}^{s} |h_{m,i}(t)|^2 \right\}. \tag{15.10}$$

We highlight that since the channel gains across the devices are i.i.d., the probability of selecting each device at each iteration is $1/M$. Let $\bar{P}_m(t)$ denote the transmit power at device $m$ during iteration $t$. We have $\bar{P}_m(t) = 0$, if $m \neq m^*(t)$, and

$$\frac{1}{MT} \sum_{t=1}^{T} \bar{P}_m(t) \leq \bar{P}, \quad \text{for } m \in [M]. \tag{15.11}$$

In the following, we provide an upper bound on the capacity of the selected device, which specifies the maximum number of bits transmitted from that device that can be received at the PS with no error. The $i$th entry of the channel output at iteration $t$ is as follows:

$$y_i(t) = h_{m^*(t),i}(t) x_{m^*(t),i}(t) + z_i(t), \quad i \in [s]. \tag{15.12}$$

This corresponds to a fast fading channel with $s$ channel uses, where the CSI is known at both the transmitter and the receiver. We place an upper bound on the capacity of the above link assuming coding across infinitely many realizations of this $s$-dimensional channel. Given a transmit power $P_{m^*(t)}(t)$, the capacity of a parallel fading channel is the solution of the optimization problem [30, Section 5.4.6]

$$\max_{P_1,\dots,P_s} \sum_{i=1}^{s} \log_2 \left(1 + P_i \left|h_{m^*(t),i}(t)\right|^2\right),$$

$$\text{subject to } \sum_{i=1}^{s} P_i = \bar{P}_{m^*(t)}(t). \qquad (15.13)$$

This problem is solved through water-filling, and the optimal power allocation is given by

$$P_i^* = \max\left\{\frac{1}{\zeta} - \frac{1}{\left|h_{m^*(t),i}(t)\right|^2}, 0\right\}, \qquad (15.14)$$

where $\zeta$ is set such that $\sum_{i=1}^{s} P_i^* = \bar{P}_{m^*(t)}(t)$. Accordingly, the upper bound on the capacity of the wireless channel in Eq. (15.12) is given by

$$C(t) = \sum_{i=1}^{s} \log_2 \left(1 + P_i^* \left|h_{m^*(t),i}(t)\right|^2\right). \qquad (15.15)$$

We emphasize that this may be a loose upper bound when $s$ is relatively small.

## 15.3.2   Model Compression

Next, we introduce the quantization technique for compression of the local model updates at the selected devices for digital transmission. We adopt the scheme introduced in [31], where error accumulation is employed after the quatization to compensate for the quantization error. We denote the accumulated error at device $m$ at time $t$ by $\delta_m(t)$, where we set $\delta_m(0) = 0$, for $m \in [M]$. The quantization technique is employed at device $m$ to compress the error compensated local model update $\Delta\boldsymbol{\theta}_m(t) + \delta_m(t-1)$, for $m \in [M]$.

Given a sparsity level $q(t)$, where typically $q(t) \ll d$, device $m$ sets all but $q(t)$ largest positive entries and $q(t)$ smallest negative entries of $\Delta\boldsymbol{\theta}_m(t) + \delta_m(t-1)$ to zero, $m \in [M]$. We denote the average of the $q(t)$ nonzero positive and negative entries with $\Delta\theta_m^+(t)$ and $\Delta\theta_m^-(t)$, respectively. If $\Delta\theta_m^+(t) \geq \left|\Delta\theta_m^-(t)\right|$, device $m$ sets those $q(t)$ nonzero negative entries to zero and $q(t)$ nonzero positive entries to $\Delta\theta_m^+(t)$. On the other hand, if $\Delta\theta_m^+(t) < \left|\Delta\theta_m^-(t)\right|$, the $q(t)$ nonzero negative and positive entries are set to $\Delta\theta_m^-(t)$ and zero, respectively. We denote the resultant sparse vector by $\Delta\widehat{\boldsymbol{\theta}}_m(t)$ and update the error accumulation vector $\delta_m(t)$, which maintains those entries that are not transmitted due to the sparsification as follows:

$$\delta_m(t) = \begin{cases} \Delta\boldsymbol{\theta}_m(t) + \delta_m(t-1) - \Delta\widehat{\boldsymbol{\theta}}_m(t), & \text{if } m = m^*(t), \\ \Delta\boldsymbol{\theta}_m(t) + \delta_m(t-1), & \text{otherwise.} \end{cases}$$

We note that, when device $m$ is selected for participating in the training, the error accumulation vector corresponds to the gap between $\Delta\boldsymbol{\theta}_m(t) + \boldsymbol{\delta}_m(t-1)$ and its sparsified estimate. However, when device $m$ is not scheduled, we keep the message at device $m$, $\Delta\boldsymbol{\theta}_m(t) + \boldsymbol{\delta}_m(t-1)$ to compensate it at the next iteration. We highlight that the compression technique requires transmission of

$$R(q(t)) = \log_2 \binom{d}{q(t)} + 33, \qquad (15.16)$$

bits from the participating device, where $\log_2 \binom{d}{q(t)}$ is to represent the locations of nonzero entries, and 33 bits represent the real number with its sign. Having device $m^*(t)$ scheduled for transmission over the MAC, we set $q(t)$ as the largest integer satisfying $R(q(t)) \leq C(t)$.

*Remark 15.1*   With this approach, a single device is scheduled at each iteration, and the PS updates the global model after receiving the quantized local update from that device. Alternatively, we can select all the devices, or a subset of the devices, and allocate distinct subchannels to each scheduled device. It was shown in [23] that providing all the subchannels to a single device as described in the digital FEEL approach performs significantly better than sharing the subchannels among all the devices in an orthogonal fashion.

## 15.4    FEEL with OAC

We note that, in the FL framework under consideration, the PS is only interested in the average of the local model updates computed at the devices and not each individual update. The fact that the underlying wireless MAC provides the PS with a noisy version of the average of the messages sent from the devices motivates transmission of the local updates in an uncoded/analog manner. In this approach, we exploit OAC to have a noisy estimate of the average of the local model updates at the PS. Algorithm 15.1 summarizes the steps of FEEL with OAC.

In the proposed FEEL with OAC scheme, devices reduce the dimension of the local model updates to $\tilde{s} = 2sN$ through linear projection and send the resultant compressed vector of dimension $\tilde{s}$ in an uncoded fashion over $N$ time slots, for some $N \in [\lceil d/2s \rceil]$. Devices first sparsify their local model updates after error compensation. Let $\rho(t)$ denote the sparsification error accumulated at time $t$, where we set $\rho(0) = 0$. Device $m$ first sparsifies $\Delta\boldsymbol{\theta}_m(t) + \boldsymbol{\rho}_m(t-1)$ by setting all but $k$ entries with the largest magnitudes to zero, for some $k \leq \tilde{s}, m \in [M]$. We denote the resultant sparse vector at device $m$ by $\Delta\boldsymbol{\theta}_m^{\mathrm{sp}}(t)$, and the error is accumulated as follows:

$$\boldsymbol{\rho}_m(t) = \Delta\boldsymbol{\theta}_m(t) + \boldsymbol{\rho}_m(t-1) - \Delta\boldsymbol{\theta}_m^{\mathrm{sp}}(t), \quad \text{for } m \in [M]. \qquad (15.17)$$

Devices perform random linear projection, described in the following, to transmit a compressed version of the sparse vector over the MAC.

---

**Algorithm 15.1** FEEL with OAC.

1: **Initialize** $\rho_1(0) = \cdots = \rho_M(0) = 0$
2: **for** $t = 1, \ldots, T$ **do**
 • **devices do:**
3:     **for** $m = 1, \ldots, M$ in parallel **do**
4:         $\Delta\boldsymbol{\theta}_m^{\mathrm{sp}}(t) = \mathrm{sparse}_k\left(\Delta\boldsymbol{\theta}_m(t) + \boldsymbol{\rho}_m(t-1)\right)$
5:         $\boldsymbol{\rho}_m(t) = \Delta\boldsymbol{\theta}_m(t) + \boldsymbol{\rho}_m(t-1) - \Delta\boldsymbol{\theta}_m^{\mathrm{sp}}(t)$
6:         $\Delta\widetilde{\boldsymbol{\theta}}_m(t) = \boldsymbol{A}\,\Delta\boldsymbol{\theta}_m^{\mathrm{sp}}(t)$
7:         **for** $n = 1, \ldots, N$ **do**
8:             $\boldsymbol{x}_m^n(t) = \boldsymbol{\alpha}_m^n(t) \circ \Delta\widetilde{\boldsymbol{\theta}}_m^n(t)$
9:         **end for**
10:     **end for**
 • **PS does:**
11:     **for** $n = 1, \ldots, N$ and $i = 1, \ldots, s$ **do**
12:         $\hat{y}_{2(n-1)s+i}(t) = \begin{cases} \dfrac{\mathrm{Re}\{y_i^n(t)\}}{\bar{\gamma}_m^n(t)\left|\mathcal{M}_i^n(t)\right|}, & \text{if } \left|\mathcal{M}_i^n(t)\right| \neq 0, \\ 0, & \text{otherwise} \end{cases}$
13:         $\hat{y}_{(2n-1)s+i}(t) = \begin{cases} \dfrac{\mathrm{Im}\{y_i^n(t)\}}{\bar{\gamma}_m^n(t)\left|\mathcal{M}_i^n(t)\right|}, & \text{if } \left|\mathcal{M}_i^n(t)\right| \neq 0, \\ 0, & \text{otherwise} \end{cases}$
14:     **end for**
15:     **if** $\widehat{\boldsymbol{y}}(t) \neq \boldsymbol{0}$ **then**
16:         $\Delta\widehat{\boldsymbol{\theta}}(t) = \mathrm{AMP}_{\boldsymbol{A}}\left(\widehat{\boldsymbol{y}}(t)\right)$
17:         $\boldsymbol{\theta}(t+1) = \boldsymbol{\theta}(t) + \Delta\widehat{\boldsymbol{\theta}}(t)$
18:     **else**
19:         $\boldsymbol{\theta}(t+1) = \boldsymbol{\theta}(t)$
20:     **end if**
21: **end for**

---

A pseudo-random matrix $\boldsymbol{A} \in \mathbb{R}^{\tilde{s} \times d}$, with each entry i.i.d. according to $\mathcal{N}(0, 1/\tilde{s})$ is generated and shared with the devices and the PS before the training, where we remind that $\tilde{s} = 2sN$, for an arbitrary $N \in \left[\lceil d/2s \rceil\right]$. At iteration $t$, device $m$ computes

$$\Delta\widetilde{\boldsymbol{\theta}}_m(t) \triangleq \boldsymbol{A}\,\Delta\boldsymbol{\theta}_m^{\mathrm{sp}}(t) \in \mathbb{R}^{\tilde{s}}, \quad \text{for } m \in [M], \tag{15.18}$$

and transmits it over $N = \tilde{s}/(2s)$ time slots in an uncoded fashion. We define, for $n \in [N], m \in [M]$,

$$\Delta\widetilde{\boldsymbol{\theta}}_{m,\mathrm{re}}^n(t) \triangleq [\Delta\tilde{\theta}_{m,2(n-1)s+1}(t), \ldots, \Delta\tilde{\theta}_{m,(2n-1)s}(t)]^T, \tag{15.19a}$$

$$\Delta\widetilde{\boldsymbol{\theta}}_{m,\mathrm{im}}^n(t) \triangleq [\Delta\tilde{\theta}_{m,(2n-1)s+1}(t), \ldots, \Delta\tilde{\theta}_{m,2ns}(t)]^T, \tag{15.19b}$$

$$\Delta\widetilde{\boldsymbol{\theta}}_m^n(t) \triangleq \Delta\widetilde{\boldsymbol{\theta}}_{m,\mathrm{re}}^n(t) + j\,\Delta\widetilde{\boldsymbol{\theta}}_{m,\mathrm{im}}^n(t), \tag{15.19c}$$

where $\Delta\tilde{\theta}_{m,i}(t)$ is the $i$th entry of $\Delta\tilde{\theta}_m(t)$, $i \in [\tilde{s}]$. At the $n$th time slot of the communication round during iteration $t$, device $m$ transmits $x_m^n(t) = \alpha_m^n(t) \circ \Delta\tilde{\theta}_m^n(t)$, where $\alpha_m^n(t) \in \mathbb{C}^s$ is the power control vector, whose $i$th entry is set as follows:

$$\alpha_{m,i}^n(t) = \begin{cases} \frac{\gamma_m^n(t)}{h_{m,i}^n(t)}, & \text{if } |h_{m,i}^n(t)|^2 \geq \lambda(t), \\ 0, & \text{otherwise}, \end{cases} \tag{15.20}$$

where $\gamma_m^n(t), \lambda(t) \in \mathbb{R}$ are set so that the average transmit power constraint is satisfied. We define, for $i \in [s], n \in [N]$,

$$\mathcal{M}_i^n(t) = \left\{ m \in [M] : |h_{m,i}^n(t)|^2 \geq \lambda(t) \right\}, \tag{15.21}$$

which includes the indices of $x_m^n(t)$ with nonzero entries.

Here we analyze the average transmit power of the FEEL approach with OAC given the previously described power allocation technique. The average transmit power at device $m$ at the $n$th time slot of the communication round during iteration $t$ is given by, $m \in [M], n \in [N]$,

$$\bar{P}_m^n(t) = \mathbb{E}\left[ \|x_m^n(t)\|_2^2 \right] \tag{15.22}$$

$$= \sum_{i=1}^s \mathbb{E}\left[ |\alpha_{m,i}^n(t)|^2 \left| \Delta\tilde{\theta}_{m,2(n-1)s+i}(t) + j\Delta\tilde{\theta}_{m,(2n-1)s+i}(t) \right|^2 \right]. \tag{15.23}$$

We highlight that the entries of $\Delta\tilde{\theta}_m^n(t)$ is independent of the channel realizations at time slot $n$. Thus, it follows that

$$\bar{P}_m^n(t) = \sum_{i=1}^s \left| \Delta\tilde{\theta}_{m,2(n-1)s+i}(t) + j\Delta\tilde{\theta}_{m,(2n-1)s+i}(t) \right|^2 \mathbb{E}\left[ |\alpha_{m,i}^n(t)|^2 \right]. \tag{15.24}$$

We note that $|h_{m,i}^n(t)|^2$ follows an exponential distribution with mean $\sigma^2$, $\forall i, n, m$. Accordingly, it follows that

$$\mathbb{E}\left[ |\alpha_{m,i}^n(t)|^2 \right] = \left( \frac{\gamma_m^n(t)}{\sigma} \right)^2 \mathrm{E}_1(\lambda(t)), \tag{15.25}$$

where $\mathrm{E}_1(x) \triangleq \int_x^\infty \frac{e^{-\tau}}{\tau} d\tau$. Plugging Eq. (15.25) into Eq. (15.24) yields, for $m \in [M]$, $n \in [N]$,

$$\bar{P}_m^n(t) = \left( \frac{\gamma_m^n(t)}{\sigma} \right)^2 \mathrm{E}_1(\lambda(t)) \|\Delta\tilde{\theta}_m^n(t)\|_2^2. \tag{15.26}$$

Having set $\bar{P}_m^n(t)$, $\forall m, n, t$, such that

$$\frac{1}{NT} \sum_{t=1}^T \sum_{n=1}^N \bar{P}^n(t) \leq \bar{P}, \tag{15.27}$$

we have

$$\gamma_m^n(t) = \sigma \left( \frac{\bar{P}_m^n(t)}{E_1(\lambda^e(t)) \left\| \Delta \widetilde{\boldsymbol{\theta}}_m^n(t) \right\|_2^2} \right)^{1/2}. \tag{15.28}$$

We assume that device $m$ first transmits the scaling factor $\gamma_m^n(t)$ to the PS in an error-free fashion before sending $x_m^n(t)$, according to which the PS computes

$$\bar{\gamma}_m^n(t) \triangleq \frac{1}{M} \sum_{m=1}^{M} \gamma_m^n(t), \quad n \in [N], t \in [T]. \tag{15.29}$$

The $i$th entry of the received signal at time slot $n$ during iteration $t$ is given by, for $i \in [s], n \in [N]$,

$$y_i^n(t) = \sum_{m \in \mathcal{M}_i^n(t)} \gamma_m^n(t) \left( \Delta \tilde{\theta}_{m,2(n-1)s+i}(t) + j \Delta \tilde{\theta}_{m,(2n-1)s+i}(t) \right) + z_i^n(t)$$

$$= \boldsymbol{a}_{2(n-1)s+i}^T \sum_{m \in \mathcal{M}_i^n(t)} \gamma_m^n(t) \Delta \boldsymbol{\theta}_m^{\mathrm{sp}}(\boldsymbol{\theta}_t) \tag{15.30}$$

$$+ j \boldsymbol{a}_{(2n-1)s+i}^T \sum_{m \in \mathcal{M}_i^n(t)} \gamma_m^n(t) \Delta \boldsymbol{\theta}_m^{\mathrm{sp}}(\boldsymbol{\theta}_t) + z_i^n(t), \tag{15.31}$$

where $\boldsymbol{a}_l^T$ denotes the $l$th row of the measurement matrix $\boldsymbol{A}$, and we note that $\Delta \tilde{\theta}_{m,l}(t) = \boldsymbol{a}_l^T \Delta \boldsymbol{\theta}_m^{\mathrm{sp}}(t), l \in [\tilde{s}]$. We note that the PS is interested in $\frac{1}{M} \sum_{m=1}^{M} \Delta \boldsymbol{\theta}_m^{\mathrm{sp}}(\boldsymbol{\theta}_t)$, which is estimated from the above received signal having access to the measurement matrix $\boldsymbol{A}$ and the CSI. To this end, the PS employs a recovery algorithm to retrieve a sparse vector from its compressed version knowing the measurement matrix. We consider the approximate message passing (AMP) recovery algorithm, represented by AMP$_{\boldsymbol{A}}$ in Algorithm 15.1. To estimate $\frac{1}{M} \sum_{m=1}^{M} \Delta \boldsymbol{\theta}_m^{\mathrm{sp}}(\boldsymbol{\theta}_t)$, the PS first computes, for $i \in [s], n \in [N]$,

$$\hat{y}_{2(n-1)s+i}(t) = \begin{cases} \frac{\mathrm{Re}\{y_i^n(t)\}}{\bar{\gamma}_m^n(t) |\mathcal{M}_i^n(t)|}, & \text{if } |\mathcal{M}_i^n(t)| \neq 0, \\ 0, & \text{otherwise,} \end{cases} \tag{15.32a}$$

$$\hat{y}_{(2n-1)s+i}(t) = \begin{cases} \frac{\mathrm{Im}\{y_i^n(t)\}}{\bar{\gamma}_m^n(t) |\mathcal{M}_i^n(t)|}, & \text{if } |\mathcal{M}_i^n(t)| \neq 0, \\ 0, & \text{otherwise,} \end{cases} \tag{15.32b}$$

and then estimates

$$\Delta \widehat{\boldsymbol{\theta}}(t) = \mathrm{AMP}_{\boldsymbol{A}} \left( \widehat{\boldsymbol{y}}(t) \right), \tag{15.33}$$

where we have defined $\widehat{\boldsymbol{y}}(t) \triangleq [\hat{y}_1(t), \dots, \hat{y}_{\tilde{s}}(t)]^T$. If $\widehat{\boldsymbol{y}}(t) = \boldsymbol{0}$, the last model parameter vector is used as the new one; that is, $\boldsymbol{\theta}(t+1) = \boldsymbol{\theta}(t)$.

*Remark 15.2*    When $s = d/2$ ($N = 1$), the local model updates can be entirely transmitted over the wireless MAC at each iteration without employing any compression technique. Also, in the special case of $N = \lceil d/(2s) \rceil$, the local model updates, each of dimension $d$, are transmitted over $N$ times slots again without employing any compression.

*Remark 15.3*   The hyperparameter $k$, which denotes the sparsity level of the local model update at each device, can take different values smaller than $\tilde{s}$. For a relatively large $k$ value, the average of the sparsified local model updates, $\frac{1}{M} \sum_{m=1}^{M} \Delta \boldsymbol{\theta}_m^{\mathrm{sp}}(t)$, provides a better estimate of the average of the local model updates, $\frac{1}{M} \sum_{m=1}^{M} \Delta \boldsymbol{\theta}_m(t)$. However, recovering $\frac{1}{M} \sum_{m=1}^{M} \Delta \boldsymbol{\theta}_m^{\mathrm{sp}}(t)$ from $\frac{1}{M} \sum_{m=1}^{M} \Delta \tilde{\boldsymbol{\theta}}_m(t)$ may be inaccurate since the sparsity level of $\frac{1}{M} \sum_{m=1}^{M} \Delta \boldsymbol{\theta}_m^{\mathrm{sp}}(t)$ may exceed $\tilde{s}$. On the other hand, relatively small $k$ values impact the performance in an opposite manner. Accordingly, for i.i.d. data distribution across the devices, where the sparsity patterns of $\Delta \boldsymbol{\theta}_m^{\mathrm{sp}}(t)$, $\forall m \in [M]$, are expected to be similar, it may be better to set $k$ to a relatively large value. For non-i.i.d. data distribution, it might be better to have a relatively small $k$ value, since the sparsity patterns of $\Delta \boldsymbol{\theta}_m^{\mathrm{sp}}(t)$, $\forall m \in [M]$, are more diverse.

## 15.5    Blind FEEL

As we mentioned in Section 15.1, one of the limitations of FEEL with OAC is the requirement of CSI at all the transmitting devices. This might be harder to obtain when devices are mobile or have limited resources. Here we consider FEEL with transmission from the devices to the PS without any CSI at the transmitters (CSIT), together with the lack of perfect CSI also at the PS. We instead equip the PS with multiple antennas to mitigate the destructive effects of channel fading through receive beamforming relying on its imperfect CSI.

### 15.5.1    System Model

Similarly to the channel model described in Section 15.2, the channel from the devices to the PS is modelled as a wireless fading MAC with $s$ subchannels, where each communication round takes place over $N$ time slots. However, here we assume that the PS is equipped with $K$ antennas. The channel vector from device $m$ to the $k$th antenna at the PS during and the noise vector at the $k$th PS antenna during the $n$th time slot of iteration $t$ are denoted by $\boldsymbol{h}_{m,k}^n(t) = [h_{m,k,1}^n(t), \dots, h_{m,k,s}^n(t)]^T \in \mathbb{C}^s$ and $\boldsymbol{z}_k^n(t) = [z_{k,1}^n(t), \dots, z_{k,s}^n(t)]^T \in \mathbb{C}^s$, respectively, for $m \in [M]$, $k \in [K]$, $n \in [N]$. Each entry of the channel vector $\boldsymbol{h}_{m,k}^n(t)$ is distributed according to $\mathcal{CN}(0, \sigma^2)$, and different entries of this vector can be correlated. However, the channel gains are assumed to be i.i.d. across the devices, PS antennas, and time. Similarly, different entries of noise vector $\boldsymbol{z}_k^n(t)$, each distributed according to $\mathcal{CN}(0, 1)$, can be correlated, while it is i.i.d. across PS antennas and time. The signal received at the $k$th PS antennas during time slot $n$ of iteration $t$ is given by

$$y_k^n(t) = \sum_{m=1}^{M} \boldsymbol{h}_{m,k}^n(t) \circ \boldsymbol{x}_m^n(t) + \boldsymbol{z}_k^n(t), \quad k \in [K], n \in [N]. \tag{15.34}$$

We assume that the devices do not have access to the CSI, and the PS has only imperfect CSI about the sum of channel gains from the devices to each PS antenna; that is, $\sum_{m=1}^{M} h_{m,k}^n(t)$, $\forall k \in [K]$, for $n \in [N]$. We model the imperfect CSI at the PS as [32]

$$\widehat{h}_k^n(t) = \sum_{m=1}^{M} h_{m,k}^n(t) + \widetilde{h}_k^n(t), \quad \forall n, k, t, \tag{15.35}$$

where $\widehat{h}_k^n(t)$ denotes the estimate of $\sum_{m=1}^{M} h_{m,k}^n(t)$ at the PS, and $\widetilde{h}_k^n(t)$ is the channel estimation error with zero mean and variance $\widetilde{\sigma}^2$. At each iteration, the PS estimates $\Delta\boldsymbol{\theta}(t)$ through the signals received from all its antennas, $\boldsymbol{y}_k^n(t)$, $\forall k$, and the CSI $\widehat{h}_k^n(t), \forall n$.

*Remark 15.4* In this model, the PS only needs to estimate the sum of the channel gains from the devices to each PS antenna, $\sum_{m=1}^{M} h_{m,k}^n(t)$, $\forall n, k$, and not each individual channel gain $h_{m,k}^n(t)$. This significantly reduces the overhead of channel estimation at the PS, and is particularly compelling when number of devices $M$ and/or number of PS antennas $K$ are relatively large. We further highlight that the overhead of channel estimation at the PS does not increase with $M$.

In this scenario, we only consider uncoded transmission of local model updates from the devices with OAC. Algorithm 15.2 summarizes the FEEL scheme with OAC when the devices do not have CSI and the PS has imperfect CSI about the wireless fading MAC.

---

**Algorithm 15.2** Blind FEEL with OAC.

---

1: **for** $t = 1, \dots, T$ **do**
- **devices do:**
2:    **for** $m = 1, \dots, M$ in parallel **do**
3:      **for** $n = 1, \dots, N$ **do**
4:        $\boldsymbol{x}_m^n(t) = \alpha(t)\Delta\boldsymbol{\theta}_m^n(t)$
5:      **end for**
6:    **end for**
- **PS does:**
7:    **for** $n = 1, \dots, N$ **do**
8:      $y^n(t) = \frac{1}{K}\sum_{k=1}^{K}\left(\sum_{m=1}^{M}\widehat{h}_{m,k}^n(t)\right)^* \circ y_k^n(t)$
9:      **for** $i = 1, \dots, s$ **do**
10:        $\Delta\hat{\theta}_{2(n-1)s+i}(t) = \frac{\text{Re}\{y_i^n(t)\}}{M\sigma^2\alpha(t)}$
11:        $\Delta\hat{\theta}_{(2n-1)s+i}(t) = \frac{\text{Im}\{y_i^n(t)\}}{M\sigma^2\alpha(t)}$
12:      **end for**
13:    **end for**
14:    $\boldsymbol{\theta}(t+1) = \boldsymbol{\theta}(t) + \Delta\widehat{\boldsymbol{\theta}}(t)$
15: **end for**

---

## 15.5.2 Blind FEEL with OAC

At iteration $t$, device $m$ transmits $\Delta\boldsymbol{\theta}_m(t)$ over $N = \lceil d/2s \rceil$ time slots in an uncoded fashion, $m \in [M]$. We define

$$\Delta\boldsymbol{\theta}_m^{n,\mathrm{re}}(t) \triangleq [\Delta\theta_{m,2(n-1)s+1}(t), \dots, \Delta\theta_{m,(2n-1)s}(t)]^T, \tag{15.36a}$$

$$\Delta\boldsymbol{\theta}_m^{n,\mathrm{im}}(t) \triangleq [\Delta\theta_{m,(2n-1)s+1}(t), \dots, \Delta\theta_{m,2ns}(t)]^T, \tag{15.36b}$$

$$\Delta\boldsymbol{\theta}_m^n(t) \triangleq \Delta\boldsymbol{\theta}_m^{n,\mathrm{re}}(t) + j\Delta\boldsymbol{\theta}_m^{n,\mathrm{im}}(t), \tag{15.36c}$$

where $\Delta\theta_{m,i}(t)$ denotes the $i$th entry of $\Delta\boldsymbol{\theta}_m(t)$, for $i \in [d]$, and we zero-pad $\Delta\boldsymbol{\theta}_m(t)$ to have dimension $2sN$. Accordingly, the $i$th entry of $\Delta\boldsymbol{\theta}_m^n(t)$ is as follows:

$$\Delta\theta_{m,i}^n(t) = \Delta\theta_{m,2(n-1)s+i}(t) + j\Delta\theta_{m,(2n-1)s+i}(t), \ i \in [s], n \in [N], m \in [M], \tag{15.37}$$

and we have

$$\Delta\boldsymbol{\theta}_m(t) = \left[\Delta\boldsymbol{\theta}_m^{1,\mathrm{re}}(t), \Delta\boldsymbol{\theta}_m^{1,\mathrm{im}}(t), \dots, \Delta\boldsymbol{\theta}_m^{N,\mathrm{re}}(t), \Delta\boldsymbol{\theta}_m^{N,\mathrm{im}}(t)\right]^T, \tag{15.38}$$

where $N = \lceil d/2s \rceil$. During the $n$th time slot od iteration $t$, device $m$ transmits

$$\boldsymbol{x}_m^n(t) = \alpha(t)\Delta\boldsymbol{\theta}_m^n(t), \quad n \in [N], m \in [M], \tag{15.39}$$

where $\alpha(t)$ is a scaling factor to control the transmit power. With $\boldsymbol{x}_m^n(t)$ given in Eq. (15.39), the average transmit power constraint dictates that

$$\frac{1}{NT}\sum_{t=1}^T \alpha^2(t) \sum_{n=1}^N \left\|\Delta\boldsymbol{\theta}_m^n(t)\right\|_2^2 \leq \bar{P}, \quad \forall m \in [M]. \tag{15.40}$$

The received signal at the $k$th PS antenna is then given by

$$\boldsymbol{y}_k^n(t) = \alpha(t)\sum_{m=1}^M \boldsymbol{h}_{m,k}^n(t) \circ \Delta\boldsymbol{\theta}_m^n(t) + \boldsymbol{z}_k^n(t), \quad k \in [K], n \in [N]. \tag{15.41}$$

Having access to $\widehat{\boldsymbol{h}}_k^n(t)$ as the CSI, the PS performs the following beamforming aiming to recover $\Delta\boldsymbol{\theta}(t)$:

$$\boldsymbol{y}^n(t) \triangleq \frac{1}{K}\sum_{k=1}^K \left(\sum_{m=1}^M \widehat{\boldsymbol{h}}_{m,k}^n(t)\right)^* \circ \boldsymbol{y}_k^n(t), \quad n \in [N]. \tag{15.42}$$

By replacing $\boldsymbol{y}_k^n(t)$ from Eq. (15.41) and $\widehat{\boldsymbol{h}}_{m,k}^n(t)$ from Eq. (15.35), the $i$th entry of $\boldsymbol{y}^n(t)$ is given by, for $i \in [s], n \in [N]$,

$$y_i^n(t) = \underbrace{\alpha(t) \sum_{m=1}^{M} \left( \frac{1}{K} \sum_{k=1}^{K} |h_{m,k,i}^n(t)|^2 \right) \Delta \theta_{m,i}^n(t)}_{\text{signal term}}$$

$$+ \underbrace{\frac{\alpha(t)}{K} \sum_{m=1}^{M} \sum_{k=1}^{K} \left( \sum_{m'=1, m' \neq m}^{M} h_{m',k,i}^n(t) + \tilde{h}_{k,i}^n(t) \right)^* h_{m,k,i}^n(t) \Delta \theta_{m,i}^n(t)}_{\text{interference term}}$$

$$+ \underbrace{\frac{1}{K} \sum_{k=1}^{K} \left( \sum_{m=1}^{M} h_{m,k,i}^n(t) + \tilde{h}_{k,i}^n(t) \right)^* z_{k,i}^n(t)}_{\text{noise term}}. \tag{15.43}$$

We observe from this that $y_i^n(t)$ includes three terms, namely signal, interference, and noise terms, respectively. As $K \to \infty$, according to the law of large numbers, the signal term approaches

$$y_{i,\text{sig}}^n(t) \triangleq \alpha(t) \sigma^2 \sum_{m=1}^{M} \Delta \theta_{m,i}^n(t), \quad i \in [s], n \in [N], \tag{15.44}$$

through which the PS can recover

$$\frac{1}{M} \sum_{m=1}^{M} \Delta \theta_{m, 2(n-1)s+i}(t) = \frac{\text{Re}\left\{ y_{i,\text{sig}}^n(t) \right\}}{M \sigma^2 \alpha(t)}, \tag{15.45a}$$

$$\frac{1}{M} \sum_{m=1}^{M} \Delta \theta_{m, (2n-1)s+i}(t) = \frac{\text{Im}\left\{ y_{i,\text{sig}}^n(t) \right\}}{M \sigma^2 \alpha(t)}, \tag{15.45b}$$

which corresponds to recovering $\Delta \boldsymbol{\theta}(t)$. However, the exact recovery of $\Delta \boldsymbol{\theta}(t)$ from $\boldsymbol{y}^n(t)$ is not possible because of the interference and noise terms, as specified in Eq. (15.43). The interference term in Eq. (15.43) is represented by

$$y_{i,\text{itf}}^n(t) = \alpha(t) \sum_{m=1}^{M} \mathfrak{h}_{m,i}^n(t) \Delta \theta_{m,i}^n(t), \quad i \in [s], n \in [N], \tag{15.46}$$

where we have defined

$$\mathfrak{h}_{m,i}^n(t) \triangleq \frac{1}{K} \sum_{k=1}^{K} \left( \sum_{m'=1, m' \neq m}^{M} h_{m',k,i}^n(t) + \tilde{h}_{k,i}^n(t) \right)^* h_{m,k,i}^n(t), \quad m \in [M]. \tag{15.47}$$

It is easy to obtain the mean and the variance of $\mathfrak{h}_{m,i}^n(t)$ as

$$\mathbb{E}\left[ \mathfrak{h}_{m,i}^n(t) \right] = 0, \tag{15.48a}$$

$$\mathbb{E}\left[ |\mathfrak{h}_{m,i}^n(t)|^2 \right] = \frac{\left( (M-1)\sigma^2 + \tilde{\sigma}^2 \right) \sigma^2}{K}, \tag{15.48b}$$

respectively. We note that the local model updates at each iteration are independent of the channel realization at the same iteration. According to this analysis and Eq. (15.46), the interference term includes $M$ components, each with zero-mean and a variance scaled with $M/K$. Thus, having a fixed number of devices $M$, the impact of the interference term can be mitigated when $K \gg M^2$. In Section 15.6, we illustrate the performance of this approach for a finite number of PS antennas.

Following the analysis in Eq. (15.45), the PS estimates $\frac{1}{M} \sum_{m=1}^{M} \Delta \theta_{m,2(n-1)s+i}(t)$ and $\frac{1}{M} \sum_{m=1}^{M} \Delta \theta_{m,(2n-1)s+i}(t)$ through

$$\Delta \hat{\theta}_{2(n-1)s+i}(t) = \frac{\text{Re}\left\{y_i^n(t)\right\}}{M \sigma^2 \alpha(t)}, \tag{15.49a}$$

$$\Delta \hat{\theta}_{(2n-1)s+i}(t) = \frac{\text{Im}\left\{y_i^n(t)\right\}}{M \sigma^2 \alpha(t)}, \tag{15.49b}$$

respectively, for $i \in [s]$, $n \in [N]$. Accordingly, it obtains an unbiased estimate of the local model updates average as $\Delta \hat{\boldsymbol{\theta}}(t) \triangleq \left[\Delta \hat{\theta}_1(t), \ldots, \Delta \hat{\theta}_d(t)\right]^T$, which is used to update the global model as in Eq. (15.9).

The interested readers are referred to [25] for the convergence analysis of the algorithm for FEEL with OAC without CSI at the devices and imperfect CSI at the PS.

*Remark 15.5* The blind FEEL algorithm with OAC we have presented provides a framework for estimating the average of the signals transmitted over the wireless fading MAC with no CSI at the devices and imperfect CSI at the PS. We can employ the sparsification and linear projection techniques presented in Section 15.4 to reduce the dimension of the vector transmitted over the wireless MAC at each device.

## 15.6 Numerical Experiments

Here we evaluate the performance of the algorithms presented in this chapter for image classification experiments on the MNIST dataset [33] with 60,000 training and 1,000 test samples and the CIFAR-10 [34] dataset with 50,000 training and 1,000 test samples using ADAM optimizer [35]. We measure the performance of each experiment with respect to the test dataset, referred to as the *test accuracy*, versus the iteration count, $t$.

We consider i.i.d. and non-i.i.d. data distributions across the devices reflecting the bias in local data. In the i.i.d. case, we assume that data samples from the training dataset are distributed across the devices at random with respect to the label/classes. This provides a homogeneous data distribution where the distribution of local dataset at each device may resemble the distribution of the whole dataset. On the other hand, in the non-i.i.d. case, we distribute the training data samples such that each device has access to the samples from only two labels/classes with $B_m/2$ samples from a single label/class at device $m$, $m \in [M]$. This introduces bias in local datasets, and it may be more aligned with a practical federated learning setting.

First, we evaluate the performance of the digital and uncoded FEEL algorithms presented in Sections 15.3 and 15.4, respectively, for $M = 25$ devices in the system. For this experiment, we train a single layer neural network with the MNIST dataset, and we measure the performance as test accuracy with respect to the normalized training time captured by $Nt$. We consider local dataset size $B_1 = \cdots = B_M = B = 10^3$, where the training data samples are randomly distributed across the devices for both i.i.d. and non-i.i.d. scenarios, in which case some data samples may not be allocated to any device, and there may be overlap among the data available at different devices. We set the number of local SGD iteration at each device to $\tau = 1$ and the number of mini-batch size for each SGD local iteration to $|\xi_m^i(t)| = B, \forall m, i, t$. We consider channel variance $\sigma^2 = 1$ and $s = \lceil d/20 \rceil$ subchannels of the wireless MAC. For the FEEL algorithm with OAC, we compress the local model updates to have length $\tilde{s} = 2s = d/10$, which results in $N = 1$, and we set the sparsity level to $k = \lfloor \tilde{s}/2.5 \rfloor$ and the threshold value to $\lambda(t) = 10^{-3}$. Furthermore, we consider $\bar{P}_m^n(t) = \bar{P}, \forall m, n, t$, for the FEEL algorithm with OAC, and $\bar{P}_{m^*}(t) = \bar{P}$ for the digital FEEL algorithm. This would satisfy the average power constraint with equality, and for the OAC approach, we can determine the scaling factor $\gamma_m^n(t)$ through Eq. (15.28).

We investigate the performance of the digital FEEL approach using two alternative quantization techniques, namely, QSGD [36] and SignSGD [37], for compression of the local model update at the participating device at each iteration. With the QSGD compression technique, each entry of the local model update is mapped to one of its nearest quantization levels with a probability scaling linearly with its distance to that level. On the other hand, with the SignSGD compression technique, only the signs of the local model update entries are transmitted by the participating device. We further consider the OAC approach introduced in [21], referred to as broadband analog FEEL, where devices transmit their local model updates entirely over the wireless MAC in an uncoded fashion without employing any compression technique. We note that, in our system model where the wireless fading MAC has only $s$ subchannels, this approach requires transmission of the local model updates over $N = \lceil d/2s \rceil$ time slots at each iteration. Finally, we include the performance of the error-free shared link scenario as a benchmark, where we assume that the accurate average of the local model updates $\Delta \theta(t)$ is available at the PS after a single time slot during each communication round; that is, $N = 1$.

In Figs. 15.2 and 15.3, we illustrate the performance of different schemes for i.i.d. and non-i.i.d. data cases, respectively, for $\bar{P} = 20$. We observe that the FEEL algorithm with OAC outperforms all other algorithms under consideration for both i.i.d. and non-i.i.d. data scenarios. For the i.i.d. case, the gap between FEEL with OAC and the error-free shared link approach is relatively small, which shows the efficiency of the proposed OAC approach in providing an accurate estimate of the average local model updates at the PS despite the compression technique employed using similarity in the sparsity patterns of the local model updates across devices. This gap increases in the non-i.i.d. case, which is due to the heterogeneity in local data providing a more biased local model updates across devices impacting the convergence speed of FEEL with OAC. However, even for the non-i.i.d. case, we observe that the FEEL with OAC

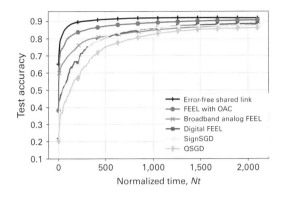

**Figure 15.2** Test accuracy for i.i.d. MNIST data for $M = 25$, $B = 10^3$, and $\bar{P} = 20$.

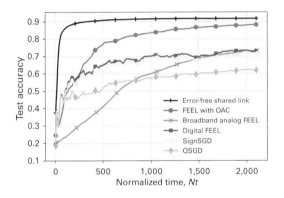

**Figure 15.3** Test accuracy for non-i.i.d. MNIST data for $M = 25$, $B = 10^3$, and $\bar{P} = 20$.

provides a final accuracy level with a relatively small difference from the error-free shared link approach, which shows the robustness of the OAC scheme against the bias in local data. We observe that the proposed OAC scheme outperforms the broadband analog FEEL approach with a significant improvement over the non-i.i.d. data case. This indicates that updating the global model more frequently using a less accurate estimate of the average of local model updates obtained as a result of the compression technique introduced together with the OAC approach provides a gain comparable to updating the global model less frequently using a more accurate estimate of the average of local model updates.

Furthermore, the OAC scheme, which benefits from the additive nature of the wireless MAC to provide the PS with the average of the local model updates, significantly improves upon the digital schemes under consideration, where the devices compress their messages separately, aiming to deliver an accurate copy of their local model updates accommodating their limited channel capacities. It can be seen that the digital FEEL scheme outperforms the SignSGD and QSGD schemes, and the performances of the digital schemes, particularly QSGD, deteriorate significantly by introducing bias in the local data. The readers are referred to [4, 23] for more experimental results

**Table 15.1.** Different layers of CNN for image classification on CIFAR-10.

| |
|---|
| $3 \times 3$ convolutional layer, 32 channels, ReLU activation, same padding |
| $3 \times 3$ convolutional layer, 32 channels, ReLU activation, same padding |
| $2 \times 2$ max pooling |
| dropout with probability 0.2 |
| $3 \times 3$ convolutional layer, 64 channels, ReLU activation, same padding |
| $3 \times 3$ convolutional layer, 64 channels, ReLU activation, same padding |
| $2 \times 2$ max pooling |
| dropout with probability 0.3 |
| $3 \times 3$ convolutional layer, 128 channels, ReLU activation, same padding |
| $3 \times 3$ convolutional layer, 128 channels, ReLU activation, same padding |
| $2 \times 2$ max pooling |
| dropout with probability 0.4 |
| softmax output layer with 10 units |

investigating the impact of various setting parameters, such as the average transmit power, number of devices, and number of MAC subchannels, as well as the case of imperfect CSI at the devices on the performance of different algorithms.

Next, we investigate the performance of the blind OAC algorithm considering the impact of the number of PS antennas on mitigating the destructive effects of channel, which cannot be cancelled at the devices due to the lack of CSI where the PS has imperfect CSI. We train a convolutional neural network (CNN), the architecture of which is outlined in Table 15.1, using the CIFAR-10 dataset, and the performance is measured as test accuracy versus iteration count, $t$, for a total of $T = 400$ iterations. We consider $M = 20$ devices in the system, and we assume i.i.d. data distribution, where the dataset is randomly split into $M$ disjoint groups of data, and each group of data is allocated to a distinct device. We further consider channel gain variance $\sigma^2 = 1$ and set $\tau = 5$ and $\left|\xi_m^i(t)\right| = 500$, $\forall m, i, t$. We set the scaling factor at the devices to $\alpha(t) = 1 + 10^{-3}t, t \in [T]$. For simplicity, we assume that $s = d/2$ resulting in $N = 1$ and that the CSI estimation error $\tilde{h}_{k,i}^n(t)$ is i.i.d. according to $(0, \tilde{\sigma}^2)$, $\forall k, i, n, t$.

In Figs. 15.4 and 15.5 we illustrate the impact of number of antennas on the performance of the blind OAC algorithm for the cases of perfect and imperfect CSI at the PS, respectively, setting $K \in \{M, 2M, 5M, 10M, 2M^2\}$. For the case of imperfect CSI at the PS, we consider the variance of the CSI estimation error as $\tilde{\sigma}^2 = M\sigma^2/2$.

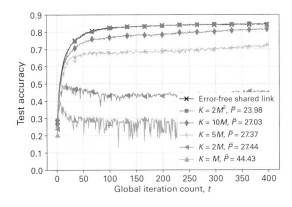

**Figure 15.4** Test accuracy of the blind OAC algorithm for i.i.d. CIFAR-10 data with different number of antennas $K \in \{M, 2M, 5M, 10M, 2M^2\}$ for $M = 20$, and perfect CSI at PS, $\tilde{\sigma}^2 = 0$

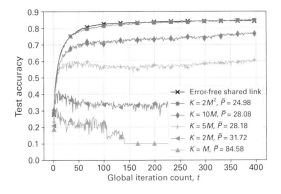

**Figure 15.5** Test accuracy of the blind OAC algorithm for i.i.d. CIFAR-10 data with different number of antennas $K \in \{M, 2M, 5M, 10M, 2M^2\}$ for $M = 20$ and imperfect CSI at PS, $\tilde{\sigma}^2 = M\sigma^2/2$

We observe that for both cases of the perfect and imperfect CSI at the PS, increasing $K$ can provide significant performance improvement without any visible gap to the error-free shared link scenario for a large enough $K$ set as $K = 2M^2$. This indicates that the negative effects of the channel due to the lack of CSIT can be alleviated by equipping the PS with multiple antennas even when the PS does not have perfect CSI. As can be seen, increasing $K$ provides more gain when having imperfect CSI at the PS. We highlight that, when all the other parameters are fixed, the average transmit power reduces with $K$. This is due to the fact that increasing $K$ provides a faster convergence rate, which leads to a faster reduction in the empirical variances of the local model updates at the devices. Please refer to [24, 25] for more experimental results considering different noise levels at the PS as well as the MNIST dataset for training.

## 15.7     Conclusion

In this chapter, we presented a new approach to distributed edge learning that can exploit the superposition property of the wireless medium, rather than mitigating it. This novel approach allows us to exploit the wireless medium to carry out computations as well as to transfer information. In particular, we studied FEEL where wireless devices use their local datasets to train a global model, which is updated at the PS and shared with the devices. We considered power- and bandwidth-limited transmissions from the devices to the PS over a wireless fading MAC and showed how OAC can be exploited to improve the efficiency and accuracy of FEEL. We also considered the digital transmission of local model updates as a benchmark for the performance of FEEL in conventional communication networks. The results presented in this chapter clearly show that FEEL with OAC outperforms the digital approach, which treats computation and communication separately.

We also observed that the local model updates at the devices can be compressed thanks to the similarity in their sparsity patterns without any significant performance degradation, which is particularly compelling for transmission over bandwidth-limited channels, which is often the case for IoT applications.

We also considered transmission of model updates to the PS over the wireless MAC without any CSI at the devices and demonstrated that receive beamforming at the multiantenna PS can mitigate the adverse effects of channel on the performance, even with imperfect CSI. The proposed OAC approach mainly exploits the fact that in distributed learning with SGD, due to the iterative nature of the optimization, we do not need accurate computations at each iteration. SGD is inherently noisy due to stochastic gradient estimation carried out at each device at each iteration. Adding noisy computation on top of the gradient estimation error does not prevent approaching the optimal solution. Note that the alternative digital approach would accurately compute the average of the transmitted gradient estimated, but would have to compress them heavily due to the limited channel resources available to each transmitting device.

Another benefit of OAC in the FEEL context is its inherent privacy preservation. The PS receives only the sum of the model updates, together with the channel noise. As it is shown in [38–41], this aspect of OAC can be used to reduce the amount of noise that the devices need to introduce on their model updates to provide differential privacy guarantees to the users, significantly improving the final accuracy of the trained models without sacrificing privacy.

## References

[1] D. Gündüz et al., "Machine learning in the air," *IEEE Journal on Selected Areas in Communications*, vol. 37, no. 10, pp. 2184–2199, 2019.

[2] B. McMahan and D. Ramage, "Federated learning: Collaborative machine learning without centralized training data," Google AI Blob, Apr. 2017; https://ai.googleblog.com/2017/04/federated-learning-collaborative.html.

[3] J. Park et al., "Wireless network intelligence at the edge," *arXiv preprint*, arXiv:1812.02858, 2018.

[4] M. M. Amiri and D. Gündüz, "Machine learning at the wireless edge: Distributed stochastic gradient descent over-the-air," *IEEE Trans. Signal Processing*, vol. 68, pp. 2155–2169, Apr. 2020.

[5] G. Zhu et al., "Toward an intelligent edge: Wireless communication meets machine learning," *IEEE Communications Magazine*, vol. 58, no. 1, pp. 19–25, 2020.

[6] M. Chen et al., "A joint learning and communications framework for federated learning over wireless networks," *arXiv preprint*, arXiv:1909.07972, Oct. 2020.

[7] H. H. Yang et al., "Scheduling policies for federated learning in wireless networks," *IEEE Trans. on Communications*, vol. 68, no. 1, pp. 317–333, 2020.

[8] M. S. H. Abad et al., "Hierarchical federated learning across heterogeneous cellular networks," in *IEEE International Conference on Acoustics, Speech and Signal Processing (ICASSP)*, pp. 8866–8870, 2020.

[9] D. Gündüz et al., "Communicate to learn at the edge," *IEEE Communications Magazine*, vol. 58, no. 12, pp. 14–19, 2020.

[10] N. H. Tran et al., "Federated learning over wireless networks: Optimization model design and analysis," in *IEEE Conf. on Computer Communications (INFOCOM)*, pp. 1387–1395, 2019.

[11] M. M. Amiri et al., "Convergence of update aware device scheduling for federated learning at the wireless edge," *arXiv preprint*, arXiv:2001.10402, 2020.

[12] M. Salehi and E. Hossain, "Federated learning in unreliable and resource-constrained cellular wireless networks," *arXiv preprint*, arXiv cs.LG.2012.05137, 2020.

[13] J. Ren et al., "Scheduling for cellular federated edge learning with importance and channel awareness," *arXiv preprint*, arXiv cs.IT.2004.00490, 2020.

[14] W.-T. Chang and R. Tandon, "Communication efficient federated learning over multiple access channels," *arXiv preprint*, arXiv cs.IT.2001.08737, 2020.

[15] B. Nazer and M. Gastpar, "Computation over multiple-access channels," *IEEE Trans. on Information Theory*, vol. 53, no. 10, pp. 3498–3516, 2007.

[16] R. Soundararajan and S. Vishwanath, "Communicating linear functions of correlated Gaussian sources over a MAC," *IEEE Trans. on Information Theory*, vol. 58, no. 3, pp. 1853–1860, 2012.

[17] B. Nazer and M. Gastpar, "Compute-and-forward: Harnessing interference through structured codes," *IEEE Trans. on Information Theory*, vol. 57, no. 10, pp. 6463–6486, 2011.

[18] D. Gündüz et al., "Multiple multicasts with the help of a relay," *IEEE Trans. on Information Theory*, vol. 56, no. 12, pp. 6142–6158, 2010.

[19] M. Nokleby and B. Aazhang, "Cooperative compute-and-forward," *IEEE Trans. on Wireless Communications*, vol. 15, no. 1, pp. 14–27, 2016.

[20] M. Goldenbaum and S. Stanczak, "Robust analog function computation via wireless multiple-access channels," *IEEE Trans. on Communications*, vol. 61, no. 9, pp. 3863–3877, Sep. 2013.

[21] G. Zhu, Y. Wang, and K. Huang, "Broadband analog aggregation for low-latency federated edge learning," *IEEE Trans. on Wireless Communications*, vol. 19, no. 1, pp. 491–506, Jan. 2020.

[22] K. Yang et al., "Federated learning via over-the-air computation," *IEEE Trans. on Wireless Communications*, vol. 19, no. 3, pp. 2022–2035, 2020.

[23] M. M. Amiri and D. Gündüz, "Federated learning over wireless fading channels," *IEEE Trans. on Wireless Communications*, vol. 19, no. 5, pp. 3546–3557, May 2020.

[24] M. M. Amiri, T. M. Duman, and D. Gündüz, "Collaborative machine learning at the wireless edge with blind transmitters," in *Proc. IEEE Global Conf. on Signal and Information Processing (GlobalSIP)*, pp. 1–5, Nov. 2019.

[25] M. M. Amiri et al., "Blind federated edge learning," *arXiv preprint*, arXiv cs.IT.2010.10030, 2020.

[26] D. G. G. Zhu, Y. Du and K. Huang, "One-bit over-the-air aggregation for communication-efficient federated edge learning: Design and convergence analysis," *IEEE Trans. on Wireless Communications*, 2021.

[27] J. Ahn, O. Simeone, and J. Kang, "Wireless federated distillation for distributed edge learning with heterogeneous data," in *IEEE Int. Symp. on Personal, Indoor and Mobile Radio Communications (PIMRC)*, pp. 1–6, 2019.

[28] Y. Sun, S. Zhou, and D. Gündüz, "Energy-aware analog aggregation for federated learning with redundant data," in *IEEE Int. Conf. on Communications (ICC)*, pp. 1–7, 2020.

[29] M. M. Amiri et al., "Convergence of update aware device scheduling for federated learning at the wireless edge," *arXiv preprint*, arXiv:2001.10402 [cs.IT], Jan. 2020.

[30] D. Tse and P. Viswanath, *Fundamentals of Wireless Communication*. Cambridge University Press, 2005.

[31] F. Sattler et al., "Sparse binary compression: Towards distributed deep learning with minimal communication," in *Proc. Int. Joint Conf. on Neural Networks (IJCNN)*, July 2019.

[32] T. Weber, A. Sklavos, and M. Meurer, "Imperfect channel-state information in MIMO transmission," *IEEE Trans. on Communications*, vol. 54, no. 3, pp. 543–552, March 2006.

[33] Y. LeCun, C. Cortes, and C. Burges, "The MNIST database of handwritten digits," database, 1998; http://yann.lecun.com/exdb/mnist/.

[34] A. Krizhevsky and G. Hinton, "Learning multiple layers of features from tiny images," tech. report, University of Toronto, 2009.

[35] D. P. Kingma and J. Ba, "Adam: A method for stochastic optimization," *arXiv preprint*, arXiv:1412.6980 [cs.LG], Jan. 2017.

[36] D. Alistarh et al., "QSGD: Communication-efficient SGD via randomized quantization and encoding," in *Proc. Advances in Neural Information Processing Systems*, pp. 1709–1720, Dec. 2017.

[37] J. Bernstein et al., "signSGD: Compressed optimisation for non-convex problems," *arXiv preprint*, arXiv:1802.04434 [cs.LG], Aug. 2018.

[38] M. Seif, R. Tandon, and M. Li, "Wireless federated learning with local differential privacy," in *2020 IEEE Int. Symp. on Information Theory (ISIT)*, pp. 2604–2609, 2020.

[39] Y. Koda et al., "Differentially private aircomp federated learning with power adaptation harnessing receiver noise," *arXiv preprint*, arXiv:2004.06337, 2020.

[40] D. Liu and O. Simeone, "Privacy for free: Wireless federated learning via uncoded transmission with adaptive power control," *arXiv preprint*, arXiv:2006.05459, 2020.

[41] B. Hasircioglu and D. Gündüz, "Private wireless federated learning with anonymous over-the-air computation," *arXiv preprint*, arXiv:2011.08579, 2020.

# 16 Federated Knowledge Distillation

Hyowoon Seo, Jihong Park, Seungeun Oh, Mehdi Bennis,
and Seong-Lyun Kim

## 16.1 Introduction

Machine learning is one of the key building blocks in 5G and beyond [1–3], spanning a broad range of applications and use cases. In the context of mission-critical applications [2, 4], machine learning models should be trained with fresh data samples that are generated by and dispersed across edge devices (e.g., phones, cars, access points, etc.). Collecting these raw data incurs significant communication overhead, which may violate data privacy. In this regard, *federated learning (FL)* [5–8] is a promising communication-efficient and privacy-preserving solution that periodically exchanges local model parameters, without sharing raw data. However, exchanging model parameters is extremely costly under modern deep neural network (NN) architectures that often have a huge number of model parameters. For instance, MobileBERT is a state-of-the-art NN architecture for on-device natural language processing (NLP) tasks, with 25 million parameters corresponding to 96 MB [9]. Training such a model by exchanging the 96 MB payload per communication round is challenging particularly under limited wireless resources.

The aforementioned limitation of FL has motivated to the development of *federated distillation (FD)* [10] based on exchanging only the local model outputs with dimensions that are commonly much smaller than the model sizes (e.g., 10 labels in the MNIST dataset). To illustrate, as shown in Fig. 16.1, consider a two-label classification example wherein each worker in FD runs local iterations with samples having either back or grey ground-truth label. For each training sample, the worker generates its prediction output distribution, termed a *local logit* that is a softmax output vector of the last NN layer activations (e.g., $\{black, grey\} = \{0.7, 0.3\}$ for a black sample). At a regular interval, the generated local logits of the worker are averaged per ground-truth label and uploaded to a parameter server for aggregating and globally averaging the *local average logits* across workers per ground-truth label. The resultant *global average logits* per ground-truth label are downloaded by each worker. Finally, to transfer the downloaded global knowledge into local models, each worker updates its model parameters by minimizing its own loss function, in addition to a regularizer that penalizes larger gap between its own logit of a given sample and the global average logit for the given sample's ground truth.

The overarching goal of this chapter is to provide a deep understanding of FD and show the effectiveness of FD as a communication-efficient distributed learning

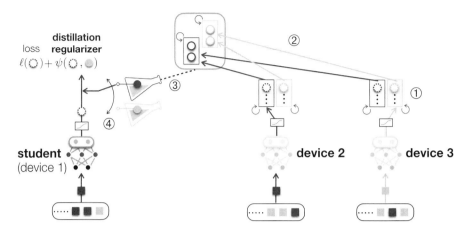

**Figure 16.1** Federated distillation (FD) with three devices and two labels in a classification task.

framework that is applicable to a variety of tasks. To this end, the rest of this chapter is organized into three parts. To demystify the operational principle of FD, by exploiting the theory of neural tangent kernel (NTK) [11], the first part in Section 16.2 provides a novel asymptotic analysis for two foundational algorithms of FD: knowledge distillation (KD) and codistillation (CD). Next, the second part in Section 16.3 elaborates on a baseline implementation of FD for a classification task and illustrates its performance in terms of accuracy and communication efficiency compared to FL. Lastly, to demonstrate the applicability of FD to various distributed learning tasks and environments, the third part presents two selected applications: FD over asymmetric uplink-and-downlink wireless channels and FD for reinforcement learning in Sections 16.4 and 16.5, respectively, followed by concluding remarks in Section 16.6.

## 16.2     Preliminaries: Knowledge Distillation and Codistillation

FD is built upon two basic algorithms. One is KD that transfers a pretrained teacher model's knowledge into a student model [12], whereas the other is an online version of KD without pretraining the teacher model, called CD [13]. Although KD has widely been used in practice since its inception, its fundamentals have not been fully understood up until now. Only a handful works [14–16] have attempted to analyze KD and its convergence, using the recently proposed NTK technique [11] as we will review in the first part of this section. Leveraging and extending this NTK framework, in the second part, we will provide a novel NTK analysis of the convergence of CD.

### 16.2.1     Knowledge Distillation

Knowledge distillation (KD), illustrated in Fig. 16.2, aims to imbue an empty student model with a teacher's knowledge [12]. In a classification task, KD is different from

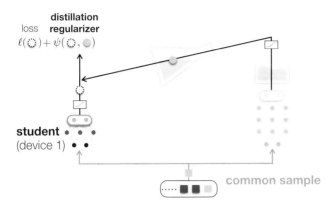

**Figure 16.2** Knowledge distillation (KD) from a pretrained teacher to a student model.

the standard model training that attempts to match a target model's one-hot prediction (e.g., [cat, dog] = [0,1]) of each unlabeled sample with its ground-truth label. Instead, KD tries to match the target model's output layer activation, that is, logit[1](e.g., [cat, dog] = [0.3, 0.7]), with the teacher's logit for the same sample. This logit contains more information than its one-hot prediction, thereby training the student model faster than the standard training with much fewer samples [15].

The teacher's knowledge of KD can be constructed in various ways. Typically, the knowledge is a pretrained teacher model's logit, which is transferred to a small-sized student model for model compression [12]. The knowledge can also be an ensemble of other student model's logits [13] in that the ensemble of predictions is often more accurate than individual predictions. Leveraging this, one can train a student model by transferring the ensemble of other student models' logits. Indeed, CD and FD utilize this key idea for enabling KD-based distributed learning without the need for any pretraining operations, to be elaborated in Sections 16.2.2 and 16.3.

Given the aforementioned teacher's knowledge, what the student model knows after KD can be clarified through the lens of NTK, a recently developed kernel method to asymptotically analyze an over-parameterized NN in an infinite width regime [11]. To illustrate, we consider a simple three-layer student NN model comprising input, hidden, and output layers with $M_i > 0$, $M_h \to \infty$ (i.e., infinite width), and $M_o = 1$ neurons, respectively. These layers are fully connected, and a nonlinear activation function is applied to the hidden layer. In a classification task, the input data tuple $\{(\mathbf{x}_i, y_i)\}_{i=1}^n$ consists of an unlabeled data sample $\mathbf{x}_i$ and its ground-truth label $y_i$. For a given input sample $\mathbf{x}_i$, the prediction output $\hat{y}_i$ of the student NN is represented by the function $f(\mathbf{x}_i)$ as follows:

---

[1] KD originally aims to match the softmax activation function of the student's logit with the temperature softmax activation function of the teacher's logit [12]. Recent KD works have also considered various activation functions of logits, such as margin rectifier linear unit (ReLU) and attention [17]. In this chapter, we consider the same activation functions as in [12], and for the sake of convenience, we hereafter call this functional output as logit.

$$\hat{y}_i = f(\mathbf{x}_i) = \frac{1}{\sqrt{M_h}} \sum_{m=1}^{M_h} a_m f_m(\mathbf{x}_i) \in \mathbb{R}, \tag{16.1}$$

where $\sigma(\cdot)$ is a real and nonlinear activation function, $f_m(\mathbf{x}_i) = \sigma(\mathbf{w}_m^T \mathbf{x}_i)$ is the $m$-th activation of the hidden layer, and $\{\mathbf{w}_m\}_{m=1}^{M_h}$ are the weights connecting the input and hidden layers. For the given NN architecture, the logit vector is the hidden layer activations $\{f_m(\mathbf{x}_i)\}_{m=1}^{M_h}$, of which the entries are linearly combined with the weight parameters $\{a_m\}$, resulting in the prediction output $\hat{y}_i$ of the student model.

In KD, the student model updates its weights $\{\mathbf{w}_m\}$ by minimizing its own loss function and a distillation regularizer that penalizes the student when the logit gap between the student and teacher is large. Applying the mean squared error function[2] to both loss function and regularizer, the problem of KD is cast as

$$\min_{\{\mathbf{w}_m\}} \underbrace{\sum_i (y_i - \hat{y}_i)^2}_{\text{loss}} + \lambda \underbrace{\sum_i \sum_m (\phi_m(\mathbf{x}_i) - f_m(\mathbf{x}_i))^2}_{\text{distillation regularizer}}, \tag{16.2}$$

where $\lambda > 0$ is a constant hyperparameter and $\{\phi_m(\mathbf{x}_i)\}_{m=1}^{M_h}$ are pretrained teacher model's logits and $\{f_m(\mathbf{x}_i)\}_{m=1}^{M_h}$ are student's logits. Note the number of logits at both teacher and student are assumed to be the same.

To solve Eq. (16.2), following the standard NTK settings [11, 14], we use the gradient descent algorithm with an infinitesimal step size. This results in the convergence of a trajectory of the discrete algorithm to a smooth curve modeled by a continuous-time differential equation as

$$\frac{d}{dt}\mathbf{w}_m(t) = \mathbf{L}_m(t)\left[ \frac{a_m}{\sqrt{M_h}}(\mathbf{y} - \hat{\mathbf{y}}(t)) + \lambda(\boldsymbol{\phi}_m - \mathbf{f}_m(t)) \right], \tag{16.3}$$

where $\mathbf{y}$ and $\hat{\mathbf{y}}(t)$ are respectively the vectors of the ground truth labels and the prediction outputs at time $t$, and $\boldsymbol{\phi}_m$ and $\mathbf{f}_m(t)$ are respectively the vectors of the teacher model's $m$th logit and the student model's $m$th logit at time $t$. The matrix $\mathbf{L}_m(t)$ consists of $\sigma'(\mathbf{w}_m^T(t)\mathbf{x}_i)\mathbf{x}_i$ as its $i$th column, where $\sigma'(\cdot)$ is the first derivative of the activation, which is also assumed to be Lipschitz continuous.

Generally, the dynamics of the weights described in Eq. (16.3) are hard to analyze, yet we can still analyze the dynamics of the logits based on the following relation:

$$\frac{d}{dt}\mathbf{f}_m(t) = \mathbf{L}_m^T(t)\frac{d}{dt}\mathbf{w}_m(t) \tag{16.4}$$

$$= \mathbf{H}_m(t)\left[ \frac{a_m}{\sqrt{M_h}}(\mathbf{y} - \hat{\mathbf{y}}(t)) + \lambda(\boldsymbol{\phi}_m - \mathbf{f}_m(t)) \right], \tag{16.5}$$

where $\mathbf{H}_m(t) = \mathbf{L}_m^T(t)\mathbf{L}_m(t)$ is often called an NTK [11].

---

[2] In KD under classification tasks, it is common to use the cross entropy functions for the loss and distillation regularizer. For the sake of the mathematical tractability, following [14], we consider the mean squared error functions for the loss and regularizer during the NTK analysis, while considering the cross entropy functions for the rest of this chapter.

Empirically, in a network with a large number of parameters, it is observed that every weight vector along the trajectory of gradient descent algorithm is static over time and stays close to its initialization. Based on such an interesting observation, the theory of NTK establishes that the over-parametrization and random initialization jointly induce a *kernel regime* (i.e., $\mathbf{H}_m(t) \approx \mathbf{H}_m(0)$ for $t \geq 0$ [11, 18]), thereby giving rise to simpler dynamics under the negligible effect of $\mathbf{H}_m(t)$ on Eq. (16.5).

*Remark 16.1* (Theorem 1 in [14])   In the kernel regime, under mild assumptions on the eigenvalues of the matrices $\{\mathbf{H}_m(0)\}_{m=1}^{M_h}$ at initialization, bounded inputs and bounded weights, it can be shown that the student NN output vector $\mathbf{f}(t)$, which is the vector of $\{f(\mathbf{x}_i)\}_{i=1}^n$, converges asymptotically as

$$\lim_{t \to \infty} \mathbf{f}(t) = \mathbf{f}_\infty = \frac{1}{a+\lambda} \left( a\mathbf{y} + \lambda \sum_{m=1}^{M_h} \frac{a_m \phi_m}{\sqrt{M_h}} \right). \tag{16.6}$$

*Proof*   Based on the observation that the behavior of gradient descent on the over-parametrized NN can be approximated by a linear dynamics of finite order, the evolution of $\mathbf{f}(t)$ can be expressed as

$$\mathbf{f}(t) = \mathbf{f}_\infty + \mathbf{u}_1 e^{-p_d t} + \mathbf{u}_1 e^{-p_d t} + \cdots + \mathbf{u}_d e^{-p_d t}, \tag{16.7}$$

where $d$ is the order of the linear system, and complex-valued vectors $\mathbf{u}_1, \ldots \mathbf{u}_d$ are determined by the dynamics. Moreover, the nonzero complex-values $p_1, \ldots, p_d$ are the poles that correspond to the singular points of the Laplace transform of $\mathbf{f}(t)$. In [14], it is shown in detail that all existing poles are positive-valued under mild assumptions, such that $\mathbf{f}(t) \to \mathbf{f}_\infty$ for $t \to \infty$. □

Consequently, as shown by Eq. (16.6), the student model after KD outputs a weighted sum of the ground truth $\mathbf{y}$ and the teacher's prediction $\sum_m a_m \phi_m$. Then, the student's prediction error compared to $\mathbf{y}$ can be represented as

$$\|\mathbf{f}_\infty - \mathbf{y}\|_2 = \frac{\lambda}{a+\lambda} \left\| \mathbf{y} - \sum_m \frac{a_m \phi_m}{\sqrt{M_h}} \right\|_2. \tag{16.8}$$

This implies that the student's prediction error decreases as the pretrained teacher's prediction $\sum_m \frac{a_m \phi_m}{\sqrt{M_h}}$ approaches to $\mathbf{y}$ (i.e., an ideally trained teacher).

## 16.2.2   Codistillation

KD postulates a pretrained teacher model that hinders distributed learning operations. However, CD, which is an online version of KD, obviates the need for the pretrained teacher model [13]. The key idea of CD, illustrated in Fig. 16.3, is to treat an ensemble of multiple models' prediction outputs as the teacher's knowledge, which is often more accurate than the individual prediction outputs [13, 19]. To this end, each worker (i.e., student model), sees the ensemble of the other $C - 1$ workers as a virtual teacher.

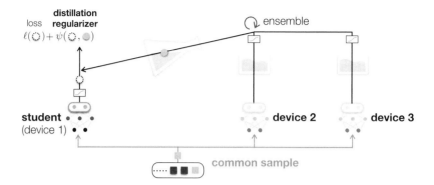

**Figure 16.3** Codistillation (CD) among three student models without any pretrained teacher model.

Consequently, the problem of CD is given by recasting Eq. (16.2) of KD as follows:

$$\min_{\{\mathbf{w}_m^1\},...,\{\mathbf{w}_m^C\}} \sum_c \left( \underbrace{\sum_i (y_i - \hat{y}_i^c)^2}_{\text{loss}} + \lambda \underbrace{\sum_i \sum_m \left( \frac{1}{C-1} \sum_{c' \neq c} f_m^{c'}(\mathbf{x}_i) - f_m^c(\mathbf{x}_i) \right)^2}_{\text{distillation regularizer}} \right),$$

(16.9)

where $\hat{y}_i^c$ is the prediction output of the $c$th worker, $\{\mathbf{w}_m^c\}$ is its weight parameters, and $\{f_m^c(\cdot)\}$ is its logits. Here, the pretrained teacher's logit $\phi_m(\mathbf{x}_i)$ of KD in Eq. (16.2) is replaced with the ensemble logit $\frac{1}{C-1}\sum_{c' \neq c} f_m^{c'}(\mathbf{x}_i)$ of $C-1$ workers in CD. Note that in Eq. (16.9) of CD is formulated for all $C$ workers, rather than considering each worker separately. This problem is more challenging than KD, in that the teacher's knowledge becomes dependent on each worker (due to exclusion) and all the other workers (due to averaging).

*Remark 16.2*   Given the aforementioned interactions across workers, based on analysis in the kernel regime, it can be shown that the output of the workers converges to the ground-truth asymptotically as

$$\lim_{r \to \infty} \mathbf{f}^c(r) = \mathbf{y},$$

(16.10)

for all $c \in \{1, \ldots, C\}$, where $\mathbf{f}^c(r)$ is the output of the worker $c$ after local training with $r$th global update (or communication round).

*Proof*   Without loss of generality, we hereafter focus only on the first worker out of $C$ workers whose models are identically structured and independently initialized. After initialization and local training for warm-up, the workers share the first updates; that is, $\mathbf{f}^1(0), \ldots, \mathbf{f}^C(0)$. Then, each worker locally and iteratively runs GD with regularization until convergence. According to the result in Eq. (16.6) from KD, the output of the worker 1 converges to

$$\mathbf{f}^1(1) = \frac{1}{a + \lambda} \left( a\mathbf{y} + \frac{\lambda}{C - 1} \sum_{c=2}^{C} \mathbf{f}^c(0) \right). \tag{16.11}$$

Thus, the output of the model $\mathcal{C}_1$ after $r$th updates will converge to

$$\mathbf{f}^1(r) = \frac{1}{a + \lambda} \left( a\mathbf{y} + \frac{\lambda}{C - 1} \sum_{c=2}^{C} \mathbf{f}^c(r - 1) \right) \tag{16.12}$$

$$= \frac{1}{a + \lambda} \left( a\mathbf{y} + \lambda \frac{\sum_{c=2}^{C} \left( a\mathbf{y} + \frac{\lambda}{C-1} \sum_{c' \neq c} \mathbf{f}^{c'}(r - 2) \right)}{(C - 1)(a + \lambda)} \right) \tag{16.13}$$

$$= \frac{1}{a + \lambda} \left( a\mathbf{y} + \lambda \frac{(C-1)a\mathbf{y} + \lambda\mathbf{f}^1(r-2) + \frac{\lambda(C-2)}{C-1} \sum_{c=2}^{C} \mathbf{f}^c(r-2)}{(C - 1)(a + \lambda)} \right). \tag{16.14}$$

By introducing $\mathbf{v}_{r+1} = \frac{\lambda}{C-1} \sum_{c=2}^{C} \mathbf{f}^c(r) = (a + \lambda)\mathbf{f}^1(r + 1) - a\mathbf{y}$, we can simplify Eq. (16.14) to a linear nonhomogeneous recurrence relation:

$$\mathbf{v}_{r+1} = \frac{(C - 2)\lambda}{(C-1)(a+\lambda)} \mathbf{v}_r + \frac{\lambda^2}{(C-1)(a+\lambda)^2} \mathbf{v}_{r-1} + \frac{\lambda^2 a\mathbf{y}}{(C-1)(a+\lambda)^2} + \frac{\lambda a\mathbf{y}}{(a+\lambda)}, \tag{16.15}$$

for $r \geq 2$. By solving the above recurrence relation [20], we obtain the closed-form solution

$$\mathbf{v}_r = \alpha \left( \frac{\lambda}{a + \lambda} \right)^r + \beta \left( -\frac{\lambda}{(C - 1)(a + \lambda)} \right)^r + \lambda\mathbf{y}, \tag{16.16}$$

where $\alpha = \frac{\lambda}{C} \sum_{c=1}^{C} \mathbf{f}^c(0) - \lambda\mathbf{y}$ and $\beta = \frac{\lambda}{C(C-1)} \sum_{c=2}^{C} \mathbf{f}^c(0) - \frac{\lambda}{C}\mathbf{f}^1(0)$. Note that for $r \to \infty$,

$$\lim_{r \to \infty} \mathbf{v}_r = \lambda\mathbf{y}, \tag{16.17}$$

since $|\frac{\lambda}{a+\lambda}| < 1$ and $|-\frac{\lambda}{(C-1)(a+\lambda)}| < 1$ for $C \geq 2$. Consequently, we can see that the output of the worker 1 converges to the ground truth as

$$\lim_{r \to \infty} \mathbf{f}^1(r) = \frac{1}{a + \lambda} (a\mathbf{y} + \lambda\mathbf{y}) = \mathbf{y}. \tag{16.18}$$

In the same way, the result in Eq. (16.18) of the worker 1 can be extended to any worker with the same conclusion. This ends the proof of Remark 16.2. □

Such a result in Remark 16.2 is remarkable in that CD achieves zero prediction error that is achievable under KD only when the teacher model is ideally pre-trained as shown in Eq. (16.8). This result highlights the importance of continual training that allows workers to reach the maximum prediction capability, as opposed to KD that is additionally guided by a pretrained yet fixed teacher model.

Lastly, it is notable that more workers yield faster convergence of CD. In essence, the convergence is achieved by eliminating the first two terms in the right-hand side

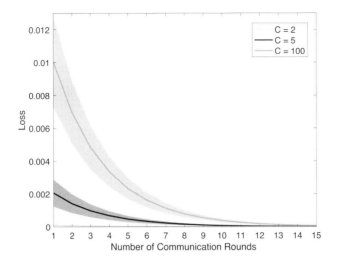

**Figure 16.4** Learning curves of CD with $C \in \{2, 5, 100\}$ workers, capturing the loss converging to zero as the number $r$ of communication rounds increases.

of Eq. (16.16). These two terms decrease not only with the number of communication rounds $r$ but also with the number of workers $C$. This implies that with more workers one needs less communications until convergence. Furthermore, we conceive that when $C \to \infty$, only one communication round can achieve convergence, enabling few-shot CD.

Figure 16.4 corroborates the aforementioned theoretical results by numerical evaluations of CD for a simple classification task considering 10 classes of samples labeled 0–9, generated with an arbitrary mapping function which is unknown to the workers. The result shows that as the number $C$ of workers grows, the convergence speed of CD increases while the variance reduces. Furthermore, as expected by the theoretical result in Eq. (16.18), numerical simulations validate that even with $C = 2$, CD is guaranteed to converge. Lastly, for $C = 100$, one can achieve convergence with only a few communication rounds, verifying the feasibility of few-shot CD.

## 16.3        Federated Distillation

CD has a great potential in enabling fast distributed learning with high accuracy as demonstrated in the previous section, yet its communication efficiency is still questionable. The fundamental reason traces back to KD that requires common training sample observations by both student and teacher models. For an online version of KD, this implies that all workers should observe the same sample per each loss calculation, requiring extensive sample exchanges that may also violate local data privacy. Eliminating such a dependency on common sample observations is the key motivation for developing FD, as elaborated next.

## 16.3.1    Federated Distillation for Classification

In a classification task, FD avoids the aforementioned problem of common sample observations in CD by grouping samples according to labels, thereby extending CD to a communication-efficient distributed learning framework. As depicted by Fig. 16.1, the operations of FD are summarized by the following four steps:

1. Each worker stores a mean logit vector per label during local training.
2. Each worker periodically uploads its *local-average logit vectors* to a parameter server averaging the uploaded local-average logit vectors from all workers separately for each label.
3. Each worker downloads the constructed *global-average logit vectors* of all labels from the server.
4. During local training based on KD, each worker selects its teacher's logit as the downloaded global-average logit associated with the same label as the current training sample's ground-truth label.

In what follows, we describe the details of FD operations. Similar to CD, we consider that the worker $c \in \{1, \ldots, C\}$ has $n$ observed samples with ground-truth label, $\{(\mathbf{x}_i^c, y_i^c)\}_{i=1}^n$, but independently observed at each worker. For the sake of simplicity, assume $\mathcal{Y} = \{1, 2, \ldots, |\mathcal{Y}|\}$ to be an alphabet of $|\mathcal{Y}|$ labels under consideration and define an index set $\mathcal{I}_\ell^c$, that is composed of $\ell$-labeled sample indices at the worker $c$, where $|\mathcal{I}_\ell^c| = n_\ell^c$ and $\sum_\ell n_\ell^c = n$. Under such circumstances, FD aims to solve the following optimization problem:

$$\min_{\{\mathbf{w}_m^1\}, \ldots, \{\mathbf{w}_m^C\}} \sum_c \sum_\ell \left( \underbrace{\sum_{i \in \mathcal{I}_\ell^c} (y_i - \hat{y}_i^c)^2}_{\text{loss}} + \lambda \underbrace{\sum_{i \in \mathcal{I}_\ell^c} \sum_m \left( \frac{1}{C-1} \sum_{c' \neq c} \bar{f}_{m,l}^{c'} - f_m^c(\mathbf{x}_i^c) \right)^2}_{\text{distillation regularizer}} \right),$$

(16.19)

where $f_m^c(\cdot)$ is the $m$-the logit of the worker $c$ as before, $\bar{f}_{m,l}^c = \frac{1}{n_l^c} \sum_{i \in \mathcal{I}_\ell^c} f_m^c(\mathbf{x}_i^c)$ is the local average of the worker $c$'s $m$th logit for the samples labeled $l$.

Following the aforementioned four-step operations, FD solves in Eq. (16.19) using Algorithm 16.1. Notations are summarized as follows. The set $\mathcal{S}^c$ denotes the training dataset of the worker $c$, and $\mathcal{B}$ represents a set of sample indices drawn as a batch per worker during the local training phase. The function $F^c(\cdot)$ is a logit vector, made by vectorizing the logits $\{f_m^c(\cdot)\}$. The function $\mathcal{L}(p, q)$ is a quadratic loss function, measuring the mean squared error between $p$ and $q$, which is used for both loss function and distillation regularizer. Note that the quadratic loss can be replaced with any other well-defined loss function, such as cross entropy. As opposed to the asymptotic analysis, we consider a constant learning rate $\eta$ for practicality, and $\lambda$ is a weighting constant for the distillation regularizer. At the $c$th worker, $\bar{F}_{\ell,r}^c$ is the local-average logit vector at the $r$th iteration when the training sample belongs to the $\ell$th ground-truth label, $\hat{F}_{\ell,r}^c$ is the global-average logit vector that equals

---

**Algorithm 16.1** Federated Distillation (FD).

---

**Require:** Prediction: $f(\mathbf{x})$, Ground-truth label: $y$, Loss function: $\mathcal{L}(f(\mathbf{x}), y)$

1: **while** not converged **do**

2:     **procedure** LOCAL TRAINING PHASE (at worker $\forall c \in \{1, \dots, C\}$)

3:         **for** $k$ steps **do** : $\mathcal{B} \leftarrow \mathcal{S}^c$

4:             **for** sample $\mathbf{x}_b$ and label $y_b$, for $b \in \mathcal{B}$ **do**

5:                 $\mathbf{w}^c \leftarrow \mathbf{w}^c - \eta \nabla \{ \mathcal{L}(f^c(\mathbf{x}_b), y_b) + \lambda \cdot \mathcal{L}(F^c(\mathbf{x}_b), \hat{F}^c_{y_b, r}) \}$

6:                 $F^c_{y_b, r} \leftarrow F^c_{y_b, r} + F^c(\mathbf{x}_b), \mathsf{cnt}^c_{y_b, r} \leftarrow \mathsf{cnt}^c_{y_b, r} + 1$

7:             **for** label $\ell = 1, 2, \cdots, |\mathcal{Y}|$ **do**

8:                 $\bar{F}^c_{\ell, r} \leftarrow F^c_{\ell, r} / \mathsf{cnt}^c_{\ell, r}$ : **return** $\bar{F}^c_{\ell, r}$ to server

9:     **procedure** GLOBAL ENSEMBLING PHASE (at the server)

10:         **for** each worker $c = 1, 2, \dots, C$ **do**

11:             **for** label $\ell = 1, 2, \dots, |\mathcal{Y}|$ **do**

12:                 $\bar{F}_{\ell, r} \leftarrow \bar{F}_{\ell, r} + \bar{F}^c_{\ell, r}$

13:         **for** each worker $c = 1, 2, \dots, C$ **do**

14:             **for** label $\ell = 1, 2, \dots, |\mathcal{Y}|$ **do**

15:                 $\hat{F}^c_{\ell, r+1} \leftarrow \bar{F}_{\ell, r} - \bar{F}^c_{\ell, r}, \hat{F}^c_{\ell, r+1} \leftarrow \frac{\hat{F}^c_{\ell, r+1}}{(C-1)}$ : **return** $\hat{F}^c_{\ell, r}$ to worker $c$

    end while

---

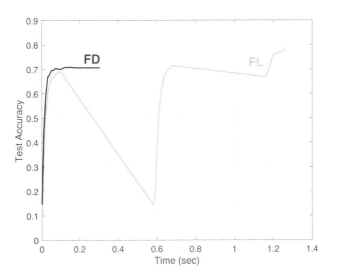

**Figure 16.5** Learning curves of FD and FL with two workers for the MNIST classification.

$\hat{F}^c_{\ell, r} = \sum_{c' \neq c} \bar{F}^{c'}_{\ell, r} / (C - 1)$ with $C$ workers, and $\mathsf{cnt}^c_{\ell, c}$ counts the number of samples that have a ground-truth label of $\ell$.

Figure 16.5 shows the numerical evaluations of FD for the MNIST (handwritten 0–9 images) classification task. The result illustrates that FD achieves 4.3× faster convergence than FL while compromising less than 10 percent accuracy, under a

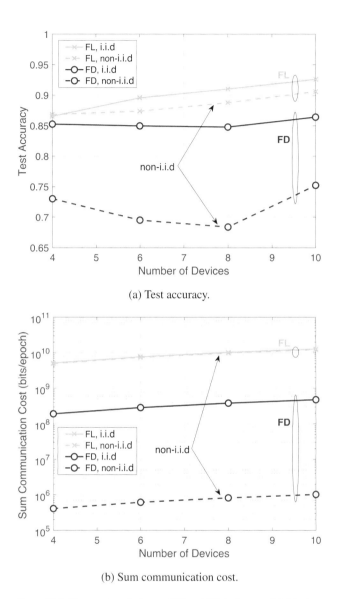

(a) Test accuracy.

(b) Sum communication cost.

**Figure 16.6** Comparison between FD and FL in terms of (a) test accuracy and (b) sum communication cost of all workers per epoch, under an i.i.d. and non-i.i.d. MNIST data.

five layer convolutional NN operated by two workers (see more details in [10]). To see the effectiveness of FD in a more generic scenario, Fig. 16.6 considers up to 10 workers, and both cases of an independent and identically distributed (i.i.d.) local dataset and a non-i.i.d. dataset with local data samples that are imbalanced across labels. The result shows that for different numbers of workers, FD can always reduce around 10,000× communication payload sizes per communication round compared to FL. Considering both fast convergence and payload size reduction, FD reduces the total communication cost until convergence by more than $40,000\times$ compared to FL. Nonetheless, FD still comes at the cost of compromising accuracy, particularly under non-i.i.d. data distributions.

## 16.3.2    Recent Progress and Future Direction

The aforementioned implementation of Vanilla FD focuses only on reducing communication payload sizes in a classification task at the cost of sacrificing accuracy. Several recent works have substantiated the communication efficiency of FD under more realistic wireless environments without compromising accuracy for applications beyond classiffication, which we review next.

- *FD Over Wireless* – FD is a communication-efficient distributed learning framework, as demonstrated by achieving a $40,000\times$ lower total communication cost than FL for an image classification task in the previous section. The communication efficiency of FD also holds under wireless fading channels [21–23]. Even with low signal-to-noise ratio and/or bandwidth, the payload size reduction of FD can be turned into more successful receptions and/or lower latency, resulting in even higher accuracy than FL [22, 23]. It could be interesting to see the effectiveness of FD under more realistic wireless environments with advanced physical-layer and multiple-access techniques such as time-varying millimeter-wave channels, reconfigurable intelligent surfaces, nonorthogonal multiple access, and many more.
- *Communication Efficiency versus Accuracy* – FD is more vulnerable to the problem of non-i.i.d. data distributions than FL. Even if a worker obtains the global average logits for all labels, when the worker lacks samples of a specific target class, the global knowledge is rarely transferred into the worker's local model. Furthermore, in many cases [10, 21, 22, 24], the communication efficiency of FD comes at the cost of compromising accuracy, yielding the trade-off between FD and FL. Given the trade-off between FD's higher communication efficiency and FL's higher accuracy, it is possible to utilize both of their strengths by taking into account the nature of uplink-downlink asymmetric channels. As shown in [21, 25], one can exploit FL in the downlink and FD in the uplink with a capacity that is much less than the downlink due to the low transmission energy at the devices, to be further discussed in Section 16.4.
- *Proxy Data Aided FD* – Recent works have overcome the aforementioned limitations of FD (i.e., accuracy degradation particularly under non-i.i.d. data distributions). The core idea is to additionally construct a common proxy dataset (e.g., a public dataset [22] or mean samples per label [23]) through which the local KD operations and the local logits to be uploaded are provided. In fact, as opposed to FL that exchanges each worker's freshest model updated right before uploading, FD is based on exchanging the locally averaged logits during which each worker's model is progressively updated. To resolve this issue, workers can collectively construct a global proxy dataset by averaging all data samples per label, referred to as global average covariate vectors in [22] or by using a prearranged public

dataset [23]. Utilizing such a proxy dataset, one can generate the local logits to be exchanged right before uploading, thereby distilling the knowledge from the freshest models. Furthermore, operating KD through the proxy dataset makes all workers observe the same samples, thereby avoiding any possible errors induced by coarse sample grouping in the original FD. Consequently, as demonstrated in [23], such proxy dataset aided FD can achieve higher accuracy than FL even under non-i.i.d. local data distributions. Extending this line of research, it could be worth investigating how to construct the proxy dataset using a coreset, a small dataset approximating the original data distribution [25, 26].

- *FD Beyond Classification* – The applicability of FD is not limited to classification tasks in supervised learning. As shown by [24], FD can be applied to an reinforcement learning (RL) application by replacing the label-wise sample grouping of the original FD with clustering based on the neighboring states (e.g., locations) of RL agents, to be further elaborated on in Section 16.5. In unsupervised learning, it could be possible to collectively train multiple conditional generative adversarial networks (cGANs) [27] using FD by exchanging their discriminators' last layer activations that are grouped based on the common conditions of cGANs. Last but not least, in self-supervised learning, one could exploit FD to train multiple bootstrap your own latent (BYOL) networks, each of which comprises a pair of online and target models [28], by constructing each target model's prediction based on an ensemble of the last layer activations of online models.

## 16.4    Application: FD Under Uplink-Downlink Asymmetric Channels

Despite the communication efficiency afforded by FD in the distributed learning framework, there still remains an accuracy issue especially under communication-limited scenarios. In a typical wireless communication network, the uplink communication is more limited by lower transmission power and smaller available bandwidth than the downlink [29], which we refer to as uplink-downlink channel asymmetry. Thus, for FD-based distributed learning built over wireless networks, a large accuracy loss of model training is inevitable, since FD goes through a number of communication rounds for model training over both uplink and downlink channels.

In this context, as an advanced form of FD, the *Mix2FLD* achieves both high accuracy and communication efficiency under the uplink-downlink channel asymmetry. As depicted in Fig. 16.7, Mix2FLD is built upon two key algorithms: *federated learning after distillation (FLD)* [24] and *Mixup* data augmentation [30]. Specifically, by leveraging FLD, each worker in Mix2FLD uploads its local model outputs as in FD and downloads model parameters as in FL, thereby coping with the uplink-downlink channel asymmetry. Between the uplink and downlink, the server runs KD. However, this output-to-model conversion requires additional training samples collected from workers, which may violate local data privacy while incurring huge communication overhead. To preserve data privacy with minimal communication overhead during

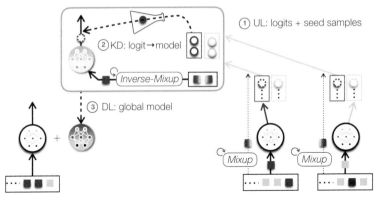

(a) **Mix2FLD**: downlink federated learning (FL) and uplink federated distillation (FD) with two-way Mixup (Mix2up) seed sample collection.

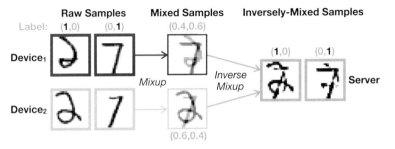

(b) **Mix2up**: mixing raw samples at workers and inversely mixing them across different workers at the server (mixing ratio $\gamma = 0.4$).

**Figure 16.7** (a) Mix2FLD operation and (b) Mix2up.

seed sample collection, Mix2FLD utilizes a *two-way Mixup algorithm (Mix2up)*, as illustrated in Fig. 16.7(b). To hide raw samples, each worker in Mix2up uploads locally superposed samples using Mixup. Next, before running KD at the server, the uploaded mixed-up samples are superposed across different workers in a way that the resulting sample labels are in the same form as raw sample labels. This inverse-Mixup provides more realistic synthetic seed samples for KD, without restoring raw samples. Furthermore, with the uploaded mixed samples from the workers, a larger number of inversely mixed-up samples can be generated, thereby enabling KD with minimal uplink cost. In the following subsections, we first elaborate on a baseline method, MixFLD that combines FLD and Mixup, and then describe Mix2FLD that integrates MixFLD with the inverse-Mixup.

## 16.4.1 Baseline: MixFLD

MixFLD integrates FLD with Mixup, within which FLD counteracts the uplink-downlink channel asymmetry as elaborated next. Following FLD, as shown in Fig. 16.7(a), at the $r$th global update, the workers upload their local average logit vectors, thereby constructing a global average logit vector at the server, as in FD.

Then, the workers download the global weight vector as in FL. To this end, the server must convert the global logit average vector into the global weight vector, since it lacks one. The key idea is to transfer the knowledge in the global average logit vector to a global model. To enable this, at the beginning of FLD, each worker uploads $n_{\text{mix}}$ seed samples randomly selected from its local dataset. By feeding the collected $Cn_{\text{mix}}$ seed samples, denoted by $\{\mathbf{x}_{s,i}\}_{i=1}^{Cn_{\text{mix}}}$, the server runs $K_s$ iterations of SGD with KD, thereby updating the global model's weight vector $\mathbf{w}_{g,k}$ as

$$\mathbf{w}_{g,k+1} = \mathbf{w}_{g,k} - \eta \cdot \nabla \left( \mathcal{L}(f_g(\mathbf{x}_{s,i}), \mathbf{y}_{s,i}) + \lambda \cdot \mathcal{L}(F_g(\mathbf{x}_{s,i}), \hat{F}_{y_{s,i},r}) \right), \qquad (16.20)$$

where $f_g(\cdot)$ is the function denoting the global NN at the server and $F_g$ is the corresponding global model logit vector. As defined earlier, $\mathcal{L}(\cdot, \cdot)$ is a well-designed loss function such as quadratic loss or cross entrophy, and $\hat{F}_{l,r}$ is the global average logit vector for $l$-labeled samples at the $r$th global update, which is obtained by averaging local logit vectors uploaded from the workers. Finally, the server yields the global model $\mathbf{w}_{g,K_s}$ that is downloaded by every worker. The remaining operations follow the same procedure of FL.

The aforementioned FLD operations include seed sample collection process that may incur nonnegligible communication overhead while violating local data privacy. To mitigate this problem, MixFLD applies Mixup before collection [25, 30] to the sample collection procedure of FLD as follows. Before uploading the seed samples, the worker $c$ randomly selects two different raw samples $\mathbf{x}_i^c$ and $\mathbf{x}_j^c$ with $i \neq j$, having the ground-truth labels $y_i^c$ and $y_j^c$, respectively. With a mixing ratio $\gamma \in (0, 0.5]$ given identically for all workers, the worker linearly combines these two samples (see Fig. 16.7(b)), thereby generating a mixed-up sample $\hat{\mathbf{x}}_{ij}^c$ as

$$\hat{\mathbf{x}}_{ij}^c = \gamma \mathbf{x}_i^c + (1 - \gamma) \mathbf{x}_j^c. \qquad (16.21)$$

In Eq. (16.21), the label is also mixed up as $\hat{\mathbf{y}}_{ij}^c = \gamma \mathbf{y}_i^c + (1 - \gamma) \mathbf{y}_j^c$. Then, each worker uploads the generated $n_{\text{mix}}$ mixed-up samples to the server without revealing raw samples.

The guaranteed privacy level can be quantified through the lens of $(\varepsilon, \delta)$-differential privacy [31], in which lower $\epsilon, \delta > 0$ preserves more privacy by making it difficult to guess with less confidence whether or not a certain data point is included in a private dataset. For the sake of the analysis, we consider that each worker selects two samples uniformly at random out of $n$ samples and mixes them with $\gamma = 0.5$, followed by inserting additive zero-mean Gaussian noises to $\hat{\mathbf{x}}_{ij}^c$ and $\hat{\mathbf{y}}_{ij}^c$ with the variances $\sigma_x^2$ and $\sigma_y^2$, respectively. When generating $n_{\text{mix}}$ samples at each worker, according to Theorem 3 in [32], the aforementioned Mixup is $(\varepsilon, \delta)$-differentially private, where

$$\varepsilon = \frac{2n_{\text{mix}}\Delta^2}{8n} \left( 1 + \sqrt{\frac{4n \log(1/\delta)}{\Delta^2 n_{\text{mix}}}} \right) + \sqrt{\frac{\Delta^2 n_{\text{mix}} \log(1/\delta)}{4n}}. \qquad (16.22)$$

The term $\Delta^2$ is given as $\Delta^2 = d_x/\sigma_x^2 + d_y/\sigma_y^2$, where $d_x$ and $d_y$ are the sample and label dimensions (e.g., for the $28 \times 28$ pixel MNIST images of handwritten

0–9 digits, $d_x = 28 \times 28 = 784$ and $d_y = 10$). As observed by $\varepsilon$ decreasing with $n$ in Eq. (16.22), Mixup can guarantee the raw sample privacy as long as the local dataset size is sufficiently large. Recall that this differential privacy analysis is based on $\gamma = 0.5$ and additive noises. For more general cases under $\gamma > 0$ without additive noise, we numerically evaluate the sample privacy by measuring the similarity between the raw and mixed-up samples in Section 16.4.3.

## 16.4.2   Proposed: Mix2FLD

While MixFLD preserves local data privacy during seed sample collection, the Mixup operations may significantly distort the collected seed samples, which may hinder achieving high accuracy. To resolve this issue, Mix2FLD additionally applies the inverse-Mixup algorithm to MixFLD, thereby not only ensuring local data privacy but also achieving high accuracy. For the sake of clear explanation, we hereafter focus on a two-worker setting, where workers $c$ and $c'$ independently mix up the following two raw samples having symmetric labels:

- Worker $c$: $\mathbf{x}_i^c$ with $\mathbf{y}_i^c = \{1, 0\}$ and $\mathbf{x}_j^c$ with $\mathbf{y}_j^c = \{0, 1\}$,
- Worker $c'$: $\mathbf{x}_{i'}^{c'}$ with $\mathbf{y}_{i'}^{c'} = \{0, 1\}$ and $\mathbf{x}_{j'}^{c'}$ with $\mathbf{y}_{j'}^{c'} = \{1, 0\}$,

where $\mathbf{y}_i^c$ is a one-hot encoded ground-truth label vector of $\mathbf{x}_i^c$, referred to as a *hard label*. Following Eq. (16.21), worker $c$ mixes up local samples $\mathbf{x}_i^c$ and $\mathbf{x}_j^c$, yielding the mixed-up sample $\hat{\mathbf{x}}_{ij}^c$ corresponding to the mixed-up label $\{\gamma, 1 - \gamma\}$, referred to as its *soft label*. Likewise, worker $c'$ superpositions $\mathbf{x}_{i'}^{c'}$ and $\mathbf{x}_{j'}^{c'}$, resulting in the mixed-up sample $\hat{\mathbf{x}}_{i'j'}^{c'}$ having the soft label $\{1 - \gamma, \gamma\}$. The workers $c$ and $c'$ upload $\hat{\mathbf{x}}_{ij}^c$ and $\hat{\mathbf{x}}_{i'j'}^{c'}$ with their soft labels to the server.

Then, the server in Mix2FLD converts the soft labels back into hard labels, such that the converted samples contain more similar features of the hard-labeled real dataset, while being still different from the raw samples. To this end, the server applies the *inverse-Mixup* that linearly combines $n_s$ mixed-up samples such that the resulting sample has a hard label. For the case of $C = 10$ workers, as depicted in Fig. 16.7(b), with the symmetric setting, the server combines $\hat{\mathbf{x}}_{ij}^c$ and $\hat{\mathbf{x}}_{i'j'}^{c'}$, such that the resulting $\tilde{\mathbf{x}}_{ij,i'j',l}^{cc'}$ has the $l$th converted hard label as the ground truth. This is described as

$$\tilde{\mathbf{x}}_{ij,i'j',l}^{cc'} = \hat{\gamma}\hat{\mathbf{x}}_{ij}^c + (1 - \hat{\gamma})\hat{\mathbf{x}}_{i'j'}^{c'}. \tag{16.23}$$

The inverse mixing ratio $\hat{\gamma}$ for $n_s = 2$ is chosen in the following way. Suppose the target hard label is $\{1, 0\}$, or $l = 1$. Applying $\{1, 0\}$ to the LHS of Eq. (16.23) and $\{\gamma, 1 - \gamma\}$ and $\{1 - \gamma, \gamma\}$ of $\hat{\mathbf{x}}_{ij}^c$ and $\hat{\mathbf{x}}_{i'j'}^{c'}$ to the right-hand side of Eq. (16.23) yields two equations:

$$1 = \hat{\gamma}\gamma + (1 - \hat{\gamma})(1 - \gamma) \tag{16.24}$$

$$0 = \hat{\gamma}(1 - \gamma) + (1 - \hat{\gamma})\gamma \tag{16.25}$$

Solving these equations yields the desired $\hat{\gamma}$. By induction, this can be generalized to $n_s > 2$.

---

**Algorithm 16.2** FLD with Mix2up (**Mix2FLD**).

**Require:** $\mathcal{S}^c$ with $c \in \{1, \ldots, C\}$, $\gamma \in (0, 1)$

1: **while** not converged **do**
2:      **procedure** LOCAL TRAINING AND MIXUP(at worker $c \in \{1, \ldots, C\}$)
3:          **if** $r = 1$ **generates** $\{\hat{\mathbf{x}}_{ij}^c\}$ via (16.21) **end if**        ▷ *Mixup*
4:          **updates** $\mathbf{w}^c$ and $\bar{F}_{l,r}^c$ for $K$ iterations as in FD (**Algorithm 16.1**)
5:          **unicasts** $\{\bar{F}_{l,r}^c\}$ (with $\{\hat{\mathbf{x}}_{ij}^c\}$ if $r = 1$) to the server
6:      **procedure** ENSEMBLING AND OUTPUT-TO-MODEL CONVERSION(at server)
7:          **if** $r = 1$ **generates** $\{\widetilde{\mathbf{x}}_{ij,i'j',l}^{cc'}\}$ via (16.23) **end if**     ▷ *Inverse-Mixup*
8:          **computes** $\{\hat{F}_{l,r}^c\}$
9:          **updates** $\mathbf{w}_{g,k}$ via (16.20) for $K_s$ iterations
10:         **broadcasts** $\mathbf{w}_{g,K_s}$ to all devices
11:      $r \leftarrow r + 1$
12:      Worker $c \in \{1, \ldots, C\}$ **substitutes** $\mathbf{w}_0^c$ with $\mathbf{w}_{g,K_s}$     ▷ *Model download*
     end while

---

Hereafter, for the sake of convenience, we explain the rest of the algorithm considering $n_s = 2$. By alternating $\hat{y}$ with $l = 1$ and 2, inversely mixing up two mixed-up samples $\hat{\mathbf{x}}_{ij}^c$ and $\hat{\mathbf{x}}_{i'j'}^{c'}$ yields two inversely mixed-up samples $\widetilde{\mathbf{x}}_{ij,i'j',1}^{cc'}$ and $\widetilde{\mathbf{x}}_{ij,i'j',2}^{cc'}$. The server generates $n_{inv}$ inversely mixed-up samples by pairing two samples with symmetric labels among $n_{mix}$ mixed-up samples. By nature, inverse-Mixup is a data augmentation scheme, so $n_{inv}$ can be larger than $n_{mix}$. Note that none of the raw samples are identical to inversely mixed-up samples. To ensure this, inverse-Mixup is applied only for the seed samples uploaded from different devices, thereby preserving data privacy. The overall operation of Mix2FLD is summarized in Algorithm 16.2.

## 16.4.3   Numerical Evaluation and Discussions

In what follows, we provide a numerical performance evaluation of Mix2FLD compared with FL, FD, and MixFLD, in terms of the test accuracy and convergence time of a randomly selected reference device, under different data distributions (i.i.d. and non-i.i.d.) and uploaded/generated seed sample configurations: $(n_{mix}, n_{inv}) \in \{(10, 10), (10, 20), (50, 50), (50, 100)\}$. The convergence time includes communication delays during the uplink and downlink, as well as the computing delays of devices and the server, measured using tic-toc elapsed time.

Every device has a three-layer convolutional NN model (two convolutional layers and one fully connected layer) having 12,544 model parameters in total. The server's global model follows the same architecture. Each worker owns its local MNIST dataset with $|\mathcal{Y}| = 10$ classes and $n = 500$ samples. For the i.i.d. case, every label has the same number of samples. For the non-i.i.d. case, two randomly selected labels have two samples, respectively, while each of the other labels has 62 samples. Other simulation parameters for model training are given as $C = 10$, $K = 6{,}400$ iterations,

$K_s = 3,200$ iterations, and $\eta = 0.01$. The simulation parameters for reflecting wireless environment is given as the same as in [21].

The *impact of channel conditions* is illustrated in Fig. 16.8. The results show that Mix2FLD achieves the highest accuracy with moderate convergence under

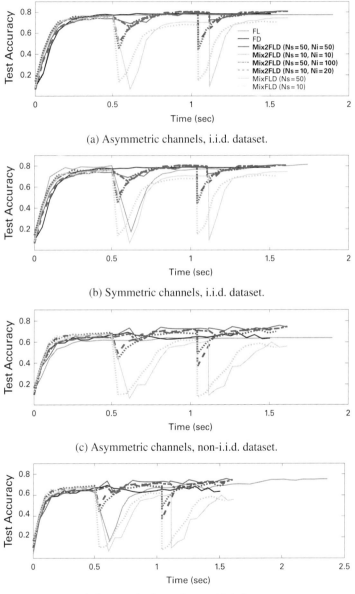

(a) Asymmetric channels, i.i.d. dataset.

(b) Symmetric channels, i.i.d. dataset.

(c) Asymmetric channels, non-i.i.d. dataset.

(d) Symmetric channels, non-i.i.d. dataset.

**Figure 16.8** Learning curves of a randomly selected device in Mix2FLD, compared to FL, FD, and MixFLD, under asymmetric and symmetric channels, when $\gamma = 0.1$ with i.i.d. and non-i.i.d. datasets.

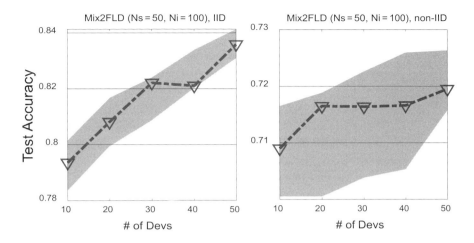

**Figure 16.9** Test accuracy distribution of Mix2FLD with respect to the number of devices, under symmetric channels with i.i.d. and non-i.i.d. datasets.

asymmetric channel conditions among others. Compared to FL uploading model weights, Mix2FLD's model output uploading reduces the uplink payload size by up to 42.4 times. Under asymmetric channels with the limited uplink capacity (Figs. 16.8(a) and 16.8(c)), this enables more frequent and successful uploading, thereby achieving up to 16.7 percent higher accuracy and 1.2 times faster convergence. Compared to FD, Mix2FLD leverages the high downlink capacity for downloading the global model weights, which often provides higher accuracy than downloading model outputs as reported in [10]. In addition, the global information of Mix2FLD is constructed by collecting seed samples and reflecting the global data distribution, rather than by simply averaging local outputs as used in FD. Thereby, Mix2FLD achieves up to 17.3 percent higher accuracy while taking only 2.5 percent more convergence time than FD. Under symmetric channels, FL achieves the highest accuracy. Nevertheless, Mix2FLD still converges 1.9 times faster than FL, thanks to its smaller uplink payload sizes and more frequent updates.

Next, the *impact of the number of devices* is observed in Fig. 16.9. When the number of devices is increased from 10 to 50, the average of test accuracy increases by 5.7 percent and the variance decreases by 50 percent with the i.i.d. dataset. In the non-i.i.d. dataset, the test accuracy gain is smaller than that of the i.i.d. dataset, but has the same tendency. This concludes that Mix2FLD is scalable under both i.i.d. and non-i.i.d. data distributions.

Furthermore, the *effectiveness of Mix2up* is depicted in Figs. 16.8(c) and 16.8(d), corroborating that Mix2FLD is particularly effective in coping with non-i.i.d. data. In our non-i.i.d. datasets, samples are unevenly distributed, and locally trained models become more biased, degrading accuracy compared to i.i.d. datasets in Figs. 16.8(a) and 16.8(c). This accuracy loss can partly be restored by additional global training (i.e., output-to-model conversion) that reflects the entire dataset distribution using few seed samples. While preserving data privacy, MixFLD attempts to realize this idea.

However, as observed in Fig. 16.8(d), MixFLD fails to achieve high accuracy as its mixed-up samples inject too much noise into the global training process. Mix2FLD resolves this problem by utilizing inversely mixed up samples, reducing unnecessary noise. Thanks to its incorporating the data distribution, even under symmetric channels (Fig. 16.8(d)), Mix2FLD achieves higher accuracy than FL. One drawback of Mix2up is its relying on an $n_s \times n_s$ matrix inversion for inverting $n_s$ linearly mixed-up samples, which may hinder the scalability of Mix2FLD for large $n_s$. Alternatively, as demonstrated in [33], one can exploit the bit-wise XOR operation and its flipping property (e.g., $(A \oplus B) \oplus B = A$) replacing mixup and inverse-mixup, respectively, thereby avoiding the matrix inversion complexity.

Lastly, the trade-offs among latency, privacy, and accuracy are illustrated in Fig. 16.8. For all the considered channel conditions and data distributions, in Mix2FLD and MixFLD, reducing the seed sample amount ($n_{\mathrm{mix}} = 10$) provides faster convergence time albeit compromising accuracy, leading to a *latency-accuracy* trade-off. The inverse-Mixup of Mix2FLD can partly resolve the trade-off by more augmenting the seed samples. Even for the same $n_{\mathrm{mix}}$, increasing $n_{inv}$ improves the accuracy by up to 1.7 percent. In doing so, the inverse-Mixup of Mix2FLD can increase the accuracy without additional communication latency. Next, to validate the data privacy guarantees of Mixup and Mix2up, we evaluate the *sample privacy*, given as the minimum similarity between a mixed-up sample and its raw sample: $\log(\min\{||\hat{\mathbf{x}}_{ij}^c - \mathbf{x}_i^c||, ||\hat{\mathbf{x}}_{ij}^c - \mathbf{x}_j^c||\})$ according to [34]. Table 16.1 shows that Mixup ($\gamma > 0$) with a single device preserves more sample privacy than the case without Mixup ($\gamma = 0$). Table 16.2 illustrates that Mix2up with two devices preserves higher sample privacy than Mixup thanks to the additional (inversely) mixing up of the seed samples across devices.

**Table 16.1.** Sample privacy, *Mixup* ($n_{\mathrm{mix}} = 100$).

| Dataset | Sample Privacy Under Mixing Ratio $\gamma$ | | | | | |
|---|---|---|---|---|---|---|
| | $\gamma = 0.001$ | 0.1 | 0.2 | 0.3 | 0.4 | 0.499 |
| MNIST | 2.163 | 4.465 | 5.158 | 5.564 | 5.852 | **6.055** |
| FMNIST | 1.825 | 4.127 | 4.821 | 5.226 | 5.514 | **5.717** |
| CIFAR-10 | 2.582 | 4.884 | 5.577 | 5.983 | 6.270 | **6.473** |
| CIFAR-100 | 2.442 | 4.744 | 5.438 | 5.843 | 6.131 | **6.334** |

**Table 16.2.** Sample privacy, *Mix2up* ($n_{\mathrm{mix}} = 100$).

| Dataset | Sample Privacy Under Mixing Ratio $\gamma$ | | | | | |
|---|---|---|---|---|---|---|
| | $\gamma = 0.001$ | 0.1 | 0.2 | 0.3 | 0.4 | 0.499 |
| MNIST | 2.557 | 4.639 | 5.469 | 6.140 | 7.007 | **9.366** |
| FMNIST | 2.196 | 4.568 | 5.410 | 6.143 | 6.925 | **9.273** |
| CIFAR-10 | 2.824 | 5.228 | 6.076 | 6.766 | 7.662 | **10.143** |
| CIFAR-100 | 2.737 | 5.151 | 6.050 | 6.782 | 7.652 | **10.104** |

It also shows that each inversely mixed-up sample does not resemble its raw sample but an arbitrary sample having the same ground-truth label. Both Tables 16.1 and 16.2 show that the mixing ratio $\gamma$ closer to 0.5 (i.e., equally mixing up two samples) ensures higher sample privacy, which may require compromising more accuracy. Investigating the privacy-accuracy trade-off could be an interesting topic for future research.

## 16.5 Application: FD for Reinforcement Learning

The original design of FD relies on grouping model outputs based on labels in classification. To demonstrate its applicability beyond classification, in this section we aim to exemplify an FD implementation under a reinforcement learning (RL) environment in which multiple interactive agents locally carry out decision making in real time. In such environments, policy distillation (PD) is a well-known solution [35], wherein multiple agents collectively train their local NNs. As illustrated in Fig. 16.10, PD is operated by (a) uploading every local *experience memory* to a server, (b) constructing a global experience memory at the server, and (c) downloading and replaying the global experience memory at each agent to train its local NN [35]. However, the local experience memory contains all local state observations and the corresponding policies (i.e., action logits). Exchanging such raw memories may thus violate the privacy of their host agents. Furthermore, the global experience memory size increases with the number of agents. The resulting ever-growing communication overhead may undermine the scalability of PD.

To obviate the aforementioned problems, by leveraging FD, we introduce *federated reinforcement distillation (FRD)* [24, 36], a communication-efficient and privacy-preserving distributed RL framework based on a *proxy experience memory*. In FRD, each agent stores a local proxy experience memory that consists of a set of prearranged *proxy states* and *locally averaged policies*. In this memory structure, the actual states are mapped into the proxy states (e.g., based on the nearest value rule), and the actual policies are averaged over time. Exchanging the local proxy memories of agents not only preserves the privacy of agents, but also avoids the continuaal increase in the communication overhead as the number of agents grows. In what follows we first elaborate the baseline PD operations and then illustrate FRD operations, followed by numerical evaluations.

## 16.5.1 Policy Distillation With Experience Memory

We consider an episodic environment modeled by a Markov decision process. The state space $\mathcal{S}$ and action space $\mathcal{A}$ are discrete. Without any prior knowledge about the environment, each agent takes an action $a \in \mathcal{A}$ at time slot $t$ and in return receives the reward $r_t \in \mathbb{R}$. The resulting policy $\pi_\theta : \mathcal{S} \rightarrow \mathcal{P}(\mathcal{A})$ (i.e., actions for given states) is stochastic, where $\mathcal{P}(\mathcal{A})$ is the set of probability measures on $\mathcal{A}$. The policy

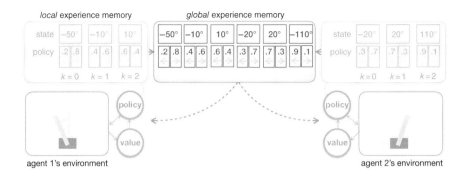

**Figure 16.10** Policy distillation (PD) with experience memory [35].

is described by the conditional probability $\pi_\theta(a|s)$ of $a \in \mathcal{A}$ for a given state $s \in \mathcal{S}$, where $\theta \in \mathbb{R}^n$ denotes the local model parameters of an agent. Hereafter, the subscript $c \in \{1, 2, \ldots, C\}$ identifies an agent out of $C$ agents, and we abuse the notations by dropping it if the relationships are clear.

In PD [35], as depicted by Fig. 16.10, the agents collectively construct a dataset named *experience memory* for training the local models. The operation of PD can be summarized by the following steps:

1. Each agent records an *local experience memory* $\mathcal{M}_c = \{(s_k, \pi_{\theta_c,k}(\mathbf{a}_k|s_k))\}_{k=1}^{K_c}$ for $E$ episodes. Note that $K_c$ is the size of local experience memory.
2. After all the agents complete $E$ episodes, the server collects the local experience memories from all agents.
3. The server constructs a *global experience memory* $\mathcal{M} = \{(s_k, \pi_{\Theta,k}(\mathbf{a}_k|s_k))\}_{k=1}^{K}$, where $K = \sum_{c=1}^{C} K_c$ and $\pi_{\Theta,k}$ is the policy collected from the clients.
4. To reflect the knowledge of other agents, the agents download the global experience memory $\mathcal{M}$ from the server.
5. Similar to the conventional classification setting, the agent $c$ optimizes the local model $\theta_c$ by minimizing the cross entropy loss $L_c(\mathcal{M}, \theta_c)$ between the policy of local model $\pi_{\theta_c}$ and the policy $\pi_\Theta$ of global experience memories $\mathcal{M}$, where $L_c(\mathcal{M}, \theta_c)$ is given as

$$L_c(\mathcal{M}, \theta_c) = -\sum_{k=1}^{K} \pi_{\Theta,k}(\mathbf{a}_k|s_k) \log\left(\pi_{\theta_c}(\mathbf{a}_k|s_k)\right). \qquad (16.26)$$

Unfortunately, under these PD operations, malicious agents and an honest-but-curious server may sneak a look at all the previously visited states and taken actions of every agent, incurring privacy leakage issues. Furthermore the global experience memory size increases with the number of agents, limiting the scalability of PD.

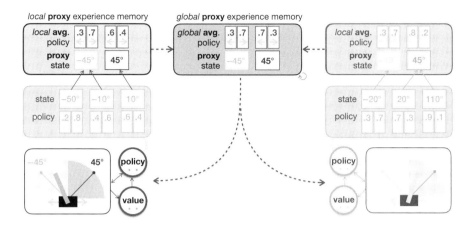

**Figure 16.11** *Federated reinforcement distillation (FRD) with proxy experience memory* [24, 36].

## 16.5.2    Federated Reinforcement Distillation with Proxy Experience Memory

Unlike PD, FRD relies on constructing and exchanging *proxy experience memories* as illustrated in Fig. 16.11, improving the communication efficiency while preserving privacy. The proxy experience memory $\mathcal{M}^P$ is comprised of *proxy state* $s^P$ and its associated *average policy* $\pi_{\Theta}^P$. A proxy state is the representative state of each *state cluster* $\mathcal{S}_j \subset \mathcal{S}$ for $i \in \{1, \dots, I\}$, where we assume $S_i \cap S_j = \emptyset$ for $i \neq j$. Given these definitions, the operations of FRD are described by the following steps:

1. Each agent categorizes the experienced policy $\pi_{\theta_c, k}(\mathbf{a}|s)$ according to the proxy state cluster that the state $s$ is included in.

2. After all the agents complete $E$ episodes, each agent constructs a *local proxy experience memory* $\mathcal{M}_c^P = \{(s_{k'}^P, \pi_{\theta_i, k'}^P(\mathbf{a}_{k'}|s_{k'}^P)\}_{k'=1}^{K_c^P}$, where $\pi_{\theta_c, k'}^P(\mathbf{a}_{k'}|s_{k'}^P)$ is the *local average policy*, obtained by averaging the policy in the same category, while $K_c^P$ is the size of local proxy experience memory describing the number of proxy state clusters that have visited by the agent. Note that the $\pi_{\theta_i}^P(\mathbf{a}_k|s_k^P)$ is not generated by the local model of agent.

3. When the local proxy experience memory of every agent is ready, the server collects it from each agent.

4. Then, the server constructs the *global proxy experience memory*

$$\mathcal{M}^P = \{(s_{k'}, \pi_{\Theta, k'}(\mathbf{a}_{k'}|s_{k'}))\}_{k'=1}^{K^P}, \tag{16.27}$$

by averaging the local average policies in the same category. The size of global proxy experience memory $K^P$ is the number of proxy state clusters that have visited by all the clients.

5. Each agent downloads the global proxy experience memory $\mathcal{M}^P$ from the server.
6. Each agent $i$ fits the local model $\theta_c$, minimizing the cross-entropy loss $L_c^P(\mathcal{M}^P, \theta_c)$ between the policy of local model $\pi_{\theta_c}(\mathbf{a}_{k'}|s_{k'})$ and the global average policy $\pi_{\Theta,k'}^P(s_{k'}^P, \mathbf{a}_{k'}|s_{k'}^P)$ of global proxy experience memory $\mathcal{M}^P$, where

$$L_c^P(\mathcal{M}^P, \theta_c) = -\sum_{k=1}^{K^P} \pi_{\Theta,k'}^P(\mathbf{a}_{k'}|s_{k'}^P) \log\left(\pi_{\theta_c,k'}(\mathbf{a}_{k'}|s_{k'}^P)\right). \tag{16.28}$$

This loss is calculated with the policy produced by the local model as the input of a proxy state.

Constructing the local and proxy experience memories can be interpreted as quantizing the memories, thereby reducing the uplink and downlink payload sizes, respectively. Notably, the downlink payload size reduction significantly benefits from sharing each global proxy experience by multiple agents. This is in stark contrast to PD, wherein the different agents' experiences are hardly overlapped with each other particularly for a large state dimension, bringing higher communication efficiency on FRD. Furthermore, exchanging proxy experience memories does not reveal any raw experiences of agents, enabling privacy-preserving distributed RL.

The local experiences are obtained by running a deep RL method at each agent. Throughout this chapter, we consider the advantage actor-critic (A2C) framework [37] in which each agent stores a pair of actor and critic NNs. The actor NN generates an action $a \in \mathcal{A}$ according to the policy $\pi_\theta$, while the critic NN evaluates the benefit of the generated action compared to other possible actions, in terms of obtaining higher expected future reward. Since the actor and critic NNs have no prior knowledge on the environment, the actor-critic pair must interact with the environment and thereby learn the optimal policy $\pi^*$ to gain the maximum expected future reward. Meanwhile, the benefit of taking an action is evaluated using the advantage function $A^\pi(s_t, a_t)$ [38], given as

$$A^\pi(s_t, a_t) = Q^\pi(s_t, a_t) - V^\pi(s_t) \tag{16.29}$$
$$= r(s_t, a_t) + \mathbb{E}_{s_{t+1} \sim \mathbb{E}}\left[V^\pi(s_{t+1})\right] - V^\pi(s_t) \tag{16.30}$$
$$\approx r(s_t, a_t) + V^\pi(s_{t+1}) - V^\pi(s_t), \tag{16.31}$$

where $V^\pi(s) = \mathbb{E}[r_0^\gamma | s_0 = s; \pi]$ is the value function, $Q^\pi(s, a) = \mathbb{E}[r_0^\gamma | s_0 = s, a_0 = a; \pi]$ is the Q-function, and $r(s_t, a_t)$ is the instant reward at learning step $t$. Note that if the output value of the advantage function is positive, it means that the selected action is not an optimal solution. Moreover, we can see from Eq. (16.31) that the advantage function is approximately described only using the value function. The critic NN who computes the value can thereby evaluate the advantage for each updating step of the actor NN. The actor NN is a policy NN who approximates the policy $\pi$ and constructs the local experience memory. Lastly, in that each agent stores a pair of actor and critic NNs, there are three possibilities of exchange: only actor NNs, critic NNs, or both actor and critic NNs across agents. As seen by several experiments [24, 36], exchanging only actor NNs (i.e., policy NNs) achieves the convergence speed as fast as exchanging both actor and critic NNs, while saving the communication cost thanks to ignoring critic NNs. Hereafter, we thus focus on an FRD implementation with the experience memory constructed by the actor NN outputs.

### 16.5.3 Experiments and Discussions

To show the effectiveness of FRD, we consider the *CartPole-v1* environment in the OpenAI gym [39], where each agent controls a cart so as to make a pole attached to the cart upright as long as possible. Each agent obtains a score of +1 for every time slot during which the pole remains upright. Playing the CartPole game with multiple episodes, the agents complete a mission when any agent first reaches an average score of 490, where the average is taken across 10 latest episodes.

The performance of FRD is evaluated in terms of the mission completion time and is compared with two baseline distributed RL frameworks: PD [35] and federated reinforcement learning (FRL) that exchanges actor NN model parameters following the standard FL operations [5–8, 36]. Each agent runs an A2C model comprising a pair of actor and critic NNs [37], each of which is a multilayer perceptron (MLP) with two hidden layers. At an interval of 25 episodes, the agents exchange their critic NN's outputs in PD and FRD or the critic NN parameters in FRL.

To construct proxy experience memories in FRD, the agent states are clustered as follows. In the *Cartpole* environment, each agent has its four-tuple state consisting of the cart location, cart velocity, pole angle, and the angular velocity of the pole. By evenly dividing each observation space into $S = 30$ subspaces, we define state clusters as the combinations of the four subspaces, resulting in $S^4$ state clusters in total. A proxy state is defined by the middle value of each state cluster, and each raw state is mapped into the proxy state based on the nearest value rule. For example, the proxy state of the pole angle is $-45°$ when the state cluster is $[-90°, 0°)$, as illustrated in Fig. 16.11. Throughout the simulations, the lines represent the median values, and the shaded areas depict the regions between the top-25 and top-75 percentiles.

In comparison with PD, FRD achieves the mission completion time as fast as PD as shown by Fig. 16.12(a), while saving the communication cost by around 50 percent as observed by Fig. 16.12(b) for two agents. In Fig. 16.12(b), the payload size gap between the uplink and downlink is due to the difference between local and global (proxy) experience memory sizes. This uplink-downlink payload size gap of PD is larger than that of FRD for two agents, which is expected to become even larger for more agents thanks to the proxy state sharing of FRD, advocating the communication efficiency and scalability of FRD.

Compared to FRL, FRD completes the mission slightly slower than FRL particularly for a small number of agents, as illustrated in Fig. 16.12(a). However, the the communication payload size of FRL increases with the actor NN model size, incurring higher payload sizes than FRD when there are over 100 neurons per layer as depicted by Fig. 16.4(b). Furthermore, due to the nature of exchanging and averaging model parameters, all the agents under FRL are forced to have an identical critic NN architecture, limiting the adoption of FRL particularly for a large-scale implementation

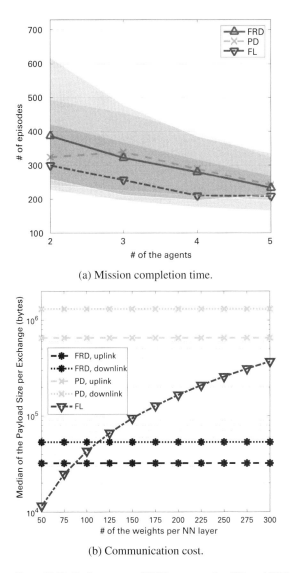

(a) Mission completion time.

(b) Communication cost.

**Figure 16.12** Performance of FRD compared to PD and FRL, in terms of (a) the mission completion time and (b) communication cost.

with heterogeneous agents. By contrast, FRD yields the communication cost upper bounded by the number of state clusters, and it does not impose any constraint on the NN architecture selection, highlighting the communication efficiency and flexibility of FRD.

## 16.6    Conclusion

In this chapter we introduced FD, a distributed learning framework that exchanges model outputs as opposed to FL based on exchanging model parameters. FD leverages key principles of CD, an online version of KD, and pushes the frontiers of its communication efficiency forward via a novel model output grouping method. To provide a deep understanding of FD, we provided a neural tangent kernel (NTK) analysis of CD in a classification task, proving that CD asymptotically achieves the convergence to the ground-truth prediction even with two workers, while more workers accelerate the convergence speed. Treating CD as the method providing the upper-bound accuracy of FD, while still effective in terms of communication efficiency, our vanilla implementation of FD is far from achieving the maximum achievable accuracy. To fill this gap, we presented several advanced FD applications that harness wireless channel characteristics and/or exploit proxy datasets, thereby achieving even higher accuracy than FL. The potential of FD is not limited to classification tasks. We partly advocated such possibilities of FD by exemplifying a reinforcement learning (RL) use case. Going beyond this, for future research, it could be worth studying the applicability of FD to unsupervised learning and self-supervised learning tasks under more realistic wireless channels and time-varying network topologies.

## References

[1] J. Park et al., "Wireless network intelligence at the edge," *Proc. IEEE*, vol. 107, no. 11, pp. 2204–2239, Oct. 2019.

[2] J. Park et al., "Extreme URLLC: Vision, challenges, and key enablers," *arXiv preprint*, arXiv:2001.09683, 2020.

[3] J. Park et al., "Communication-efficient and distributed learning over wireless networks: Principles and applications," *arXiv preprint*, arXiv:2008.02608, 2020.

[4] M. Bennis, M. Debbah, and V. Poor, "Ultra-reliable and low-latency wireless communication: Tail, risk and scale," *Proc. IEEE*, vol. 106, no. 10, pp. 1834–1853, Oct. 2018.

[5] H. B. McMahan et al., "Communication-efficient learning of deep networks from decentralized data," in *Proc. Int. Conf. on Artificial Intelligence and Statistics (AISTATS)*, April 2017.

[6] S. Samarakoon et al., "Distributed federated learning for ultra-reliable low-latency vehicular communications," *IEEE Trans. on Communications*, vol. 68, no. 2, pp. 1146–1159, 2020.

[7] H. Kim et al., "Blockchained on-device federated learning," *IEEE Communications Letters*, vol. 24, no. 6, pp. 1279–1283, 2020.

[8] P. Kairouz et al., "Advances and open problems in federated learning," *arXiv preprint*, arXiv:1912.04977, 2019.

[9] Z. Sun et al., "MobileBERT: a compact task-agnostic BERT for resource-limited devices," *arXiv preprint*, arXiv:2004.02984, 2020.

[10] E. Jeong et al., "Communication-efficient on-device machine learning: Federated distillation and augmentation under non-iid private data," in *Advances in Neural Information Processing Systems (NeurIPS) Workshop on Machine Learning on the Phone and other Consumer Devices (MLPCD)*, 2018.

[11] A. Jacot, F. Gabriel, and C. Hongler, "Neural tangent kernel: Convergence and generalization in neural networks," in *Proc. Advances in Neural Information Processing Systems (NeurIPS)*, Dec. 2018.

[12] G. Hinton, O. Vinyals, and J. Dean, "Distilling the knowledge in a neural network," in *Advances in Neural Information Processing Systems (NeurIPS) Workshop on Deep Learning and Representation Learning*, Dec. 2015.

[13] R. Anil et al., "Large scale distributed neural network training through online distillation," in *Proc. Int. Conf. on Learning Representations (ICLR)*, May 2018.

[14] A. Rahbar et al., "On the unreasonable effectiveness of knowledge distillation: Analysis in the kernel regime," *arXiv preprint*, arXiv:2003.13438, 2020.

[15] M. Phuong and C. Lampert, "Towards understanding knowledge distillation," in *Proc. Int. Conf. on Machine Learning (ICML)*, June 2019.

[16] J. Tang et al., "Understanding and improving knowledge distillation," *arXiv preprint*, arXiv:2002.03532, 2020.

[17] B. Heo et al., "A comprehensive overhaul of feature distillation," in *Int. Conf. on Computer Vision (ICCV)*, 2019.

[18] S. S. Du et al., "Gradient descent provably optimizes over-parameterized neural networks," *arXiv preprint*, arXiv:1810.02054, 2018.

[19] I. Goodfellow, Y. Bengio, and A. Courville, *Deep Learning*. MIT Press, 2016.

[20] K. H. Rosen, *Discrete Mathematics and Its Applications*, 7th ed. McGraw-Hill, 2011.

[21] S. Oh et al., "Mix2FLD: downlink federated learning after uplink federated distillation with two-way mixup," *IEEE Communications Letters*, vol. 24, no. 10 pp. 2211–2215, Oct. 2020.

[22] J.-H. Ahn, O. Simeone, and J. Kang, "Wireless federated distillation for distributed edge learning with heterogeneous data," in *Proc. IEEE Annual Int. Symp. on Personal, Indoor and Mobile Radio Communications (PIMRC)*, Sep. 2019.

[23] S. Itahara et al., "Distillation-based semi-supervised federated learning for communication-efficient collaborative training with non-iid private data," *arXiv preprint*, arXiv:2008.06180, 2020.

[24] H. Cha et al., "Federated reinforcement distillation with proxy experience memory," in *Int. Joint Conf. on Artificial Intelligence (IJCAI) Workshop on Federated Machine Learning for User Privacy and Data Confidentiality (FML)*, Aug. 2019.

[25] J. Park et al., "Distilling on-device intelligence at the network edge," *arXiv preprint*, arXiv:1908.05895, 2019.

[26] H. Lu et al., "Robust coreset construction for distributed machine learning," *IEEE Journal on Selected Areas in Communications*, vol. 38, no. 10, pp. 2400–2417, 2020.

[27] M. Mirza and S. Osindero, "Conditional generative adversarial nets," *arXiv preprint*, arXiv:1411.1784, 2014.

[28] J.-B. Grill et al., "Bootstrap your own latent: A new approach to self-supervised learning," *arXiv preprint*, arXiv:2006.07733, 2020.

[29] J. Park, S. Kim, and J. Zander, "Tractable resource management with uplink decoupled millimeter-wave overlay in ultra-dense cellular networks," *IEEE Trans. on Wireless Communications*, vol. 15, no. 6, pp. 4362–4379, 2016.

[30] H. Zhang et al., "mixup: Beyond empirical risk minimization," in *Proc. Int. Conf. on Learning Representations (ICLR)*, May 2018.

[31] C. Dwork, "Differential privacy: A survey of results," *Theory and Applications of Models of Computation*, M. Agrawal, D. Du, Z. Duan, and A. Li, eds. Springer, pp. 1–19, 2008.

[32] K. Le et al., "Synthesizing differentially private datasets using random mixing," in *Proc. IEEE Int. Symp. on Information Theory (ISIT)*, Jul. 2019.

[33] M. Shin et al., "XOR Mixup: Privacy-preserving data augmentation for one-shot federated learning," in *Int. Conf. on Machine Learning (ICML) Workshop on Federated Learning for User Privacy and Data Confidentiality (FL-ICML)*, July 2020.

[34] E. Jeong et al., "Multi-hop federated private data augmentation with sample compression," in *Int. Joint Conf. on Artificial Intelligence (IJCAI) Workshop on Federated Machine Learning for User Privacy and Data Confidentiality (FML)*, Aug. 2019.

[35] A. Rusu et al., "Policy distillation," in *Proc. Int. Conf. on Learning Representations (ICLR)*, May 2016.

[36] H. Cha et al., "Proxy experience replay: Federated distillation for distributed reinforcement learning," *IEEE Intelligent Systems*, vol. 35, no. 4, pp. 94–101, 2020.

[37] V. Mnih et al., "Asynchronous methods for deep reinforcement learning," in *Proc. Int. Conf. on Machine Learning (ICML)*, June 2016.

[38] Z. Wang et al., "Dueling network architectures for deep reinforcement learning," in *Proc. Int. Conf. on Machine Learning (ICML)*, May 2016.

[39] G. Brockman et al., "Open AI gym," *arXiv preprint*, arXiv:1606.01540, 2016.

# 17 Differentially Private Wireless Federated Learning

Dongzhu Liu, Amir Sonee, Osvaldo Simeone, and Stefano Rini

## 17.1 Introduction

With the steady increase in data generated, processed, and stored at mobile devices, current wireless networks offer new opportunities to develop intelligent applications and communication protocols in a data-driven fashion by training machine learning (ML) models [1–5]. Prevalent machine learning solutions assume central processing of the training data at a server in the cloud. In wireless networks, this approach would entail the transmission of the local datasets to the data center, increasing the communication load and affecting the privacy of the local data. Alternatively, training at each individual device may require excessive energy resources, and it generally limits the model performance due to the limited data available at the device. These facts motivate the design of distributed learning protocols that can leverage data from multiple devices via communications. This decentralized ML framework is typically referred to as federated learning (FL).

In FL, distributed agents jointly train a shared ML model by sharing information about the local ML models over multiple iterations, instead of sharing data points [6–12]. Applications include the Internet of Things (IoT), autonomous driving, remote sensing, control, and other communication technologies [13–15]. FL can not only enhance the communication efficiency as compared to the full sharing of data [16–20], but it also alleviates the information leakage of local datasets [21–26]. This chapter reviews recent work on private FL in wireless systems.

Among the original motivations for the introduction of FL are the inherent privacy properties that arise from sharing only model information. However, it has been shown that a malicious server can still infer information about local data samples from the released model information via membership inference attack [27], model inversion attack [25] and reconstruction attacks [28]. Theoretical frameworks that provide insights into FL privacy leakage include [29].

Two distinct approaches can be adopted to obtain privacy guarantees for FL. The first is cryptographic and the second is information-theoretic. The primitives of secure multiparty computation (SMC) and homomorphic encryption address the risk of privacy leakage during the learning process by enabling the server to compute statistics on encrypted local data using asymmetric cryptographic methods [30, 31]. This approach generally entails a high computation complexity, which makes it more suitable for simple learning models such as linear regression [32] and logistic regression [33]. It also requires the setting up and maintenance of a public key infrastructure.

The second approach, which is grounded in information-theoretic arguments, revolves around the notion of differential privacy (DP) [34, 35]. DP is a well-established measure that quantifies the information leaked by released statistics regarding the presence of individual data samples in the input training set [35]. DP measures the sensitivity of the released statistics to changes in any of the training data points for an arbitrary training dataset. Therefore, a mechanism achieving DP guarantees output statistics that are almost as likely to have been produced whether any individual data point is present or not. As a result, unlike the cryptographic approach, DP provides ad omnia information-theoretic guarantees that are not affected by the computational power or by any side information available at agents observing the released statistics.

Typical DP mechanisms for the case of an "honest-but-curious" recipient introduce uncertainty via random perturbations of the disclosed statistics. This perturbation can be in the form of a randomized digital mapping to one of a limited number of output levels [36–39]; or in the form of continuous additive noise following distribution among which Gaussian [40] and Laplacian [41, 42] are the most commonly used distributions.

DP mechanisms have been investigated for FL based on gradient descent [39, 43–50], alternating direction method of multipliers (ADMM) [51, 52], and distributed consensus [53]. Most prior works assume the edge server is honest-but-curious and that communication is noiseless and unconstrained. In [44], Gaussian noise is added to the local model updates after clipping. The clipping threshold determines the sensitivity of the disclosed updates, and hence it controls the power of Gaussian noise needed to ensure a target privacy requirement. The analysis in [44] demonstrates a trade-off between the convergence rate of learning and the level of privacy. Using the related privacy metric based on mutual information [54], the analysis of the convergence rate and privacy was provided in [55]. Furthermore, it was proved that running the algorithm on random mini-batches achieves a higher privacy guarantee, a principle dubbed as "privacy amplification by subsampling" [45]. An FL protocol that combines DP mechanisms, random scheduling, and homomorphic encryption was introduced in [49].

When the dimensionality of the data and the number of servers increase, it becomes imperative to reduce the size of transmitted messages using sparsification [56–58] or quantization schemes [17, 39, 50, 59–61], or to decrease the number of communication iterations [49, 62]. Among the works dealing with both communication-efficiency and DP are references [39, 50, 61], which apply different approaches. While [39] proposes adding random binomial noise on the quantized version of the gradients to achieve both DP and communication-efficiency, reference [50] proposes a vector-quantization scheme based on the convex-hull of a discrete set of points. The latter can offer DP guarantees and an improved convergence rate, while maintaining the communication cost that is sublinear in the dimension of the model parameters.

The works reviewed so far assume ideal communication. In a wireless system, the presence of a communication channel between the edge server and distributed devices offers novel challenges and opportunities [63, 64]. The challenges are related to the limited reliability of wireless links, and opportunities arise from the inherent source of randomness provided by fading and noise, which may be leveraged as a privacy mechanism. The aim of this chapter is to study FL in wireless systems under DP constraints. We consider two protocols based on distinct principles, with the first adopting digital communication and separation of source and channel coding, and the second using analog transmission with joint source and channel coding.

## Organization

The rest of the chapter is organized as follows. We first describe the system model and main assumptions in Section 17.2. This is followed by a detailed design of digital transmission-based protocols in Section 17.3 and of uncoded analog transmission in Section 17.4. Performance evaluations and comparisons of two schemes are provided in Section 17.5. Concluding remarks are given in Section 17.6.

## Notation

Henceforth, the following notations are used throughout this chapter. Boldface lower-case letters denote vectors. Calligraphic capital letters indicate sets. The set of integers between $a$ and $b$ is indicated as $\{a, \ldots, b\} \subseteq \mathbb{N}$. Matrices are indicated with capital bold letters. $\mathbf{I}_d$ is the identity matrix of dimension $d$, and the dimension is omitted when clear from the context. $\Pr[Z]$ and $\mathbb{E}[Z]$ represent the probability and expected value of a random variable $Z$. $\nabla$ is the gradient operator, and $\|.\|_\wp$ denotes the $\wp$-norm of its argument vector. Without a specified $\wp$, $\|.\|$ denotes the 2-norm. The binomial distribution with parameters $p$ and $m$ is indicated as $\mathrm{Bin}(p, m)$.

## 17.2     System Model

In this section, we set up the problem of interest by describing the learning and communication models adopted in this chapter, along with the main assumptions made in the analysis. As illustrated in Fig. 17.1, we focus on a wireless FL system comprising a single edge server and $K$ edge devices connected through it via a multiple access channel. Each device $k \in \{1, \ldots, K\}$ has a local dataset $\mathcal{D}_k$, with the overall global dataset denoted as $\mathcal{D} = \bigcup_{k=1}^{K} \mathcal{D}_k$. Local datasets are disjoint. Each $\mathcal{D}_k$ consists of labelled data samples; that is,

$$\mathcal{D}_k \triangleq \{\mathbf{z}_{k,i}\}_{i \in \{1, \ldots, N_k\}} = \left\{(\mathbf{u}_{k,i}, v_{k,i})\right\}_{i \in \{1, \ldots, N_k\}}, \tag{17.1}$$

where each example $\mathbf{z}_{k,i}$ consists of a vector of covariates $\mathbf{u}_{k,i}$ and of its associated label $v_{k,i}$, and $N_k$ is the number of available local data points.

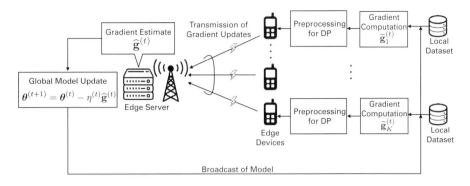

**Figure 17.1** DP-FedSGD: differentially private federated edge learning system based on distributed stochastic gradient descent.

## Differential Privacy

Consider a stochastic algorithm $M(\cdot)$ that outputs a random statistic of an input dataset. DP evaluates the information leaked by $M(\mathcal{D})$ for each data sample in the input dataset $\mathcal{D}$. Accordingly, DP is defined by bounding the difference between the distributions of the outputs $M(\mathcal{D}')$ and $M(\mathcal{D}'')$ produced by the algorithm $M(\cdot)$, where the input is one of two datasets $\mathcal{D}'$ and $\mathcal{D}''$ differing only in the value of one sample. Two such datasets are defined as *neighboring datasets*.

DEFINITION 17.1 (Neighboring datasets) Two datasets $\mathcal{D}' = \{\mathbf{z}_i'\}_{i=1}^N$ and $\mathcal{D}'' = \{\mathbf{z}_i''\}_{i=1}^N$ are said to be neighboring if there exists a single $i' \in \{1, \ldots, N\}$ such that $\mathbf{z}_{i'}' \neq \mathbf{z}_{i'}''$, while, for all other $i \in \{1, \ldots, N\}$ with $i \neq i'$, we have $\mathbf{z}_i' = \mathbf{z}_i''$.

DEFINITION 17.2 (Differential privacy [35]) A random algorithm $M(\cdot)$ is said to be $(\epsilon, \delta)$-differentially private, with $\epsilon > 0$ and $\delta \in [0, 1)$, if for any two neighboring datasets $\mathcal{D}'$ and $\mathcal{D}''$ as in Definition 17.1 it holds that

$$\Pr(M(\mathcal{D}') \in \mathcal{S}) \leq \exp(\epsilon)\Pr(M(\mathcal{D}'') \in \mathcal{S}) + \delta \tag{17.2}$$

for any subset $\mathcal{S}$ of the output space.

The definition of DP in Eq. (17.2) can be equivalently given in terms of the *DP loss*, which is defined as the (random) log-likelihood ratio

$$(\text{DP loss}) \quad L_{\mathcal{D}', \mathcal{D}''}(\xi) = \ln \frac{\Pr(M(\mathcal{D}') = \xi)}{\Pr(M(\mathcal{D}'') = \xi)}, \tag{17.3}$$

where $\xi$ is drawn according to mechanism $M(\mathcal{D}')$. It can be proved that the $(\epsilon, \delta)$-DP condition in Eq. (17.2) can also be expressed as [35, Lemma 3.17]

$$\Pr\left(|L_{\mathcal{D}', \mathcal{D}''}(\xi)| \leq \epsilon\right) \geq 1 - \delta. \tag{17.4}$$

The condition in Eq. (17.4) states that the privacy loss is bounded by $\epsilon$ with probability at least $1 - \delta$. For small values of $\epsilon$ and $\delta$, DP hence guarantees that any observer of the output statistics is unable to accurately predict the presence or absence of any single data input, even when all the other elements in a dataset are known for any possible input dataset.

## Federated Learning

The goal of FL is to collaboratively train a common machine learning model parameterized by vector $\boldsymbol{\theta} \in \mathbb{R}^d$ across multiple devices, by exchanging the local information via edge server. Each local device $k$ evaluates the performance of the model according to a local loss function defined as

$$\text{(Local loss function)} \quad L_k(\boldsymbol{\theta}) = \frac{1}{N_k} \sum_{i=1}^{N_k} \ell(\boldsymbol{\theta}; \mathbf{z}_{k,i}), \tag{17.5}$$

where $\ell(\boldsymbol{\theta}; \mathbf{z}_{k,i})$ is the sample-wise loss function quantifying the prediction error of the model $\boldsymbol{\theta}$ on the training sample $\mathbf{z}_{k,i}$; The global loss is defined as

$$\text{(Global loss function)} \quad L(\boldsymbol{\theta}) = \frac{1}{\sum_{k=1}^{K} N_k} \cdot \sum_{k=1}^{K} N_k L_k(\boldsymbol{\theta}). \tag{17.6}$$

Equation (17.6) is the regularized empirical loss on the global data set $\mathcal{D}$. The training process aims to determine the model $\theta^*$ which minimizes the global loss function:

$$\boldsymbol{\theta}^* = \arg \min L(\boldsymbol{\theta}). \tag{17.7}$$

To solve the optimization problem in Eq. (17.7), we consider the gradient descent method in the context of FL via distributed stochastic gradient averaging, which we will refer to as federated stochastic gradient descent (FedSGD). A similar idea would also apply to federated averaging [8, 14, 44], but we will not study this case here. During the $t$th iteration, the edge server first broadcasts the current model $\boldsymbol{\theta}^{(t)}$ to the edge devices via downlink ideal communication such that the model is received at each device without distortion.

Having received model $\boldsymbol{\theta}^{(t)}$, each device computes the local gradient with respect to $\boldsymbol{\theta}^{(t)}$ as

$$\text{(Local gradient)} \quad \mathbf{g}_k^{(t)} = \frac{1}{N_k} \sum_{i=1}^{N_k} \nabla \ell(\boldsymbol{\theta}^{(t)}; \mathbf{z}_{k,i}). \tag{17.8}$$

By using the received signals, the edge server produces an estimate $\widehat{\mathbf{G}}^{(t)}$ of the global gradient

$$\text{(Global gradient)} \quad \mathbf{G}^{(t)} = \frac{1}{N} \sum_{k=1}^{K} N_k \mathbf{g}_k^{(t)}, \tag{17.9}$$

and it carries out the model updating via

$$\text{(Model updating)} \quad \boldsymbol{\theta}^{(t+1)} = \boldsymbol{\theta}^{(t)} - \eta^{(t)} \widehat{\mathbf{G}}^{(t)}, \tag{17.10}$$

where $\eta^{(t)}$ denotes the learning rate. The steps in Eqs. (17.8) and (17.10) are iterated until convergence.

The performance of the FedSGD for a given iteration time $T$ can be evaluated in terms of the optimality gap, defined as the difference between the loss of the model estimated at iteration $T$ and its optimal value $L^*$; that is,

$$\text{(Optimality gap)} \quad \mathbb{E}\big[L(\boldsymbol{\theta}^{(T)}) - L^*\big], \tag{17.11}$$

where the expectation is over the random estimates of the gradients obtained as a result of the stochastic communication channels to be discussed next.

## Communication Model

In this chapter, we assume that the users communicate toward the server over a fading multiaccess channel (MAC) to exchange the information of local gradient $\mathbf{g}_k^{(t)}$ with the edge server. The communication phase of each iteration $t$ takes place over $M$ channel uses. We assume a block-fading model, so that the channel coefficients are constant within an iteration. Furthermore, perfect channel state information (CSI) is assumed at both the edge devices and the server.

We consider two transmission strategies, namely, orthogonal and non-orthogonal multiple access, which are referred to as OMA and NOMA, respectively.

**Orthogonal multiple access (OMA):** Assuming time division multiple access (TDMA), the $M$ channel uses are split into $K$ orthogonal slots, each with $M_o = M/K$ channel uses. Only one of the devices is scheduled for communication in each slot with the edge server. We will adopt round-robin scheduling, so that the received signal in any of the channel use of the $k$th slot is given as

$$y_k^{(t)} = h_k^{(t)} x_k^{(t)} + w_k^{(t)}, \quad \text{for each slot } k \in \{1, \ldots, K\}, \tag{17.12}$$

where $x_k^{(t)}$ is the channel input of device $k$. At each remote device $k$, channel input is subject to an average per-symbol power constraint:

$$\text{(Power constraint)} \quad \mathbb{E}\big[\big(x_k^{(t)}\big)^2\big] \le P. \tag{17.13}$$

**Nonorthogonal multiple access (NOMA):** Nonorthogonal multiple access enables all devices to transmit simultaneously over the $M$ channel uses available for each iteration $t$. The signal received by the edge server for any channel use allocated at iteration $t$ is given as

$$y^{(t)} = \sum_{k=1}^{K} h_k^{(t)} x_k^{(t)} + w^{(t)}. \tag{17.14}$$

Note that NOMA generally requires stronger assumptions of symbol-level synchronization among the devices for ensuring successful decoding.

According to the quasi-static fading assumption, the fading coefficients $\{h_k^{(t)}\}_{k=1}^K$ are independent and identically distributed (i.i.d.) and drawn from a given fading distribution. Moreover, the additive noise term $w^{(t)}$ is i.i.d. drawn from the Gaussian distribution $\mathcal{N}(0, \sigma^2)$. Under the assumption of perfect CSI, devices can compensate for the phase of the channels, which we can hence consider to be real without loss of generality.

We consider two transmission approaches:

- **Digital transmission:** Gradients are first quantized, then perturbed by an integer-valued artificial noise to enhance privacy, and then encoded before transmission.
- **Analog transmission:** Gradients estimates are transmitted uncoded, and channel noise is leveraged for privacy.

Privacy is defined from the view of any device with respect to the edge server. The server is honest-but-curious, meaning it follows the prescribed FL protocol but may attempt to infer information about the local datasets. In contrast, the other remote devices are implicitly trusted.

## 17.2.1    Assumptions on the Loss Functions

As is customary in the convergence analysis of optimization schemes, we shall assume that the loss function is smooth and that it satisfies the Polyak-Lojasiewicz (PL) condition.

ASSUMPTION 17.3 (Smoothness)    The global loss function $L(\boldsymbol{\theta})$ is smooth with constant $\nu > 0$; that is, it is continuously differentiable and the gradient $\nabla L(\boldsymbol{\theta})$ is Lipschitz continuous with constant $\nu$:

$$\left\| \nabla L(\boldsymbol{\theta}) - \nabla L(\boldsymbol{\theta}') \right\| \leq \nu \left\| \boldsymbol{\theta} - \boldsymbol{\theta}' \right\|, \quad \text{for all } \boldsymbol{\theta}, \boldsymbol{\theta}' \in \mathbb{R}^d. \tag{17.15}$$

The inequality in Eq. (17.15) implies that the global loss function $L(\boldsymbol{\theta})$ can be upper bounded by the following quadratic approximation:

$$L(\boldsymbol{\theta}) \leq L(\boldsymbol{\theta}') + \nabla L(\boldsymbol{\theta}')^T (\boldsymbol{\theta} - \boldsymbol{\theta}') + \frac{\nu}{2} \left\| \boldsymbol{\theta} - \boldsymbol{\theta}' \right\|^2, \quad \text{for all } \boldsymbol{\theta}, \boldsymbol{\theta}' \in \mathbb{R}^d. \tag{17.16}$$

ASSUMPTION 17.4 (Polyak-Lojasiewicz inequality)    Assuming that the optimization Eq. (17.7) has a nonempty solution set with its optimal value denoted as $L^*$, the global loss function $L(\boldsymbol{\theta})$ satisfies the PL condition with constant $\mu > 0$ if the following inequality holds

$$\frac{1}{2} \left\| \nabla L(\boldsymbol{\theta}) \right\|^2 \geq \mu \left[ L(\boldsymbol{\theta}) - L^* \right]. \tag{17.17}$$

As detailed in [65], the PL condition is more general than the assumption of strong convexity: strong convexity with constant $\mu > 0$ implies the PL inequality with same parameter $\mu$ [66]. Furthermore, for a convex sample-wise loss function $\ell(\boldsymbol{\theta}; \mathbf{z}_{k,i})$ (e.g., for least squares and logistic regression), its global loss function $L(\boldsymbol{\theta})$ can be modified to be strongly convex with a constant $\mu$ by adding a quadratic regularization term with multiplicative constant $\lambda$.

## 17.3    Digital Transmission

Digital transmission is implemented for the FL model by quantizing the analog-valued gradients prior to transmission [63]. In the literature, various quantization schemes have been proposed that operate either coordinate-wise [59, 60, 67–69] or at the vector level [50]. The relative performance of these schemes can be evaluated in terms of computational complexity, convergence of the FedSGD protocols, as well as privacy, where the latter generally requires the inclusion of additional privacy-preserving mechanisms in the encoding of the gradients.

Existing works on coordinate-wise quantization schemes include signSGD [60], TernGrad [67], and uniform multilevel stochastic quantization techniques [39]. The signSGD scheme makes use of two quantization levels that allow for the encoding of the sign of each coordinate in the gradient vector. Similarly, TernGrad applies ternary quantization, thus making it possible to distinguish small values of the entries in the gradient vector. In [63], the authors consider multilevel quantization of gradients and argued that multilevel quantization is particularly useful when the quantization levels are assigned according to the channel conditions.

Privacy-preserving mechanisms for quantized SGD (QSGD) methods have been first studied in [39], where the authors propose adding an artificial binomial noise for enhancing the privacy of coordinate-wise quantization. In [50], the privacy mechanism is studied in conjunction with vector quantization jointly addressing quantization and privacy.

In the rest of this section, we review and expand upon the protocol proposed in [39, 70], which employs QSGD with multilevel quantization and binomial noise addition as a DP mechanism.

### 17.3.1    Quantization, DP Mechanism, and Channel Coding

In this section, we describe the quantization, DP mechanism, and channel-coding schemes. We start this section by providing the following assumption on the elements in the local gradient.

ASSUMPTION 17.5 (Bounded element-wise gradient)    At any iteration $t$, for any training sample $\mathbf{z}$, each element in the local gradient, $[\mathbf{g}_k^{(t)}]_j$, is upper bounded by a given constant $C_e^{(t)}$; that is, for all possible $\mathbf{z}$ (not limited to those in dataset $\mathcal{D}$), we have the inequality

$$[\mathbf{g}_k^{(t)}]_j \le C_e^{(t)}, \quad \forall j \in \{1, \ldots, d\}, \forall k \in \{1, \ldots, K\}. \tag{17.18}$$

This assumption indicates that the local gradient is bounded as $\|\mathbf{g}_k^{(t)}\| \le dC_e^{(t)}$. Furthermore, the value of $C_e^{(t)}$ is independent of the local datasets, and it is typically achieved by clipping.

**Quantization:** Let us begin this section by introducing the notation necessary to describe the stochastic quantizer with element-wise clipping bound $C_e^{(t)}$ and rate $r_k^{(t)}$ employed for analog-to-digital conversion by device $k$ at iteration $t$. From a high-level perspective, the quantizer performs an unbiased uniform stochastic quantization followed by randomness ensuring DP. To detail the quantization operation, the subscript $k$ and superscript $(t)$ denoting the device and iteration, respectively, are dropped.

Consider an arbitrary local gradient $\mathbf{g} \in \mathbb{R}^d$, and clip this vector so that each coordinate $[\mathbf{g}]_j$ in $\mathbf{g}$, $j \in \{1, \ldots, d\}$, is limited in value between $-C_e$ and $C_e$ (i.e., $-C_e \le [\mathbf{g}]_j \le C_e$). This can be done by scaling the coordinate as $\min\{1, C_e/[\mathbf{g}]_j\}[\mathbf{g}]_j$. Then, divide the interval $[-C_e, C_e]$ into $l - 1$ levels with $l = 2^r$. The quantization step size is

$$\gamma = \frac{2C_e}{l - 1}, \tag{17.19}$$

and for $i \in \{1, \ldots, l - 1\}$, the reconstruction points of the quantizer are

$$V(i) = -C_e + i\frac{\gamma}{l - 1}. \tag{17.20}$$

As a final step, let us describe the mapping between each coordinate $[\mathbf{g}]_j$ in $\mathbf{g}$ and the reconstruction points in Eq. (17.20). Assume $i$ is such that $[\mathbf{g}]_j \in [V(i), V(i + 1))$, then reconstruction point $[\bar{\mathbf{g}}]_j$, as illustrated in Fig. 17.2, is given as

$$[\bar{\mathbf{g}}]_j = Q_{C_e, l}([\mathbf{g}]_j) = \begin{cases} V(i + 1) & \text{w.p.} \quad \frac{[\mathbf{g}]_j - V(i)}{\gamma} \\ V(i) & \text{otherwise.} \end{cases} \tag{17.21}$$

As in [39, 63], the stochastic mapping in Eq. (17.21) is such that the quantization is unbiased:

$$\mathbb{E}\left[[\bar{\mathbf{g}}]_j\right] = [\mathbf{g}]_j, \tag{17.22}$$

and the variance is bounded as

$$\mathbb{V}\mathrm{ar}\left[[\bar{\mathbf{g}}]_j\right] \le \frac{C_e^2}{(l - 1)^2}. \tag{17.23}$$

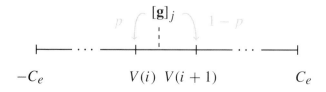

**Figure 17.2** A graphical representation of the stochastic quantization in Eq. (17.21).

Quantization is applied coordinate-wise. Accordingly, we define the quantized vector

$$\overline{\mathbf{g}} = Q_{C_e,l}(\mathbf{g}) = \left[[\overline{\mathbf{g}}]_1, \dots, [\overline{\mathbf{g}}]_d\right]^T. \tag{17.24}$$

When employing the stochastic quantizer $Q_{C_e,l}(\mathbf{g})$, we assume that at each iteration $t$, the clipping bound $C_e^{(t)}$ is communicated to the devices by the edge server ahead of transmission.

**DP mechanism:** Following quantization, the discrete-valued quantized gradient is further perturbed by adding a binomial distributed noise to ensure privacy. As the quantization step, the privacy-preserving mechanism is applied coordinate-wise, producing

$$[\widetilde{\mathbf{g}}]_j = M_{p,m}([\overline{\mathbf{g}}]_j) = [\overline{\mathbf{g}}]_j + \gamma\left([\mathbf{n}]_j - mp\right), \tag{17.25}$$

where $[\mathbf{n}]_j \sim \text{Bin}(p,m)$ are independent, and we define the output of the privacy mechanism as

$$\widetilde{\mathbf{g}} = M_{p,m}(\overline{\mathbf{g}}) = \left[[\widetilde{\mathbf{g}}]_1, \dots, [\widetilde{\mathbf{g}}]_d\right]^T. \tag{17.26}$$

**Channel Encoding Scheme:** As a last step before transmission, the quantized and noisy gradient in Eq. (17.26) is encoded into a codeword of block-length $n$ with $n = M$ for NOMA and $n = M/K$ for OMA satisfying the per-symbol average power constraint in Eq. (17.13). Note that $\widetilde{\mathbf{g}}$ in Eq. (17.26) can be represented with at most $d \log_2 (1 + m)$ bits. These bits are encoded into a codeword $\mathbf{x}$ of $n$ symbols. This operation is expressed as

$$\mathbf{x} = X_R(\widetilde{\mathbf{g}}), \tag{17.27}$$

where we assume that $\mathbf{x}$ satisfies the power constraint in Eq. (17.13) and the channel coding rate is obtained as

$$R = \frac{d \log_2 (1 + m)}{n}. \tag{17.28}$$

**Model Update:** Having described the quantization, DP, and channel encoding schemes for a generic gradient, we are now ready to detail the overall algorithm. Each node $k$ at time $t$ produces the channel input vector $\mathbf{x}_k^{(t)}$ as

$$\mathbf{x}_k^{(t)} = X_{R_k^{(t)}} \left( M_{p_k^{(t)}, m_k^{(t)}} \left( Q_{C_e^{(t)}, l_k^{(t)}} \left(\mathbf{g}_k^{(t)}\right) \right) \right). \tag{17.29}$$

Note that the clipping value $C_e^{(t)}$ is chosen to be equal at all users. Furthermore, given $l_k^{(t)}$, the binomial noise parameters $m_k^{(t)}$ and $p_k^{(t)}$, the channel rate is given as

$$R_k^{(t)} = \frac{d \log_2 \left(l_k^{(t)} + m_k^{(t)}\right)}{n}. \tag{17.30}$$

When NOMA is employed, the edge server can decode the codeword transmitted by the users when the set of transmission rates $R_k^{(t)}$ are to within the capacity of the Gaussian MAC in Eq. (17.14).[1] This condition can be expressed as [71],

$$\sum_{k=1}^{K} R_k^{(t)} \leq \log\left(1 + \sum_{k=1}^{K} \left(h_k^{(t)}\right)^2 P/\sigma^2\right), \qquad \text{for NOMA.} \qquad (17.31)$$

In contrast, when OMA is employed, the conditions for correct decoding follows from the capacity of point-to-point channel:

$$R_k^{(t)} \leq \log\left(1 + \left(h_k^{(t)}\right)^2 P/\sigma^2\right), \qquad \forall k \in \{1, \ldots, K\}, \text{ for OMA.} \qquad (17.32)$$

Either way, once the codewords sent by devices are decoded into $\widehat{\mathbf{g}}_k^{(t)}$, the edge server updates the global model via Eq. (17.10) by using the estimate

$$\widehat{\mathbf{G}}^{(t)} = \frac{1}{N} \sum_{k=1}^{K} N_k \widehat{\mathbf{g}}_k^{(t)}. \qquad (17.33)$$

## 17.3.2 Performance Analysis

**Differential privacy analysis:** The standard results on DP provide guidelines on how to choose the parameters of additive noise in a mechanism, which should be designed according to the sensitivity of the function being disclosed without the noise.

In the case of digital transmission, at iteration $t$, each device $k$ releases the composition of gradient $\mathbf{g}_k^{(t)}$ and of the stochastic quantization function $Q_{C_e^{(t)}, l_k^{(t)}}$ specified in Eq. (17.21). For this randomized function, following [39], we introduce probabilistic sensitivity bounds $\Delta_{k,\wp}^{(t)}$ with respect to the $\wp$-norm. This is an upper bound holding for a set of norms with probability at least $1 - \delta'$. Formally, the probabilistic sensitivity bounds are defined as

$$\Pr\left(\bigcup_{\wp \in \mathcal{S}} \left(\left\|U^{(t)}\left(\mathcal{D}_k'\right) - U^{(t)}\left(\mathcal{D}_k''\right)\right\|_{\wp} \leq \Delta_{k,\wp}^{(t)}\right)\right) \geq 1 - \delta' \qquad (17.34)$$

if there exist random variables $U^{(t)}(\mathcal{D}_k')$ and $U^{(t)}(\mathcal{D}_k'')$ that have the same marginal distributions of the respective released functions and arbitrary joint distribution [39]. We specifically define the quantities $\Delta_{k,\wp}^{(t)}$ as the maximum values for which the respective inequality in Eq. (17.34) holds. We note that the analysis of other DP mechanisms, such as the Gaussian mechanism assumed by analog transmission rely on the deterministic sensitivity to be discussed in the next section.

For the binomial mechanism as adopted in digital transmission, the set of norms to be considered is $\mathcal{S} = \{1, 2, \infty\}$. It can be shown that for any $\delta' \in (0, 1)$, the deterministic sensitivities are obtained as in [39]

---

[1] For simplicity of analysis, finite-block length effects are ignored here.

$$\Delta_{k,\infty}^{(t)} = l_k^{(t)} + 1,$$

$$\Delta_{k,1}^{(t)} = \frac{\sqrt{d}C^{(t)}}{C_e^{(t)}}\left(l_k^{(t)} - 1\right) + \sqrt{2\frac{\sqrt{d}C^{(t)}}{C_e^{(t)}}\left(l_k^{(t)} - 1\right)\ln\left(\frac{2}{\delta'}\right)} + \frac{4}{3}\ln\left(\frac{2}{\delta'}\right),$$

$$\Delta_{k,2}^{(t)} = \frac{C^{(t)}}{C_e^{(t)}}\left(l_k^{(t)} - 1\right) + \sqrt{\Delta_{k,1}^{(t)} + 2\frac{\sqrt{d}C^{(t)}}{C_e^{(t)}}\left(l_k^{(t)} - 1\right)\ln\left(\frac{2}{\delta'}\right)}. \tag{17.35}$$

With the definition of probabilistic sensitivity in Eq. (17.34) and the parameters $\Delta_{k,S}^{(t)}$ in Eq. (17.35) with $\delta = \delta'$, the composed binomial mechanism $M_{p,m_k^{(t)}}$ $(Q_{C_e^{(t)},l_k^{(t)}}(\mathbf{g}_k^{(t)}))$ for each iteration $t$ is $(\Xi, 2\delta)$-DP for

$$\Xi(\delta, m_k^{(t)}, l_k^{(t)}) = \frac{\Delta_{k,2}^{(t)}a(\delta, p)}{\gamma_k^{(t)}\sqrt{m_k^{(t)}}} + \frac{\Delta_{k,2}^{(t)}c(\delta, p) + \Delta_{k,1}^{(t)}b(\delta, p)}{\gamma_k^{(t)}m_k^{(t)}} + \frac{\Delta_{k,\infty}^{(t)}d(\delta, p)}{\gamma_k^{(t)}m_k^{(t)}}, \tag{17.36}$$

under the condition that $p(1 - p)m_k^{(t)} \geq \max\{23\ln(10d/\delta), 2\Delta_{k,\infty}^{(t)}/\gamma_k^{(t)}\}$. In Eq. (17.36), $a(\delta, p)$, $b(\delta, p)$, $c(\delta, p)$ and $d(\delta, p)$ are decreasing functions of $\delta$ and are minimized over $p$ by setting $p = 1/2$. Further details are available in [39]. Note that, the case of $p = 1$ yields $\Xi = \infty$, which corresponds to the scenario without DP constraints. It can be inferred from Eq. (17.36) that better privacy is achieved via either reducing the quantization levels or increasing the noise levels at the devices. As we will show next, this is in contrast with the goal of achieving better learning performance, which motivates the optimal solution proposed in Section 17.3.3.

In both OMA and NOMA, the edge server decodes each local gradient separately; thereby the disclosed information are identical to both access schemes, so as with the sensitivity analysis. By using the composition theorem [35, Theorem 3.16], the proposed scheme achieves $(\epsilon, 2\delta)$-DP over $T$ iterations if per iteration $(\epsilon/T, 2\delta/T)$-DP is attainable for all the devices; that is,

$$\text{(DP guarantees)} \quad \Xi(\delta/T, m_k^{(t)}, l_k^{(t)}) \leq \epsilon/T, \quad \forall k \in \{1, \ldots, K\}. \tag{17.37}$$

Without claim of optimality in terms of learning performance, the DP constraint is allocated uniformly across all time instants $t$. In Section 17.4, we study a more general form for analog transmission in which the DP constraint across all $T$ iterations is enforced, while allowing different time instants to have potentially distinct contributions to the privacy loss.

**Convergence analysis:** The convergence rate of FL models is mostly analyzed in terms of the optimality gap (i.e., the difference between the expected value of the loss function obtained after $T$ iterations and the optimal value). In the literature, it is known that, for a $\mu$-strongly convex, and $\nu$-smooth loss function, the optimality gap using SGD with learning rate $\eta^{(t)} = 1/\mu t$ is upper bounded by the summation of the second-order moments of the estimated global gradient over $T$ iterations (i.e., $\sum_{t=1}^{T} \mathbb{E}[\|\widehat{\mathbf{G}}^{(t)}\|^2]$). This term can be further decomposed into the mean squared error (MSE) of the gradient estimate

$$\text{(MSE)} \quad \mathbb{E}[\|\widehat{\mathbf{G}}^{(t)} - \mathbf{G}^{(t)}\|^2] = \frac{d}{K^2} \sum_{k=1}^{K} \frac{\left(C_e^{(t)}\right)^2 \left(1 + 4m_k^{(t)} p(1-p)\right)}{\left(l_k^{(t)} - 1\right)^2}, \tag{17.38}$$

and the norm of (true) global gradient (i.e., $\|\mathbf{G}^{(t)}\|^2$), which can be bounded via Assumption 17.5 with $dC_e^{(t)}$. From these classic results, it follows that one can bound the optimality gap under the assumptions in Section 17.2.1. In digital transmission, we assume the error free channel that the gradients are only corrupted by the binomial noise added locally. Accordingly, the convergence analyses in the OMA and NOMA case are equivalent, and the optimality gap in Eq. (17.11) can be upper bounded as

$$\mathbb{E}\left[L(\boldsymbol{\theta}^{(T)}) - L^*\right] \leq \frac{2\nu}{\mu^2 T^2} \sum_{t=1}^{T} \left(\mathbb{E}[\|\widehat{\mathbf{G}}^{(t)} - \mathbf{G}^{(t)}\|^2] + (dC_e^{(t)})^2\right). \tag{17.39}$$

Note that the optimality gap in Eq. (17.39) depends on the clipping bound and the levels of quantization and binomial noise. For a given clipping bound, increasing the quantization levels $l_k^{(t)}$ or decreasing the binomial noise levels $m_k^{(t)}$ help reduce the optimality gap. However, this degrades the level of DP that motivates the rate allocation problem as formulated in the next section, to balance the trade-off between privacy and learning performance.

### 17.3.3    Rate Allocation Policy

We now consider rate allocation policy via optimization over the number of quantization levels $l_k^{(t)}$ and noise levels $m_k^{(t)}$. The problems are formulated as minimizing the optimality gap in Eq. (17.39), under the privacy constraint in Eq. (17.37), and the transmission rate constraint in Eq. (17.31) for NOMA and Eq. (17.32) for OMA, respectively. Furthermore, the optimization variables $l_k^{(t)}$ and $m_k^{(t)}$ should be positive integer, and the quantization levels $l_k^{(t)}$ should allow for at least one bit representation (i.e., $l_k^{(t)} \geq 2$).

For a constant clipping threshold $C_e^{(t)} = C_e$ for all $t$, the optimization can be reduced to minimize MSE in Eq. (17.38) of each iteration, since the weight factors of MSE in Eq. (17.39) are identical over $t$. Therefore, the noise introduced by DP mechanism degrades the learning performance equally across the iterations. In OMA, the optimization can be tackled at distributed devices in parallel, where each device minimizes its individual MSE; that is, $[1 + 4m_k^{(t)} p(1-p)]/(l_k^{(t)} - 1)^2$ in Eq. (17.38), under the DP constraint in Eq. (17.37) and channel capacity in Eq. (17.32). However, as for NOMA, the constraint on channel capacity in Eq. (17.31) requires jointly optimization over different devices, and thus the optimization problem should be generally tackled in a centralized manner at the edge server.

The optimization at hand is an integer nonlinear programming (INLP) problem, which can be addressed by noting the objective function is decreasing as the number of signal quantization levels $m_k^{(t)}$ increase and the artificial noise levels $l_k^{(t)}$ decrease. Therefore, for OMA, we can initialize the algorithm for all devices at the maximum value of the number of quantization levels allowed by the individual capacity constraint in Eq. (17.32) as $l_{k,\,\text{max}}^{(t)} = (1 + (h_k^{(t)})^2 P/\sigma^2)^{M/Kd} - 1$ and at the minimum value $m_k^{(t)} = 1$. If the DP constraint in Eq. (17.37) is satisfied, the algorithm returns the optimal solution. Otherwise, at each round, the value of $l_k^{(t)}$ is decreased by one, and the values of $m_k^{(t)} \leq l_{k,\,\text{max}}^{(t)} - l_k^{(t)}$ are successively tested until the DP constraint per iteration is satisfied or the maximum value is reached. In the former case, the procedure terminates, while in the latter, a new round is initiated. For NOMA, without claiming optimality, the same approach is used by fixing the rates to the maximum equal-rate point in the capacity region.

## 17.4   Analog Transmission

In this section, we consider uncoded analog transmission of local gradients. A key novel aspect of this solution is that channel noise can be leveraged as a privacy mechanism. This is in contrast to digital schemes that mitigate the effect of channel noise via coding. This way, uncoded transmission may allow each device to obtain privacy "for free" (i.e., to achieve no performance loss with respect to a counterpart uncoded scheme with no privacy constraints). Under NOMA, analog communication also enables over-the-air computing, whereby one can estimate the global gradient in Eq. (17.9) from superimposed transmission of the local gradient [72–76]. In this section, we follow reference [77] and focus on the optimization of the power allocation (PA) for both OMA and NOMA, with the aim of minimizing the learning optimality gap under privacy and power constraints. The general problem formulation is similar to that used in the analysis of digital schemes in the previous section, although tools and conclusions differ, as we review next.

## 17.4.1   Description of the Transmission Scheme

To start, we design the transmitted signal at each iteration $t$ for $M$ channel uses as a vector $\mathbf{x}_k^{(t)} \in \mathbb{R}^M$.

$$\text{(Transmit signal)} \quad \mathbf{x}_k^{(t)} = \sqrt{P_k^{(t)}} \mathbf{A}_k \frac{N_k \mathbf{g}_k^{(t)}}{\|N_k \mathbf{g}_k^{(t)}\|}, \tag{17.40}$$

where $P_k^{(t)} \geq 0$ is the transmitted power to be optimized, and $\mathbf{A}_k \in \mathbb{R}^{M \times d}$ is encoding matrix due to the constraint on the channel uses. Specifically, we consider $M = dK$ for OMA and $M = K$ for NOMA. In OMA, we adopt round-robin scheduling from devices 1 to $K$, and the resultant $\mathbf{A}_k = \mathbf{J}_{kk} \otimes \mathbf{I}^d$. $\mathbf{J}_{kk}$ represents the $K \times K$ single-entry matrix where 1 at $(k,k)$ and zero elsewhere, and $\otimes$ is Kronecker product. For NOMA,

$\mathbf{A}_k = \mathbf{I}^d$. Furthermore, the power control for NOMA is subject to the additional condition of gradient alignment

$$\text{(Gradient alignment for NOMA)} \quad \sqrt{P_k^{(t)}} \frac{h_k^{(t)}}{\|N_k \mathbf{g}_k^{(t)}\|} = \beta^{(t)}, \tag{17.41}$$

for a constant $\beta^{(t)}$. This condition, widely adopted for over-the-air computation [72–76], ensures that in the absence of noise, the edge server can recover the global gradient $\mathbf{G}^{(t)}$, scaled by a constant $\beta^{(t)}$, from the aggregated signals in Eq. (17.14) after vectorization for each iteration, denoted as $\mathbf{y}^{(t)}$.

As we discussed, in the DP literature, an artificial noise term is added to $\mathbf{g}_k^{(t)}$ before transmission. However, in uncoded transmission we can make direct use of the channel noise by adapting the transmit power to the DP requirements without the need to add noise.

The $t$th iteration in OMA is comprised by $K$ orthogonal slots, and the received signal at $k$th slot is vectorized as $\mathbf{y}_k^{(t)}$. The edge server estimates each scaled local gradient $N_k \mathbf{g}_k^{(t)}$ as $\mathbf{y}_k^{(t)} (P_k^{(t)})^{-\frac{1}{2}} \|N_k \mathbf{g}_k^{(t)}\|_2 / h_k^{(t)}$ separately, and then the global gradient is obtained as

$$\widehat{\mathbf{G}}^{(t)} = \frac{1}{N} \sum_{k=1}^{K} N_k \mathbf{g}_k^{(t)} + \left( \sqrt{P_k^{(t)}} \frac{h_k^{(t)}}{\|N_k \mathbf{g}_k^{(t)}\|_2} \right)^{-1} \mathbf{w}_k^{(t)}, \tag{17.42}$$

where $\mathbf{w}_k^{(t)}$ is the vectorized channel noise corresponding to $\mathbf{y}_k^{(t)}$.

In contrast, with NOMA, at the $t$th iteration, by using gradient alignment, the edge server estimates the global gradient as

$$\widehat{\mathbf{G}}^{(t)} = \frac{1}{N} \sum_{k=1}^{K} N_k \mathbf{g}_k^{(t)} + \left( \beta^{(t)} \right)^{-1} \mathbf{w}^{(t)}, \tag{17.43}$$

where $\mathbf{w}^{(t)}$ is the vectorized channel noise corresponding to $\mathbf{y}^{(t)}$.

In the following, we are interested in optimizing over the power control sequences $\{P_k^{(1)}, \ldots, P_k^{(T)}\}_{k=1}^{K}$ so as to maximize the learning performance under privacy and power constraints. To this end, we first introduce the DP and convergence analysis for $T$ iterations, followed by problem formulations and solutions for OMA and NOMA separately. For simplicity, without compromising the fairness, in this section, we consider energy constraint for each iteration to surrogate power constraints on the transmit symbols.

$$\text{(Energy constraint)} \quad \|\mathbf{x}_k^{(t)}\|^2 \leq PM. \tag{17.44}$$

## 17.4.2  Performance Analysis

**Differential privacy analysis:** We start DP analysis by making the following common assumption (see, e.g., [42, 78, 79]) on the sample-wise gradient.

ASSUMPTION 17.6 (Bounded sample-wise gradient)   At any iteration $t$, for any training sample $\mathbf{z}$, the gradient is upper bounded by a given constant $C^{(t)}$; that is, for all possible $\mathbf{z}$ (not limited to those in data sets $\{\mathcal{D}\}$), we have the inequality

$$\|\nabla\ell(\boldsymbol{\theta}^{(t)};\mathbf{z})\| \leq C^{(t)}. \tag{17.45}$$

This assumption indicates that the value of $C^{(t)}$ is independent of the local data sets, and it is typically achieved by clipping the per-sample gradient; that is, $\min\{1, C^{(t)}/\|\nabla\ell(\boldsymbol{\theta}^{(t)};\mathbf{z})\|\}\nabla\ell(\boldsymbol{\theta}^{(t)};\mathbf{z})$. Furthermore, with this assumption, one can also bound local gradient as $\|\mathbf{g}_k^{(t)}\| \leq C^{(t)}$.

For uncoded analog transmission, we exploit channel noise for DP, corresponding to the Gaussian mechanism. As discussed in Section 17.3.2, the analysis of DP depends on the sensitivity of the disclosed function, namely, the noiseless vectorized received signals. The noiseless disclosed signal for in OMA is $h_k^{(t)}\sqrt{P_k^{(t)}}N_k\mathbf{g}_k^{(t)}/\|N_k\mathbf{g}_k^{(t)}\|_2$ for device $k$; while for NOMA, it is identical across all devices as $\sum_k h_k^{(t)}\sqrt{P_k^{(t)}}N_k\mathbf{g}_k^{(t)}/\|N_k\mathbf{g}_k^{(t)}\|_2$.

Mathematically, the sensitivity $\Delta_k^{(t)}$ of $k$th device is defined as maximum 2-norm difference between noiseless disclosed signal for two neighboring datasets $\mathcal{D}_k'$ and $\mathcal{D}_k''$, while the other local datasets stay the same. This can be bounded by triangular inequality and Assumption 17.6 as [77]

$$\text{(Sensitivity)} \quad \Delta_k^{(t)} \leq 2C^{(t)}\sqrt{P_k^{(t)}}\frac{h_k^{(t)}}{\|N_k\mathbf{g}_k^{(t)}\|}. \tag{17.46}$$

With the upper bound in Eq. (17.46), we can apply the advanced composition theorem [35, Theorem 3.20] to derive the following condition under which FedSGD via either OMA or NOMA guarantees $(\epsilon, \delta)$-DP after $T$ iterations [77, Appendix A]. We use $\Gamma_{dp}(\epsilon, \delta)$ to denote the privacy level, which is a decreasing function of $\epsilon$ and $\delta$. A smaller value of $\Gamma_{dp}$ represents a stricter DP requirement. The result shows that the privacy level is determined by the summation of per-iteration ratios between the powers of disclosed data point $(C^{(t)})^2$ and the effective noise $\sigma^2\|N_k\mathbf{g}_k^{(t)}\|^2/(P_k^{(t)}(h_k^{(t)})^2)$. This suggests that power control should not only adapt to the noise variance and channel gains as for the conventional power allocation in wireless networks, but it should also adapt to the sample-wise gradient norm $(C^{(t)})^2$, which accounts for DP constraints. This requirement can be written as

$$\text{(DP guarantees)} \quad \sum_{t=1}^{T}\frac{P_k^{(t)}(C^{(t)}h_k^{(t)})^2}{\sigma^2\|N_k\mathbf{g}_k^{(t)}\|^2} \leq \Gamma_{dp}(\epsilon, \delta), \text{ for all } k. \tag{17.47}$$

**Convergence analysis:** Following the standard results on noisy gradient descent [66], under Assumptions 17.3 and 17.4, we can bound the average optimality gap at the end of iteration $T$, for a constant learning rate $\eta^{(t)} = 1/\nu$ for all $t$.

(Optimality gap bound for OMA)

$$
\mathrm{E}\left[L\left(\boldsymbol{\theta}^{(T)}\right) - L^*\right] \le \left(1 - \frac{\mu}{\nu}\right)^T \left[L\left(\boldsymbol{\theta}^{(0)}\right) - L^*\right] + \frac{d}{2\nu N^2} \sum_{t=1}^{T}\left(1 - \frac{\mu}{\nu}\right)^{T-t}
$$
$$
\times \sum_{k=1}^{K} \frac{\sigma^2}{(h_k^{(t)}\alpha_k^{(t)})^2}. \quad (17.48)
$$

(Optimality gap bound for NOMA)

$$
\mathrm{E}\left[L\left(\boldsymbol{\theta}^{(T)}\right) - L^*\right] \le \left(1 - \frac{\mu}{\nu}\right)^T \left[L\left(\boldsymbol{\theta}^{(0)}\right) - L^*\right] + \frac{d}{2\nu N^2} \sum_{t=1}^{T}\left(1 - \frac{\mu}{\nu}\right)^{T-t} \frac{\sigma^2}{(\beta^{(t)})^2}.
$$
$$
(17.49)
$$

The first term in Eqs. (17.48) and (17.49) is common also to the standard gradient descent without noise, and it indicates a geometric decay of initial optimality gap $L(\boldsymbol{\theta}^{(T)}) - L^*$ as $T$ increases. The results of OMA and NOMA differ in the second term, which represents the impact of DP requirement in Eq. (17.47). The second term in OMA is a summation of individual noises from all users over $k$. This is replaced by a single noise term in NOMA due to the efficiency of over-the-air aggregation. The opportunities for PA arise from the fact that the effect of the channel noise at iteration $t$ is weighted by a factor $(1 - \mu/\nu)^{T-t}$. Therefore, the noise added in the later iterations is more harmful to the optimality gap than the noise added in the initial. This can be leveraged to optimize PA, as discussed next.

## 17.4.3   Power Allocation

Before detailing adaptive PA, we introduce a reference suboptimal approach, namely, static power allocation for DP-FedSGD [76, 80]. This naive design divides up the DP constraint equally across all iterations, without taking into account the learning performance. That is, to satisfy the DP constraint in Eq. (17.47), it requires the condition $P_k^{(t)}(C^{(t)}h_k^{(t)})^2/(\sigma\|N_k\mathbf{g}_k^{(t)}\|)^2 < \Gamma_{\mathrm{dp}}/T$ for all $t = 1, \ldots, T$. By including the power constraint in (17.13) thus yields

$$
\text{(Static PA in OMA)} \quad P_k^{(t)} = \min\left\{\frac{\Gamma_{dp}\left(\sigma\|N_k\mathbf{g}_k^{(t)}\|\right)^2}{T\left(C^{(t)}h_k^{(t)}\right)^2}, PdK\right\}. \quad (17.50)
$$

In contrast, to achieve gradient alignment Eq. (17.41), each user in NOMA should scale down their power toward the one with minimum $h_k^{(t)}/\|N_k\mathbf{g}_k^{(t)}\|$. Thereby, the static power allocation in NOMA is given as

$$
\text{(Static PA in NOMA)} \quad P_k^{(t)} = \min\left\{\frac{\Gamma_{dp}\left(\sigma\|N_k\mathbf{g}_k^{(t)}\|\right)^2}{T\left(C^{(t)}h_k^{(t)}\right)^2}, \frac{\min_k\left(h_k^{(t)}/\|N_k\mathbf{g}_k^{(t)}\|\right)^2}{\left(h_k^{(t)}/\|N_k\mathbf{g}_k^{(t)}\|\right)^2}Pd\right\}.
$$
$$
(17.51)
$$

Adaptive power allocation aims to minimize the optimality bound Eq. (17.48) for OMA and Eq. (17.49) for NOMA, respectively, under $(\epsilon, \delta)$-DP constraint Eq. (17.47) and power constraints Eq. (17.13), for all $K$ devices across $T$ iterations.

We consider this here as an off-line optimization by assuming the parameters $\{h_k^{(t)}, C^{(t)}, \|\mathbf{g}_k^{(t)}\|\}$ are known beforehand. With this assumption, optimization problems over the power control parameters $\{P_k^{(1)}, \ldots, P_k^{(T)}\}$ are convex programs. Specifically, for OMA, the problem for each devices $k$ can be carried out in parallel, and the closed form solution is provided in [77] as follows.

**Adaptive PA in OMA:**

- If condition $\frac{Pd}{\sigma^2} \sum_{t=1}^{T} (C^{(t)} h_k^{(t)})^2 / \|N_k \mathbf{g}_k^{(t)}\|^2 < \Gamma_{\mathrm{dp}}$ holds,

$$(P_k^{(t)})_{\mathrm{opt}} = Pd. \tag{17.52}$$

- Otherwise, the optimal solution is

$$(P_k^{(t)})_{\mathrm{opt}} = \min \left\{ \frac{\left(\sigma \|N_k \mathbf{g}_k^{(t)}\|\right)^2 (\lambda_k)^{-\frac{1}{2}}}{\left(h_k^{(t)} \sqrt{C^{(t)}}\right)^2} \left(1 - \frac{\mu}{\nu}\right)^{-t/2}, Pd \right\}, \tag{17.53}$$

where the value of parameter $\lambda_k$ can be obtained by bisection to satisfy the constraint

$$\sum_{t=1}^{T} \min \left\{ \frac{(1 - \mu/\nu)^{-t/2} C^{(t)}}{\sqrt{\lambda_k}}, \frac{Pd (C^{(t)} h_k^{(t)})^2}{\left(\sigma \|N_k \mathbf{g}_k^{(t)}\|\right)^2} \right\} = \Gamma_{\mathrm{dp}}(\epsilon, \delta).$$

The solution of the first case is identical as that of without DP constraint, since it requires the power budget $Pd$ to be fully allocated for the transmission of the local gradient, thus minimizing the noise power that is harmful to the convergence rate. In this case, the privacy is obtained "for free" without degrading the learning performance of the system. However, this approach is not optimal if a strict DP constraint is enforced in the optimization since the larger power of noise benefits privacy. As shown in the second case, the transmitted power has to be scaled down in order to leverage the channel noise to ensure $(\epsilon, \delta)$-DP. The solutions of the two cases indicate that it is generally suboptimal to use static power allocation as introduced at the beginning of this section. Another interesting observation is that devices with larger datasets can attain privacy "for free" over a broader range of signal to noise (SNR) levels, as indicated by the condition that is less strict as $N_k$ increases.

**Adaptive PA in NOMA:** Unlike the optimization problem for OMA that can be tackled in parallel, in NOMA, the optimization over power control parameters are coupled by different users, since there is an additional constraint for gradient alignment. The problem can be formulated as an optimization over the alignments points $\{\beta^{(1)}, \ldots, \beta^{(T)}\}$. We refer to [77] for details.

A heuristic solution was also proposed in [77] for online PA that does not require a prior knowledge of parameters $\{h_k^{(t)}, C^{(t)}, \|\mathbf{g}_k^{(t)}\|\}$ across all communication blocks $t = 1, \ldots, T$. The design is built on iterative one-step-ahead optimization by using the predictions for the future parameters $\{h_k^{(t)}, C^{(t)}, \|\mathbf{g}_k^{(t)}\|\}$. We illustrate the key idea per $t$th iteration. At the $t$th iteration, we have the predicted values $\{\widehat{h}_k^{(t')}, \widehat{C}^{(t')}, \widehat{g}_k^{(t')}\}$ for $t' = t, t+1, \ldots, T$ and the accumulated DP cost for the past iterations given by $\Gamma_k^{(t-1)} = \sum_{t'=1}^{t-1} P_k^{(t')} (C^{(t')} h_k^{(t')})^2 / (\sigma \|N_k \mathbf{g}_k^{(t')}\|)^2$. The off-line solution is then applied to the interval $(t, t+1, \ldots, T)$ by replacing the true parameters $\{h_k^{(t)}, C^{(t)}, \|\mathbf{g}_k^{(t)}\|\}$ with their estimates $\{\widehat{h}_k^{(t')}, \widehat{C}^{(t')}, \widehat{g}_k^{(t')}\}$ and the DP constraint with its residual $\Gamma_{dp}(\epsilon, \delta) - \Gamma_k^{(t-1)}$. The solutions of power control $\{P_k^{(t)}\}_{k=1}^{K}$ are then applied for transmission, and the procedure is repeated for iteration $t+1$. The prediction of parameters $\{h_k^{(t)}, C^{(t)}, \|\mathbf{g}_k^{(t)}\|\}$ is detailed in [77].

## 17.5    Numerical Results

In this section, we evaluate the performance of digital and analog implementations of FedSGD to illustrate the impact of DP constraints on the learning performance. We first use a synthetic, randomly generated dataset that is comprised of $N = 10,000$ pairs $(\mathbf{u}, v)$, where the covariates $\mathbf{u} \in \mathbb{R}^{10}$ are drawn i.i.d. following distribution $\mathcal{N}(0, \mathbf{I})$, and the label $v$ for each vector $\mathbf{u}$ is obtained as $v = [\mathbf{u}]_2 + 3[\mathbf{u}]_5 + 0.2w_o$, where $[\mathbf{u}]_j$ is the $j$th entry in vector $\mathbf{u}$ and the observation noise $w_o \sim \mathcal{N}(0, 1)$ is i.i.d. across the samples. We consider ridge regression, and the sample-wise loss function is given as $\ell(\boldsymbol{\theta}; \mathbf{u}, v) = 0.5\|\boldsymbol{\theta}^T \mathbf{u} - v\|^2 + \lambda\|\boldsymbol{\theta}\|^2$ with $\lambda = 5 \times 10^{-3}$. The PL parameter $\mu$ and smoothness parameter $\nu$ are computed as the smallest and largest eigenvalues of the data Gramian matrix $\mathbf{U}^T \mathbf{U}/N + 2\lambda\mathbf{I}$, where $\mathbf{U} = [\mathbf{u}_1, \ldots, \mathbf{u}_N]^T$ is the data matrix of the training dataset. The initial value of $\boldsymbol{\theta}$ is set as an all-zero vector. We note that the (unique) optimal solution to the joint learning problem in Eq. (17.7) is $\boldsymbol{\theta}^* = (\mathbf{U}^T \mathbf{U} + 2N\lambda\mathbf{I})^{-1} \mathbf{U}^T \mathbf{v}$, where $\mathbf{v} = [v_1, \ldots, v_N]^T$ is the label vector. The channel model is considered to be additive white Gaussian noise (AWGN), where $h_k^{(t)} = 1$ for all $k$ and $t$, and the variance of channel noise is set as $\sigma = 1$. The performance is evaluated by using the normalized optimality gap $\mathbb{E}[L(\boldsymbol{\theta}^{(T)}) - L^*]/L^*$ after $T = 20$ iterations, and all the results are averaged over 100 experiments.

To evaluate the performance of the digital transmission, a setting with $K = 2$ devices is considered where each device has a disjoint subset of $N_k = 5,000$ samples. The gradient, quantization, and global model clipping parameters are set as $C^{(t)} = 4$, $C_e^{(t)} = 4$, and $C_\theta^{(t)} = 5$, respectively. On the communication phase, we consider a total $M = 25$ channel uses in NOMA transmission. The artificial noise is generated according to the binomial distribution with probability $p = 1/2$, and we set $\epsilon/T = 4$ per iteration with $\delta = 10^{-2}$.

In Fig. 17.3, NOMA is seen to outperform OMA due to the larger rate region. The larger transmission rates increase the number of available signal levels, which

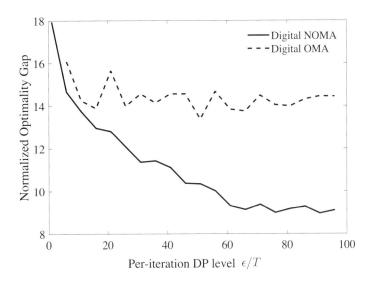

**Figure 17.3** Digital transmission: optimality gap versus *per-iteration* DP level $\epsilon/T$ for SNR $= 10$, $M = 40$, $\delta = 10^{-2}$.

can be used to improve both signal resolution and privacy. Note, however, that the privacy guarantees are hardly practical since the value of $\epsilon/T$ shown in the figure is per iteration and should be multiplied by the number of iterations $T$ to obtain the total DP level. This problem is addressed by analog transmission, as discussed next.

We now turn to evaluate the performance of analog transmission in a setting with $K = 10$ devices, with each device having a disjoint subset of $N_k = 1,000$ samples. By default, we set $\delta = 10^{-2}$ and $C^{(t)} = 4$. Furthermore, we consider the limited searching space $\|\boldsymbol{\theta}\| \leq 3.5$ to enhance the robustness of DP-FedSGD under large noise perturbation. To this end, the updated global model is clipped as $\boldsymbol{\theta} = \min\{1, 3.5/\|\boldsymbol{\theta}\|\}\boldsymbol{\theta}$ if its norm beyond the threshold. Figure 17.4 plots the normalized optimality gap versus the *total* privacy level $\epsilon$ by using NOMA with $M = 10$. The figure shows that the performance gain due to adaptive PA is more pronounced in the more challenging regime with stricter DP constraints, while the performance under both static and adaptive PA solutions converges to the one without DP constraint as $\epsilon$ increases. The figure also shows that for $\epsilon \geq 10$, adaptive PA achieves DP "for free" in the sense that no performance loss is measured in the presence of a DP constraint.

The impact of SNR, $P/\sigma^2$, for analog NOMA with $M = 10$ is shown in Fig. 17.5. The normalized optimality gap of all schemes is seen to decrease with the value of SNR until the DP constraint limits the learning performance. Furthermore, the performance of static PA diverges from that of the scheme without DP constraint for SNR $> 0$ dB, while such divergence happens to adaptive PA for SNR $> 15$ dB. This confirms the advantage of adaptive PA that is able to attain privacy "for free" in a broader SNR regime.

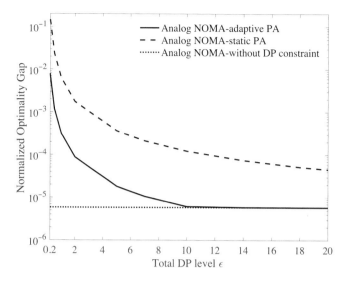

**Figure 17.4** Analog transmission: optimality gap versus the *total* DP level $\epsilon$ (for $\delta = 10^{-2}$) for different power allocation (PA) schemes and for the scheme without DP constraint (SNR $= 10$ dB).

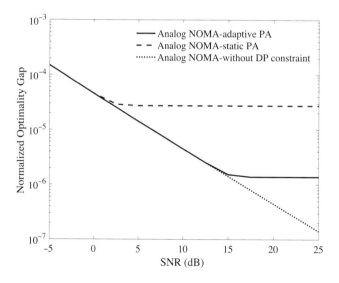

**Figure 17.5** Analog transmission: optimality gap versus SNR for different power allocation (PA) schemes and for the scheme without DP constraint (total DP level $\epsilon = 30$, $\delta = 10^{-2}$).

## 17.6    Concluding Remarks

This chapter has presented a review of differentially private FL over wireless networks. The presence of a noisy channel between the transmitters and server was seen to provide a challenge for digital communication methods, while offering new opportunities for privacy via analog transmission. Under digital transmission, the privacy mechanism we considered was composed of quantization and additive binomial noise. Within the limited channel capacity, one hence needs to allocate both signal levels, carrying useful information and noise levels necessary to ensure privacy. In contrast, for uncoded analog transmission, the additive channel noise can be harnessed as a privacy mechanism. Given a targeted DP level, we have seen how power control can be useful to meet privacy constraints while improving convergence.

Both digital and analog implementations reviewed in this chapter can be improved and generalized. As an example, the analog transmission scheme can be extended to operate under any ratio between the size of the model and the number of channel uses available per iteration by applying linear or nonlinear gradient compression, or expansion, techniques.

From the perspective of modelling DP, the threat model could also include honest-but-curious edge devices, which would generally incur a larger DP loss. The study could be further generalized to other network topologies including multihop device-to-device (D2D) networks as studied in [81, 82]. Finally, the considered design approaches could be extended to other forms of federated learning that are more challenging to analyze, such as Federated Averaging.

# References

[1] J. Park et al., "Wireless network intelligence at the edge," *Proc. IEEE*, vol. 107, no. 11, pp. 2204–2239, 2019.

[2] Z. Zhou et al., "Edge intelligence: Paving the last mile of artificial intelligence with edge computing," *Proc. IEEE*, vol. 107, no. 8, pp. 1738–1762, 2019.

[3] G. Zhu et al., "Toward an intelligent edge: wireless communication meets machine learning," *IEEE Communications Magazine*, vol. 58, no. 1, pp. 19–25, 2020.

[4] C. Zhang, P. Patras, and H. Haddadi, "Deep learning in mobile and wireless networking: A survey," *IEEE Communications Surveys and Tutorials*, vol. 21, no. 3, pp. 2224–2287, 2019.

[5] M. Chen et al., "A joint learning and communications framework for federated learning over wireless networks," *arXiv preprints*, arXiv:1909.07972, 2020.

[6] T. Li et al., "Federated learning: Challenges, methods, and future directions," *IEEE Signal Processing Magazine*, vol. 37, no. 3, pp. 50–60, 2020.

[7] P. Kairouz et al., "Advances and open problems in federated learning," *arXiv preprint*, arXiv:1912.04977, 2019.

[8] H. B. McMahan et al., "Federated learning of deep networks using model averaging," *arXiv preprint*, arXiv:1602.05629, 2016.

[9] Y. Qiang et al., "Federated machine learning: Concept and applications," *ACM Trans. on Intelligent Systems and Technology*, vol. 10, no. 2, pp. 12:1–12:19, Jan. 2019.

[10] F. Hanzely and P. Richtárik, "Federated learning of a mixture of global and local models," *arXiv preprint*, arXiv:2002.05516, 2020.

[11] S. Dhakal et al., "Coded federated learning," *arXiv preprint*, arXiv:2002.09574, 2020.

[12] S. J. Reddi et al., "Adaptive federated optimization," *arXiv preprint*, arXiv:2003.00295, 2020.

[13] A. M. Elbir and S. Coleri, "Federated learning for vehicular networks," *arXiv preprint*, arXiv:2006.01412v1, 2020.

[14] S. Wang et al., "Adaptive federated learning in resource constrained edge computing systems," *IEEE Journal on Selected Areas in Communications*, vol. 37, no. 6, pp. 1205–1221, 2019.

[15] X. Wang et al., "In-edge AI: Intelligentizing mobile edge computing, caching and communication by federated learning," *IEEE Network*, vol. 33, no. 5, pp. 156–165, Sep. 2019.

[16] H. B. McMahan and D. Ramage, "Communication-efficient learning of deep networks from decentralized data," in *20th Int. Conf. on Artificial Intelligence and Statistics (AISTATS)*, pp. 1–10, 2017.

[17] J. Konečný et al., "Federated learning: Strategies for improving communication efficiency," *arXiv preprint*, arXiv:1610.05492, 2017.

[18] U. Mohammad and S. Sorour, "Adaptive task allocation for asynchronous federated mobile edge learning," *arXiv preprint*, arXiv:1905.01656, 2020.

[19] N. H. Tran et al., "Federated learning over wireless networks: Optimization model design and analysis," in *IEEE Conf. Computer Communications*, pp. 1387–1395, 2019.

[20] H. H. Wang et al., "Scheduling policies for federated learning in wireless networks," *IEEE Trans. on Communications*, vol. 68, no. 1, pp. 317–333, 2020.

[21] R. Shokri and V. Shmatikov, "Privacy-preserving deep learning," in *22nd ACM SIGSAC Conf. Computers and Communications Security (CCS)*, pp. 1310–1321, 2015.

[22] M. Hao et al., "Towards efficient and privacy-preserving federated deep learning," in *53rd IEEE Int. Conf. on Communications (ICC)*, pp. 1–6, 2019.

[23] L. Zhao et al., "Privacy-preserving collaborative deep learning with unreliable participants," *IEEE Trans. on Information Forensics and Security*, vol. 15, pp. 1486–1500, Sep. 2020.

[24] L. Melis et al., "Exploiting unintended feature leakage in collaborative learning," in *IEEE Symp. Security and Privacy (SP)*, pp. 691–706, 2019.

[25] M. Fredrikson, S. Jha, and T. Ristenpart, "Model inversion attacks that exploit confidence information and basic countermeasures," in *Proc. ACM SIGSAC Conf. on Computers and Communications Security (CCS)*, Oct. 2015.

[26] A. Bhowmick et al., "Protection against reconstruction and its applications in private federated learning," *arXiv preprint*, arXiv:1812.00984v2, 2018.

[27] R. Shokri et al., "Membership inference attacks against machine learning models," in *IEEE Symp. on Security and Privacy (SP)*, pp. 3–18, 2017.

[28] M. Al-Rubaie and J. M. Chang, "Reconstruction attacks against mobile-based continuous authentication systems in the cloud," *IEEE Trans. on Information Forensics and Security*, vol. 11, no. 12, pp. 2648–2663, 2016.

[29] W. Wei et al., "A framework for evaluating gradient leakage attacks in federated learning," *arXiv preprint*, arXiv:2004.10397, 2020.

[30] C. Zhao et al., "Secure multi-party computation: Theory, practice and applications," *Information Sciences*, vol. 476, pp. 357–372, 2019.

[31] O. Goldreich, "Secure multi-party computation," unpublished manuscript, 1998; https://www.wisdom.weizmann.ac.il/~oded/pp.html.

[32] V. Nikolaenko et al., "Privacy-preserving ridge regression on hundreds of millions of records," in *IEEE Symp. on Security and Privacy*, pp. 334–348, 2013.

[33] P. Mohassel and Y. Zhang, "Secureml: A system for scalable privacy-preserving machine learning," in *IEEE Symp. on Security and Privacy*, pp. 19–38, 2017.

[34] C. Dwork, "Differential privacy: A survey of results," in *Int. Conf. Theory and Applications of Models of Computation (TAMC)*, vol. 4978, pp. 1–19, 2008.

[35] C. Dwork et al., "The algorithmic foundations of differential privacy," *Foundations and Trends in Theoretical Computer Science*, vol. 9, no. 3–4, pp. 211–407, 2014.

[36] J. Duchi, M. J. Wainwright, and M. I. Jordan, "Local privacy and minimax bounds: Sharp rates for probability estimation," in *Advances in Neural Information Processing Systems*, pp. 265–284, 2013.

[37] M. Al-Rubaie and J. M. Chang, "Minimax optimal procedures for locally private estimation," *Journal of the American Statistical Association*, vol. 113, no. 521, pp. 82–201, 2018.

[38] N. Wang et al., "Collecting and analyzing multidimensional data with local differential privacy," in *IEEE 35th Int. Conf. Data Engineering (ICDE)*, pp. 638–649, 2019.

[39] N. Agarwal et al., "cpSGD: Communication-efficient and differentially-private distributed SGD," in *Advances in Neural Information Processing Systems (NIPS)*, Dec. 2018.

[40] M. Abadi et al., "Deep learning with differential privacy," in *Proc. ACM SIGSAC Conf. Computers and Communications Security (CCS)*, Oct. 2016.

[41] C. Dwork et al., "Calibrating noise to sensitivity in private data analysis," in *Theory of Cryptography Conf. (TCC)*, pp. 265–284, 2006.

[42] N. Wu et al., "The value of collaboration in convex machine learning with differential privacy," *arXiv preprint*, arXiv:1906.09679, 2019.

[43] R. C. Geyer, T. Klein, and M. Nabi, "Differentially private federated learning: A client level perspective," *arXiv preprint*, arXiv:1712.07557, 2017.

[44] K. Wei et al., "Federated learning with differential privacy: Algorithms and performance analysis," *IEEE Trans. on Information Forensics and Security*, 2020.

[45] B. Balle, G. Barthe, and M. Gaboardi, "Privacy amplification by subsampling: Tight analyses via couplings and divergences," in *Advances in Neural Information Processing Systems (NIPS)*, Dec. 2018.

[46] Y. Zhao et al., "Local differential privacy based federated learning for internet of things," *arXiv preprint*, arXiv:2004.08856, 2020.

[47] S. Truex et al., "A hybrid approach to privacy-preserving federated learning," *arXiv preprint*, arXiv:1812.03224, 2018.

[48] Y. Li et al., "Asynchronous federated learning with differential privacy for edge intelligence," *arXiv preprint*, arXiv:1912.07902, 2019.

[49] R. Hu, Y. Gong, and Y. Guo, "CPFed: Communication-efficient and privacy-preserving federated learning," *arXiv preprint*, arXiv:2003.13761, 2020.

[50] V. Gandikota, R. K. Maity, and A. Mazumdar, "vqSGD: Vector quantized stochastic gradient descent," *arXiv preprint*, arXiv:1911.07971v4, 2019.

[51] Z. Huang et al., "Dp-admm: Admm-based distributed learning with differential privacy," *IEEE Trans. on Information Forensics and Security*, vol. 15, pp. 1002–1012, 2019.

[52] A. Elgabli et al., "Harnessing wireless channels for scalable and privacy-preserving federated learning," *arXiv preprint*, arXiv:2007.01790v2, Jul. 2020.

[53] Z. Huang, S. Mitra, and G. Dullerud, "Differentially private iterative synchronous consensus," in *Proc. ACM Workshop Privacy in the Electronic Society*, pp. 81–90, 2012.

[54] L. Yu and P. Cuff, "Differential privacy as a mutual information constraint," in *23rd ACM SIGSAC Conf. on Computers and Communications Security (CCS)*, pp. 43–54, 2016.

[55] S. Yagli, A. Dytso, and H. V. Poor, "Information-theoretic bounds on the generalization error and privacy leakage in federated learning," in *Proc. IEEE Intl. Workshop Signal Process. Advances Wireless Comm. (SPAWC)*, May 2020.

[56] D. Alistarh et al., "The convergence of sparsified gradient methods," in *Advances in Neural Information Processing Systems*, pp. 5973–5983, 2018.

[57] N. S. Shai Shalev-Shwartz and T. Zhang, "Trading accuracy for sparsity in optimization problems with sparsity constraints," *SIAM Journal on Optimization*, 2010.

[58] H. Wang et al., "Atomo: Communication-efficient learning via atomic sparsification," in *Advances in Neural Information Processing Systems*, pp. 9850–9861, 2018.

[59] F. Seide et al., "1-bit stochastic gradient descent and its application to data-parallel distributed training of speech dnns," in *INTERSPEECH*, pp. 9850–9861, 2014.

[60] J. Bernstein et al., "signsgd: Compressed optimization for non-convex problems," in *Advances in Neural Information Processing Systems*, pp. 560–569, 2018.

[61] T. Li et al., "Privacy for free: Communication efficient learning with differential privacy using sketches," *arXiv preprint*, arXiv:1911.00972, 2019.

[62] H. B. McMahan, D. R. K. Talwar, and L. Zhang, "Learning differentially private recurrent language models," in *Int. Conf. Learning Representations (ICRL)*, 2018.

[63] W.-T. Chang and R. Tandon, "Communication efficient federated learning over multiple access channels," *arXiv preprint*, arXiv:2001.08737v1, 2020.

[64] M. M. Amiri and D. Gündüz, "Federated learning over wireless fading channels," *IEEE on Wireless Communications*, 2020.

[65] H. Karimi, J. Nutini, and M. Schmidt, "Linear convergence of gradient and proximal-gradient methods under the Polyak-Lojasiewicz condition," in *Joint European Conf. on Machine Learning and Knowledge. Discovery in Databases (ECML KDD)*, 2016.

[66] L. Bottou, F. E. Curtis, and J. Nocedal, "Optimization methods for large-scale machine learning," *SIAM Review*, vol. 60, no. 2, pp. 223–311, 2018.

[67] W. Wen et al., "Terngrad: Ternary gradients to reduce communication in distributed deep learning," in *Advances in Neural Information Processing Systems*, pp. 1509–1519, 2017.

[68] D. Alistarh et al., "QSGD: Communication-efficient SGD via gradient quantization and encoding," in *Advances in Neural Information Processing Systems (NIPS)*, Dec. 2017.

[69] J. Bernstein et al., "signSGD: Compressed optimisation for non-convex problems," in *Proc. Int. Conf. Machine Learning (PMLR)*, July 2018.

[70] A. Sonee and S. Rini, "Efficient federated learning over multiple access channel with differential privacy constraints," *arXiv preprint*, arXiv:2005.07776v2, 2020.

[71] A. El-Gamal and Y.-H. Kim, *Network Information Theory*. Cambridge University Press, 2012.

[72] G. Zhu, Y. Wang, and K. Huang, "Broadband analog aggregation for low-latency federated edge learning," *IEEE Trans. on Wireless Communications*, 2019.

[73] K. Yang et al., "Federated learning via over-the-air computation," *IEEE Trans. on Wireless Communications*, vol. 19, no. 3, pp. 2022–2035, 2020.

[74] M. M. Amiri and D. Gündüz, "Machine learning at the wireless edge: Distributed stochastic gradient descent over-the-air," *IEEE Trans. on Signal Processing*, vol. 68, pp. 2155–2169, 2020.

[75] T. Sery and K. Cohen, "On analog gradient descent learning over multiple access fading channels," *IEEE Trans. on Signal Processing*, vol. 68, pp. 2897–2911, 2020.

[76] M. Seif, R. Tandon, and M. Li, "Wireless federated learning with local differential privacy," *arXiv preprint*, arXiv:2002.05151v1, 2020.

[77] D. Liu and O. Simeone, "Privacy for free: Wireless federated learning via uncoded transmission with adaptive power control," *arXiv preprint*, arXiv:2006.0545, Jun. 2020.

[78] M. P. Friedlander and M. Schmidt, "Hybrid deterministic-stochastic methods for data fitting," *SIAM Journal on Scientific Computing*, vol. 34, no. 3, pp. A1380–A1405, 2012.

[79] P. Zhao and T. Zhang, "Stochastic optimization with importance sampling for regularized loss minimization," in *Intl. Conf. Machine Learning (ICML)*, July 2015.

[80] Y. Koda et al., "Differentially private aircomp federated learning with power adaptation harnessing receiver noise," *arXiv preprint*, arXiv:2004.06337, 2020.

[81] H. Xing, O. Simeone, and S. Bi, "Decentralized federated learning via SGD over wireless D2D networks," *arXiv preprint*, arXiv:2002.12507, 2020.

[82] E. Ozfatura, S. Rini, and D. Gunduz, "Decentralized SGD with over-the-air computation," *IEEE Global Communications Conf. (GLOBECOM)*, 2020.

# 18 Timely Wireless Edge Inference

Sheng Zhou, Wenqi Shi, Xiufeng Huang, and Zhisheng Niu

## Notation

| | |
|---|---|
| AFM | atomic force microscope |
| AKPZ | anisotropic KPZ equation |
| $a_0$ | lattice constant |
| $c_q(\ell)$ | qth order correlation function |
| $d_E$ | embedding dimension |
| $d_f$ | fractal dimension |
| $L$ | system size |
| $\equiv$ | *defined* to be equal |
| $\sim$ | *asymptotically* equal (in scaling sense) |
| $\approx$ | *approximately* equal (in numberical value) |

## 18.1 Introduction

With recent breakthroughs in artificial intelligence (AI), people are witnessing explosive growth in AI-based intelligent services and applications. Deep neural networks (DNNs), as an essential AI technology, have achieved state-of-the-art performance in many areas, including computer vision [1], natural language processing [2], audio recognition [3], and wireless channel estimation [4]. Once properly trained, these DNN models have strong prediction and fitting capabilities and the ability to generalize predictions on unseen data (known as inference) [5]. As a result, DNN models can be trained for various purposes, and deploying them on mobile devices brings the possibility of plentiful intelligent end-user applications, such as intelligent voice assistants, smart wearable devices, and autonomous vehicles. These have profound impacts on our lifestyles and will continuously bring benefits to the people and society.

However, most existing DNN models are memory and computation intensive. For instance, the state-of-the-art natural language processing model, GPT-3 produced by OpenAI [2], has about 175 billion parameters. These gigantic DNN models bring stringent requirements on the end devices' computation and memory resources, which

The authors of this chapter would like to acknowledge the support of the National Key R&D Program of China 2018YFB1800804, the Nature Science Foundation of China (no. 61871254, no. 91638204, and no. 61861136003), and Hitachi Ltd.

makes it hardly possible for everyone to enjoy high-quality intelligent services. On the other hand, real-time intelligent applications and services usually have timeliness requirements; that is, the results must be returned within low latencies. For instance, voice assistants like Apple Siri and Microsoft Cortana must response very fast (e.g., within 1 second) after receiving requests from the user. To ensure safety, autonomous driving vehicles must frequently perceive the surrounding environments (e.g., at least 30 frames per second), and then accordingly the inference latency of the environment perception DNN should be no more than 33.3 ms. Moreover, if DNN is applied to estimate channel state information of fast fading channels, the estimation results must be returned within the channel coherence time, which can be as short as several milliseconds. Although the mobile devices have increasing computation capabilities, they can hardly keep up with the increasing complexity of DNN models, and battery life is also a major limitation on mobile devices. Therefore, it is challenging to directly deploy state-of-the-art DNN models on resources constrained mobile devices, while satisfying timeliness requirements.

To tackle these issues, traditional wisdom from the cloud computing and the deep learning communities suggests two different approaches. An intuitive way of reducing the inference latency of mobile devices is (cloud-based) task offloading, where the task input is uploaded to the powerful cloud data center, and the inference result is sent back after the cloud server completes the computation. However, the data volume can be large for images and videos, and the transmission latency may be unacceptable when the channel is bad or the network is congested. Also, uploading raw input data to the cloud may increase the risk of privacy leakage. An alternative approach is to derive lightweight DNN models, so they can meet the latency requirements when being deployed on mobile devices. Existing methods of deriving lightweight DNN models can be divided into two categories: model architecture design and model compression. Efficient DNN architectures can be either designed by human experts or searched by computers. While in model compression, technologies like low-rand approximation, knowledge distillation, pruning, and quantization can be leveraged to produce a compact model from a complex model. (For details, please refer to [6, 7] and references therein). Although the derived lightweight DNN models are far more computation- and energy-efficient, there will be a notable accuracy loss compared to the state-of-the-art DNN models. Therefore, both these approaches cannot afford reliable and low-latency DNN inference for mobile devices, which is in particular critical for applications like autonomous driving.

Recently, mobile edge computing (MEC) has been emerging as an essential element of 5G and beyond radio access networks [8, 9]. In order to provide computing services in proximity to end users, the computation resources can be deployed at the edge of networks, such as gateways, base stations, vehicles, and roadside units. As a result of proximal services, MEC has two main advantages over traditional cloud computing: (a) ultra-low latency and high reliability and (b) increased privacy and security. Since the edge server is closer to the end user, high and fluctuating latency of transmitting data over the Internet can be avoided. Also, by avoiding uploading the users' data to the remote cloud, the risk of hijack and extraction of the private data

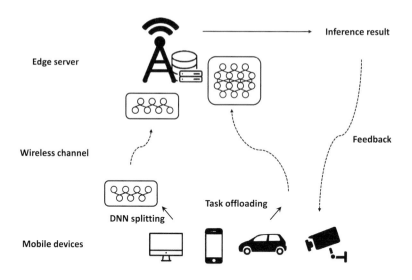

**Figure 18.1** Edge inference.

can be reduced. These advantages make MEC a promising candidate for addressing the key challenges of timely inference, especially for resource constrained mobile devices. A new paradigm, edge inference, where the DNN inference task at the mobile device can be fully or partially offloaded to the edge server, has been recently proposed [10, 11].

As shown in Fig. 18.1, in edge inference systems, mobile devices and edge servers can complete inference tasks cooperatively in order to meet the timeliness requirements. Mobile devices offload inference tasks to the edge server, and inference results are sent back to the device upon completion. It is also possible to split the DNN model and deploy the lower part at the mobile device and the rest on the edge server. The intermediate features or data are uploaded via the wireless channel. Since inference results are often lightweight, the latency of result feedback can be ignored. As a result, the total inference latency, defined as the time elapsed from task arrival or generation at the mobile device to the reception of inference result, mainly consists of following parts: (a) communication latency for uploading the data via the wireless channel, (b) mobile computation latency for the local computations; and (c) edge computation latency at the edge server. Since the computation resources and the wireless bandwidth are always limited, the communication and computation in edge-inference systems should be carefully managed and optimized to satisfy stringent timeliness requirements.

Although many existing works on MEC have studied latency-optimized task offloading, general computation tasks and models are mainly considered. While considering the specific DNN inference, new research opportunities and challenges will emerge, and they are the focus of this chapter. In the following parts of this chapter, state-of-the-art edge-inference technologies with the aim of reducing inference latency are reviewed, followed by two case studies on DNN splitting and inference task offloading.

## 18.2        Device-Edge Coinference

### 18.2.1        Neural Network Splitting and Pruning

Most DNNs consist of several types of basic structure and module, such as, fully connected layer, convolutional layer, and long short-term memory (LSTM) layer. Therefore, a large DNN model can be intrinsically split into parts. As shown in Fig. 18.1, in edge-inference systems, the mobile device can compute the inference task up to an intermediate layer of the whole DNN model (i.e., the lower part of the DNN), upload the intermediate data (the output of the last layer of the lower part DNN) to the edge server, and download the inference result after the edge server completes the rest part of the DNN (i.e., the higher part of the DNN). The optimal split point depends not only on various system factors, including the wireless channel conditions and the computation capabilities of the mobile device and edge server, but also the DNN model itself. As a result, how to select a split point in different scenarios needs to be considered.

For this purpose, the Neurosurgeon framework has been proposed [12]. A general DNN splitting process has been proposed, which consists of (a) measuring or estimating the cost of running different DNN layers and transmitting the intermediate data between layers; (b) predicting the total cost of all candidate DNN split points; and (c) selecting the optimal split points according to the objective, such as, inference latency and mobile energy consumption. A regression-based estimation method has been proposed in [12] to estimate the inference latency of each layer. By measuring the DNN inference latency via runtime profiling, the authors of [13] formulate the split-point selection as a shortest path problem, which is proved to be NP-hard, and they proposed an approximation solution. In [14], the authors study the DNNs that are characterized by a directed acyclic graph (DAG) rather than a chain. They prove that optimizing the DNN splitting for these kinds of DNNs is NP-hard and proposed two approximation methods for both heavy and light workload scenarios.

Furthermore, as shown by [12, 13], the intermediate data can be much larger than the input data for some kind of DNN models, making DNN splitting inefficient. Therefore, intermediate data volume reduction technologies have been proposed. Both lossy and lossless JPEG coding for the intermediate data have been introduced in [15], while PNG coding is studied in [13]. More aggressively, [16, 17] add side branch classifiers to the intermediate data in order to leverage the intermediate data to get early inference results (known as early exit). If an early inference result has a high degree of confidence, the rest of the inference can be saved, and thus there is no need to transmit the intermediate data.

Recently, some researchers propose combining model compression technologies with DNN splitting [18–20]. In one of the pioneering works [18], structured pruning technology was leveraged, so that some unimportant parts in the original DNN model (e.g., neurons or convolutional filters) can be removed, resulting in a smaller and faster DNN model. This can reduce the computation workload of the inference task and the volume of the intermediate data as well. A two-step-pruning algorithm has been proposed to jointly optimize the DNN pruning and splitting to minimize inference latency, while preserving the accuracy of the model. Details about the two-step-pruning algorithm will be provided in the next section of this chapter. The advantage of using pruning to reduce the volume of intermediate data is two-fold. First, pruning can be viewed as a learning-driven feature compression method, since the DNN can learn which part is unimportant and needs to be removed. As a result, the DNN model and the feature compression are jointly optimized via pruning, leading to a better accuracy performance compared to separately designed feature compression technologies (e.g., [13, 15]). Second, the output of the pruning process is still a DNN model and thus can be combined with existing methods, like early exit technology [17], to further reduce the inference latency. Due to these advantages, the idea of joint DNN pruning and splitting is leveraged and extended in [19, 20].

## 18.2.2     Joint Source and Channel Coding for Coinference

In device-edge coinference systems, the intermediate data as the result of DNN splitting is transmitted between mobile devices and the edge server. To ensure timely edge inference, an important point of optimization is reducing the cost of transmitting the intermediate layer outputs of the neural network. Traditional method focus on reducing the size of the neural network. On the one hand, light-weight architectures of neural network model like MobileNet [21] and ShuffleNet [22] are proposed. On the other hand, techniques like gradient compression with quantization [23] and sparsification [24], model prunning [25], and network distillation [26] can further reduce the size of original models for edge-inference systems.

However, these methods do not consider the noisy wireless channel. Bottlenet++ [27] proposes a feature compression scheme that takes the channel state into account. An end-to-end trainable architecture is introduced for intermediate result compression, which takes the noisy wireless channel as a nontrainable neural network layer. Based on that, the authors exploit the joint source and channel coding. Simulation results show that Bottlenet++ can achieve aggreseive compression while guaranteeing high inference performance. With joint source and channel coding, [20] proposes an efficient communication framework for device-edge coinference, including split-point selection, communication-aware model compression, and task-oriented encoding of intermediate data or features.

To apply joint source and channel coding for practical scenarios, [28] optimizes communication schemes for wireless image retrieval systems. In practical scenarios, retrieval tasks cannot be performed locally at the mobile devices due to their limited computational power and lack of database for retrieval. Therefore, coinference is needed for image retrieval, and they propose a neural network for retrieval-oriented image compression. For the transmission of analog signals over wireless channels, the authors exploit joint source and channel coding that maps the feature vector directly to channel inputs. Simulation results show that the joint source and channel coding scheme can significantly increase the end-to-end accuracy and speed up the encoding process. The work [19] considers the image classification task with device-edge coinference. To deal with the errors that may be introduced over the noisy wireless channel, the authors propose an autoencoder-based network for intermediate feature map transmission.

### 18.2.3    Inference-Aware Scheduling

Besides reducing the model sizes and data volume, another important aspect of saving communication resources and providing timely inference is scheduling. A good scheduling scheme can allocate communication and computation resources to the most proper task, so as to optimize the metrics like energy efficiency and latency, according to the state of queued tasks and resources.

In this regard, [29] considers a mobile-edge coinference system with multiple mobile devices where the objective is minimizing their energy consumption, while completing the inference tasks by given deadlines. Offloading tasks to the edge server can potentially bring timley inference and save precious computation resources of mobile devices. However, limited wireless resources may fail to support the transmissions of a large amount of task offloading. Therefore, a scheduling scheme that allocates communication resources to mobile devices according to the usage of bandwidth and the number of queued tasks on mobile devices is needed. The scheduling problem is formulated as a convex optimization problem, and an offloading priority function depending on the channel gain and local computing energy consumption is derived for making scheduling decisions. Similar to this work, [30] also formulates the energy consumption minimization problem as a convex optimization, considering the local computing cost and transmission cost. This work exploits an important feature that in some multiuser systems there are data samples shared by different users in proximity, which can be exploited to save the transmission costs. Based on this insight, they propose a joint computation offloading and communications resource allocation scheme that outperforms the baseline without considering shared data property.

The dual problem of minimizing energy consumption under the latency constraint is minimizing the total delay with given energy. The work [31] considers an edge-inference system with random task arrivals and renewable energy arrivals. The scheduling problem is formulated as a Markov decision process, and the minimized long-term cost is the average inference latency. The authors exploit deep Q-network to learn the optimal offloading policy without having a priori knowledge of the dynamic statistics. Moreover, a mixed metric that is the weighted sum of energy consumption and inference latency is considered in [32]. By taking the channel conditions and the computation task queue as the system state, the scheduling problem is also modeled as Markov decision process, and the authors exploit Q-learning to balance the trade-off between the energy consumption and inference latency.

### 18.3    Pruning-Based Dynamic Neural Network Splitting for Edge Inference

As explained in Section 18.1, the inference latency of DNN splitting mainly consists of three parts: mobile computation latency, transmission latency of the intermediate data, and edge computation latency. The first two parts become the bottleneck of latency performance because of the limited wireless bandwidth and mobile computation resources. As such, pruning can be leveraged to reduce inference latency of both the mobile computation and wireless transmission. In this section, we will propose a two-step pruning framework to enable dynamic DNN splitting for timely edge inference. In the proposed framework, the original DNN is trained off-line and pruned first. Then the output pruned DNN models are profiled based on the optimal split point, and the corresponding DNN model is selected by a proposed algorithm.

### 18.3.1    Neural Network Pruning

Recently, network pruning has become an important direction among DNN model compression methods. Network pruning technologies can be divided into two main categories: structured pruning and unstructured pruning. For unstructured pruning, unimportant connections in the DNN are discarded by setting corresponding parameters to zero, which causes the sparsity and irregularity of the DNN to increase. As a result, although unstructured pruning yields high efficiency and low inference latency on some specific DNN inference engines, it may increase the actual inference latency when deployed on commercial hardware platforms (e.g., a CPU or GPU) due to the increased irregularity [33]. On the other hand, structured pruning discards convolutional filters (also called filter pruning or channel pruning) or entire DNN layers, so as to avoid increasing irregularity of the DNN. Therefore, structured pruning is able to reduce inference latency when deployed on general commercial hardware and is increasingly attracting attention [34, 35]. Because some convolutional filters are removed during the structured pruning, the volume of intermediate data in the resulting network is also reduced. This byproduct of pruning makes it suitable for edge inference, because the mobile computation latency and transmission latency can be reduced simultaneously.

A general filter pruning process is shown in Fig. 18.2. First, the original DNN model and the pruning range (i.e., which part of the DNN model is going to be pruned) will be given as the input of the pruning process. Then, all filters in the pruning range are ranked according to some given criteria, such as the magnitudes of their weights [25], their impacts on the energy consumption [36], or their significance on the loss function [37]. Here we use the last criteria, which accounts for the importance of each filter for the accuracy of the DNN. After that, we can remove unimportant filters and fine tune the pruned DNN to recover the accuracy loss caused by pruning. Subsequently, the accuracy of the pruned DNN after fine-tuning is tested on the test

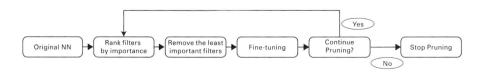

**Figure 18.2**  General filter pruning process.

dataset. The aforementioned pruning iteration can be continued, until the accuracy drops below a given accuracy threshold.

## 18.3.2    Two-Step Pruning

The goal of conventional pruning methods is to accelerate the inference or reduce the computation energy consumption. Nevertheless, reducing the volume of intermediate data, which is also critical in edge-inference systems, has not been considered yet. Therefore, we propose a two-step pruning method as shown in Fig. 18.3.

In *pruning step 1*, the pruning range is the entire DNN. Since filters in any layer can be pruned in this step, the pruned DNN has a similar shape (i.e., the distribution of the number of filters in each layer) compared to the original DNN. As a result, the volume of intermediate data between layers of the pruned DNN may still be larger than the original input data. The mobile computation latency is reduced, while the transmission latency may still stay high when partitioning at the front end of the DNN.

The intermediate data to be transmitted is the output of the last layer in the front-end part of the DNN. Based on this observation, we can further introduce the second-step pruning, which prunes the last layer in the front-end part of the DNN (i.e., the layer right before the split point) to reduce the transmission load. However, the split point should be determined before second-step pruning, because pruning can change the structure of the DNN and further affect the selection of the split point. Therefore, there is a tight coupling between split-point selection and second-step pruning, and it is hard to decouple them due to the difficulty in predicting the resultant DNN structure after pruning. As a result, we propose a brute force search process to derive good pruned DNN models for each potential split point, as follows.

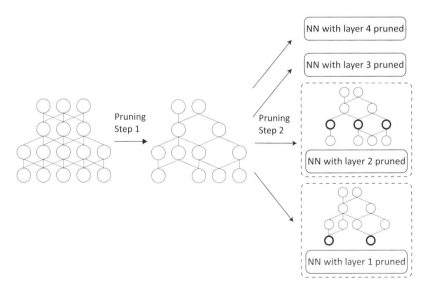

**Figure 18.3**  Proposed two-step pruning.

In *pruning step 2*, we individually apply the pruning method to each layer in the pruned DNN output of pruning step 1 by restricting the pruning range to each layer. A series of pruned DNNs will be generated for each layer that corresponds to the split point right after this layer. Hence, all DNNs as outputs of pruning step 2 have only one different layer as compared to the result of pruning step 1. After the pruning is completed, all pruned DNNs produced by pruning step 2 are profiled and stored for the profiling and split-point selection stage.

## 18.3.3    Profiling and Split-Point Selection

In order to select the optimal split point and its corresponding pruned DNN model, some profiles are needed:

1. $D_{i,j}$: the output data volume of the $i$th layer in the pruned DNN corresponding to the split point right after the $j$th layer;
2. $t_{i,j}^{\text{edge}}$: the edge computation latency of the $i$th layer in the pruned DNN corresponding to the split point right after the $j$th layer;
3. $A_i$: the accuracy of pruned DNN corresponding to the split point right after the $i$th layer;
4. $R$: the transmission rate of the wireless channel; and
5. $t_{i,j}^{\text{mobile}}$: the mobile computation latency of the $i$th layer in the pruned DNN corresponding to the split point right after the $j$th layer.

Except the mobile computation latency, all the parameters can either be profiled offline (by running the pruned DNNs at edge server) or estimated in real time. Since profiling all pruned DNNs at mobile device is impossible, we introduce $\gamma = \frac{t_{i,j}^{\text{mobile}}}{t_{i,j}^{\text{edge}}}$, as the ratio of the computation capability between the mobile device and the edge server. Further, we assume $\gamma$ remains the same for all $i$ and $j$, and one can estimate the mobile computation latency by $t_{i,j}^{\text{mobile}} = \gamma t_{i,j}^{\text{edge}}$. In real edge-inference systems, $\gamma$ for popular mobile device CPUs can be profiled and recorded off-line as a lookup table. When a typical mobile device requests edge inference, an appropriate $\gamma$ can be selected according to its CPU type.

Given all profiles and assuming the original DNN has a total of $M$ layers, we can write the inference latency given the split point right after the $i$th layer as

$$t_i^{\text{infer}} = \sum_{k=1}^{i} t_{k,i}^{\text{mobile}} + \frac{D_{k,k}}{R} + \sum_{k=i+1}^{M} t_{k,i}^{\text{edge}}. \tag{18.1}$$

where the first part is the mobile computation latency, the second part is the transmission latency, and the third part is the edge computation latency. Then, given the lowest tolerable accuracy $A$, one can select the optimal split point $x$ and corresponding pruned DNN, achieving the lowest inference latency while preserving accuracy, as follows:

$$x = \underset{i=1,2,\ldots,M \text{ and } A_i \geq A}{\arg\min} t_i^{\text{infer}}. \tag{18.2}$$

Then, a traversal split-point selection algorithm can be used to find the solution of Eq. (18.2). Since $x$ has at most $M$ different choices, and the complexity of computing Eq. (18.1) for each choice is $\mathcal{O}(M)$, the complexity of the split-point selection algorithm is $\mathcal{O}(M^2)$, which is low because most DNNs deployed at mobile devices are not extremely deep.

After determining the optimal split-point and corresponding pruned DNN model, it can be deployed into edge-inference systems. The front-end part before the split point of the selected DNN is first transmitted to the mobile device, by which the front-end part computation can be locally performed, and the intermediate data is uploaded to the edge server for computing the rest part. Note that all profiling can be done before deployment, and only the low-complexity split-point selection algorithm needs to run in real time. Therefore, the proposed two-step pruning framework is flexible and can adapt to system dynamics.

## 18.3.4    Experiments

We use PyTorch [38], a Python-based deep learning framework, in the following experiments. Our server platform is shown in Table 18.1. GPUs are used in off-line training and pruning, while a CPU is used in profiling the pruned DNN models. VGG [39], a well-known convolutional neural network (CNN) for image classification, is used as an example. Our dataset is CIFAR-10 [40], a widely used image classification dataset with 10 classes of objects. Typical average *upload* rates $R$ of 3G, 4G, and Wi-Fi networks are 1.1 Mbps, 5.85 Mbps, and 18.88 Mbps respectively [41, 42]. We range $\gamma$ from 0.1 to 100 to simulate various computation capabilities of mobile devices. We allow at most 2 percent accuracy loss in each pruning step, resulting in at most 4 percent accuracy loss.

We first show the transmission and computation workload reduction of the proposed two-step pruning. Figure 18.4 shows the transmission workload and cumulative computation time at each layer for the original VGG model, VGG model after pruning step 1, and VGG model after pruning step 2, respectively. The left bars in the histogram in Fig. 18.4 represent the transmission workload in the original VGG at each layer. As the number of feature maps at the front-end part of VGG increases, the intermediate data gets an order of magnitude larger than the input, leading to high transmission

**Table 18.1.** Server platform specifications.

| Hardware | Specifications |
| --- | --- |
| System | Supermicro SYS-7048GR-TR, 4 × PCIe 3.0 × 16 slots |
| CPU | 2 × Intel Xeon E5-2640 V4, 2.4GHz |
| Memory | 128GB DDR4 2400MHz |
| GPU | 4 × NVIDIA TITAN Xp |

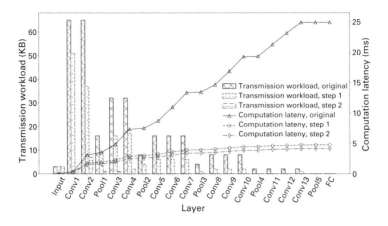

**Figure 18.4** Layer level transmission and computation characteristics of the original, step 1 pruned, and step 2 pruned VGG.

latency if we split the original VGG in the front-end part. The middle bars in the histogram are for the pruned VGG model after pruning step 1. Since filters in the back-end part of VGG get a higher probability to be pruned due to the pruning algorithm we used [37], splitting in the front-end part will face even more severe transmission latency issues (Fig. 18.6(b)). Nevertheless, the curves show that pruning step 1 can reduce the overall computation time by up to 5.35 times.

The right bars in the histogram are for the pruned VGG models after pruning step 2. Note that each bar stands for a specific pruned VGG. For example, the right bar with index conv1 represents a pruned VGG with conv1 layer pruned by step 2, and other layers are the same as those in the pruned VGG after pruning step 1. Since pruning step 2 only prunes one layer, it can significantly reduce the transmission workload but with only little computation latency reduction. Combining two pruning steps, we can get up to 25.6 times transmission workload reduction and 6.01 times computation acceleration as compared to the original model.

Then we use an average upload transmission rate $R = 137.5$ KB/s (3G network) and a computation capability ratio $\gamma = 5$ as an example system. Figure 18.5 shows the edge-inference latency and accuracy in this setting. The histograms in the figure are the edge-inference latency with different split points, and we use textures to distinguish different latency components. Figure 18.5(c) shows that the accuracy loss varies over different split-points due to pruning step 2, and thus the split-point can be adaptively selected according to the accuracy and latency requirements. If the system allows low-accuracy (e.g., 88 percent accuracy), the split point Pool 3 can achieve the best latency performance. When an accurate network (e.g., 90 percent accuracy) is needed, the split point should be Pool 4.

Consequently, optimal split-points given by the split-point selection algorithm under various system configurations are shown in Fig. 18.6. Figure 18.6(a) shows the optimal split points under various average uplink rates with $\gamma = 5$, and Fig. 18.6(c) is for various computation capability ratios with $R = 137.5$ KB/s, respectively.

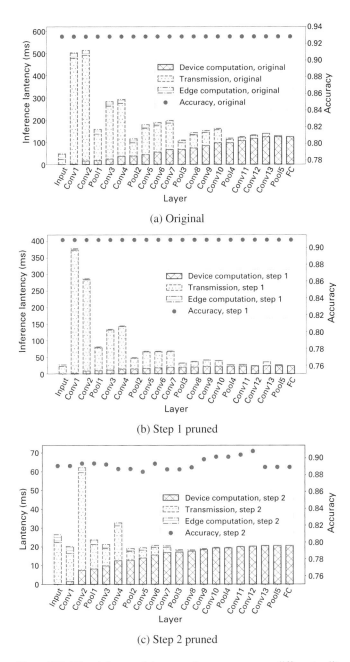

(a) Original

(b) Step 1 pruned

(c) Step 2 pruned

**Figure 18.5** Edge-inference latency and accuracy versus different split points. The average upload transmission rate is $R = 137.5$ KB/s, and the computation capability ratio is $\gamma = 5$.

Figure 18.6(b) and (d) shows the corresponding edge-inference latency. Due to the high transmission latency for the intermediate data, the original VGG prefers completing the entire inference either at the mobile device (split point 0) or at the edge server (split point 18). The VGG after step 1 pruning is similar, while it has relatively

**Table 18.2.** Edge-inference latency improvements under three typical mobile networks with $\gamma = 5$.

| Network | Original (ms) | Step 2 pruned (ms) | Improvement |
|---------|---------------|--------------------|-------------|
| 3G      | 46.64         | 17.84              | 2.61 ×      |
| 4G      | 28.50         | 7.73               | 3.69 ×      |
| Wi-Fi   | 25.61         | 5.32               | 4.81 ×      |

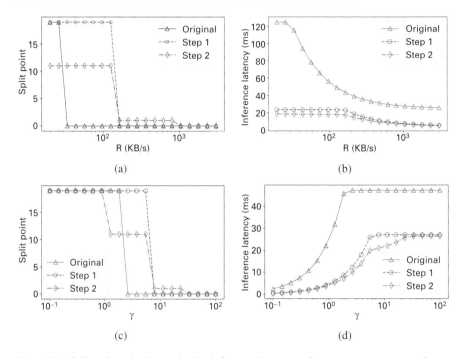

**Figure 18.6** Split-point selection and edge-inference latency performance versus system factors. (a) and (b) show the optimal split point and corresponding edge-inference latency versus average upload transmission rate $R$. (c) and (d) show the optimal split point and corresponding edge-inference latency versus computation capability raito $\gamma$.

less computation workload and thus prefers local computation over edge computation in bad wireless channel conditions or under limited edge computation capabilities. However, the system can still benefit from splitting after pruning step 2. Table 18.2 shows the edge-inference latency improvements for three typical mobile networks (3G, 4G, and Wi-Fi) with $\gamma = 5$. By applying the two-step pruning framework, 4.81× acceleration can be achieved in Wi-Fi environments.

## 18.4     Dynamic Compression for Edge Inference with Hard Deadlines

An edge-inference system consisting of an edge server (attached to a base station) and several edge devices is considered as shown in Fig. 18.7. The edge server has a

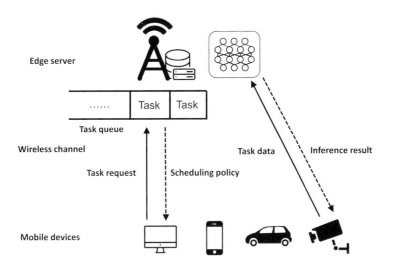

**Figure 18.7** The edge-inference system with a task queue.

well-trained learning model so that mobile devices can offload the inference tasks to the edge server. The considered system is time slotted. When an inference task arrives at a device, the device sends a request to the edge server for performing inference and waits for the scheduling decision from the server for uploading its task data. All tasks have the same maximum latency deadline, denoted by $\tau$, meaning that every task should be completed before $\tau$ time slots after its arrival at the device; otherwise the task is considered failed. The time used for processing tasks at the server side and sending requests as well as scheduling decisions are constants for different tasks. Therefore, it can be considered by subtracting the total latency constraint with the corresponding overhead; that is, with larger time consumption required for task processing, sending requests, and scheduling decisions, the system is faced with smaller equivalent $\tau$. The edge server maintains a task queue with task requests from all edge devices. The request arrival process of the task queue is assumed to be a Bernoulli process so that there is at most one request arriving at the edge server in every time slot with probability $p$. The edge server adopts the first-come, first-serve (FCFS) scheduling principle because all tasks have the same priority.

For edge inference, there may be a large amount of raw data for transmission. To reduce the transmission latency, lossy compression is applied. However, the compression of data can possibly bring performance degradation. To balance the trade-off between the inference timliness and accuracy, the edge server should select the appropriate compression ratio for every task according to the state of the task queue and the task arrival process. The objective is successfully completing most tasks under given deadline constraints, and thus inference accuracy can be regarded as the reward of performing inference. We assume that all tasks have the same raw data size and the reward is a function of compression ratio. Compression ratio $r \in [1, +\infty)$ is defined as the ratio of the size of raw data to the size of compressed data (i.e., larger

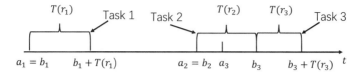

**Figure 18.8** Example of the arrivals and transmissions of inference tasks.

compression ratio leads to less amount of data size for transmission). The accuracy of the machine learning model with compression ratio $r$ (i.e., the reward function) is denoted by $\rho(r) \in [0, 1]$, which can be obtained off-line by performing inference on a validation dataset and stored as a lookup table at the server side. For communication, the edge devices use fixed transmission rate, and the time slots used for transmission using compression ratio $r$ are denoted by $T(r) \in \mathbb{N}$. $\rho(r)$ and $T(r)$ are both decreasing functions of $r$. The fading of the wireless channel is assumed to be independent and identically distributed (i.i.d.) among different slots. In every time slot, the transmission can fail with probability $p_e$. As a result, transmission of task using $T(r)$ time slots will fail with probability

$$P_e(T(r)) = 1 - (1 - p_e)^{T(r)}, \tag{18.3}$$

which is also called as packet error ratio (PER).

The objective of the transmission scheme is to maximize the expected number of successfully completed tasks within the deadline $\tau$. Consider $M$ tasks, which have an arrival time of $a_1 < a_2 < \cdots < a_M$. The $i$th task is scheduled to transmit the data *from* time slot $b_i$ with compression ratio $r_i$. Figure 18.8 shows an example with three tasks. Successful completion means the inference result is correct and the task is completed before meeting deadline. The optimization problem is formulated as

$$\max_{r_i} \sum_{i=1}^{M} \rho(r_i)(1 - P_e(T(r_i))) \tag{18.4}$$

$$\text{s.t. } b_1 = a_1, \tag{18.5}$$

$$b_{i+1} = \max\{b_i + T(r_i), a_{i+1}\}, i = 1, 2, \ldots, M - 1, \tag{18.6}$$

$$b_i + T(r_i) \le a_i + \tau, i = 1, 2, \ldots, M, \tag{18.7}$$

where Eqs. (18.5) and (18.6) indicate that the $(i + 1)$th task will be transmitted when the transmission of the $i$th task is completed (if the $(i + 1)$th task has not arrived, then it will be transmitted when it arrives). Equation (18.7) is the deadline constraint. Notice that $T(r_i)$ can equal to 0 and then $\rho(r_i)$ will also equal to 0 (it means that the task is failed since it cannot be delivered before the deadline), which ensures that Eq. (18.7) can be satisfied for all tasks.

## 18.4.1    Off-Line Dynamic Compression Algorithm

As a benchmark, off-line means that the arrival time slots of all the tasks are known beforehand, so that the optimal algorithm can be estanblished through dynamic

---

**Algorithm 18.1** Off-line DP algorithm.

**Require:**

   Number of tasks $M$; Maximum waiting time of tasks $\tau$; Arriving time of tasks $a_m$;

**Ensure:**

   Maximum number of successfully completed task $F(m,t)$;

   Optimal policy $r(m)$;

1: set $F(0,t) = 0 \ (1 \leq t \leq a_M + \tau)$

2: **for** $m = 1$ to $M$ **do**

3:   **for** $t = a_m + 1$ to $a_m + \tau$ **do**

4:     $F(m,t) = \max_{a_m \leq i \leq t-1} \left( F(m-1,i) + \rho\left(T^{-1}(t-i)\right)\left(1 - P_e(t-i)\right) \right)$

5:     $G(m,t) = \arg\max_{a_m \leq i \leq t-1} \left( F(m-1,i) + \rho\left(T^{-1}(t-i)\right)\left(1 - P_e(t-i)\right) \right)$

6:   **end for**

7: **end for**

8: set $t = a_M + \tau$

9: set $m = M$

10: **while** $m > 0$ **do**

11:   $r(m) = T^{-1}(G(m,t))$

12:   $t = \min(t - G(m,t), a_{m-1} + \tau)$

13:   $m = m - 1$

14: **end while**

---

programming (DP). We use $F(m,t), (1 \leq m \leq M, 1 \leq t \leq a_M + \tau)$ to denote the maximum number of successfully completed tasks when processing the first $m$ tasks by time $t$. To achieve $F(m,t)$, the number of time slots used for transmitting the $m$th task is denoted by $G(m,t)$. Then we can get the optimal off-line policy (i.e., the selected compression ratio for every task $r(m)$) after calculating $F(m,t)$ and $G(m,t)$ $(1 \leq m \leq M, 1 \leq t \leq a_M + \tau)$. The algorithm is expressed in Algorithm 18.1.

## 18.4.2  Online Dynamic Compression Algorithm

In practice, the arrival times of tasks cannot be known beforehand. Therefore, an online algorithm is needed to select the compression ratios for the tasks in the queue without the knowledge of future task arrivals. Assume that the arrival process is known, which is a Bernoulli process with probability $p$. We can formulate the probem as Markov decision process (MDP), where we map the task queue to the MDP state and find the optimal action based on the state to select the optimal compression ratio for the head-of-line (HoL) task. The detailed definition of the state, action, and reward are provided as follows.

**State:** The arrival times of all tasks in the queue are represented by $s = \{a_1, a_2, \ldots, a_N\}$, where $N$ is the number of tasks in the waiting queue and $a_1 < a_2 < \cdots < a_N$. Because of the deadline constraint, $N$ is at most $\tau$. Note that the difference between the deadline of tasks and current time is sufficient for making a decision. Accordingly, the state can be transformed to $s = \{a_1 + \tau - t, a_2 + \tau - t, \ldots, a_N + \tau - t\}$, where $\tau$ is the latency constraint of tasks and $t$ is the current time slot. Because $0 < a_1 + \tau - t < a_2 + \tau - t < \cdots < a_N + \tau - t \leq \tau$ (there is at most one task request arriving at the edge server in one time slot), the state $s$ can be encoded to a binary number of $\tau$ digits

$$s = \sum_{i=1}^{N} 2^{a_i + \tau - t - 1}. \tag{18.8}$$

**Action**: The action of the MDP is the selected compression ratio of the HoL task in the queue. The action space is the set of compression ratio options, denoted by $\mathcal{R}$.

**Reward**: The reward of taking action $r \in \mathcal{R}$ (selecting compression ratio $r$) is the probability of getting the correct inference result

$$W(r) = \rho(r)(1 - P_e(T(r))), \tag{18.9}$$

which is the product of the expected accuracy of the machine learning model under compression ratio $r$ and the probability of successful transmission.

**State-transition probability**: The state-transition probability depends on the action and the arrival process of tasks. With action $r$, state $s$ will transit to state $s' = 2^{\tau - T(r)} i + \left\lfloor \frac{s}{2^{T(r)}} \right\rfloor$ ($i = 0, 1, \ldots, 2^{T(r)} - 1$) with probability

$$\mathbf{P}_{ss'}(r) = p^{B(i)}(1 - p)^{T(r) - B(i)}, \tag{18.10}$$

where $B(i)$ is the number of 1's in the binary expression of nonnegative integer $i$.

Generally, the space complexity and time complexity of calculating the state-transition probability matrix $\mathbf{P}$ and performing value iteration are both $\mathcal{O}(S \times S \times |\mathcal{R}|)$, where $S = 2^\tau - 1$ is the number of states. When $\tau$ is large, the complexity can be unacceptable. To this end, we use the similarity of state transitions and the fact that there are many zero elements in the state-transition probability matrix; then we can substantially reduce the complexity of the algorithm as follows. First, the value iteration equation of MDP is

$$V^{k+1}(s) = \min_{r \in \mathcal{R}(s)} \left[ W(r) + \sum_{i=1}^{2^\tau - 1} \mathbf{P}_{si}(r) V^k(i) \right], \tag{18.11}$$

where $V^k(i)$ is the value of state $i$ of the $k$th iteration and $R(s)$ is the set of optional action of state $s$. To reduce the complexity, we use Eq. (18.12) for value iteration.

$$V^{k+1}(s) = \min_{r \in \mathcal{R}(s)} \left[ W(r) + \sum_{i=0}^{2^{T(r)} - 1} \mathbf{P}'(r, i) V^k \left( 2^{\tau - T(r)} i + \left\lfloor \frac{s}{2^{T(r)}} \right\rfloor \right) \right]. \tag{18.12}$$

For Eq. (18.12), the state-transition probability is $\mathbf{P}'(r, i) = p^{B(i)}(1 - p)^{T(r) - B(i)}$. For efficient value iteration, $\mathbf{P}'$ is calculated and stored before performing the value iteration. Therefore one only needs to store $\mathbf{P}'$, with $S \times |\mathcal{R}|$ elements, instead of $\mathbf{P}$.

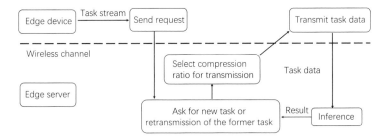

**Figure 18.9** Workflow of information augmentation scheme.

In addition, Eq. (18.12) only sums up $2^{T(r)} - 1$ terms when considering action $r$, instead of $2^T - 1$ terms in Eq. (18.11). As a result, the space complexity is now $\mathcal{O}(S \times |\mathcal{R}|)$, and the time complexity is $\mathcal{O}(S \times S \times |\mathcal{R}| \times \frac{1}{\tau})$.

## 18.4.3   Information Augmentation

Notice that some task samples can get correct inference results with a high compression ratio because of the information redundancies in the raw data. Inspired by this observation, we can spend fewer communication resources on these tasks, leaving more communication resources for tasks that require lower compression ratios in the hopes of completing more tasks within the deadline. Therefore, we propose an information augmentation scheme to find the proper compression ratio for every task, and the workflow is shown in Fig. 18.9. Here we first assume that there is a method to judge whether the inference result is correct or not and the PER is $p_e = 0$. The edge server can first ask the edge devices to transmit data with a high compression ratio (not necessarily the highest one, and the compression ratio for the first attempt is also subject to our optimization). If the result is wrong, the edge server can decide whether to ask the device to transmit the task data with lower compression ratio, according to the system state. In short, the key to the information augmentation scheme is to exploit the state of the task queue so as to decide the compression ratio of the first transmission attempt and later on augmentation transmissions if needed.

The proposed information augmentation scheme is also based on MDP. In the online algorithm, we use state $s = \{a_1, a_2, \ldots, a_N\}$ for MDP and the action space $R$ is the set of optional compression ratios. Notice that one state transition corresponds to one transmission and the objective of MDP is to maximize the average reward per state transition. However, for the information augmentation scheme, the times of transmission of different tasks may be different. If one state transition of MDP still corresponds to one transmission over the wireless channel, different tasks will have different weights (depending on the times of retransmissions) on the reward, and thus we cannot directly apply the MDP formulation from the previous algorithm.

To address the issue, we should ensure that different tasks have the same number of state transitions, and thus we consider a new state space and corresponding state transition formulation. For every task, the edge server will *virtually* consider all optional compression ratios from high to low for its transmission even if not selected for transmission. (If the task can be completed with a high compression ratio, the transmission with lower compression is not needed.) Every state transition corresponds to one time of virtual consideration. There are only two actions for MDP, transmission (or retransmission) with the current compression ratio under consideration or no transmission. If the action is no transmission, the edge server will consider the next compression ratio. If the action is to transmit, the task will be transmitted with the considered compression ratio. Given the inference result and the state of task queue, the edge server will continue to consider the next compression ratio for possible information augmentation. In this way, every task will have $|R|$ times of state transitions. The corresponding state, action, reward, and state-transition probability is shown as follows.

**State**: The key here is that the compression ratio under consideration is put into the state, and the state becomes $s = \{a_1, a_2, \ldots, a_N, r_L, r, f\}$, where $r_L$ is the compression ratio for the last transmission for current task (if the task has not been transmitted, $r_L = +\infty$), $r$ is the compression ratio considered for current transmission, and $f \in \{0, 1\}$ indicates whether the last transmission results in correct inference (1 for correct and 0 for wrong). For simplicity, the state can be transformed to

$$s = \left\{ \sum_{i=1}^{N} 2^{a_i + \tau - t - 1}, r_L, r, f \right\} = \{a, r_L, r, f\}, \tag{18.13}$$

where $a = \sum_{i=1}^{N} 2^{a_i + \tau - t - 1}$ is the state of the task queue.

**Action**: Because the compression ratio under consideration is formulated into states, the actions for MDP become transmitting or not, denoted by $r_I$, $r_I = 1$ for transmission and $r_I = 0$ for no transmission. If the inference of last transmission is correct ($f = 1$) or the time to the deadline is not enough for additional transmission, the action can only be no transmission.

**Reward**: For the state transition from $s = \{a, r_L, r, f\}$ to $s' = \{a', r'_L, r', f'\}$ with action $r_I$, the reward is

$$W_I(s, s', r_I) = r_I f'. \tag{18.14}$$

**State-transition probability**: The state-transition probability depends on the task arrival process and the accuracy of inference conditioned on the inference correctness of the last transmission. If the action is no transmission, the state $s = \{a, r_L, r, f\}$ will transit to state $s' = \{a, r_L, r', f\}$ with probability 1, where $r'$ is the next considered compression ratio after $r$. If the action is to transmit, then $s = \{a, r_L, r, f\}$ will transit to state $s' = \{\lfloor \frac{a}{2^{T(r)}} \rfloor + i \times 2^{\tau - T(r)}, r, r', f\}$ ($i \in [0, 2^{T(r)} - 1] \cap \mathbb{N}$) and the state-transition probability is

$$\mathbf{P}(s, s') = p^{B(i)}(1 - p)^{T(r) - B(i)} \times \mathbf{P}_a(r, f | r_L). \tag{18.15}$$

$\mathbf{P}_a(r, 0 | r_L)$ is the probability of getting the wrong inference result with compression ratio $r$, conditioned on that the last transmission is with compression ratio $r_L$ and the inference result was wrong (otherwise there is no need to transmit again). $\mathbf{P}_a(r, 1|r_L)$ is the probability of getting the correct inference result with compression ratio $r$, conditioned on that the last transmission is with compression ratio $r_L$. $\mathbf{P}_a$ is obtained by performing inference on the validation dataset with optional compression ratios.

The number of states is $S_R = 2(2^\tau - 1)|R|^2$ and the size of action space is 2. The memory usage is the simplified state transition matrix $\mathbf{P}'$ as explained in Eq. (18.12) and $\mathbf{P}_a$. The space complexity is $\mathcal{O}(S_R)$ and the time complexity is $\mathcal{O}(S_R \times S \times \frac{1}{\tau})$.

We have assumed that there is a method to judge whether the inference result is correct or not, which cannot be satisfied in many scenarios. Therefore, we further introduce uncertainty to estimate the confidence of the inference result if one does not know whether the inference is correct. The uncertainty $\mathcal{U}$ of the output of the learning model is defined as

$$\mathcal{U} = -\sum_{i=1}^{n} X_i \log X_i, \tag{18.16}$$

where $X = (X_1, X_2, \ldots, X_n)$ is the normalized output of learning model ($\sum_{i=1}^{n} X_i = 1$) and $n$ is number of elements of the model output. The lower uncertainty $\mathcal{U}$ indicates higher probability of correct inference. For example, the relation between correctness and uncertainty of data samples in the handwritten digit images dataset (MNIST) [43] is shown in Fig. 18.10. It shows the histogram of uncertainty of correct results and wrong results with different resolutions. The true results have lower uncertainty, and the wrong results have higher uncertainty, meaning that it is reasonable to use uncertainty to estimate the confidence of inference results.

We now adapt the information augmentation scheme with uncertainty by changing the definitions of state, action, and reward.

**State**: Because we use uncertainty of the inference result to estimate the correctness, the uncertainty should be added into the state for decision making, and thus the state of the MDP becomes $s = \{a, r_L, r, \mathcal{U}\}$, where $\mathcal{U}$ is the uncertainty of the result of last transmission of the HoL task. To get a discrete state, uncertainty $\mathcal{U}$ should be quantized, and we apply uniform quantization for uncertainty.

**Action**: There are still two actions for the MDP, transmit or not transmit, denoted by $r_U$.

**Reward**: If the action is no transmission, the reward is 0. If the action is to transmit, the reward of state transition from state $s = \{a, r_L, r, \mathcal{U}\}$ to state $s' = \{a', r, r', \mathcal{U}'\}$ is

$$W_U(s, s') = \rho(r|r_L, \mathcal{U}, \mathcal{U}') - \rho(r_L|\mathcal{U}). \tag{18.17}$$

Here, $\rho(r|r_L, \mathcal{U}, \mathcal{U}')$ is the accuracy of the learning model with compression ratio $r$, conditioned on the compression ratio of last transmission $r_L$, the uncertainty of the result of last transmission $\mathcal{U}$ and the uncertainty of the newest result $\mathcal{U}'$. $\rho(r_L|\mathcal{U})$ is the accuracy of the learning model with compression ratio $r_L$, conditioned on the uncertainty of the result $\mathcal{U}$. Again, $\rho(r|r_L, \mathcal{U}, \mathcal{U}')$ and $\rho(r_L|\mathcal{U})$ can be obtained by performing inference on the validation dataset with optional compression ratios.

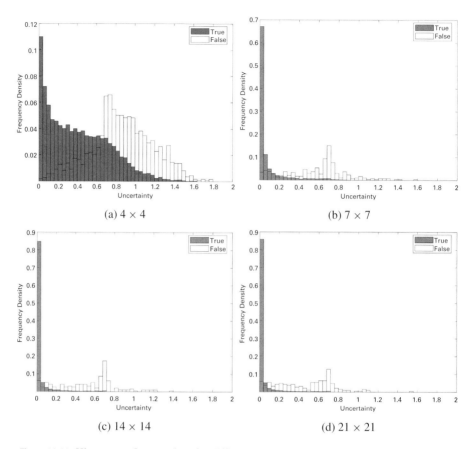

**Figure 18.10** Histogram of uncertainty for different resolutions of figures from MNIST.

**State-transition probability**: The state-transition probability depends on the arrival process of tasks and the probability distribution of uncertainty $\mathcal{U}$ of learning model. If the action is no transmission, the state $s = \{a, r_L, r, \mathcal{U}\}$ will transit to state $s' = \{a, r_L, r', \mathcal{U}\}$ with a probability of 1, where $r'$ is the next considered compression ratio after $r$. If the action is to transmit, state $s = \{a, r_L, r, \mathcal{U}\}$ will transit to state $s' = \{\lfloor \frac{a}{2^{T(r)}} \rfloor + i \times 2^{\tau - T(u)}, r, r', \mathcal{U}'\}$ ($i \in [0, 2^{T(r)} - 1] \cap \mathbb{N}$) and the state transition probability is

$$\mathbf{P}(s, s') = p^{B(i)}(1 - p)^{T(r) - B(i)} \times \mathbf{P}_U(r, \mathcal{U}'|r_L, \mathcal{U}), \tag{18.18}$$

where $\mathbf{P}_U(r, \mathcal{U}'|r_L, \mathcal{U})$ is the probability that the model output has uncertainty $\mathcal{U}'$ with compression ratio $r$ conditioned on that the model output of the last transmission with compression ratio $r_L$ has uncertainty $\mathcal{U}$.

The number of states is now $S_{UR} = (2^\tau - 1)U|R|^2$ and the size of action space is 2, where $U$ is the number of the quantization levels of uncertainty $\mathcal{U}$. The memory usage is the simplified state transition matrix $\mathbf{P}'$ as explained in Eq. (18.12) and $\mathbf{P}_U$. The space complexity is $\mathcal{O}(S_{UR})$ and the time complexity of the MDP algorithm is $\mathcal{O}(S_{UR} \times S \times \frac{1}{\tau})$.

## 18.4.4    Packet-Loss-Aware Evolutionary Retransmission

Due to the unreliable wireless channels, the transmission of data samples may fail with packet loss. If the transmission fails, the edge server can keep this task in the task queue for retransmissions and update the queue state with new task arrivals. Then the edge server can still use the online dynamic compression algorithm to select the compression ratios for the transmission/retransmissions of this task according to the new queue state. This can be regarded as the baseline and is called the original retransmission scheme.

Notice that the algorithm does not consider the packet error when designing MDP. Therefore, we incorporate the packet error and retransmissions into the state transition of MDP to obtain an *evolutionary* retransmission scheme. In the considered scenario, one transmission of the task needs at least one time slot, so the edge device can transmit the data of a task for at most $\tau$ times due to the hard deadline $\tau$. Therefore, we can use the same method as in information augmentation to ensure that different tasks have the same number of state transitions (i.e., the edge server virtually considers $\tau$ times of transmissions for every task, including whether transmitting and the corresponding compression ratio). Therefore, the definitions of state, action, and reward are adapted as follows.

**State**: The state of MDP is $s = \{a, r_L, \tau_s\}$, where $a$ is the queue state, $r_L$ is compression ratio of last successful transmission (no packet error) and $\tau_s$ is number of considered transmissions.

**Action**: The action of MDP is the compression ratio for transmission or no transmission for the HoL task.

**Reward**: For state $s = \{a, r_L, \tau_s\}$, if the action is to transmit with compression ratio $r$ and the transmission succeeds, the reward is

$$W_R(s, r) = \rho(r) - \rho(r_L), \tag{18.19}$$

otherwise the reward is 0.

**State-transition probability**: The state-transition probability depends on the arrival process of tasks, the accuracy of learning model and the PER. With action $r$, the state $s = \{a, r_L, \tau_s\}$ will transit to state $s' = \{\lfloor \frac{a}{2^{T(r)}} \rfloor + i \times 2^{\tau - T(r)}, r, \tau_s + 1\}$ if transmission succeeds, or state $s'' = \{\lfloor \frac{a}{2^{T(r)}} \rfloor + i \times 2^{\tau - T(r)}, r_L, \tau_s + 1\}$ if transmission fails. The state transition probability is

$$\mathbf{P}(s, s') = p^{B(i)}(1 - p)^{T(r) - B(i)} \times (1 - P_e(T(r))), \tag{18.20}$$

and

$$\mathbf{P}(s, s'') = p^{B(i)}(1 - p)^{T(r) - B(i)} \times P_e(T(r)). \tag{18.21}$$

Here the space complexity is $\mathcal{O}(S \times |R| \times \tau)$ and the time complexity of the MDP algorithm is $\mathcal{O}(S \times S \times |R|)$.

## 18.4.5     Experiments

The dataset used in experiments is handwritten digit images dataset MNIST, and the inference task of the edge learning system is the number recognition for images in the testing dataset. In MNIST, there are 60,000 images in the training dataset and 10,000 images in the testing dataset. Each image has $28 \times 28$ pixels, and corresponds to an integer number from 0 to 9. The machine learning model deployed on the edge server is a multilayer perceptron (MLP) with one hidden layer and 700 hidden units and is trained with the training dataset. The output of the model is vector $X$ with 10 elements and $\sum_{i=1}^{10} X_i = 1$. The $i$th element of $X$ represents the probability that the input image belongs to number $i - 1$.

The compression algorithm used in the experiment is downsampling. The optional resolution of images for transmission include $4 \times 4$, $7 \times 7$, and $14 \times 14$, for which the compression ratio $r$ is 49, 16, and 4. A time slot is normalized to be the time used for transmitting an image with resolution $4 \times 4$. Hence the number of time slots used for transmitting data with these optional resolutions are 1, 3, and 10 slots, respectively. The inference accuracy with different compression ratios is 0.89, 0.97, and 0.98, for compression ratio 49, 16, and 4, respectively.

First, we provide the performance under different task arrival rates. Here we suppose that there is no packet loss, which means PER $p_e = 0$. The latency deadline is $\tau = 12$ slots. The number of quantization level of $\mathcal{U}$ of information augmentation scheme is 10. Figure 18.11 shows the performance of the online algorithm and information augmentation using MDP. It is shown that online algorithm has almost the same performance as the off-line algorithm. The information augmentation brings significant improvement, especially when the arrival rate is high, and it is more robust to different arrival rates. When the inference correctness is unknown, exploiting uncertainty for confidence estimation can still bring benefits. The improvement of the information augmentation is marginal when the arrival rate is low. This is because tasks

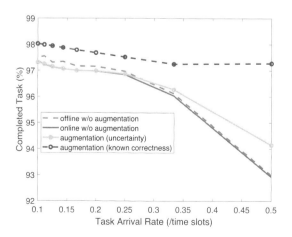

**Figure 18.11**  Performance of the proposed online algorithms with dynamic compression ratio selection under different arrival rates.

can be transmitted with low compression ratio (i.e., with high resolutions) for the first transmission when the arrival rate is low, and in this case, the information augmentation is unnecessary for many tasks.

Next, in Fig. 18.12 we show the performance with different latency requirements, where the arrival rate is set to $p=0.11$. The gaps between off-line algorithm and online algorithm become small when $\tau$ increases, and at the same time, the information augmentation scheme shows more advantages. This is because the information augmentation scheme can reduce the average communication cost but results in extra latency of some tasks. Some of tasks may fail due to this extra latency. With larger $\tau$, there is more time for these extra transmissions and the advantage of information augmentation scheme, which reduces the average communication cost, becomes more significant.

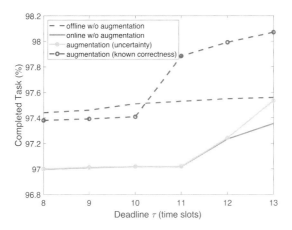

**Figure 18.12** Performance of the proposed online algorithms with dynamic compression ratio selection under different deadlines.

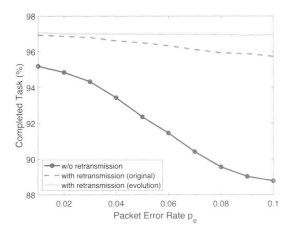

**Figure 18.13** Performance of the proposed online retransmission schemes without information augmentation.

Finally, we consider the packet loss in the edge learning system and compare the performance of the retransmission scheme under different values of PER. Figure 18.13 shows the performance of three schemes, including scheme without retransmission, baseline of retransmission scheme (original MDP without considering the packet losses), and the evolutionary retransmission scheme that considers packet losses. The performance of the scheme without retransmission indicates that the packet loss brings severe performance degradation. The evolutionary retransmission scheme is much more robust to various values of PER and improves the performance for over 1 percent as compared to the baseline when $p_e = 0.1$.

## 18.5    Conclusion

In this chapter, we introduced the concept of wireless edge inference and emphasized the importance and challenges of guaranteeing timely inference for mobile applications. Representative approaches from the aspects of DNN pruning, joint source and channel coding, and inference-aware scheduling are reviewed, and their benefits in terms of low latency inference are discussed. Our initial research efforts on DNN pruning and inference-aware scheduling are presented. We have shown that pruning the DNN not only for the whole network, but also on the specific splitting points, can better utilize the limited wireless bandwidth while at the same time enjoy the computation resources on edge servers, so that the overall inference latency is substantially reduced. We also proposed a way of removing the redundancies in the data via compression and showed how compression can interact with transmission scheduling and retransmissions to allow more inference tasks completed within hard deadlines. As for future works, the proposed pruning-based DNN splitting framework can incorporate dynamic compression and other joint source and channel coding schemes, as its extension to supporting multiple mobile devices is nontrivial and requires cooptimization with scheduling. Also, existing work mainly looks at multiple access channels, while purely distributed networks, where nodes can cooperate to perform inference, deserve further study.

## References

[1]  A. Voulodimos et al., "Deep learning for computer vision: A brief review," *Computational Intelligence and Neuroscience*, vol. 2018, 2018.

[2]  T. B. Brown et al., "Language models are few-shot learners," *arXiv preprint*, arXiv:2005.14165, 2020.

[3]  O. Plchot et al., "Audio enhancing with DNN autoencoder for speaker recognition," in *IEEE Int. Conf. on Acoustics, Speech and Signal Processing (ICASSP)*, pp. 5090–5094, 2016.

[4]  H. Ye, G. Y. Li, and B.-H. Juang, "Power of deep learning for channel estimation and signal detection in OFDM systems," *IEEE Wireless Communications Letters*, vol. 7, no. 1, pp. 114–117, 2017.

[5] I. Goodfellow et al., *Deep Learning*, vol. 1. MIT Press, 2016.

[6] Y. Cheng et al., "Model compression and acceleration for deep neural networks: The principles, progress, and challenges," *IEEE Signal Processing Magazine*, vol. 35, no. 1, pp. 126–136, 2018.

[7] T. Elsken, J. H. Metzen, and F. Hutter, "Neural architecture search: A survey," *arXiv preprint*, arXiv:1808.05377, 2018.

[8] S. Kekki et al., "MEC in 5G networks," white paper, ETSI, 2018.

[9] Q.-V. Pham et al., "A survey of multi-access edge computing in 5G and beyond: Fundamentals, technology integration, and state-of-the-art," *IEEE Access*, vol. 8, pp. 116 974–117 017, 2020.

[10] Z. Zhou et al., "Edge intelligence: Paving the last mile of artificial intelligence with edge computing," *Proc. IEEE*, vol. 107, no. 8, pp. 1738–1762, 2019.

[11] X. Wang et al., "Convergence of edge computing and deep learning: A comprehensive survey," *IEEE Communications Surveys & Tutorials*, vol. 22, no. 2, pp. 869–904, 2020.

[12] Y. Kang et al., "Neurosurgeon: Collaborative intelligence between the cloud and mobile edge," *ACM SIGARCH Computer Architecture News*, vol. 45, no. 1, pp. 615–629, 2017.

[13] A. E. Eshratifar, M. S. Abrishami, and M. Pedram, "JointDNN: An efficient training and inference engine for intelligent mobile cloud computing services," *IEEE Trans. on Mobile Computing*, 2019.

[14] C. Hu et al., "Dynamic adaptive DNN surgery for inference acceleration on the edge," in *IEEE Conf. on Computer Communications (INFOCOM)*, pp. 1423–1431, 2019.

[15] J. H. Ko et al., "Edge-host partitioning of deep neural networks with feature space encoding for resource-constrained Internet-of-Things platforms," in *15th IEEE Int. Conf. on Advanced Video and Signal Based Surveillance (AVSS)*, pp. 1–6, 2018.

[16] E. Li, Z. Zhou, and X. Chen, "Edge intelligence: On-demand deep learning model co-inference with device-edge synergy," in *Proc. Workshop on Mobile Edge Communications*, pp. 31–36, 2018.

[17] E. Li et al., "Edge AI: On-demand accelerating deep neural network inference via edge computing," *IEEE Trans. on Wireless Communications*, vol. 19, no. 1, pp. 447–457, 2019.

[18] W. Shi et al., "Improving device-edge cooperative inference of deep learning via 2-step pruning," *arXiv preprint*, arXiv:1903.03472, 2019.

[19] M. Jankowski, D. Gündüz, and K. Mikolajczyk, "Joint device-edge inference over wireless links with pruning," in *IEEE 21st Int. Workshop on Signal Processing Advances in Wireless Communications (SPAWC)*, pp. 1–5, 2020.

[20] J. Shao and J. Zhang, "Communication-computation trade-off in resource-constrained edge inference," *arXiv preprint*, arXiv:2006.02166, 2020.

[21] A. G. Howard et al., "Mobilenets: Efficient convolutional neural networks for mobile vision applications," *arXiv preprint*, arXiv:1704.04861, 2017.

[22] X. Zhang et al., "Shufflenet: An extremely efficient convolutional neural network for mobile devices," in *Proc. IEEE Conf. on Computer Vision and Pattern Recognition (CVPR)*, pp. 6848–6856, June 2018.

[23] Y. Lin et al., "Deep gradient compression: Reducing the communication bandwidth for distributed training," *arXiv preprint*, arXiv:1712.01887, 2017.

[24] S. U. Stich, J.-B. Cordonnier, and M. Jaggi, "Sparsified SGD with memory," in *Advances in Neural Information Processing Systems*, pp. 4447–4458, 2018.

[25] S. Han et al., "Learning both weights and connections for efficient neural network," in *Advances in Neural Information Processing Systems*, pp. 1135–1143, 2015.

[26] R. Anil et al., "Large scale distributed neural network training through online distillation," *arXiv preprint*, arXiv:1804.03235, 2018.

[27] J. Shao and J. Zhang, "Bottlenet++: An end-to-end approach for feature compression in device-edge co-inference systems," in *IEEE Int. Conf. on Communications Workshops (ICC)*, pp. 1–6, 2020.

[28] M. Jankowski, D. Gündüz, and K. Mikolajczyk, "Wireless image retrieval at the edge," *arXiv preprint*, arXiv:2007.10915, 2020.

[29] C. You et al., "Energy-efficient resource allocation for mobile-edge computation offloading," *IEEE Trans. on Wireless Communications*, vol. 16, no. 3, pp. 1397–1411, 2016.

[30] X. He et al., "Energy-efficient mobile-edge computation offloading for applications with shared data," in *IEEE Global Communications Conf. (GLOBECOM)*, pp. 1–6, 2018.

[31] X. Chen et al., "Performance optimization in mobile-edge computing via deep reinforcement learning," in *IEEE 88th Vehicular Technology Conf. (VTC-Fall)*, pp. 1–6, 2018.

[32] X. Liu, Z. Qin, and Y. Gao, "Resource allocation for edge computing in iot networks via reinforcement learning," in *IEEE Int. Conf. on Communications (ICC)*, pp. 1–6, 2019.

[33] W. Wen et al., "Learning structured sparsity in deep neural networks," in *Advances in Neural Information Processing Systems*, pp. 2074–2082, 2016.

[34] Y. He, X. Zhang, and J. Sun, "Channel pruning for accelerating very deep neural networks," in *Proc. IEEE Int. Conf. on Computer Vision*, pp. 1389–1397, 2017.

[35] J.-H. Luo and J. Wu, "Autopruner: An end-to-end trainable filter pruning method for efficient deep model inference," *Pattern Recognition*, p. 107461, 2020.

[36] T.-J. Yang, Y.-H. Chen, and V. Sze, "Designing energy-efficient convolutional neural networks using energy-aware pruning," in *Proc. IEEE Conf. on Computer Vision and Pattern Recognition*, pp. 5687–5695, 2017.

[37] P. Molchanov et al., "Pruning convolutional neural networks for resource efficient inference," *arXiv preprint*, arXiv:1611.06440, 2016.

[38] A. Paszke et al., "PyTorch: Tensors and dynamic neural networks in Python with strong GPU acceleration," open-source project, 2017; www.findbestopensource.com/product/pytorch-pytorch.

[39] K. Simonyan and A. Zisserman, "Very deep convolutional networks for large-scale image recognition," *arXiv preprint*, arXiv:1409.1556, 2014.

[40] A. Krizhevsky, V. Nair, and G. Hinton, "The CIFAR-10 dataset," dataset, 2014; www.cs.toronto.edu/kriz/cifar.html.

[41] OpenSignal.com, "State of mobile networks: USA," tech. report, Jul. 2018; https://opensignal.com/reports/2018/07/usa/state-of-the-mobile-network.

[42] Speedtest, "United States speedtest market report," tech. report; http://www.speedtest.net/reports/united-states/2018/Mobile/.

[43] Y. Lecun et al., "Gradient-based learning applied to document recognition," *Proc. IEEE*, vol. 86, no. 11, pp. 2278–2324, Nov. 1998.

# Index